Birkhäuser

ISNM
International Series of Numerical Mathematics

Volume 168

More information about this series at www.birkhauser-science.com/series/4819

Hervé Le Dret · Brigitte Lucquin

Partial Differential Equations: Modeling, Analysis and Numerical Approximation

Hervé Le Dret
Laboratoire Jacques-Louis Lions
Université Pierre et Marie Curie—Paris VI
Paris
France

Brigitte Lucquin
Laboratoire Jacques-Louis Lions
Université Pierre et Marie Curie—Paris VI
Paris
France

ISSN 0373-3149 ISSN 2296-6072 (electronic)
International Series of Numerical Mathematics
ISBN 978-3-319-27065-4 ISBN 978-3-319-27067-8 (eBook)
DOI 10.1007/978-3-319-27067-8

Library of Congress Control Number: 2015955864

Mathematics Subject Classification (2010): 35J20, 35J25, 35K05, 35K20, 35L03, 35L05, 65M06, 65M08, 65M12, 65N30

Springer Cham Heidelberg New York Dordrecht London

Printed on acid-free paper

Springer International Publishing AG Switzerland is part of Springer Science+Business Media
(www.birkhauser-science.com)

Preface

This book is devoted to the study of partial differential equation problems both from the theoretical and numerical points of view. A partial differential equation (PDE) is a relation between the partial derivatives of an unknown function u in several variables to be satisfied by this function, for example $\frac{\partial u}{\partial t} - \frac{\partial u}{\partial x} = 0$, where u is a function in the two variables t and x. Partial differential equations constitute a major field of study in contemporary mathematics. They also arise in other fields of mathematics, such as differential geometry or probability theory for example. In addition, partial differential equations appear in a wide variety of contexts in many applied fields, not only in the traditional fields of physics, mechanics and engineering, but also more recently in chemistry, bioscience, medicine, meteorology, climatology, finance and so on.

In all of these applied fields, numerical simulation is playing an increasingly prominent role, because even though solutions to such PDE problems can be shown to exist, there is in general no closed form solution. Therefore, quantitative information about the solutions can only be obtained by means of numerical approximation methods. It is very important not to take numerical simulation results at face value since they inherently present errors. In order to be able to do this, a comprehensive knowledge of all the steps involved is necessary, starting from modeling to mathematical existence theorems, to numerical approximation methods. This is one of the main purposes of this book.

A numerical approximation method is a procedure in which the original continuous unknowns are replaced by a finite number of discrete, computable approximate unknowns. It is thus important to be able to properly understand on the one hand the properties of partial differential equations and, on the other hand, the properties of the numerical methods that are used to define approximations of their solutions and effectively compute such approximations. In particular, it is essential to quantify the approximation by appropriately defining the error between approximate and continuous unknowns, and to show that this error goes to zero, preferably at a known rate, in the continuous limit when the number of approximated unknowns goes to infinity.

The goal of this book is to try and illustrate this program through a rather wide spectrum of classical or not so classical examples. We thus present some modeling aspects, develop the theoretical analysis of the partial differential equation problems thus obtained for the three main classes of partial differential equations, and analyze several numerical approximation methods adapted to each of these examples.

We have selected three broad families of numerical methods, finite difference, finite element and finite volumes methods. This is far from being exhaustive, but these three families of methods are the most widely used and constitute the core skills for anyone intending to work with numerical simulation of partial differential equations. There are many other numerical methods that we have chosen not to develop, in order to keep the size of the book within reasonable bounds.

Parts of the book are accessible to Bachelor students in mathematics or engineering. However, most of the book is better suited to Master students in applied mathematics or computational engineering. We put emphasis on mathematical detail and rigor for the analysis of both continuous and discrete problems.

The book is structured globally according to the three major types of partial differential equations: elliptic equations, parabolic equations and hyperbolic equations. We mainly consider linear equations, except for a few nonlinear examples. Each one of the above three types of equations requires a specific set of mathematical techniques for theoretical analysis, and a specific set of numerical methods, i.e., specific discretization procedures and convergence analyses for approximation. We follow a path of progressive difficulty in each case inasmuch as possible. We begin with the most elementary approaches either theoretical or numerical, which also happen to be the earliest ones from the historical viewpoint. We then continue to more advanced topics that require a more sophisticated mathematical background, both from the theoretical and numerical points of view, and that are also more recent than the previous ones. We also give along the way several numerical illustrations of successes as well as failures of numerical methods, using free software such as Scilab and FreeFem++.

The book is divided into ten chapters. Chapter 1 is devoted to mathematical modeling. We give examples ranging from mechanics and physics to finance. Starting from concrete situations, we try to present the various steps leading to a mathematical model involving partial differential equations to the extent possible. In some cases, it is possible to start from first principles and entirely derive the equations and boundary conditions. In other, more complicated cases, we just give a few indications concerning the modeling approach and sometimes a historical account. For some of the examples considered, we also give a short elementary mathematical study: existence, uniqueness, maximum principle.

In Chap. 2, we present the simplest possible and earliest numerical method, the so-called finite difference method. We first use the example of a one-dimensional elliptic problem, already introduced in Chap. 1, which is elementary enough not to require sophisticated mathematical machinery. We then consider a few generalizations: one-dimensional problems with other boundary conditions, two-dimensional problems, still on a rather elementary mathematical level.

The first two chapters are accessible at the Bachelor level. The remainder of the book calls for a higher level of mathematics, with the possible exception of parts of Chaps. 8 and 10. Chapter 3 is thus devised as a toolbox of more advanced mathematical analysis techniques that are required to study more general partial differential equations, especially in more than one dimension: a review of Hilbert space theory, usual function spaces, properties of open sets in $\mathbb{R}^d$, multidimensional integration by parts, distributions, and Sobolev spaces. The results are standard and we sometimes refer to classical references. We have however detailed some of the proofs for those results that are not always easy to find in the literature. Readers who are more interested in the numerical aspects can leaf through this chapter and keep it for future reference. For the reader's convenience, the chapter is concluded by a summary of the most important results that are used in the sequel.

Chapter 4 is concerned with the variational formulation of multidimensional elliptic boundary value problems, which is a very powerful way of rewriting certain partial differential equation problems in an abstract Hilbertian setting, in order to prove the existence and uniqueness of solutions. We provide many examples of such problems. Most of these examples are problems of second order, i.e., the maximum order of derivatives appearing in the partial differential equation is two, with one fourth order example. We also consider a variety of boundary conditions.

The variational formulation also makes it possible to devise numerical approximation methods in a very natural and unified fashion. We introduce such methods in Chap. 5. The approximate problems are set in the same framework as the continuous problem. This framework also provides a fundamental error estimate. Of particular interest to us is the finite element method introduced and analyzed in detail on simple one-dimensional examples.

In Chap. 6, we study the generalization of the finite element method to two-dimensional elliptic problems. We start by a detailed presentation of approximation using rectangular finite elements. We provide a convergence estimate in the case of the Lagrange rectangular element of lowest possible degree and give indications about convergence for higher degrees. We then introduce the concept of barycentric coordinates and use them to describe Lagrange triangular finite elements of degree up to three.

The elliptic problems considered so far correspond to the modeling of static or equilibrium situations, with no time evolution. We then turn to evolution problems in Chap. 7 with the theoretical study of the heat equation, which is the prototypical parabolic equation: maximum principle, existence and uniqueness of regular solutions, energy estimates, variational formulation and weak solutions. New mathematical tools are needed, mainly Hilbert space-valued function spaces, which we introduce as the need arises. We also mention and discuss the heat kernel in the case of the heat equation in the whole space.

We next consider the numerical approximation of the heat equation in Chap. 8. We focus on the finite difference method, already seen for static problems in Chap. 2. We consider several finite difference schemes of various precision. The convergence of such schemes rest on their consistency and stability. The analysis of these schemes is significantly more complicated in the evolution case than in the static case, due in particular to rather subtle stability issues, which we analyze in detail and from several different viewpoints. We also mention other methods such as the finite difference-finite element method, in which time is discretized using finite differences and space is discretized using finite elements.

Chapter 9 is devoted to both theoretical and numerical analyses of another classical evolution problem, the wave equation. This equation is the prototypical hyperbolic equation of second order. We first study the continuous problem and prove the existence and uniqueness of regular solutions and then of weak solutions. We next consider finite difference schemes for the wave equation. The stability issues are again significantly subtler than in the case of the heat equation, and take up most of the exposition concerning numerical methods.

Finally, we present the finite volume method in Chap. 10, again on examples. Finite volume methods are the most recent of the numerical methods covered in the book. They are currently widely used in certain areas of applications such as computational fluid dynamics. We start with the one-dimensional elliptic problem of Chap. 2 with a description of the finite volume discretization and a complete convergence analysis. We then consider the one-dimensional transport equation, which is the prototypical hyperbolic equation of first order. This equation is solved via the method of characteristics. We then introduce several linear finite volume schemes and study their properties of consistency, stability and convergence. We also present the method on a few examples in the nonlinear case and for the two-dimensional transport equation.

The contents of this book are significantly expanded from a series of first year Master degree classes, taught by the authors at UPMC (Université Pierre et Marie Curie) in Paris, France, over several years. It is intended to be as self-contained as possible. It consequently provides more than enough material for a one semester Master class on the subject. The book can also serve as a wide spectrum monograph for more advanced readers.

We are indebted to the students who have attended our classes and asked many questions that contributed to making the text more readable. We would like to thank our colleagues Muriel Boulakia, Edwige Godlewski, Sidi Mahmoud Kaber and Gérard Tronel, for carefully reading parts of the manuscript and making numerous suggestions that improved it significantly. We also thank the anonymous referees for making several constructive comments.

Paris, France

Hervé Le Dret
Brigitte Lucquin

Contents

Chapter 1
Mathematical Modeling and PDEs

In this chapter, we consider several concrete situations stemming from various areas of applications, the mathematical modeling of which involves partial differential equation problems. We will not be rigorous mathematically speaking. There will be quite a few rather brutal approximations, not always convincingly justified. This is however the price to be paid if we want to be able to derive mathematical models that aim to describe the complex phenomena we are dealing with in a way that remains manageable. At a later stage, we will study some of these models with all required mathematical rigor.

The simplest examples arise in mechanics. Let us start with the simplest example of all.

1.1 The Elastic String

Let us first consider the situation depicted in Fig. 1.1.

What is an elastic string in real life? The term can refer to several different objects, such as a stretched rubber band, the strings of a musical instrument made of nylon or steel, or again a cabin lift cable. Up to a certain level of approximation, all these objects are modeled in the same way. What they all have in common is a very small aspect ratio: they are three-dimensional, however much thinner in two directions than in the third one. Thus, the first step toward a simple mathematical model is to simply declare them to be one-dimensional at the onset. Points in a string will thus be labeled by a single real-valued variable x belonging to a segment $[0, L]$, embedded in $\mathbb{R}^2$ or $\mathbb{R}^3$. Another implicit assumption used here is the assumption of continuum mechanics. We assume that matter is a continuum which can be divided indefinitely. This is obviously untrue, but it is true enough at the macroscopic scale to be an extremely effective modeling hypothesis.

H. Le Dret and B. Lucquin, *Partial Differential Equations: Modeling, Analysis and Numerical Approximation*, International Series of Numerical Mathematics 168, DOI 10.1007/978-3-319-27067-8_1

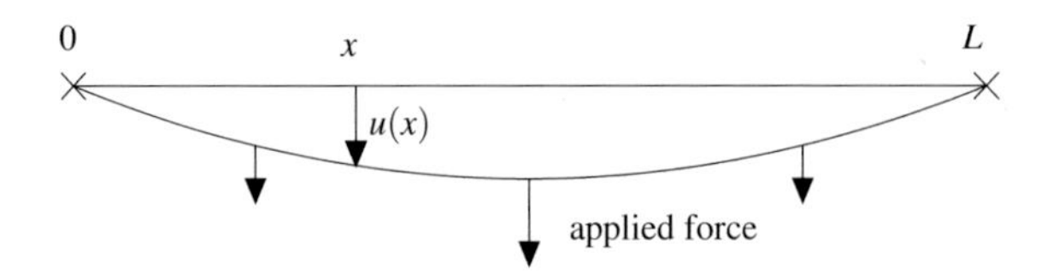

Fig. 1.1 An elastic string stretched between two points and pulled by some vertical force. The point initially located at x moves by a vertical displacement $u(x)$ to an equilibrium position

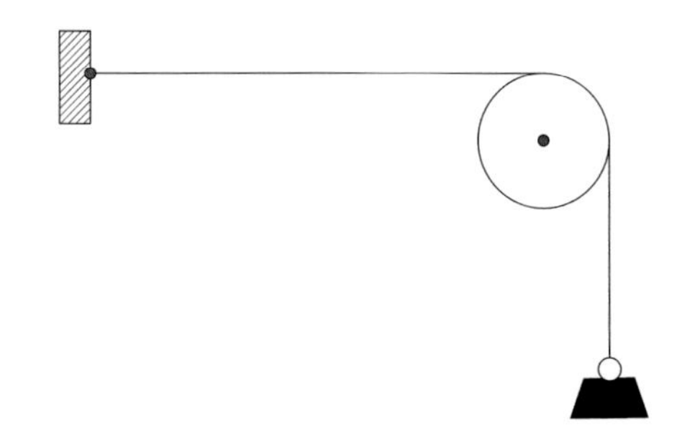

Fig. 1.2 Stretching a string with a wheel and a weight

Fig. 1.3 A piece of string cut at x and y, then kept in equilibrium

We assume that the string is stretched with a tension $T > 0$. The tension is just the force that is applied at both extremities 0 and L in order to make the string taut, for instance by working the tuners of a guitar or by passing the string on a wheel and suspending a weight at the end with the other end attached to a wall, see Fig. 1.2. In terms of physical units, the tension is measured in Newton (N). If the only force acting on the string is the tension, then the string settles in an equilibrium position which is none but the segment $[0, L]$.

We now perform a thought experiment by cutting the string at two points of $[0, L]$, located at two arbitrary abscissae x and y with $x < y$. If the piece of string that has just been cut is going to stay in the same place, it is quite clear that we need to apply horizontal forces $T_-(x) < 0$ and $T_+(y) > 0$ at the two newly created extremities, in order to compensate for the disappearance of the rest of the string, see Fig. 1.3.

As a matter of fact, $T_-(x)$ is the force that was exerted by the $[0, x]$ part of the string on the segment $[x, y]$ at point x before the string was cut, and $T_+(y)$ is likewise the force formerly exerted by the $[y, L]$ part of the string at point y. The action-reaction principle immediately shows that we have $T_-(x) = -T_+(x)$. Let us thus just set $T(x) = T_+(x)$.

Since the cut out piece of string stays in equilibrium and the only forces acting on it are the above two tensions, Newton's law of motion implies that the resultant force vanishes, that is to say that

$$T(y) - T(x) = 0.$$

Now this holds true for all x and y, therefore the tension $T(x)$ inside the string is constant. Taking $x = 0$ or $x = L$, we see that this constant is equal to T, the string tension applied at the extremities. This is an important—even if obvious—result, because it holds true irrespective of the physical nature of the string. Whether a string is made of rubber, nylon or steel, the tension inside is constant and equal to the applied tension. This is quite remarkable.

From now on, we will work with planar deformations, that is to say, we assume that the string lives in $\mathbb{R}^2$. Of course, a similar model can be derived in three dimensions. Let us apply other forces to the string, for example its weight or the weight of an object that is suspended to the string. For simplicity, we assume that this extra force is perpendicular to the segment—we will say that it is vertical whereas the segment is considered to be horizontal—and described by a lineic density f. This means that we are given a function $f\colon [0, L] \to \mathbb{R}$ such that the vertical component of the force applied to a portion $[a, b]$ of the string is equal to the integral $\int_a^b f(x)\,dx$. Such is the case of the weight of the string. Assuming the string is homogeneous, then its weight is represented by the function $f(x) = -\rho g$, where ρ is the string mass per unit length and g is the gravitational acceleration. If we suspend a weight P to a device occupying a segment $[\alpha, \beta]$ of the string, we may take $f(x) = -P\mathbf{1}_{[\alpha,\beta]}(x)$, where $\mathbf{1}_E$ denotes the characteristic function of a set E: $\mathbf{1}_E(x) = 1$ if $x \in E$, $\mathbf{1}_E(x) = 0$ otherwise.

Due to the extra applied force, the string deforms and settles in a new, unknown equilibrium position that we wish to determine. We assume that any point initially situated at $(x, 0)$ moves vertically and reaches an equilibrium position $(x, u(x))$, as in Fig. 1.1. This verticality hypothesis is again a *modeling hypothesis*. It is not strictly speaking true. In reality, the point in question also settles a little bit to the left or to the right of the vertical position. However, this hypothesis is reasonable when the force is vertical and the displacement is small. In this case, it can be justified, and we just admit it here, that the horizontal displacement is negligible in comparison with the vertical displacement $u(x)$. If the force was slanted, or the displacement large, it would be an entirely different story.

The deformed string is at this point described by a parametric curve in $\mathbb{R}^2$, $x \mapsto (x, u(x))$ where u is an unknown function to be determined. We now make another modeling hypothesis, which is that we are only interested in those situations where the derivative $u'(x)$ has small absolute value compared to 1. In this case, the length element of the deformed string satisfies

$$\sqrt{1 + u'(x)^2} \approx 1 + \frac{1}{2}u'(x)^2 \approx 1,$$

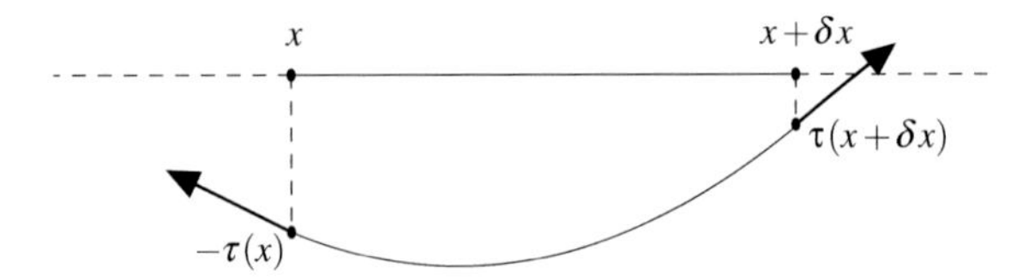

Fig. 1.4 Cutting out a piece of the deformed string

since if $u'(x)$ is small, then $u'(x)^2$ is negligible.[1] We are thus dealing with situations in which the string is approximately inextensional, i.e., there are basically no length changes compared to the initial straight stretched string.

Let us pick up our thought experiment scissors again and cut the string between abscissae x and $x + \delta x$, $\delta x > 0$. This time, the string is no longer straight. When we think about the forces exerted by the rest of the string on the extremities of the part that was cut, it appears reasonable that these forces should be tangent to the deformed string at the cut points, see Fig. 1.4. This is yet another modeling hypothesis, which can be justified by a more refined mechanical analysis.

The thought experimenter thus applies a tension force of the form $-T(x)\tau(x)$ at point $(x, u(x))$, and a tension force $T(x+\delta x)\tau(x+\delta x)$ at point $(x+\delta x, u(x+\delta x))$, where τ is the unit tangent vector

$$\tau(x) = \frac{1}{\sqrt{1+u'(x)^2}}\begin{pmatrix}1\\u'(x)\end{pmatrix} \approx \begin{pmatrix}1\\u'(x)\end{pmatrix},$$

in order to keep the piece that was cut out in equilibrium. The above approximation of the tangent vector is legitimate in view of our decision to neglect terms involving $u'(x)^2$ and other quantities of higher order.

We apply Newton's law of motion again, which yields the vector equation

$$T(x+\delta x)\tau(x+\delta x) - T(x)\tau(x) + \begin{pmatrix}0\\ \int_x^{x+\delta x} f(s)\,ds\end{pmatrix} = \begin{pmatrix}0\\0\end{pmatrix}.$$

The horizontal component of the equation implies that $T(x) = T$ is the same constant as before. The vertical component then reads

$$T\big(u'(x+\delta x) - u'(x)\big) + \int_x^{x+\delta x} f(s)\,ds = 0,$$

using the above approximation for the tangent vector. Dividing everything by δx, we obtain

[1] As a general rule, we neglect all terms of order strictly higher than one with respect to $u'(x)$. This leads to a simplified *linearized* model. A model that would take into account such higher order terms would be by nature *nonlinear*, and thus a lot more difficult to study from the point of view of mathematics.

$$-T\frac{u'(x+\delta x)-u'(x)}{\delta x}=\frac{1}{\delta x}\int_x^{x+\delta x}f(s)\,ds.$$

Since one of our modeling hypotheses is that matter can be indefinitely divided, the above relation holds true for any value of δx, no matter how small. We thus let δx tend to 0. The left-hand side is a differential quotient, the right-hand side is an average over a small interval, and we thus obtain in the limit $\delta x \to 0$,

$$-Tu''(x)=f(x),$$

which can be rewritten as the first equation of the following *string problem:*

$$\begin{cases} -u''(x)=\dfrac{1}{T}f(x) \text{ for all } x \text{ in }]0,\ L[, \\ u(0)=u(L)=0. \end{cases} \tag{1.1}$$

The second line of (1.1) expresses the fact that the string is fixed at the endpoints $x = 0$ and L. These points never move and the displacement is zero there. This condition is called a *boundary condition*. Problem (1.1) consists of a differential equation in an open set (here an ordinary differential equation, since we are in dimension one), together with a condition on the boundary of the open set. This type of problem is called a *boundary value problem*, and we will see many more of them.

If we are somehow capable of solving this problem, then we will have determined the deformed shape of the string under the action of the applied forces. Indeed, it is easily checked by following the computations backward, that any solution of problem (1.1) yields an equilibrium position for the string, at least within the range of approximations that have been made.

Remark 1.1 In order to appease natural suspicions that it does not feel right to neglect terms before differentiating them, we can note that

$$\Big(\frac{u'(x)}{\sqrt{1+u'(x)^2}}\Big)'=\frac{u''(x)}{\sqrt{1+u'(x)^2}}+\frac{u'(x)^2u''(x)}{(1+u'(x)^2)^{3/2}}\approx u''(x),$$

therefore it was not so bad, a posteriori. □

Remark 1.2 Let us admit for the time being that problem (1.1) has a unique solution for given f and T. If we consider the same string subjected to the same force, but with different tensions, we see that the displacement is inversely proportional to the tension: the tauter the string, the more rigid it behaves and conversely. This is in agreement with daily life.

Let us emphasize again that the string model is independent of the physical nature of the string, which can be counterintuitive. A rubber string and a steel string stretched with the same tension have the same behavior insofar as the model is concerned.

Once the model is derived and solved, a natural question arises: how far apart is the model solution from real life or how large is the modeling error? This is a difficult question in all generality. However, a few rules of thumb can be useful. For instance, if a given modeling hypothesis is clearly not satisfied by the model solution, then its validity is dubious at best. For example, in the case of a string, if we find a solution which is such that $|u'(x)|$ is large for some values of x, in the sense that $u'(x)^2$ can no longer be neglected, then we know that a more precise model is needed. Of course, this is still rather vague. An ideal situation would be one in which it was possible to have an explicit quantitative estimate of the difference in the solutions of models of various accuracy. Unfortunately, this is rarely possible. □

It is important to understand that, even though the string model derived here involves an ordinary differential equation because we are in dimension one, a boundary value problem such as problem (1.1) has *strictly nothing to do* with the Cauchy problem for the same ordinary differential equation, either in terms of theory or in terms of numerical approximation.

In particular, the numerical schemes used for the Cauchy problem, such as the forward and backward Euler methods or the Runge-Kutta method (see for example [8, 16, 23, 60, 64]), are of no use to compute approximations of the solution of problem (1.1) (except in the shooting method, see below). Different, more adapted numerical methods are needed, and we will discuss several of them later on.

To illustrate this, let us introduce a slightly generalized version of the string problem. We thus consider the following boundary value problem:

$$\begin{cases} -u''(x) + c(x)u(x) = f(x) \text{ in }]0, L[, \\ u(0) = A, u(L) = B, \end{cases} \tag{1.2}$$

where f and c are two given functions defined on $]0, L[$ and A, B are two given constants. The function c has no specific mechanical interpretation in the context of the elastic string. It just adds generality without costing any extra complexity. The boundary condition in (1.2) is called a *Dirichlet boundary condition*. In general, imposing a Dirichlet condition means ascribing given values to the unknown function u on the boundary. In the case when $A = B = 0$, it is called a *homogeneous* Dirichlet condition.

Let us now see some of the fundamental differences between a boundary value problem and a seemingly similar Cauchy problem. The Cauchy problem consists in replacing the boundary conditions in (1.2) by initial conditions of the form $u(0) = \alpha$, $u'(0) = \beta$ and no condition at $x = L$. We know from the classical theory of ordinary differential equations (see [16, 23, 60] for example) that this Cauchy problem has one and only one solution if for example c and f are continuous. The boundary value problem (1.2) may however not have any solution at all under the same hypotheses!

Let us take the following apparently innocuous example: $L = 1$, $A = B = 0$, $c(x) = -\pi^2$ and $f(x) = 1$. We can treat the first equation in (1.2) as a linear ordinary differential equation of second order with constant coefficients that admits the general solution

$$u(x) = \lambda \cos(\pi x) + \mu \sin(\pi x) - \frac{1}{\pi^2},$$

where λ and μ are two real parameters. If we write down the homogeneous Dirichlet boundary conditions, we obtain two relations

$$\lambda - \frac{1}{\pi^2} = 0, \quad -\lambda - \frac{1}{\pi^2} = 0,$$

which are impossible to satisfy simultaneously. Consequently, there is no solution to this particular boundary value problem.

More generally, let us assume that the problem

$$\begin{cases} -u''(x) - \pi^2 u(x) = f(x) \text{ in }]0, 1[, \\ u(0) = u(1) = 0, \end{cases}$$

has a regular solution that is for example of class C^2. We multiply the differential equation by $\sin(\pi x)$, which yields

$$-u''(x) \sin(\pi x) - \pi^2 u(x) \sin(\pi x) = f(x) \sin(\pi x).$$

We now integrate this equality between 0 and 1. We obtain

$$-\int_0^1 u''(x) \sin(\pi x)\, dx - \pi^2 \int_0^1 u(x) \sin(\pi x)\, dx = \int_0^1 f(x) \sin(\pi x)\, dx. \quad (1.3)$$

Now, if we integrate the first integral by parts twice, we see that

$$\int_0^1 u''(x) \sin(\pi x)\, dx = [u'(x) \sin(\pi x)]_0^1 - \pi [u(x) \cos(\pi x)]_0^1 - \pi^2 \int_0^1 u(x) \sin(\pi x)\, dx.$$

The first two terms vanish because the sine function vanishes at $x = 0$ and $x = 1$ for the first one, and u vanishes at the same points for the second one by the homogeneous Dirichlet condition. The remaining integral cancels out with the second integral in Eq. (1.3). Finally, we find that if the problem has a solution u, then

$$\int_0^1 f(x) \sin(\pi x)\, dx = 0,$$

which is therefore a necessary condition for existence. By contraposition, if f is such that $\int_0^1 \sin(\pi x) f(x)\, dx \neq 0$, for instance $f = 1$, then the problem cannot have any regular solution.

Concerning uniqueness, we note that for $f = 0$, we have the infinite family of solutions $u(x) = \mu \sin(\pi x)$, $\mu \in \mathbb{R}$. Hence, there is no uniqueness either. □

The above trick of multiplying the equation by certain well-chosen functions and integrating the result by parts will be at the heart of the existence and uniqueness theory using variational formulations, as well as the basis of such variational approximation methods as the finite element method that we will encounter later on.

We can already prove a uniqueness result. The problem in the previous example is that the function c is negative (and take the specific value $-\pi^2$).

Theorem 1.1 *If c is a continuous, nonnegative function, then problem* (1.2) *has at most one solution of class $C^2([0, L])$.*

Proof Let u_1 and u_2 be two solutions of class $C^2([0, L])$, and set $w = u_2 - u_1$. It is easily checked that w solves the homogeneous boundary value problem:

$$\begin{cases} -w''(x) + c(x)w(x) = 0 \text{ in }]0, L[, \\ w(0) = w(L) = 0. \end{cases} \tag{1.4}$$

We multiply the differential equation by w and integrate between 0 and L. This yields

$$-\int_0^L w''(x)w(x)\,dx + \int_0^L c(x)w(x)^2\,dx = 0.$$

Integrating the first term by parts once, we obtain

$$\int_0^L [w'(x)^2 + c(x)w(x)^2]\,dx = 0,$$

since $[w'w]_0^L = 0$, given the boundary conditions satisfied by w. The integrand is a continuous function which is nonnegative due to the sign hypothesis for c. Its integral is zero, hence it is identically zero. In particular, $w'(x) = 0$, which implies $w(x) = w(0) = 0$ for all x, hence $u_1 = u_2$, which is the uniqueness result. □

Let us say a few words about the *shooting method* alluded to above, applied to boundary value problem (1.2). The idea is to use the Cauchy problem for the same ordinary differential equation, $-u''(x) + c(x)u(x) = f(x)$ on $]0, L[$, with initial values $u(0) = A$, $u'(0) = \lambda$ and to try and adjust the parameter λ so as to reach the target B for $x = L$, i.e., $u(L) = B$, see Fig. 1.5.

This Cauchy problem has one and only one solution, which we denote by u_λ, and we consider the shooting mapping $S\colon \mathbb{R} \to \mathbb{R}$, $S(\lambda) = u_\lambda(L)$. The boundary value problem (1.2) thus has a solution for all B if and only if the mapping S is onto. Let us study this mapping.

Lemma 1.1 *The mapping S is affine.*

Proof Let $\lambda \in \mathbb{R}$. We claim that $u_\lambda = \lambda u_1 + (1 - \lambda)u_0$. Indeed, if we set $v = \lambda u_1 + (1 - \lambda)u_0$, we see that $v(0) = \lambda A + (1 - \lambda)A = A$, $v'(0) = \lambda u_1'(0) + (1 - \lambda)u_0'(0) = \lambda$, and

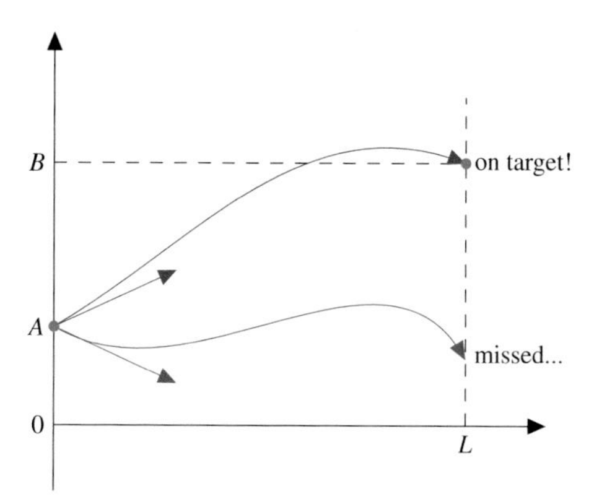

Fig. 1.5 The shooting method

$$-v''(x)+c(x)v(x) = \lambda(-u_1''(x)+c(x)u_1(x))+(1-\lambda)(-u_0''(x)+c(x)u_0(x)) = f(x)$$

in $]0, L[$, hence the claim by uniqueness of the Cauchy problem. The Lemma follows then by taking $x = L$, which shows that $S(\lambda) = \lambda S(1) + (1-\lambda)S(0)$. □

Theorem 1.2 *If c is a continuous, nonnegative function, then problem* (1.2) *has one and only one solution of class* $C^2([0, L])$.

Proof The mapping $S : \mathbb{R} \to \mathbb{R}$ is affine by Lemma 1.1. Let us show that it is one-to-one and onto when $c \geq 0$. Since $S(\lambda) = (S(1) - S(0))\lambda + S(0)$, it is enough to show $S(0) \neq S(1)$. Let thus assume that $S(0) = S(1)$. We set $w = u_1 - u_0$, which is thus a solution of problem (1.4). By Theorem 1.1, it follows that $w = 0$, therefore $1 = u_1'(0) = u_0'(0) = 0$, contradiction. □

We have seen that existence and uniqueness fail in the particular case $L = 1$ and $c = -\pi^2 < 0$.

The shooting method can also be used in a numerical context, and in the present case, it is particularly simple. Indeed, since S is affine, it is sufficient to compute its values for two different values of λ to determine it entirely. This can be done numerically using ordinary differential equation schemes such as the forward Euler scheme or the fourth-order Runge-Kutta scheme for better precision.

The shooting method also applies to nonlinear boundary value problems in one dimension. In this case, the mapping S is no longer affine, and computing two of its values would be the initialization step in the application of a bisection method or a secant method, for example. Now of course, there is no analogue of the shooting method in dimensions higher than one, which severely restricts its interest.

Such boundary value problems as (1.2) have an important property called the *maximum principle* [35, 40]. As we will see shortly, the proof is banal in dimension one. However, the maximum principle is a profound property in dimensions strictly greater than one.

Theorem 1.3 (Maximum Principle) *Assume that* $c \geq 0$ *and that problem* (1.2) *has a solution* u *of class* C^2. *If* $f \geq 0$ *in* $]0, L[$, $A \geq 0$ *and* $B \geq 0$, *then we have* $u \geq 0$ *in* $[0, L]$.

Proof We argue by contradiction by assuming that there exists a point x_0 such that $u(x_0) < 0$. Since $u(0) = A \geq 0$ and $u(L) = B \geq 0$, it follows that $x_0 \in]0, L[$. Now u is continuous, therefore there is an interval $[\alpha, \beta]$ such that $x_0 \in [\alpha, \beta] \subset [0, L]$ and $u \leq 0$ on $[\alpha, \beta]$. We may assume that $u(\alpha) = u(\beta) = 0$ by the intermediate value theorem.

On the interval $[\alpha, \beta]$, c and f are positive and u is nonpositive, therefore

$$u''(x) = c(x)u(x) - f(x) \leq 0.$$

We deduce from this that the function u is concave on $[\alpha, \beta]$.

Now as $x_0 \in [\alpha, \beta]$, there exists $\lambda \in [0, 1]$ such that $x_0 = \lambda\alpha + (1 - \lambda)\beta$. Consequently, the concavity of u implies that

$$u(x_0) \geq \lambda u(\alpha) + (1 - \lambda)u(\beta) = 0,$$

which is a contradiction. □

Remark 1.3 Under the form given above, it is a little hard to see where the maximum of the principle is... because it is hiding. Anyway, one way of understanding Theorem 1.3 is to see it as a monotonicity result. Indeed, the function that maps the data triple (f, A, B) to the solution u is monotone. Thus, in the case of the elastic string, when $A = B = 0$ and $f \geq 0$, in other words when we pull upwards on the string, then $u \geq 0$, which means that the string bends upwards too. So we see a very natural physical interpretation of the maximum principle that is in agreement with our intuition. This is also the reason why, in mathematics, the operator $-u''$, or more generally $-\Delta u = -\sum_{i=1}^{d} \frac{\partial^2 u}{\partial x_i^2}$ in higher dimension d, is preferred to the operator u'', which has the opposite behavior. □

1.2 The Elastic Beam

Our second example is also an example taken from mechanics. However, the mathematical modeling of this example is considerably more complicated than that of the string, and we will not explain it here. We are again dealing with essentially one-dimensional objects that are a lot more rigid than the previous ones, such as a metal rod, a concrete pillar or a wooden beam. Such objects exhibit a strong resistance to bending, as opposed to strings.

If we assume that our beam is clamped in a rigid wall at both ends, see Fig. 1.6, then the following boundary value problem is found for the vertical displacement u:

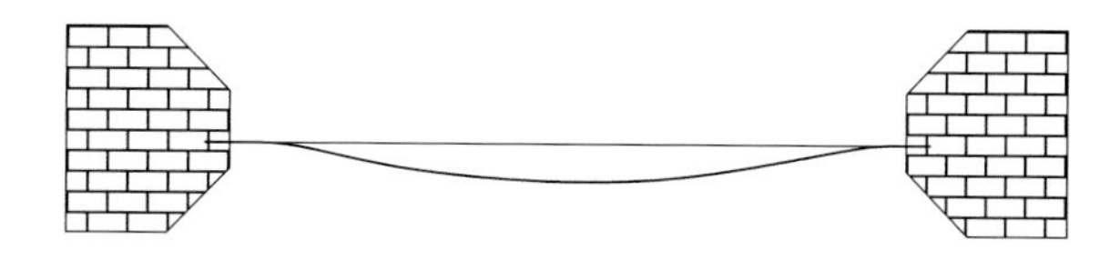

Fig. 1.6 A beam clamped at both ends

$$\begin{cases} EI\,u^{(4)}(x) = f(x) \text{ in }]0, L[, \\ u(0) = u'(0) = u(L) = u'(L) = 0, \end{cases}$$

where f is again the density of the applied vertical force, $E > 0$ is a coefficient which is characteristic of the material of the beam[2] and I is a geometric coefficient depending on the shape of the cross-section of the beam.[3] This is in striking contrast with the string model in which the nature of the string material and the shape of the cross-section of the string play no role whatsoever. Note also that there is no tension in a beam.

The differential equation is a fourth order equation ($u^{(4)}$ denotes the fourth derivative of u), as opposed to a second order equation in the case of the string, and accordingly, the Dirichlet boundary conditions involve both u and u'.

We can generalize the equation in the same spirit as before by considering the boundary value problem

$$\begin{cases} u^{(4)}(x) - (a(x)u'(x))' + c(x)u(x) = f(x) \text{ in }]0, L[, \\ u(0) = u'(0) = u(L) = u'(L) = 0, \end{cases} \tag{1.5}$$

where the given functions a and c still have no particular mechanical meaning. We also have uniqueness of C^4 solutions when a and c are nonnegative. Indeed, if $w = u_2 - u_1$ is the difference between two solutions, multiplying by w the differential equation satisfied by w, which is the first equation in (1.5) with zero right-hand side, and integrating by parts as many times as needed, we obtain

$$\int_0^L [(w''(x))^2 + a(x)(w'(x))^2 + c(x)w(x)^2]\,dx = 0,$$

whence $w''(x) = 0$. Consequently, w is an affine function of the form $w(x) = \alpha x + \beta$. Since it vanishes at both ends, we deduce that $w = 0$.

Remark 1.4 A word of warning: there is no maximum principle for such problems as (1.5) in general. The maximum principle is a property of second order boundary value problems that does not extend to fourth order problems. Physically, this means that some mechanical systems modeled by a fourth order equation may exhibit the strange behavior that when pulled downwards, part of such systems may move upwards. □

---

[2] The coefficient E is called the Young modulus of the material. It is measured in units of pressure. The higher the coefficient, the more rigid the material.

[3] I is an inertia momentum of the cross-section.

1.3 The Elastic Membrane

Let us now switch to actual PDEs in more than one dimension. The first example is still taken from mechanics. It is the two-dimensional version of the elastic string, and it is called the elastic membrane. As we will see, many of the characteristics of the elastic string carry over to the elastic membrane.

To get a feeling for what an elastic membrane is, think of plastic wrap suitable for food contact that you can find in your favorite supermarket. Stretch the film up to the sides of some container in order to seal it before you store it in the fridge. In the beginning, the stretched part of the plastic film is planar. Then, as the temperature of the air inside the container goes down, the inside air pressure diminishes. At the same time, the atmospheric pressure inside the fridge remains more or less constant (you are bound to open the door every once in a while). The pressure differential thus created pushes on the film which bends inwards as a result. We wish to determine the final shape of the film in three-dimensional space.

This kitchen example above is by far not the only one. There are many instances of elastic membranes around: the skin of a drum, a biological cell membrane, the sails of a boat, a party balloon, and so on.

To model this situation, let us be given an open set Ω of $\mathbb{R}^2$, whose boundary $\partial\Omega$ represents the edge of the container opening. Each point x of the closure $\bar{\Omega}$ of Ω represents a material point of the membrane when it is stretched without any other applied force. Again, we identify a small aspect ratio, three-dimensional object with a two-dimensional object filling $\bar{\Omega}$.

We now subject the membrane to a given force density, such as the above pressure differential, which is orthogonal to its plane, and is represented by a given function $f : \Omega \to \mathbb{R}$. This time, f is a surfacic force density and the resultant force applied to a part ω of Ω is given by the integral $\int_\omega f(x_1, x_2)\, dx_1 dx_2$.

As in the case of the elastic string, we make the reasonable albeit approximative hypothesis that point x is displaced by a quantity $u(x)$ perpendicularly to the membrane (vertically in Fig. 1.7). The displacement u is thus now a function in two variables $u : \bar{\Omega} \to \mathbb{R}$, and the shape of the membrane at equilibrium is a parametric surface in $\mathbb{R}^3$ given by $(x_1, x_2) \mapsto (x_1, x_2, u(x_1, x_2))$.

Since we assume that the membrane sticks to the opening of the container, we get at once a homogeneous Dirichlet boundary condition to be satisfied by the displacement u of the membrane

$$u(x) = 0 \text{ on } \partial\Omega. \tag{1.6}$$

This condition is the exact analogue in two dimensions of the Dirichlet boundary condition for an elastic string that is attached at both of its ends.

We next need to obtain an equation that will determine the function u in Ω, and based on our previous one-dimensional experience, we can expect partial differential equations to play the leading role here.

Figure 1.7 represents the graph of the function $u(x_1, x_2) = (1 - x_1^2)(1 - x_2^2)$ on the square $\Omega =]-1, 1[^2$, which is the solution of the as yet unwritten membrane

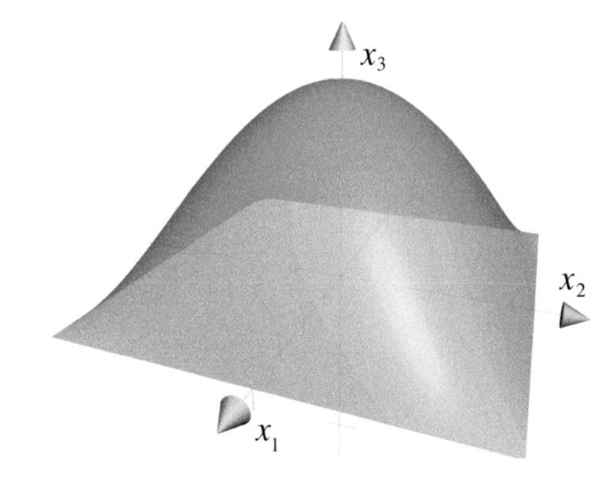

Fig. 1.7 An elastic membrane stretched with tension $T = 1$ on the square $]-1, 1[^2$

problem (1.8) for the vertical force $f(x_1, x_2) = 4 - 2(x_1^2 + x_2^2)$ and homogeneous Dirichlet boundary conditions.

The two vectors

$$a_1 = \begin{pmatrix} 1 \\ 0 \\ \frac{\partial u}{\partial x_1}(x_1, x_2) \end{pmatrix} \text{ and } a_2 = \begin{pmatrix} 0 \\ 1 \\ \frac{\partial u}{\partial x_2}(x_1, x_2) \end{pmatrix}$$

form a basis of the tangent plane to the surface at point $(x_1, x_2, u(x_1, x_2))$.

As in the case of the elastic string, we will only consider situations in which $\|\nabla u\| = \sqrt{\left(\frac{\partial u}{\partial x_1}\right)^2 + \left(\frac{\partial u}{\partial x_2}\right)^2}$ is small (which is not exactly the case in Fig. 1.7!). This hypothesis leads us to neglect all quantities that are at least quadratic in the partial derivatives of u. In particular, when we normalize the above tangent basis vectors, we obtain the approximation

$$\frac{a_i}{\|a_i\|} = \frac{1}{\sqrt{1 + \left(\frac{\partial u}{\partial x_i}\right)^2}} a_i \approx a_i,$$

which is analogous to the approximate normalization of the tangent vector to the deformed elastic string used earlier.

Let us now explain what the word tension means in the case of a membrane. Because we are in a two-dimensional setting, the situation is a bit more complicated than for the elastic string. The general principle remains however the same. Let us consider a part A of the membrane and isolate this part as if it was cut out of the membrane. Just like the cut piece of string before, what keeps the part A in place must be forces exerted by the rest of the membrane. It seems reasonable to assume that these forces are exerted exactly on the boundary Γ_A of A relative to the membrane, since the membrane cannot act at a distance. Now the boundary in question is a curve, so that the force in question must be given by a lineic density distributed on

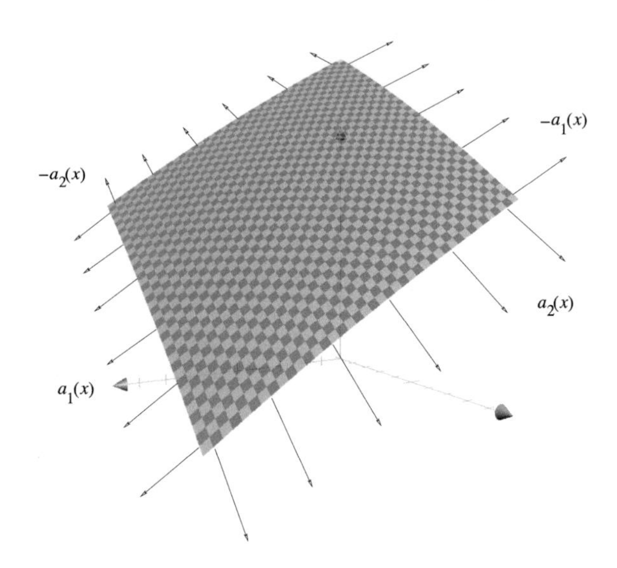

Fig. 1.8 Magnified view of a small square cut out of the membrane. A few of the normal vectors are drawn. The force density exerted by the rest of the membrane is equal to T times these vectors. We can see it pulling to stretch the piece of membrane

Γ_A, the resultant force being the integral of the density on Γ_A. This is general for all two-dimensional continuum mechanics models.

In the case of an elastic membrane, as in the case of a string, we assume that the above force density lies in the tangent plane to the deformed surface, and furthermore, that it is normal to Γ_A in the tangent plane and pointing outwards, see Fig. 1.8. Actually, this assumption can be seen as the very definition of an elastic membrane. The tension $T > 0$ is the norm of this force density—we admit here for simplicity that this norm is a constant, independent of the point[4]—it measures the tautness of the membrane. The physical unit for T is the Newton per meter (N/m).

Let us thus take the scissors out again and cut a small square out of the membrane around an arbitrary point $(\bar{x}, u(\bar{x}))$ with $\bar{x} = (\bar{x}_1, \bar{x}_2)$. More precisely, we consider the square

$$C_{\bar{x},\delta x} =]\bar{x}_1 - \delta x, \bar{x}_1 + \delta x[\times]\bar{x}_2 - \delta x, \bar{x}_2 + \delta x[,$$

in $\mathbb{R}^2$, with $\delta x > 0$, and cut out its image by the mapping $x \mapsto (x, u(x))$ in $\mathbb{R}^3$, see Fig. 1.8. We also make no distinction between the boundary of the image of $C_{\bar{x},\delta x}$ in $\mathbb{R}^3$ and its boundary as a subset of $\mathbb{R}^2$ in the computation of the integrals. This is because $\|\nabla u\|$ is assumed to be small. We already made this approximation in the case of the string, without mentioning it... The exact computations can of course be performed in order to make sure that this approximation is really justified.

In order to compute the integral, we number the four sides of the square counter-clockwise: $\gamma^1_{\bar{x},\delta x} =]\bar{x}_1-\delta x, \bar{x}_1+\delta x[\times\{\bar{x}_2-\delta x\}, \gamma^2_{\bar{x},\delta x} = \{\bar{x}_1+\delta x\}\times]\bar{x}_2-\delta x, \bar{x}_2+\delta x[$, and so on for $\gamma^3_{\bar{x},\delta x}$ and $\gamma^4_{\bar{x},\delta x}$. According to the normal vectors depicted in Fig. 1.8, Newton's law of motion for the vertical force component then reads

[4]This can of course be proved with a little more work.

$$T\Big[\int_{\gamma^1_{\bar{x},\delta x}} -[a_2(x)]_3\, d\gamma + \int_{\gamma^2_{\bar{x},\delta x}} [a_1(x)]_3\, d\gamma + \int_{\gamma^3_{\bar{x},\delta x}} [a_2(x)]_3\, d\gamma + \int_{\gamma^4_{\bar{x},\delta x}} -[a_1(x)]_3\, d\gamma\Big] + \int_{C_{\bar{x},\delta x}} f(x)\, dx = 0, \quad (1.7)$$

where $[z]_3$ denotes the vertical component of vector z. It is a simple exercise to check that the horizontal components already satisfy Newton's law. Let us write each integral separately. We have

$$\int_{\gamma^1_{\bar{x},\delta x}} [a_2(x)]_3\, d\gamma = \int_{\bar{x}_1-\delta x}^{\bar{x}_1+\delta x} \frac{\partial u}{\partial x_2}(x_1, \bar{x}_2 - \delta x)\, dx_1,$$

$$\int_{\gamma^2_{\bar{x},\delta x}} [a_1(x)]_3\, d\gamma = \int_{\bar{x}_2-\delta x}^{\bar{x}_2+\delta x} \frac{\partial u}{\partial x_1}(\bar{x}_1 + \delta x, x_2)\, dx_2,$$

$$\int_{\gamma^3_{\bar{x},\delta x}} [a_2(x)]_3\, d\gamma = \int_{\bar{x}_1-\delta x}^{\bar{x}_1+\delta x} \frac{\partial u}{\partial x_2}(x_1, \bar{x}_2 + \delta x)\, dx_1,$$

$$\int_{\gamma^4_{\bar{x},\delta x}} [a_1(x)]_3\, d\gamma = \int_{\bar{x}_2-\delta x}^{\bar{x}_2+\delta x} \frac{\partial u}{\partial x_1}(\bar{x}_1 - \delta x, x_2)\, dx_2.$$

Formula (1.7) can thus be rewritten as

$$T\Big[\int_{\bar{x}_1-\delta x}^{\bar{x}_1+\delta x} \Big(\frac{\partial u}{\partial x_2}(x_1, \bar{x}_2 + \delta x) - \frac{\partial u}{\partial x_2}(x_1, \bar{x}_2 - \delta x)\Big)\, dx_1 + \int_{\bar{x}_2-\delta x}^{\bar{x}_2+\delta x} \Big(\frac{\partial u}{\partial x_1}(\bar{x}_1 + \delta x, x_2) - \frac{\partial u}{\partial x_1}(\bar{x}_1 - \delta x, x_2)\Big)\, dx_2\Big] + \int_{C_{\bar{x},\delta x}} f(x)\, dx = 0.$$

The situation is less transparent than in dimension one, but the idea is the same. We divide everything by $4(\delta x)^2$. This yields

$$-T\frac{1}{2\delta x}\Big[\int_{\bar{x}_1-\delta x}^{\bar{x}_1+\delta x} \frac{\frac{\partial u}{\partial x_2}(x_1, \bar{x}_2 + \delta x) - \frac{\partial u}{\partial x_2}(x_1, \bar{x}_2 - \delta x)}{2\delta x}\, dx_1 + \int_{\bar{x}_2-\delta x}^{\bar{x}_2+\delta x} \frac{\frac{\partial u}{\partial x_1}(\bar{x}_1 + \delta x, x_2) - \frac{\partial u}{\partial x_1}(\bar{x}_1 - \delta x, x_2)}{2\delta x}\, dx_2\Big] = \frac{1}{4(\delta x)^2}\int_{C_{\bar{x},\delta x}} f(x)\, dx.$$

Now the length of each of the segments on which differential quotients of the partial derivatives $\partial u/\partial x_i$ are integrated is exactly $2\delta x$, and the area of the square is exactly $4(\delta x)^2$. We thus see that all the above quantities are averages over small segments or squares, which is good in view of letting δx tend to 0 later.

Let us assume that u is of class C^2. We can thus write the following Taylor-Lagrange expansion at $\bar{x}$

$$\frac{\partial u}{\partial x_2}(x) = \frac{\partial u}{\partial x_2}(\bar{x}) + \frac{\partial^2 u}{\partial x_2 \partial x_1}(\bar{x})(x_1 - \bar{x}_1) + \frac{\partial^2 u}{\partial x_2^2}(\bar{x})(x_2 - \bar{x}_2) + r(x)$$

where $r(x)/\|x - \bar{x}\| \to 0$ when $\|x - \bar{x}\| \to 0$. Therefore

$$\frac{\frac{\partial u}{\partial x_2}(x_1, \bar{x}_2 + \delta x) - \frac{\partial u}{\partial x_2}(x_1, \bar{x}_2 - \delta x)}{2\delta x} = \frac{\partial^2 u}{\partial x_2^2}(\bar{x}) + r_1(x_1, \delta x)$$

where $r_1(x_1, \delta x) \to 0$ when $|x_1 - \bar{x}_1| \leq \delta x$ and $\delta x \to 0$. Integrating with respect to x_1, we obtain

$$\frac{1}{2\delta x}\int_{\bar{x}_1-\delta x}^{\bar{x}_1+\delta x} \frac{\frac{\partial u}{\partial x_2}(x_1, \bar{x}_2 + \delta x) - \frac{\partial u}{\partial x_2}(x_1, \bar{x}_2 - \delta x)}{2\delta x}\, dx_1 = \frac{\partial^2 u}{\partial x_2^2}(\bar{x}) + r_2(\delta x)$$

where $r_2(\delta x) \to 0$ when $\delta x \to 0$.

We treat the remaining integral on the boundary in the same fashion and we obtain

$$\frac{1}{2\delta x}\int_{\bar{x}_2-\delta x}^{\bar{x}_2+\delta x} \frac{\frac{\partial u}{\partial x_1}(\bar{x}_1 + \delta x, x_2) - \frac{\partial u}{\partial x_1}(\bar{x}_1 - \delta x, x_2)}{2\delta x}\, dx_2 \to \frac{\partial^2 u}{\partial x_1^2}(\bar{x})$$

when $\delta x \to 0$. Finally, assuming that f is continuous, we have

$$\frac{1}{4(\delta x)^2}\int_{C_{\bar{x},\delta x}} f(x)\, dx \to f(\bar{x}) \quad \text{when} \quad \delta x \to 0,$$

since the left-hand side is the average of f over the square.

We have thus obtained in the $\delta x \to 0$ limit

$$\forall \bar{x} \in \Omega, \quad -\Delta u(\bar{x}) = \frac{1}{T} f(\bar{x}). \tag{1.8}$$

The differential operator $\Delta = \frac{\partial^2}{\partial x_1^2} + \frac{\partial^2}{\partial x_2^2}$ is called the *Laplacian* or *Laplace operator*. Equation (1.8) is called the *elastic membrane equation*. It is a second order equation since it only involves second derivatives, the order of a partial differential equation being the highest order of derivatives that appear in the equation. The equation must naturally be complemented by some boundary conditions, such as the homogeneous Dirichlet condition (1.6).

The mechanical remarks made in the case of the elastic string also apply to the elastic membrane and we will not repeat them.

More generally, the boundary value problem in any dimension, $\Omega \subset \mathbb{R}^d$, $d \geq 1$,

$$\begin{cases} -\Delta u = f \text{ in } \Omega, \\ \quad u = 0 \text{ on } \partial\Omega, \end{cases} \tag{1.9}$$

with $\Delta u = \sum_{i=1}^{d} \frac{\partial^2 u}{\partial x_i^2}$, is called the *Poisson equation*. The Poisson equation shows up in a surprising number of different areas of mathematics and its applications. For example, for $d = 3$, if f represents the density of electrical charge present in Ω and the boundary of Ω is covered by a perfectly conducting material, then $-u$ is the electric potential[5] inside Ω. The gradient of $-u$ is the electric field. More generally, the Poisson equation is central in all questions relating to the Newtonian potential, e.g. in electromagnetism, in classical gravity.

There are many other interpretations. Thus, if f represents a density of heat sources in Ω, say the distribution of radiators in a room and how much heat they give off, then u is the equilibrium temperature in Ω when the walls of the room $\partial\Omega$ are somehow kept at temperature $0°$. This is why the Poisson equation is sometimes referred to as the diffusion equation, as it also models the diffusion of heat (and of other things that may want to diffuse). We will return to the heat equation later on.

There is also a probabilistic interpretation of the Poisson equation, not unrelated to the diffusion interpretation. For $f = 2$, $u(x)$ is the expectation of the first exit time from Ω of a standard Brownian motion starting from point x. Roughly speaking, a particle moving randomly in $\mathbb{R}^d$ and starting from a point x in $\bar{\Omega}$ will reach $\partial\Omega$ for the first time in an average time $u(x)$.

Finally, when $f = 0$, the equation is known as the *Laplace equation* whose solutions are the harmonic functions (it is clearly better to impose a nonzero boundary condition to have $u \neq 0$, or no condition at all). Harmonic functions are of course extremely important in many areas.

The Poisson equation satisfies the maximum principle, see [20, 25, 61] among others.

Theorem 1.4 (Maximum Principle) *Let Ω be a bounded open subset of $\mathbb{R}^d$ and $u \in C^2(\Omega) \cap C^0(\bar{\Omega})$ be a solution of the Poisson equation* (1.9) *with $f \geq 0$ in Ω. Then we have $u \geq 0$ in $\bar{\Omega}$.*

This is again a monotonicity result. For instance, in Fig. 1.7, the membrane is pulled upward by the applied force, and consequently bends upwards.

Let us close this section by rapidly mentioning the plate equation. A plate is to a membrane what a beam is to a string: sheet iron, concrete wall, wood plank. The clamped plate problem reads

[5]The minus sign is due to the physical convention that goes contrary to the mathematical convention in this case.

$$\begin{cases} \Delta^2 u = f \text{ in } \Omega, \\ u = \dfrac{\partial u}{\partial n} = 0 \text{ on } \partial\Omega, \end{cases} \tag{1.10}$$

where the operator $\Delta^2 = \Delta \circ \Delta = \frac{\partial^4}{\partial x_1^4} + 2\frac{\partial^4}{\partial x_1^2 \partial x_2^2} + \frac{\partial^4}{\partial x_2^4}$ is called the *bilaplacian* and $\frac{\partial u}{\partial n} = \nabla u \cdot n = \frac{\partial u}{\partial x_1} n_1 + \frac{\partial u}{\partial x_2} n_2$ is the *normal derivative* of u on the boundary, n denoting the unit exterior normal vector (we will go back to this later). This is a fourth order boundary value problem. A real plate model also contains mechanical constants which depend on the nature of the material as well as the thickness of the plate.

All the problems considered up to now are stationary problems in which time plays no role and only model equilibrium situations. Let us now talk about problems where time intervenes, that is to say evolution problems.

1.4 The Transport Equation

Let us imagine a kind of gas composed of particles moving in an infinite straight tube $\mathscr{T}$ in $\mathbb{R}^3$ of the form $\mathbb{R} \times D$, where D is a disk of unit area in the (x_2, x_3) plane. Instead of tracking each particle individually, which would be impossible in practice due to their huge number, we can describe the gas by using a function $u : \mathscr{T} \times \mathbb{R}_+ \to \mathbb{R}_+$, where $u(x, t)$ measures the quantity of particles, or rather their density at point x and instant t. This is called a *kinetic description*. The initial density of particles at $t = 0$ is denoted by $u_0(x) = u(x, 0)$. We assume it to be given, it is called an *initial condition*.

Let us count the total number of particles in a section $[y, y + \delta y] \times D$ of the tube. We disregard the fact that this number should be an integer. In fact, we consider cases in which this integer is so large as to appear like a continuous quantity at the macroscopic scale. Think of the Avogadro number and the fact that quantities of matter are actually measured in moles. By definition of a density, at time t, this quantity is equal to $Q(y, \delta y, t) = \int_y^{y+\delta y} \int_D u(x, t)\, dx_1 dx_2 dx_3$. For simplicity, we assume that the density u only depends on x_1, and we will henceforth drop the subscript 1, so that $Q(y, \delta y, t) = \int_y^{y+\delta y} u(x, t)\, dx$, since the area of D is 1.

Now the question is how does the gas evolve in time? We clearly need to make hypotheses on the individual motions of particles in order to answer this question. For maximum simplicity, we assume here that all the particles move at the same constant speed ae_1, where $a \in \mathbb{R}$, is given. If $a > 0$, they all move to the right on Fig. 1.9, if $a < 0$, they all move to the left, and if $a = 0$, they do not move at all.

Since all particles move as a group at speed ae_1, all the particles that were in the tube section situated between $\{y\} \times D$ and $\{y + \delta y\} \times D$ at time 0, are going to be located between $\{y + at\} \times D$ and $\{y + \delta y + at\} \times D$ at time t, and no other particle will be there at the same time. Therefore, we have a *conservation law*: for all y, δy and t

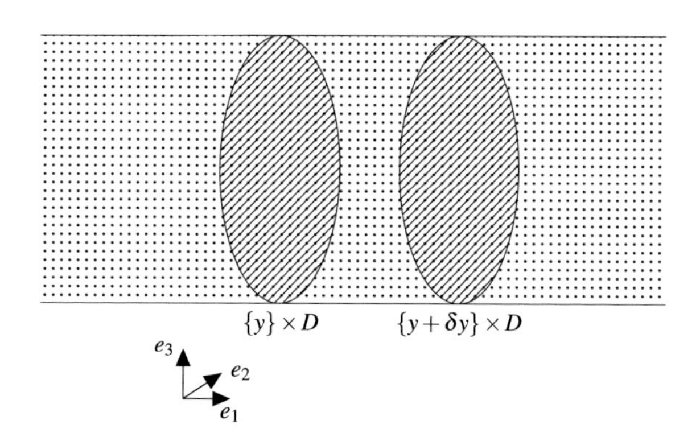

Fig. 1.9 Particles in the tube

$$Q(y + at, \delta y, t) = Q(y, \delta y, 0). \tag{1.11}$$

Let us differentiate relation (1.11) with respect to t. We obtain

$$\begin{aligned} 0 = \frac{d}{dt} Q(y + at, \delta y, t) &= \frac{d}{dt}\Big(\int_{y+at}^{y+\delta y+at} u(x, t)\, dx\Big) \\ &= a[u(y + \delta y + at, t) - u(y + at, t)] + \int_{y+at}^{y+\delta y+at} \frac{\partial u}{\partial t}(x, t)\, dx. \end{aligned}$$

Here again we find a relation that begs to be divided by δy. So we oblige, then let δy tend to 0 so that

$$\frac{\partial u}{\partial t}(y + at, t) + a\frac{\partial u}{\partial x}(y + at, t) = 0.$$

Now y and t are arbitrary, therefore we can perform the change of variables $x = y+at$ and obtain the following PDE problem:

$$\begin{cases} \dfrac{\partial u}{\partial t}(x, t) + a\dfrac{\partial u}{\partial x}(x, t) = 0 \text{ for } (x, t) \in \mathbb{R} \times \mathbb{R}_+, \\ u(x, 0) = u_0(x) \text{ for } x \in \mathbb{R}. \end{cases}$$

The PDE above is the *transport equation* (at velocity a), together with an initial condition. The conjunction of the two form an *initial value problem*. There is no boundary condition here because the space variable x ranges over the whole of $\mathbb{R}$. The PDE itself is of first order in time and space.

Let us now proceed to solve the transport equation. Since the particles all move at the same velocity a, we can look at the variation of u on the trajectory of one particle $t \mapsto x + at$ with x fixed. We thus compute the derivative

$$\frac{d}{dt}\big[u(x + at, t)\big] = a\frac{\partial u}{\partial x}(x + at, t) + \frac{\partial u}{\partial t}(x + at, t) = 0.$$

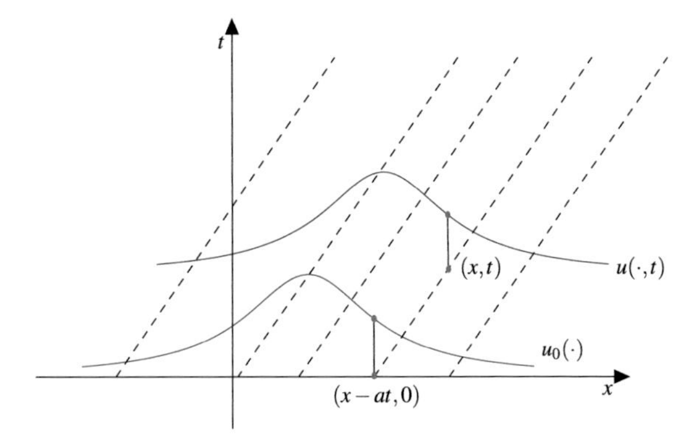

Fig. 1.10 The characteristics are the dashed straight lines with slope $1/a$. If $a = 0$, they are vertical and there is no propagation

In other words, u is constant on the trajectories. In particular,

$$u(x + at, t) = u(x, 0) = u_0(x). \tag{1.12}$$

The curves $t \mapsto (x + at, t)$ in space-time $\mathbb{R} \times \mathbb{R}_+$—which are here straight lines—are called the *characteristics* of the equation, and their use to solve the equation is accordingly called the *method of characteristics*. They are often drawn in a space-time diagram as shown on Fig. 1.10.

To determine the value of u at a point (x, t) in space-time, it is enough to look at the unique characteristic going through this point, take the point where it intersects the $t = 0$ axis and take the value of u_0 at that point, see Fig. 1.10. This construction simply amounts to rewriting formula (1.12) in the form

$$u(x, t) = u_0(x - at), \tag{1.13}$$

which proves the uniqueness of the solution, due to an *explicit* formula![6]

We have established the uniqueness of the solution, but have not yet established its existence. Fortunately, we have an explicit formula, therefore we just need to check that it actually is a solution. Let us compute the partial derivatives of u given by formula (1.13), assuming u_0 smooth enough. We have

$$\frac{\partial u}{\partial x}(x, t) = u_0'(x - at) \text{ and } \frac{\partial u}{\partial t}(x, t) = -au_0'(x - at),$$

where u_0' is the ordinary derivative of u_0. The PDE is thus clearly satisfied. Moreover, the initial condition is also trivially satisfied by setting $t = 0$ in formula (1.13). Hence, we have found the unique solution.

It is apparent that u propagates or transports the initial data at constant speed a, hence the name of the equation. This transport of the density is the macroscopic manifestation of the individual microscopic behavior of the gas particles.

[6]Explicit solutions are very rare in PDE problems.

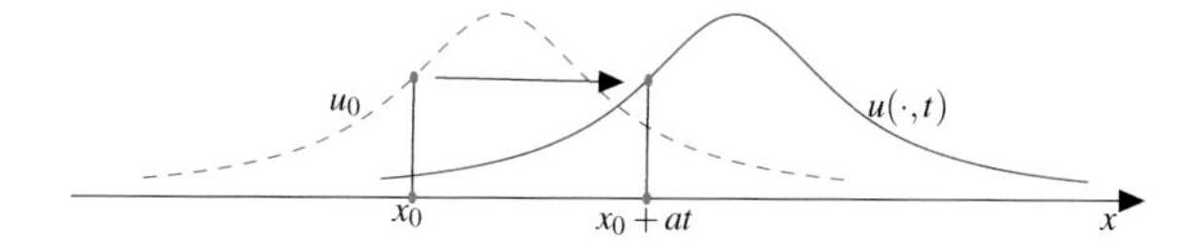

Fig. 1.11 Propagation of the initial data u_0

If it was possible to animate Fig. 1.11 on paper, the solid curve would be seen to glide to the right at a steady pace ($a > 0$ in the picture) without changing shape, after having coincided with the dashed curve at $t = 0$.

The transport equation has higher dimensional versions, which are much more complicated than the one-dimensional version. It can also be set in open sets of $\mathbb{R}^d$ instead of on the whole of $\mathbb{R}^d$, see Chap. 10. In this case, boundary conditions must be added in addition to the initial condition, which makes it an *initial-boundary value problem*. The boundary value question is delicate depending on whether the transport velocity, which is then a vector, points inwards or outwards of the open set. Let us illustrate this on a simple one-dimensional example. To be specific, we suppose that the constant speed a is strictly positive and we consider the problem

$$\begin{cases} \dfrac{\partial u}{\partial t}(x,t) + a\dfrac{\partial u}{\partial x}(x,t) = 0 \text{ for } x \in]0,1[, t > 0, \\ u(x,0) = u_0(x) \text{ for } x \in [0,1], \\ u(0,t) = g(t) \text{ for } t > 0, \end{cases}$$

where g is a given Dirichlet boundary condition at $x = 0$ such that $g(0) = u_0(0)$ and $g'(0) = -au_0'(0)$. This problem has the explicit solution for $t > 0$

$$u(x,t) = \begin{cases} u_0(x - at) & \text{for } at \le x \le 1, \\ g(t - \frac{x}{a}) & \text{for } 0 \le x \le \min(at, 1). \end{cases} \quad (1.14)$$

This means that the initial condition is transported along the characteristics in the region $at \le x \le 1, t > 0$, whereas it is the boundary condition at $x = 0$ that is transported, still along the characteristics, in the region $0 \le x \le \min(at, 1)$, see Fig. 1.12.

In particular, expression (1.14) imposes the value $u(1,t) = u_0(1 - at)$ for $t < \frac{1}{a}$ and $u(1,t) = g\big(t - \frac{1}{a}\big)$ for $t \ge \frac{1}{a}$ at $x = 1$, so that it is not possible to ascribe a Dirichlet boundary condition at point $x = 1$. Note that in the case $a < 0$, it would be the other way around: a boundary condition would be expected at $x = 1$ and no condition at $x = 0$. We will go back to the transport equation in higher dimension in Chap. 10.

We can also consider periodic boundary conditions $u(0,t) = u(1,t)$ for all t. In that case, we consider a periodic initial data u_0, which we extend to $\mathbb{R}$ by 1-periodicity.

Let us take $a > 0$. We can use the previous formulas for the solution for $t < \frac{1}{a}$. We have $u(x,t) = u_0(x - at)$ in the lower right triangle of Fig. 1.12, that is to say

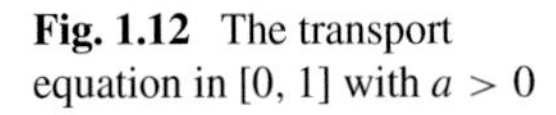
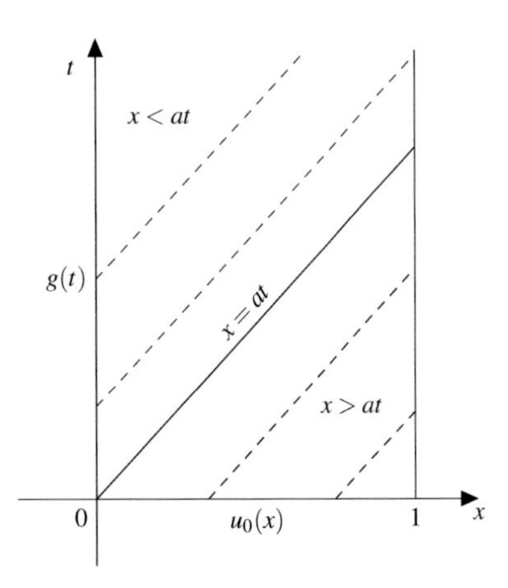

Fig. 1.12 The transport equation in $[0, 1]$ with $a > 0$

for $at < x \leq 1$, $0 < t \leq \frac{1}{a}$. Therefore, $u(1, t) = u_0(1 - at)$ for $0 < t \leq \frac{1}{a}$. The periodic condition $u(0, t) = u(1, t)$ then gives a Dirichlet boundary data at $x = 0$ for $0 < t \leq \frac{1}{a}$, more precisely $g(t) = u_0(1 - at)$. We thus obtain an expression for $u(x, t)$ in the region $0 \leq x \leq at$, $t \leq \frac{1}{a}$

$$u(x, t) = g\Big(t - \frac{x}{a}\Big) = u_0(x + 1 - at) = u_0(x - at),$$

since u_0 has been extended by 1-periodicity. For $t = \frac{1}{a}$, we thus have $u(x, \frac{1}{a}) = u_0(x - 1) = u_0(x)$, and we can continue by periodicity in time on $[\frac{1}{a}, \frac{2}{a}]$ and so on.

Note that considering the transport problem on $\mathbb{R}$ with the extended initial data u_0 directly gives the periodic solution $u(x, t) = u_0(x - at)$ by restriction to $[0, 1]$. However, the above argument establishes the uniqueness of the solution of the problem with periodic boundary condition.

The transport equation is relevant in many areas, whenever a spatially distributed quantity u_0 is transported by a velocity field, think of a concentration of pollutants carried away by the wind. A diffusion term of second order is often added, yielding what are called convection-diffusion problems.

1.5 The Vibrating String Equation

Let us return to the elastic string in the context of dynamics. The displacement u of the string is now a function of space x and time t. The analysis of applied forces is exactly the same as in the static case, except that Newton's law of motion says that the resultant of the applied forces is equal to the time derivative of the momentum for each piece that can be cut out of the whole string. There is no point in going through all the detail again—it is actually a good exercise—and the result is

$$T\Big(\frac{\partial u}{\partial x}(x+\delta x,t)-\frac{\partial u}{\partial x}(x,t)\Big)+\int_x^{x+\delta x} f(s,t)\,ds=\int_x^{x+\delta x}\rho\frac{\partial^2 u}{\partial t^2}(s,t)\,ds,$$

where T is still the constant tension, ρ is the mass of the string per unit length, which we assume to be a constant independent of x, i.e., that the string is homogeneous, and $\frac{\partial^2 u}{\partial t^2}(x,t)$ is the acceleration of the string at point x and time t. Note that the applied force f can now depend on time as well. Dividing by δx and letting δx tend to 0, we obtain

$$T\frac{\partial^2 u}{\partial x^2}(x,t)+f(x,t)=\rho\frac{\partial^2 u}{\partial t^2}(x,t),$$

which is best rewritten as

$$\frac{\partial^2 u}{\partial t^2}(x,t)-c^2\frac{\partial^2 u}{\partial x^2}(x,t)=\frac{1}{\rho}f(x,t), \tag{1.15}$$

with $c=\sqrt{\frac{T}{\rho}}$. This partial differential equation, which is also called the one-dimensional wave equation, is attributed to Jean le Rond d'Alembert. The constant c is the propagation speed. We will see later that this equation propagates waves to the right at speed c and to the left at speed $-c$. This is easily seen experimentally on a long rope held by two persons. In fact, the vibrating string differential operator is a composition of two transport operators

$$\frac{\partial^2}{\partial t^2}-c^2\frac{\partial^2}{\partial x^2}=\Big(\frac{\partial}{\partial t}-c\frac{\partial}{\partial x}\Big)\Big(\frac{\partial}{\partial t}+c\frac{\partial}{\partial x}\Big)=\Big(\frac{\partial}{\partial t}+c\frac{\partial}{\partial x}\Big)\Big(\frac{\partial}{\partial t}-c\frac{\partial}{\partial x}\Big),$$

hence the two propagation directions. Note that the propagation speed increases with the tension and decreases with the mass of the string.

Equation (1.15) must be complemented by initial conditions that prescribe the initial shape and initial velocity of the string (this is a problem of second order in time)

$$u(x,0)=u_0(x),\ \frac{\partial u}{\partial t}(x,0)=u_1(x) \text{ for all } x\in\,]0,L[, \tag{1.16}$$

and by boundary conditions, meaning here that the string is fixed at both endpoints

$$u(0,t)=u(L,t)=0,\ \text{ for all } t\in\mathbb{R}_+. \tag{1.17}$$

It should be noted that if a regular solution is expected, then a certain compatibility between initial data (1.16) and boundary conditions (1.17) must be imposed

$$u_0(0)=u_0(L)=0 \text{ and } u_1(0)=u_1(L)=0,$$

otherwise a discontinuity in the displacement or velocity will arise at $t=0$.

We will return in Chap. 9 to a more in-depth study of the wave equation. For the time being, let us consider a particular case: *harmonic vibrations* (see [71, 74] for example). We are looking for solutions to Eq. (1.15) with right-hand side $f = 0$ and by separation of variables, i.e., solutions of the special form $u(x,t) = \phi(x)\psi(t)$, non identically zero and satisfying the boundary condition (1.17). Obviously, in the case of harmonic vibrations, we cannot impose an arbitrary initial condition.

Let us rewrite the problem in this setting:

$$\begin{cases} \phi(x)\psi''(t) - c^2\phi''(x)\psi(t) = 0 \text{ for all } x \in]0,L[, t \in \mathbb{R}_+, \\ \phi(0)\psi(t) = \phi(L)\psi(t) = 0, \text{ for all } t \in \mathbb{R}_+. \end{cases}$$

Naturally, if $\psi = 0$ then $u = 0$, which is not a very interesting solution. We thus assume that there exists t_0 such that $\psi(t_0) \neq 0$. It is therefore legal to divide by $\psi(t_0)$, so that

$$\begin{cases} -\phi''(x) + \dfrac{\psi''(t_0)}{c^2\psi(t_0)}\phi(x) = 0 \text{ for all } x \in]0,L[, \\ \phi(0) = \phi(L) = 0. \end{cases}$$

This a boundary value problem in the variable x of a kind we have already encountered, and we know that if $\frac{\psi''(t_0)}{c^2\psi(t_0)} \geq 0$, then $\phi = 0$ is the unique solution. This again means that $u = 0$, which is definitely not interesting. Let us thus consider the case when $\lambda = -\frac{\psi''(t_0)}{c^2\psi(t_0)} > 0$ and see under which conditions there could exist a nonzero solution.

Forgetting the boundary conditions for an instant, we recognize a second order linear differential equation with constant coefficients, the general solution of which is of the form

$$\phi(x) = A\sin\big(\sqrt{\lambda}x\big) + B\cos\big(\sqrt{\lambda}x\big).$$

The boundary condition $\phi(0) = 0$ implies that $B = 0$. The boundary condition $\phi(L) = 0$ then either imposes $A = 0$, but then we are back to $\phi = 0$, hence a trivial solution $u = 0$, or $\sin(\sqrt{\lambda}L) = 0$, that is to say

$$\sqrt{\lambda}L = k\pi \text{ for some } k \in \mathbb{Z},$$

or again

$$\lambda = \frac{k^2\pi^2}{L^2} \text{ and } \phi(x) = A\sin\Big(\frac{k\pi}{L}x\Big),$$

where k is an integer. This gives a nontrivial solution, at last.

Without loss of generality, we take $A = 1$ and plug $u(x,t) = \sin\big(\frac{k\pi}{L}x\big)\psi(t)$ back into the original wave equation, which gives an equation for ψ

$$\psi''(t) + c^2 \frac{k^2\pi^2}{L^2}\psi(t) = 0,$$

that we solve immediately

$$\psi(t) = \alpha \sin\Big(\frac{ck\pi}{L}t\Big) + \beta \cos\Big(\frac{ck\pi}{L}t\Big),$$

where α and β are arbitrary constants. Finally, we have found

$$u(x,t) = \Big[\alpha \sin\Big(\frac{ck\pi}{L}t\Big) + \beta \cos\Big(\frac{ck\pi}{L}t\Big)\Big] \sin\Big(\frac{k\pi}{L}x\Big),$$

and it is easily checked that all these functions solve the wave equation with the homogeneous Dirichlet condition. We thus have found all the separated variable solutions.

These solutions are harmonic vibrations of frequency $\nu_k = \frac{ck}{2L} = \sqrt{\frac{T}{\rho}}\frac{k}{2L}$ indexed by the integer k. The lowest possible frequency is obtained for $k = 1$. It is called the fundamental and is the note that is heard from that string. The following integers correspond to the harmonics of this note: $k = 2$ double frequency, one octave above the fundamental, $k = 3, k = 4$ two octaves above the fundamental, etc., see Fig. 1.13. Naturally, the actual vibration of an ideal musical string is never a separated variable solution, but a superposition of harmonics. This superposition gives the note its timbre. From the point of view of mathematics, this is a question of Fourier series, but we will not pursue this angle here, see Chap. 9, Theorem 9.2.

To close this section, we deduce from the formula for the frequency that a longer string will ring a lower note, hence the relative lengths of the necks of a guitar and

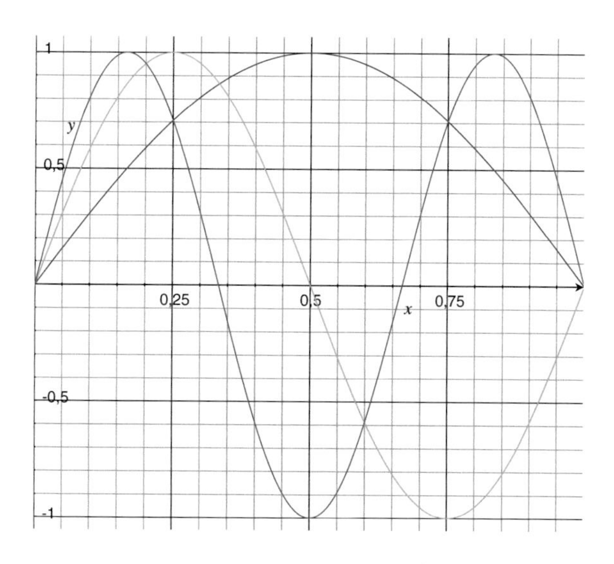

Fig. 1.13 Three successive harmonics: functions ϕ for $k = 1, 2, 3$

a bass and the different notes played on the same string on the frets, that a heavier string also rings a lower note, hence the mass differences between the strings of a guitar or piano, and that a higher tension yields a higher note, which is how such instruments are tuned.

1.6 The Wave Equation

This is the higher dimensional analogue of the vibrating string equation. If we consider a vibrating membrane in dimension two, we easily obtain the problem

$$\begin{cases} \dfrac{\partial^2 u}{\partial t^2}(x,t) - c^2 \Delta u(x,t) = f(x,t) \text{ in } \Omega \times \mathbb{R}_+, \\ u(x,t) = 0 \text{ on } \partial\Omega \times \mathbb{R}_+, \\ u(x,0) = u_0(x),\ \dfrac{\partial u}{\partial t}(x,0) = u_1(x) \text{ in } \Omega, \end{cases}$$

with $c = \sqrt{\frac{T}{\rho}}$, T is the tension and ρ the membrane mass per unit area. Note that compatibility conditions between the boundary and initial conditions must again be imposed if we expect a regular solution.

The harmonic vibration problem consists in looking for a solution of the form $u(x,t) = e^{i\lambda t}\phi(x)$ (we no longer need to pretend that we do not know what $\psi(t)$ must be...), hence the problem

$$\begin{cases} -\Delta\phi(x) = \dfrac{\lambda^2}{c^2}\phi(x) \text{ in } \Omega, \\ \phi(x) = 0 \text{ on } \partial\Omega, \end{cases} \tag{1.18}$$

with $\phi \neq 0$.

Problem (1.18) is an eigenvalue problem for the linear operator $-\Delta$, that is to say an infinite dimensional spectral problem. This was already the case in dimension one, but there was no need for the whole apparatus of self-adjoint compact operator spectral theory since everything could be done by hand.

What we need to know for now is that the eigenvalues, i.e., the possible values for $\mu = \frac{\lambda^2}{c^2}$ for which problem (1.18) admits a nonzero solution, form an infinite increasing sequence $0 < \mu_1 < \mu_2 \leq \mu_3 \leq \cdots$, with $\mu_k \to +\infty$ when $k \to +\infty$, which depends on the shape of Ω, see for example [5, 26, 28]. The situation is thus a lot more complex than in dimension one, where the shape of Ω is just characterized by its length L and we have an explicit formula for the eigenvalues. In particular, the vibration frequencies $\frac{\lambda_k}{2\pi} = c\frac{\sqrt{\mu_k}}{2\pi}$ are no longer proportional to successive integers. If the first eigenvalue still gives the fundamental tone, the following harmonics are not in rational proportion to each other, and the timbre of the sound is entirely different.

This explains why a drum produces a sound that has nothing in common with the sound produced by a guitar. It is all a question of dimensionality of the source of vibrations.

A classical problem that was only solved fairly recently was formulated as "Can you hear the shape of a drum?" The meaning of the question was to know whether the knowledge of the spectrum, that is to say of the entire sequence $(\mu_k)_{k\in\mathbb{N}^*}$, made it possible to determine Ω up to a rigid motion. The answer is negative. There are open sets in $\mathbb{R}^2$ of different shapes with exactly the same spectrum. Drums of these different shapes would thus nonetheless sound exactly the same (up to the modeling errors).

In higher dimensions, the wave equation is used to model the propagation of sound waves in the air, the propagation of light waves in the void (the wave equation in this case is deduced from Maxwell's equations, the system of PDEs of electromagnetism). There are all sorts of different kinds of waves, such as seismic waves or oceanic waves, the propagation of which is governed by equations that are more complex than the wave equation.

1.7 The Heat Equation

The heat equation is yet another evolution equation, of a totally different nature from the previous ones. For example, time is reversible in the wave equation: changing t to $-t$ does not change the equation. The heat equation describes the evolution of temperature. It thus has a connection with thermodynamics and time can only flow from the past to the future. From the point of view of mathematics, changing t into $-t$ modifies the equation and leads to problems with no solution in general.

The heat equation is as follows:

$$\begin{cases} \dfrac{\partial u}{\partial t}(x,t) - \Delta u(x,t) = f(x,t) \text{ in } \Omega \times \mathbb{R}_+, \\ u(x,t) = 0 \text{ on } \partial\Omega \times \mathbb{R}_+, \\ u(x,0) = u_0(x) \text{ in } \Omega. \end{cases}$$

Here $\Omega \subset \mathbb{R}^d$ is an open set that represents a material body (for $d = 1, 2$ and 3) and $u(x,t)$ is its temperature at point x and time t. We have set all physical constants to the value 1, as is customary in mathematics, which can in any case be achieved by a change of units. The equation is of first order in time and second order in space with a boundary condition (of Dirichlet type here) and an initial condition. When $f = 0$, the effect of the heat equation is to diffuse the initial condition.

The heat equation was discovered by Joseph Fourier, based on arguments of exchange of heat between small particles. Let us rephrase in modern terms some of

Fourier's quite remarkable modeling arguments, published in his famous 1822, 639 pages long memoir, *Théorie analytique de la chaleur*[7] [38].

Fourier starts by observing that two particles of the same substance at the same temperature have no mutual effect on each other. However, if one of the particles is hotter than the other, then there is a transfer of heat, which is a form of energy, from the warmer to the cooler particle. He then states that the quantity of heat that is transferred depends on the duration of the exchange, on the distance between the two particles, which is assumed to be very small, on both temperatures and on the nature of the substance.

Fourier goes on to assert that the exchange of heat is furthermore proportional to the temperature difference, based on very precise experiments at the time. He concludes that the amount of heat exchanged between particle m at temperature u and particle m' at temperature u' during the period of time δt (small enough so that there is no appreciable change of temperature during that time) is of the form $(u - u')\varphi(\delta x)\delta t$, where δx is the distance between the particles and φ is a rapidly decreasing function that tends to 0 when $\delta x \to +\infty$, with the intended meaning that heat exchange between particles is localized in space. This function depends on the nature of the substance. He says nothing else about it.

Fourier next considers an infinite homogeneous medium contained between two parallel planes A and B so that A is kept at constant temperature a and B at constant temperature b. Assume that $A = \{x = 0\}$ and $B = \{x = e\}$ where $e > 0$ is the distance between the two planes.[8] Fourier claims to show that the equilibrium temperature profile in the medium is given by $u = a + \frac{b-a}{e}x$ (it is reasonable that the temperature should only depend on x). What he actually explains is that this profile is consistent with equilibrium, in the sense that the heat flux[9] through all planes $C = \{x = c\}$ is independent of c. Hence, considering a slab of material contained between two arbitrary planes C and C', there is a perfect balance of incoming heat on one plane and outgoing heat on the other plane, so that the temperature must stay stationary.

His argument is as follows. Assume that the former temperature profile is present in the material, and consider two very close particles m and m' situated on both sides of C, say at $x = c + \zeta$ and $x = c - \zeta$, and two other very close particles n and n' situated on both sides of C' at $x = c' + \zeta$ and $x = c' - \zeta$, such that the distance d between m and m' is equal to the distance between n and n'. The exchanges of heat between m and m', and between n and n', are then both equal to $2\frac{b-a}{e}\zeta\varphi(d)\delta t$, and independent of c and c'. Fourier argues that since the total heat flux going through C (and C') results from all possible configurations of such pairs of particles, it is likewise independent of c (and c').

Using the same kind of reasoning with two slabs of different thicknesses e and e', left temperatures a and a' and right temperatures b and b', Fourier determines that

---

[7] In which, not only is the heat equation derived and solved in special cases, but Fourier series are invented, the heat kernel appears, etc.

[8] e stands for *épaisseur* in French, i.e. thickness.

[9] The heat flux is the quantity of heat going through a unit area in the plane during a unit of time.

the corresponding heat fluxes F and F' are such that

$$\frac{F}{F'} = \frac{\frac{a-b}{e}}{\frac{a'-b'}{e'}}.$$

Calling K the heat flux necessary to have a slab of unit thickness support a unit temperature difference, he thus obtains the following law, which is now called the Fourier law,

$$F = K\frac{a-b}{e} = -K\frac{du}{dx}.$$

The constant K is characteristic of the material, it is called its *heat conductivity*. Fourier shows that the flux law actually holds for general, time-dependent temperature repartitions, replacing the derivative $\frac{du}{dx}$ with what we would call today the gradient ∇u.

For simplicity, let us stay in the one-dimensional case, i.e., an infinite medium between two parallel planes, and derive the evolution equation in the manner of Fourier. The temperature is thus an unknown function u of x and t.

Let C be the *specific heat* of the material, that is to say, the amount of heat needed to heat up a unit of mass of the material from temperature 0 to temperature 1. Consider a parallelepiped bounded by the planes $x = x_0$ and $x = x_0 + \delta x$, the section of which has unit area. The total heat balance coming from the outside during a small period of time δt is thus the flux on the left minus the flux on the right multiplied by the duration

$$Q = \big(F(x,t) - F(x+\delta x,t)\big)\delta t = K\Big(\frac{\partial u}{\partial x}(x_0+\delta x,t) - \frac{\partial u}{\partial x}(x_0,t)\Big)\delta t.$$

This heat input, which can be positive or negative, results in a change of temperature

$$u(x_0, t+\delta t) - u(x_0,t) = \frac{Q}{\rho \delta x C},$$

where ρ is the mass density of the material. Indeed, $\rho\delta x$ is the mass of the parallelepiped under consideration. Therefore, we obtain

$$\frac{u(x_0,t+\delta t) - u(x_0,t)}{\delta t} = \frac{K}{\rho C}\,\frac{\frac{\partial u}{\partial x}(x_0+\delta x,t) - \frac{\partial u}{\partial x}(x_0,t)}{\delta x}.$$

Letting δx and δt go to 0, we thus obtain the heat equation

$$\frac{\partial u}{\partial t}(x,t) = \frac{K}{\rho C}\frac{\partial^2 u}{\partial x^2}(x,t).$$

Of course, the presence of an external heat source is easily taken into account in the above balance of fluxes argument and results in a nonzero right-hand side in the heat equation. In the case of a stationary heat distribution, $\frac{\partial u}{\partial t} = 0$, we recover the Poisson equation.

Other arguments used by Fourier are actually very close to the finite difference method that we will see later on. It is quite remarkable, and Fourier points it out himself, that it is not necessary to know the ultimate nature of heat, which remained mysterious at the time but which we now know to be the kinetic energy corresponding to the random vibrations of molecules, and how it propagates precisely, in order to be able to derive an extremely accurate macroscopic evolution equation.

1.8 The Schrödinger Equation

The Schrödinger equation is another evolution equation of an again totally different nature. This time, u is a wave function in the sense of quantum mechanics. It is complex-valued. The domain is the whole of $\mathbb{R}^3$. The equation reads[10]

$$i\frac{\partial u}{\partial t}(x,t) + \Delta u(x,t) = 0 \text{ in } \mathbb{R}^3 \times \mathbb{R}_+.$$

The Schrödinger equation is the basic equation of quantum mechanics that governs the evolution of the wave function of one particle in the absence of any potential, that is in the void. Physical constants are missing (set to 1), such as Planck's constant $\hbar$ and the mass of the particle. We need to add an initial condition $u(x,0) = u_0(x)$ on $\mathbb{R}^3$.

Since the square of the modulus of the wave function is interpreted as a presence probability density, we need to impose

$$\int_{\mathbb{R}^3} |u(x,t)|^2\,dx = 1.$$

Actually, if the initial condition satisfies this normalization condition, then the solution satisfies it automatically at all times.

Even though the Schrödinger equation presents a formal similarity with the heat equation—first order in time, second order in space—the presence of the imaginary factor i gives it radically different properties. In particular, the Schrödinger equation propagates waves, also not at all in the same way as the wave equation, whereas the heat equation does not propagate waves (heat waves notwithstanding!).

[10] Here $i^2 = -1$.

Let us note that in the Schrödinger equation for a system of N particles, the variable x must belong to $\mathbb{R}^{3N}$, which becomes rapidly difficult for practical purposes when N is large...[11]

As a general rule, physics is a nearly inexhaustible source of partial differential equations problems. Let us cite the Dirac equation, a first order system of equations and relativistic version of the Schrödinger equation; Einstein's equations of general relativity, a system of nonlinear PDEs; the Boltzmann equation for the kinetic description of gases, all the equations of fluid mechanics, Euler, Stokes, Navier-Stokes, and so on, and so forth. We refer for example to [21, 22, 24, 34] for other physical models leading to the study of PDEs.

1.9 The Black and Scholes Equation

Physics is not by far the only source of PDEs. PDEs are also playing an increasing role in diverse areas, such as biology, chemistry, material science, meteorology, climatology, road traffic modeling, crowd movement modeling, economy, finance, among many others. Let us give a famous example in the latter area, the Black and Scholes equation.

The question is to set the price of a call option. A call option is a contract between a seller and a buyer, drawn at time $t = 0$. The contract gives the buyer the right to buy an asset belonging to the seller, not right away but later and at a price K, the strike, that is agreed on in advance. The contract has a price paid by the buyer to the seller at $t = 0$, otherwise the seller would have no real reason to agree to it. For the buyer, it is an insurance against future price fluctuations since the strike is fixed.

The price C must be computed in such a way that the game should be fair on average, or at least seem to be fair... The possibility of option pricing hinges on a modeling of the market and on a hypothesis called no arbitrage opportunity (no free lunch) meaning that it is impossible to make sure gains without taking risks.

To make things a little more precise, the price of the asset at instant t is denoted S_t. It is a continuous time stochastic process. In the case of an american call, the buyer acquires the right to exercise the option, that is to say to buy the asset for the price K, at any moment $t \in [0, T]$, where T is an expiration date agreed on in advance. The buyer is under no obligation to do so, and after time T, the option disappears.

Of course, the buyer has no interest in exercising the option at time t if $S_t < K$. In this case, it is better to buy at the market price or not buy at all. On the other hand, the buyer could also have invested the amount C at a fixed interest rate r without risk. Therefore, a profit would only be made by exercising the option if $S_t > e^{rt}C + K$, which is the decision criterion. The buyer bets this situation will occur before time T, in which case he or she buys the asset for a price K and sells it back immediately on the market at price S_t, thus pocketing the difference $S_t - K$. The global balance

[11] Think Avogadro's number.

of the operation is either $-C$ if the option is not exercised or $s_t - K - C$ if it is exercised.

The seller always gains C and loses $S_t - K$ if the buyer exercises the option, in the sense that he or she could have sold at time t at the market price to somebody else. Therefore, the bet is that the buyer will not exercise the option. The seller must also seek to cover losses in case the buyer exercises the option. The price C is meant to compensate for such potential losses.

The option price C is naturally a function of the asset price, which is represented by a variable $x \in \mathbb{R}_+$, because a price is nonnegative. It is also useful to introduce the price at instant t, that is to say, the price the option would have if it was bought at instant t with the same strike K and expiry date T. The option price is thus a function in two variables $C(x, t)$ (let us emphasize again that the space variable x is actually a price).

We want to determine $C(x, 0)$ as a function of x in order to define the terms of the contract, since at $t = 0$, the price of the asset S_0 is known and the price of the option is then $C(S_0, 0)$. The price of the option at $t = T$ is obviously $C(x, T) = (x - K)_+$ since the option is exercised at T only if the price of the asset is larger than K, and there is no time left to invest $C(x, T)$ at a fixed interest rate (the notation $C_+ = \max(C, 0)$ denotes the positive part of C).

At this point, stochastic modeling is needed in order to describe the evolution of asset prices and to ensure a viable game, which we do not describe in detail. Anyway, hypotheses are made concerning the S_t process. As recent world events have made quite clear, such hypotheses are not always satisfied in real life, but let us proceed anyway. At the end of this stochastic modeling phase, we end up with a deterministic PDE for the function $C(x, t)$ of the form

$$\frac{\partial C}{\partial t}(x,t) + \frac{1}{2}\sigma^2 x^2 \frac{\partial^2 C}{\partial x^2}(x,t) + \mu x \frac{\partial C}{\partial x}(x,t) - rC(x,t) = 0 \text{ in } \mathbb{R}_+ \times [0, T],$$

with the final condition

$$C(x, T) = (x - K)_+.$$

This is the Black and Scholes equation. It has a final condition and not an initial condition because of modeling reasons, as we have seen, in fact the initial value is the unknown quantity of interest. Another reason is that the principal part of the differential operator is basically similar to a backward heat equation. We have seen that the heat equation is incapable of going back in time. Therefore, a backward heat equation needs a final condition in order to be well-posed. There is an additional difficulty since the coefficients of the space derivatives are functions of the space variables that vanish for $x = 0$. There is thus a degeneracy at the boundary, so whether or not boundary conditions are in order it is not so clear. The constant σ is called the asset volatility, a measure of the more or less erratic behavior of the asset price, and μ is the trend, a sort of average growth rate.

These oddities of the Black and Scholes equation are mostly corrected by a simple change of variable. Let us set $u(y, \tau) = C(e^y, T - \tau)$, then

$$\frac{\partial u}{\partial \tau} - \frac{1}{2}\sigma^2 \frac{\partial^2 u}{\partial y^2} - \left(\mu - \frac{1}{2}\sigma^2\right)\frac{\partial u}{\partial y} + ru = 0 \text{ in } \mathbb{R} \times [0, T],$$

with the initial (since time has been reversed) condition

$$u(y, 0) = (e^y - K)_+.$$

We are comfortably back with an ordinary heat equation with the right time direction, whose effect is to diffuse the price, corrected by a transport term whose effect is to make the price drift (in backward time) at speed $-(\mu - \frac{1}{2}\sigma^2)$. The ru term is an updating term with respect to the interest rate which can be eliminated by a further change of variables.

To conclude, let us remark that the Black and Scholes equation for one asset is a two-dimensional equation, one space dimension and one time dimension. The analogous equation for a portfolio of N assets is in $N + 1$ dimensions, which is a source of difficulty for numerical approximation even for N moderately large.

1.10 A Rough Classification of PDEs

We give a rather informal classification of PDEs which is neither very precise, nor exhaustive, but which has the advantage of giving a general idea of their properties.

Let us start with the Laplace operator $\Delta = \frac{\partial^2}{\partial x_1^2} + \frac{\partial^2}{\partial x_2^2}$, and replace $\frac{\partial}{\partial x_i}$ by multiplication by a variable ξ_i (which is more or less what the Fourier transform does). The equation $\Delta u = f$ is thus replaced by an equation of the type $\|\xi\|^2 = g$ which is the equation of a circle in $\mathbb{R}^2$, a special case of an ellipse. We say that the Poisson equation is *elliptic*. More generally, if we repeat the same operation on the principal part $\sum_{i,j=1}^d a_{ij}\frac{\partial^2}{\partial x_i x_j}$ of a general second order linear operator, we obtain $\sum_{i,j=1}^d a_{ij}\xi_i\xi_j = g$. If this yields the equation of an ellipsoid in $\mathbb{R}^d$, then we say that the equation is elliptic. This is the case if the matrix (a_{ij}) is positive definite.

The same game played on the heat equation, replacing $\partial/\partial t$ by ξ_0, yields $\xi_0 - \xi_1^2 = g$, which is the equation of a parabola, or a paraboloid in higher dimension. We say that the heat equation is *parabolic*.

Finally, in the case of the wave equation, we obtain $\xi_0^2 - \xi_1^2 = g$, the equation of a hyperbola. We say that the wave equation is *hyperbolic*.

In the two-dimensional case, $d = 2$, the above classification reduces to the discriminant criterion, which is as follows. First we can always assume that $a_{12} = a_{21}$ and let $D = a_{12}^2 - a_{11}a_{22}$. Then the equation is elliptic if and only if $D < 0$, it is parabolic if and only if $D = 0$ and it is hyperbolic if and only if $D > 0$.

It is possible to give more precise definitions [24], but this is not useful here. The important idea is that an elliptic equation (resp. a parabolic equation, resp. a hyperbolic equation) has more or less the same properties as the Poisson equation (resp. the heat equation, resp. the wave equation). The transport equation is considered to be hyperbolic.

1.11 What Comes Next

Up to now, we have presented a few examples of diverse mathematical models that involve partial differential equations. We have also obtained existence and uniqueness results for the simplest of these models, in Sect. 1.1. The rest of the book is devoted to the mathematical analysis and numerical approximation of the three main types of partial differential equations, elliptic, parabolic and hyperbolic.

General elliptic problems exemplified by the elastic membrane equation are treated in Chap. 4 for their mathematical analysis and Chaps. 5 and 6 for their numerical approximation. Parabolic equations are the subject of Chaps. 7 and 8 for the theoretical and numerical aspects respectively. Finally, we consider hyperbolic equations in Chaps. 9 and 10 from both points of view.

In the next chapter, Chap. 2, we take the simplest possible elliptic problem, i.e., the elastic string problem (1.2). Since the mathematical analysis of this problem has already been performed, we skip directly to its numerical approximation. We use this example to present the finite difference method. This is the earliest and simplest numerical approximation method and it only requires elementary mathematical tools. This method has drawbacks for more general elliptic problems, as we will see. We will however return to the finite difference method later when dealing with evolution equations of parabolic or hyperbolic type.

Chapter 2
The Finite Difference Method for Elliptic Problems

Even though it can be shown that the boundary value problems introduced in Chap. 1 admit solutions, they do not admit explicit solutions as a general rule, i.e., solutions that can be written in closed form. Therefore, in order to obtain quantitative information on the solutions, it is necessary to define approximation procedures that are effectively computable. We present in this chapter the simplest of all approximation methods, the finite difference method. We apply the method to the numerical approximation of a model problem, the Dirichlet problem for the Laplacian in one space dimension. We then give some indications for the extension of the method to Neumann boundary conditions and to two-dimensional problems.

2.1 Approximating Derivatives by Differential Quotients

The finite difference method is based on an elementary concept which is directly connected to the definition of differentiation of functions. The idea is simply to approximate any derivative by a differential quotient. As opposed to more sophisticated methods, such as the finite element method that we will present later on, we can directly treat the boundary value problem instead of a more abstract formulation thereof, the variational formulation, see Chap. 4. This explains the popularity of the finite difference method, in spite of its shortcomings in particular in higher dimensions.

Let us explain the method in one dimension. Using the definition of the derivative of a function u at point $x \in \mathbb{R}$, we can write

$$u'(x) = \lim_{h \to 0} \frac{u(x+h) - u(x)}{h},$$

H. Le Dret and B. Lucquin, *Partial Differential Equations: Modeling, Analysis and Numerical Approximation*, International Series of Numerical Mathematics 168, DOI 10.1007/978-3-319-27067-8_2

and deduce that, when $h \neq 0$ is small, the differential quotient $[u(x+h) - u(x)]/h$ should be a good approximation of the first order derivative $u'(x)$, in the sense that the error induced by this approximation goes to 0 when h goes to 0. If the function is regular (in a neighborhood of x), we can make a precise estimate of this error by using a Taylor–Lagrange expansion. Namely, if u is C^2 in a neighborhood of x, we have

$$u(x+h) = u(x) + hu'(x) + \frac{h^2}{2}u''(\xi), \tag{2.1}$$

where ξ lies between x and $x+h$. Therefore, taking $h_0 > 0$ and setting $C = \sup_{y \in [x-h_0, x+h_0]} |u''(y)|/2$, we see that

$$\left| \frac{u(x+h) - u(x)}{h} - u'(x) \right| \leq C|h|,$$

for all $h \neq 0$ such that $|h| \leq h_0$, In other words, the error made by replacing the derivative $u'(x)$ by the differential quotient $\frac{u(x+h)-u(x)}{h}$ is of the order of h. We say that we have a *consistent approximation* of order one of u' at point x. More generally, we say that we have a consistent approximation of order p ($p > 0$) of $u'(x)$ if there exists a constant C, which does not depend on h, such that this error is bounded by Ch^p.

Other consistent approximations are possible. For example, the differential quotient $[u(x) - u(x-h)]/h$ is also a consistent approximation of order one of $u'(x)$.

One way of improving the accuracy is to center the approximation, by using the points $x+h$ and $x-h$ to form the differential quotient $\frac{u(x+h)-u(x-h)}{2h}$. In fact, if u is C^3 in a neighborhood of x, we can write

$$u(x+h) = u(x) + hu'(x) + \frac{h^2}{2}u''(x) + \frac{h^3}{6}u^{(3)}(\xi^+),$$
$$u(x-h) = u(x) - hu'(x) + \frac{h^2}{2}u''(x) - \frac{h^3}{6}u^{(3)}(\xi^-),$$

where ξ^+ and ξ^- are located between $x-h$ and $x+h$. Subtracting the two formulas above and using the intermediate value theorem, we obtain

$$\frac{u(x+h) - u(x-h)}{2h} = u'(x) + \frac{h^2}{6}u^{(3)}(\xi),$$

for some ξ between $x-h$ and $x+h$. We deduce the following estimate of the error: For all h, $0 < |h| \leq h_0$, we have

$$\left| \frac{u(x+h) - u(x-h)}{2h} - u'(x) \right| \leq Ch^2,$$

where

$$C = \frac{1}{6} \sup_{y \in [x-h_0, x+h_0]} |u^{(3)}(y)|.$$

We have then defined a consistent approximation of order two of $u'(x)$. Note that the above error estimate is a priori not valid if u is less regular than C^3.

The model problem (1.2) of Chap. 1 or problem (2.4) below involve second derivatives so that we also need to approximate second derivatives by differential quotients.

Lemma 2.1 *Let us suppose that u is C^4 on the interval $[x - h_0, x + h_0]$ ($h_0 > 0$). Then there exists a constant C such that, for all h, $0 < |h| \leq h_0$, we have*

$$\left| \frac{u(x+h) - 2u(x) + u(x-h)}{h^2} - u''(x) \right| \leq Ch^2. \tag{2.2}$$

In other words, the differential quotient $\frac{u(x+h)-2u(x)+u(x-h)}{h^2}$ is a consistent approximation of order two of the second derivative of u at point x.

Proof The proof is again based on Taylor–Lagrange expansions. We have

$$u(x+h) = u(x) + hu'(x) + \frac{h^2}{2}u''(x) + \frac{h^3}{6}u^{(3)}(x) + \frac{h^4}{24}u^{(4)}(\xi^+),$$
$$u(x-h) = u(x) - hu'(x) + \frac{h^2}{2}u''(x) - \frac{h^3}{6}u^{(3)}(x) + \frac{h^4}{24}u^{(4)}(\xi^-),$$

where ξ^+ and ξ^- are between $x - h$ and $x + h$. Adding the two formulas above and using the intermediate value theorem, we thus obtain

$$\frac{u(x+h) - 2u(x) + u(x-h)}{h^2} = u''(x) + \frac{h^2}{12}u^{(4)}(\xi), \tag{2.3}$$

for some ξ between $x - h$ and $x + h$. It follows that estimate (2.2) holds with

$$C = \sup_{y \in [x-h_0, x+h_0]} \frac{|u^{(4)}(y)|}{12},$$

for $0 < |h| \leq h_0$, which completes the proof. □

Remark 2.1 Let us remark that this error estimate also depends on the regularity of u. For example, if u is only C^3, the Taylor expansion cannot be carried out up to fourth order, and gives an error estimate of order h. Conversely, any regularity higher than C^4 will not improve the accuracy, which remains of order h^2. There is thus no reason to use a higher order Taylor–Lagrange expansion, see Remark 8.1 of Chap. 8.
□

Remark 2.2 We remark that

$$\frac{u(x+h)-2u(x)+u(x-h)}{h^2} = \frac{1}{h}\left[\frac{u(x+h)-u(x)}{h} - \frac{u(x)-u(x-h)}{h}\right]$$
$$= D_h^+ D_h^- u(x) = D_h^- D_h^+ u(x),$$

where the operators D_h^+ and D_h^- are defined by

$$D_h^+ u(x) = \frac{u(x+h)-u(x)}{h}, \quad D_h^- u(x) = \frac{u(x)-u(x-h)}{h}.$$

These are precisely the operators, respectively called *forward difference operator* and *backward difference operator*, of order one that we have already met for the approximation of first order derivatives. □

2.2 Application to a One-Dimensional Model Problem

We consider the model problem consisting in finding $u\colon [0,1] \to \mathbb{R}$ such that

$$\begin{cases} -u''(x) + c(x)u(x) = f(x), & x \in]0,1[, \\ u(0) = g_0, \quad u(1) = g_1, \end{cases} \tag{2.4}$$

where c and f are two given continuous functions, defined on $[0,1]$, and g_0 and g_1 are given constants. We know that this problem has a unique solution if $c \geq 0$, by the shooting method, see Theorem 1.2 of Chap. 1. We will thus take $c \geq 0$ in the sequel.

In the particular case $c = 0$, u has the following expression

$$u(x) = \int_0^1 G(x,y)f(y)dy + g_0 + x(g_1 - g_0), \tag{2.5}$$

where G is the *Green function*, given by

$$G(x,y) = x(1-y) \text{ if } y \geq x, \quad G(x,y) = y(1-x) \text{ if } y \leq x.$$

In the general case (i.e., $c \neq 0$), formula (2.5) still holds with a Green function which is not explicitly known. The idea is thus to define an approximation of u.

More precisely, we are going to look for an approximation of u at specific points x_i of the interval $[0,1]$ in finite number. These points are called *discretization points* or *grid points*. For simplicity, we assume here that these points are uniformly distributed, see Fig. 2.1, i.e., of the form $x_i = ih$, $i \in \{0,\ldots,N+1\}$, where N is a given positive

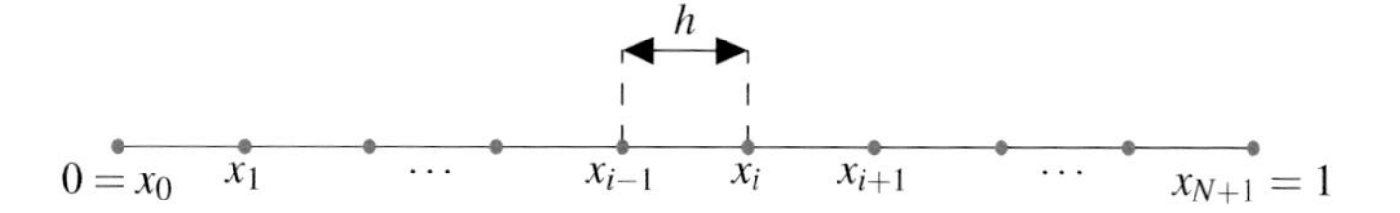

Fig. 2.1 A uniform 1d grid and grid points

integer and $h = 1/(N+1)$ is the *grid space step*, or grid step in short. Even though the notation does not make it clear, we see that x_i not only depends on i, but also on h or equivalently on N. Note that $0 \leq h \leq 1$ and that h goes to 0 when the number $N+2$ of grid points goes to infinity.

We have at the ends of the interval $x_0 = 0$ and $x_{N+1} = 1$. The other grid points x_i, for $i \in \{1, \ldots, N\}$, are called *internal points*.

We would like to compute an approximated value u_i of the unknown exact value $u(x_i)$ at each point x_i, with $i \in \{1, \ldots, N\}$. We naturally set at the ends $u_0 = g_0$ and $u_{N+1} = g_1$, i.e., we enforce the exact boundary condition at the discrete level. For the internal points, the idea is to start from the first equation in system (2.4) expressed at point x_i, and approximate $u''(x_i)$ based on the central differential quotient of Lemma 2.1.

The unknowns of the discrete problem are thus only $u_1, \ldots, u_N$. Just like x_i, we remark that u_i also depends on h or N. Let us denote by U_h the vector in $\mathbb{R}^N$ with components u_i, for $i \in \{1, \ldots, N\}$.

Recall that $c \in C^0([0,1])$ and $f \in C^0([0,1])$. We start from problem (2.4). We replace each exact value $u(x_i)$ at each grid point by the corresponding approximated value u_i, and the second order derivative $u''(x_i)$ at each internal point x_i, $i \in \{1, \ldots, N\}$, of the grid by the central difference quotient $\frac{u_{i+1}-2u_i+u_{i-1}}{h^2}$. We thus get the following discrete system expressed solely in terms of the approximated discrete values u_i:

$$\begin{cases} -\dfrac{u_{i+1} - 2u_i + u_{i-1}}{h^2} + c(x_i)u_i = f(x_i), \quad i \in \{1, \ldots, N\}, \\ u_0 = g_0, \quad u_{N+1} = g_1. \end{cases} \tag{2.6}$$

At this point, it must be emphasized that there is no indication that $u_i \approx u(x_i)$, in any sense whatsoever. We say that we have discretized the problem by a finite difference method, using the *three point central finite difference scheme* or simply three point scheme.

The discrete problem has the following equivalent vector formulation:

$$A_h U_h = F_h, \tag{2.7}$$

where A_h is the $N \times N$ matrix defined by

$$A_h = A_h^0 + \begin{pmatrix} c(x_1) & 0 & \cdots & 0 \\ 0 & c(x_2) & \ddots & \vdots \\ \vdots & \ddots & \ddots & 0 \\ 0 & \cdots & 0 & c(x_N) \end{pmatrix}, \tag{2.8}$$

with

$$A_h^0 = \frac{1}{h^2} \begin{pmatrix} 2 & -1 & 0 & \cdots & 0 \\ -1 & 2 & -1 & \ddots & \vdots \\ 0 & \ddots & \ddots & \ddots & 0 \\ \vdots & \ddots & -1 & 2 & -1 \\ 0 & \cdots & 0 & -1 & 2 \end{pmatrix}, \tag{2.9}$$

and

$$F_h = \begin{pmatrix} f(x_1) + \frac{g_0}{h^2} \\ f(x_2) \\ \vdots \\ f(x_{N-1}) \\ f(x_N) + \frac{g_1}{h^2} \end{pmatrix}. \tag{2.10}$$

In order to compute the discrete solution U_h, we thus have to solve the linear system (2.7). The first point to check is whether or not this system has a solution for any right-hand side, in other words, whether or not the matrix A_h is invertible. This is a consequence of the following proposition.

Proposition 2.1 *The matrix A_h is symmetric. Furthermore, when $c \geq 0$, it is positive definite.*

Proof The matrix A_h is clearly symmetric. Let us show it is positive definite when $c \geq 0$. Let V be a vector in $\mathbb{R}^N$ with components v_i, $i \in \{1, \ldots, N\}$. Let us compute $V^T A_h V$. First, thanks to the assumption $c \geq 0$, we get

$$V^T A_h V = V^T A_h^0 V + \sum_{i=1}^{N} c(x_i) v_i^2 \geq V^T A_h^0 V.$$

It is thus sufficient to show that A_h^0 is positive definite. A simple computation gives

$$h^2\, V^T A_h^0 V = v_1^2 + (v_2 - v_1)^2 + (v_3 - v_2)^2 + \cdots + (v_{N-1} - v_N)^2 + v_N^2,$$

from which we deduce that $V^T A_h^0 V \geq 0$. Moreover, if $V^T A_h^0 V = 0$, then each term in the sum above has to be zero, which implies that $0 = v_1 = v_2 - v_1 = \cdots = v_{N-1} - v_N = v_N$. Thus $v_1 = v_2 = \cdots = v_N = 0$, i.e. $V = 0$, which completes the proof. □

It is well-known that positive definite matrices are invertible, since $A_h V = 0$ implies that $V^T A_h V = 0$, thus

Corollary 2.1 *The discrete system* (2.7) *has one and only one solution* U_h *for any* N *and* F_h.

It follows that the finite difference scheme (2.6) is well defined. Note that the matrix of the linear system (2.7) is *sparse*, i.e., it has many zero elements. More precisely, A_h is a tridiagonal matrix in the sense that the only nonzero elements are either on the diagonal or on the super- or sub-diagonal.

The next question concerns in which sense the discrete system solution U_h is actually an approximation of the exact solution u of the boundary value problem (2.4). The two quantities are of different nature since U_h is a vector in $\mathbb{R}^N$ and u is a function defined on $[0, 1]$. The only reasonable way of comparing the two is to compare the ith component u_i of U_h with the value $u(x_i)$ of u at the corresponding grid point, since this is what the finite difference scheme is intended for. In particular we would like to know if $u_i - u(x_i) \to 0$ when the grid step h goes to zero and at which rate, in a sense that is made precise in the next section.

2.3 Convergence of the Finite Difference Method

We study the convergence of the method in the case of the model problem (2.4). The analysis below applies to more general situations.

In order to compare the discrete solution $U_h \in \mathbb{R}^N$ with the exact solution u, we introduce the *grid sampling operator* and what is meant by convergence in this context.

Definition 2.1 The *grid sampling operator* $S_h : C^0([0, 1]) \to \mathbb{R}^N$ is defined by

$$S_h(v) = \begin{pmatrix} v(x_1) \\ v(x_2) \\ \vdots \\ v(x_N) \end{pmatrix}.$$

We say that the method (2.7) is *convergent* if

$$\|U_h - S_h(u)\|_\infty \to 0 \text{ when } h \to 0.$$

Moreover, we say that the scheme is *convergent of order p* if there exists $p > 0$ and a constant C which do not depend on h such that, for all $h > 0$ we have

$$\|U_h - S_h(u)\|_\infty \le Ch^p.$$

Remark 2.3 The norm $\|X\|_\infty = \max_{i\in\{1,\dots,N\}} |X_i|$ is defined on $\mathbb{R}^N$, and thus depends on N. This specific choice is made for simplicity. We will see more general choices of norms in Chap. 8 in the study of the numerical approximation of the heat equation. With the present choice of norm, convergence thus means that $\max_{i\in\{1,\dots,N\}} |u_i - u(x_i)| \to 0$ when $N \to +\infty$, which is quite natural. Remember that $h \to 0$ is equivalent to $N \to +\infty$ and that u_i and x_i also depend on N and h. □

In order to evaluate the error $U_h - S_h(u)$ of the method, we next introduce the *truncation error* of the scheme (2.6) or equivalently (2.7).

Definition 2.2 The truncation error of the scheme $A_h U_h = F_h$ is the vector in $\mathbb{R}^N$ denoted by $\varepsilon_h(u)$ and defined by

$$\varepsilon_h(u) = A_h(S_h(u)) - F_h.$$

We say that the scheme is *consistent* if

$$\lim_{h\to 0} \|\varepsilon_h(u)\|_\infty = 0.$$

Moreover, we say that the scheme is *consistent of order p* if there exists $p > 0$ and a constant C which do not depend on h such that, for all $h > 0$ we have the following error estimate

$$\|\varepsilon_h(u)\|_\infty \le Ch^p.$$

Remark 2.4 The truncation error is not to be confused with the error of the method itself. The evaluation of the truncation error is however an important intermediate step in the evaluation of the error estimate. □

Let us thus evaluate the truncation error of the scheme. We assume that u is C^4 on $[0, 1]$, which means that f is C^2. Using formula (2.3), we obtain

$$\varepsilon_h(u) = -\frac{h^2}{12}\begin{pmatrix} u^{(4)}(\xi_1) \\ u^{(4)}(\xi_2) \\ \vdots \\ u^{(4)}(\xi_N) \end{pmatrix},$$

where each point ξ_i is such that $\xi_i \in]x_{i-1}, x_{i+1}[$. We deduce that

$$\|\varepsilon_h(u)\|_\infty \le \frac{h^2}{12} \max_{y\in[0,1]} |u^{(4)}(y)|. \tag{2.11}$$

We have thus shown the

Proposition 2.2 *Let us suppose that the solution u of problem (2.4) is C^4 on $[0, 1]$. Then the scheme (2.6) is consistent of order two.*

Remark 2.5 We remark that, if the exact solution u is such that its derivative of order 4 is zero, then the truncation error $\varepsilon_h(u)$ is zero, so that $A_h(U_h - S_h(u)) = 0$. Since the matrix A_h is invertible, it follows that $U_h = S_h(u)$, which means that for any $i \in \{0, \dots, N+1\}$, we have $u_i = u(x_i)$. The discrete solution is thus equal to the exact solution at each grid point (internal or not)! This is very specific to the case when the exact solution u happens to be a polynomial function of degree at most 3, see the proof of Proposition 2.5 where this property is used. Of course, in general $u_i \neq u(x_i)$. □

Let us now compute the error of the method. By definition of the scheme on the one hand, and of the truncation error on the other hand, we have

$$A_h U_h = F_h \quad \text{and} \quad A_h S_h(u) = F_h + \varepsilon_h(u).$$

Subtracting one from the other, we deduce that

$$A_h(U_h - S_h(u)) = -\varepsilon_h(u).$$

Since A_h is invertible, this implies that the error is given by

$$U_h - S_h(u) = -(A_h)^{-1}\varepsilon_h(u). \tag{2.12}$$

We thus have a decomposition of the error into two independent terms, the truncation error $\varepsilon_h(u)$ and the matrix term $(A_h)^{-1}$. We now proceed to show the following convergence result:

Theorem 2.1 *Let us suppose that $c \geq 0$. If the solution u of problem (2.4) is C^4 on $[0, 1]$, then the scheme (2.6) is convergent of order two. More precisely, we have*

$$||U_h - S_h(u)||_\infty \leq \frac{h^2}{96} \max_{x\in[0,1]} |u^{(4)}(x)|. \tag{2.13}$$

In order to show this result, we need to briefly introduce a few concepts, and in particular the definition of matrix norms. In this chapter all matrices are real. We refer for example to [6, 18, 53] for details.

Let $\|\cdot\|$ be a given norm on $\mathbb{R}^N$. For any $N \times N$ matrix A, we denote by $|||A|||$ the associated *induced matrix norm* or *operator norm* defined by

$$|||A||| = \sup_{X\in\mathbb{R}^N, X\neq 0} \frac{\|AX\|}{\|X\|} = \sup_{X\in\mathbb{R}^N, \|X\|=1} \|AX\| = \sup_{X\in\mathbb{R}^N, \|X\|\leq 1} \|AX\|.$$

The mapping $|||\cdot|||$ satisfies the usual properties of norms, i.e.,

1. $|||A||| \geq 0$ and $|||A||| = 0$ if and only if $A = 0$,
2. for all real λ, $|||\lambda A||| = |\lambda|\, |||A|||$,
3. the triangle inequality $|||A + B||| \leq |||A||| + |||B|||$, for all matrices A and B.

It also satisfies the following additional property: For all matrices A and B, we have

$$|||AB||| \leq |||A|||\ |||B|||.$$

We say that a matrix norm is *submultiplicative*. Finally, by definition, we have for all $N \times N$ matrices A and for all vectors X in $\mathbb{R}^N$,

$$\|AX\| \leq |||A|||\ \|X\|. \tag{2.14}$$

We note that $|||I||| = 1$ for any induced matrix norm. In this chapter, we only use the $\|\cdot\|_\infty$ norm on $\mathbb{R}^N$. The induced matrix norm is given below.

Proposition 2.3 *Let A be a $N \times N$ matrix with entries A_{ij}. We have*

$$|||A|||_\infty = \max_{i\in\{1,\ldots,N\}} \sum_{j=1}^{N} |A_{ij}|. \tag{2.15}$$

Proof Let us set $M = \max_{i\in\{1,\ldots,N\}} \sum_{j=1}^{N} |A_{ij}|$. Let X be an arbitrary vector in $\mathbb{R}^N$ with $\|X\|_\infty = 1$. By definition of $\|\cdot\|_\infty$, we have

$$\|AX\|_\infty = \max_{i\in\{1,\ldots,N\}} |\sum_{j=1}^{N} A_{ij}X_j| \leq \max_{i\in\{1,\ldots,N\}} \sum_{j=1}^{N} |A_{ij}|\,|X_j| \leq M,$$

because for any $j \in \{1,\ldots N\}$, $|X_j| \leq \|X\|_\infty = 1$. We deduce that $|||A|||_\infty \leq M$.

In order to show (2.15), it is thus sufficient to find a vector $X \in \mathbb{R}^N$, with $\|X\|_\infty = 1$, such that $\|AX\|_\infty = M$. By definition of M, there exists an index $i_0 \in \{1,\ldots,N\}$ such that $M = \sum_{j=1}^{N} |A_{i_0 j}|$. Let us consider the vector $X \in \mathbb{R}^N$ defined by: $X_j = 1$, if $A_{i_0 j} \geq 0$ and $X_j = -1$ otherwise. We clearly have $\|X\|_\infty = 1$ and $(AX)_{i_0} = \sum_{j=1}^{N} A_{i_0 j}X_j = \sum_{j=1}^{N} |A_{i_0 j}| = M \geq 0$. This shows that $|||A|||_\infty \geq M$, which completes the proof. □

Let us now introduce another useful concept, inverse nonnegative matrices.

Definition 2.3 We say that a vector $X \in \mathbb{R}^N$ is nonnegative, and we write $X \geq 0$, if all its components X_i are nonnegative, and that a $N \times N$ matrix A is nonnegative, and we write $A \geq 0$, if all its entries A_{ij} are nonnegative. A matrix A is said to be *inverse nonnegative* or *monotone* if it is invertible and its inverse matrix is nonnegative.

We also note $X \leq 0$ or $A \leq 0$ whenever $-X \geq 0$ or $-A \geq 0$. We have the following characterization of inverse nonnegative matrices:

Lemma 2.2 *A matrix A is inverse nonnegative if and only if for all vectors X in $\mathbb{R}^N$, we have*

$$\textit{if } AX \geq 0 \textit{ then } X \geq 0. \tag{2.16}$$

Proof Let us suppose that A is inverse nonnegative. Let X be a vector in $\mathbb{R}^N$ such that $AX \geq 0$. The matrix A^{-1} is nonnegative, therefore the vector $X = A^{-1}(AX)$ is nonnegative by the usual formula for matrix-vector products.

Conversely, let A satisfy (2.16). We first show that A is invertible. Let X be a vector in $\mathbb{R}^N$ such that $AX = 0$. In particular $AX \geq 0$, so that, thanks to (2.16), we have $X \geq 0$. Likewise $A(-X) \geq 0$, so that $-X \geq 0$. It follows that $X = 0$, which shows that A is invertible. Let us next show that the matrix A^{-1} is nonnegative. It is sufficient to show that each column of A is nonnegative. Let C_i be one of these columns. By definition, $C_i = A^{-1}E_i$, where E_i is the ith vector of the canonical basis of $\mathbb{R}^N$ (i.e., all its components are zero, except the component of index i which is 1). Since $E_i \geq 0$ and $AC_i = E_i$, we deduce from (2.16) that $C_i \geq 0$, which completes the proof. □

We now apply the above definitions and properties to the discrete problem (2.7).

Proposition 2.4 *Let us suppose that $c \geq 0$. The matrix A_h defined by (2.8)–(2.9) is inverse nonnegative.*

Proof We use the above characterization of inverse nonnegative matrices. Let X be a vector in $\mathbb{R}^N$ such that $A_hX \geq 0$. Let us show that $X \geq 0$. Let i_0 be the smallest index such that $X_{i_0} \leq X_i$, for all $i \in \{1, \dots, N\}$. We will show that $X_{i_0} \geq 0$, which implies the result.

Let us first consider the case when $i_0 = 1$. The first component of A_hX is nonnegative, therefore

$$[2 + h^2c(x_1)]X_1 - X_2 = (X_1 - X_2) + [1 + h^2c(x_1)]X_1 \geq 0.$$

As $X_1 - X_2 \leq 0$ and $1 + h^2c(x_1) \geq 1$ we deduce that $X_1 \geq 0$. The same proof applies in the case $i_0 = N$.

Let us now consider the case when $i_0 \in \{2, \dots, N-1\}$. Since $(A_hX)_{i_0} \geq 0$, we obtain

$$-X_{i_0-1} + [2 + h^2c(x_{i_0})]X_{i_0} - X_{i_0+1} = (X_{i_0} - X_{i_0-1}) + (X_{i_0} - X_{i_0+1}) + h^2c(x_{i_0})X_{i_0} \geq 0, \tag{2.17}$$

with $X_{i_0} - X_{i_0-1} \leq 0$ and $X_{i_0} - X_{i_0+1} \leq 0$ by definition of i_0. If $c(x_{i_0}) = 0$, we get $X_{i_0} - X_{i_0-1} = X_{i_0} - X_{i_0+1} = 0$. In other words, $i_0 - 1$ is also an index where the components of X achieve their minimum, which is a contradiction since $i_0 - 1 < i_0$. On the other hand, if $c(x_{i_0}) > 0$, then by (2.17), we see that $X_{i_0} \geq 0$. □

Remark 2.6 The above proof shows that if we have $A_hU_h = F_h$ with $F_h \geq 0$, then $U_h \geq 0$. This is the *discrete maximum principle*, which is the discrete analogue of the maximum principle of Theorem 1.3 of Chap. 1. □

Remark 2.7 We note that the matrix A_h in addition to being inverse nonnegative or monotone, is also such that its off-diagonal coefficients are nonpositive. Such matrices are called *M-matrices*. These matrices have applications in many fields in addition to the discretization of differential operators, notably in probability theory and economics among others, see [54, 79]. □

We now are in a position to prove Theorem 2.1. Starting from (2.12) and using formula (2.14), we deduce the following estimate:

$$\|U_h - S_h(u)\|_\infty \leq \|\!|(A_h)^{-1}|\!\|_\infty \ \|\varepsilon_h(u)\|_\infty.$$

We have already estimated the second term $\|\varepsilon_h(u)\|_\infty$, see (2.11). We thus need an estimate of the quantity $\|\!|(A_h)^{-1}|\!\|_\infty$, which is called a *stability estimate*.

Proposition 2.5 *Let us suppose that $c \geq 0$. We have the following estimate, for all $h = \frac{1}{N+1}$,*

$$\|\!|(A_h)^{-1}|\!\|_\infty \leq \frac{1}{8}. \tag{2.18}$$

Proof We first remark that $A_h^{-1} - (A_h^0)^{-1} = A_h^{-1}[A_h^0 - A_h](A_h^0)^{-1}$. Since $c \geq 0$, the matrix $A_h^0 - A_h$ is diagonal with nonpositive entries. By Proposition 2.4, $A_h^{-1} \geq 0$ and $(A_h^0)^{-1} \geq 0$. Therefore $A_h^{-1} - (A_h^0)^{-1} \leq 0$ by the usual formula for matrix products. Since both A_h^{-1} and $(A_h^0)^{-1}$ are nonnegative, formula (2.15) then implies that $\|\!|A_h^{-1}|\!\|_\infty \leq \|\!|(A_h^0)^{-1}|\!\|_\infty$. In order to get estimate (2.18), it is thus sufficient to show that

$$\|\!|(A_h^0)^{-1}|\!\|_\infty \leq \frac{1}{8}.$$

The matrix $(A_h^0)^{-1}$ is nonnegative, therefore by formula (2.15)

$$\|\!|(A_h^0)^{-1}|\!\|_\infty = \|(A_h^0)^{-1}E\|_\infty,$$

where E is the vector in $\mathbb{R}^N$ all the components of which are equal to 1. We notice that $(A_h^0)^{-1}E$ is precisely the discrete solution U_h of system (2.7) in the particular case $c = 0$ and for $F_h = E$. In other words, it is the discrete solution associated with the following boundary value problem:

$$\begin{cases} -\bar{u}''(x) = 1, & x \in]0, 1[, \\ \bar{u}(0) = 0, & \bar{u}(1) = 0. \end{cases}$$

This problem clearly has the explicit solution $\bar{u}(x) = x(1-x)/2$, a second degree polynomial. Thanks to Remark 2.5, p. 43, we know that, in this particular case, the discrete solution coincides with the exact solution at each grid point. We thus have $((A_h^0)^{-1}E)_i = \bar{u}(x_i) = x_i(1-x_i)/2$. Consequently,

$$\|(A_h^0)^{-1}E\|_\infty \leq \sup_{x\in[0,1]} |\bar{u}(x)| = \bar{u}\Big(\frac{1}{2}\Big) = \frac{1}{8},$$

which completes the proof. □

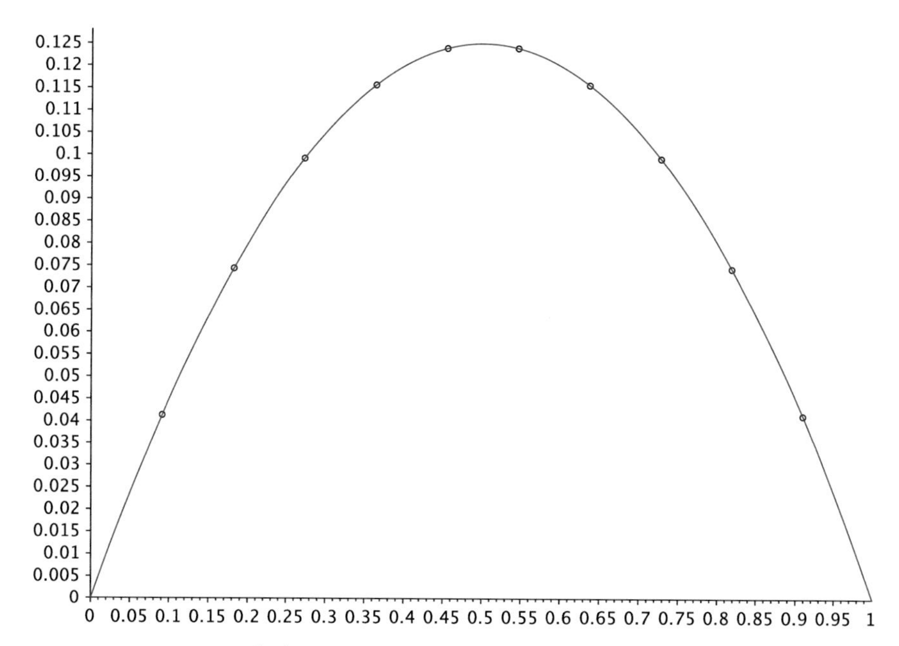

Fig. 2.2 Plot of $\bar{u}(x) = \frac{x(1-x)}{2}$ in *solid line* and $U_h = (A_h^0)^{-1}E$ for $N = 10$ with $\circ$ marks

Remark 2.8 Note that the convergence proof relies on two fundamental properties: consistency and stability. We will encounter a very similar idea in the study of numerical approximations of the heat and wave equations in Chaps. 8 and 9. □

We plot in Fig. 2.2 the function u and the discrete solution $U_h = (A_h^0)^{-1}E$ in the particular case used above in the stability estimate when the discrete solution happens to coincide with the exact solution at the grid points.

Let us illustrate the preceding results with a numerical example. We take $c(x) = 1000\sin(10\pi x)^2$ and $f(x) = 1$, and we compute the finite difference approximations of the corresponding boundary value problem for $N = 19$, 29, 49, 99, and 199, i.e., $h = 0.05$, 0.033, 0.02, 0.01, and 0.005, by solving the associated linear systems. To compare the results, we plot in Fig. 2.3 the computed discrete values u_i against x_i for these five cases on the same plot, with different marks for each value of N and linear interpolations[1] in between points. The solid curve with no marks is a so-called "reference" solution that is meant to represent the exact solution, but that we actually also computed using finite differences with $N = 8000$, since no explicit formula seems to be available. The computation is performed with Scilab, a free, general purpose scientific computing package (http://www.scilab.org/).

[1]This is for visualization purposes only. The finite difference method does not compute a function on [0, 1].

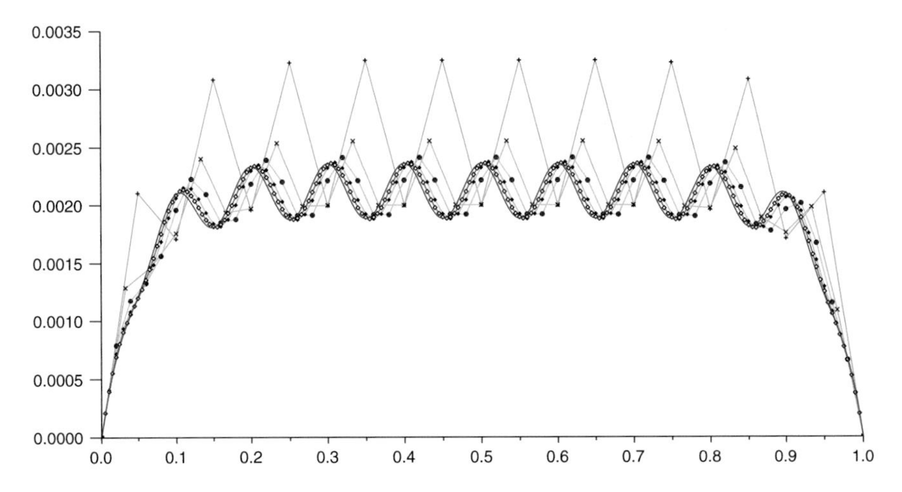

Fig. 2.3 Convergence of the finite difference method

2.4 Neumann Boundary Conditions

Let us briefly describe the method in the case of Neumann boundary conditions at both ends of the interval. The Neumann boundary conditions are different from the Dirichlet conditions seen up to now. For a one-dimensional second order problem, they concern the values of the first derivative of the unknown function at the ends of the interval, instead of the values of the function itself. In terms of modeling, in the heat equation interpretation, they consist in imposing the heat flux at the boundary instead of imposing the temperature.

We thus consider the problem: Find $u\colon [0,1] \to \mathbb{R}$ solution of

$$\begin{cases} -u''(x) + c(x)u(x) = f(x), & x \in]0,1[, \\ u'(0) = g_0, \quad u'(1) = g_1, \end{cases} \tag{2.19}$$

where c and f are two given functions ($c \in C^0([0,1])$ and $f \in C^0([0,1])$), and g_0 and g_1 are two given constants. It is easy to prove by the shooting method that such a function u exists if we suppose, for example, that there exists a constant $c_0 > 0$ such that $c \geq c_0$ (see also the variational theory in Chap. 4). More generally, if $c \geq 0$ is not identically 0, we have existence and uniqueness. On the other hand, if $c = 0$ there is no uniqueness, hence no existence in general. Indeed, if a solution exists, then we can add any arbitrary constant and still find another solution of problem (2.19). This is one important difference with the Dirichlet problem (2.4) studied before. For simplicity, we will assume a bound from below $c \geq c_0 > 0$ from now on.

We now proceed to define a finite difference approximation of this problem. We use the same uniform grid in $[0,1]$ as before, i.e., with grid points $x_i = ih$, $i \in \{0, \dots, N+1\}$, where $h = \frac{1}{N+1}$ and N is a given positive integer. There will be a discrete unknown u_i attached to each grid point, including for $i = 0$ and $i = N + 1$, since the values of u at the ends of the interval are not prescribed by the continuous problem as in the case of the Dirichlet problem.

For the internal grid points, i.e. points x_i, with $i \in \{1, \dots, N\}$, we will use the same three point scheme as before to approximate the second order derivative at this point. The problem is now to approximate the Neumann boundary condition, since it cannot be satisfied exactly—contrarily to the Dirichlet case. The condition does not even make sense at the discrete level. There are several possibilities. The simplest one consists in approximating the first order derivative by one of the two decentered difference schemes introduced in Remark 2.2. More precisely, we can use the forward difference to approximate $u'(0)$ and the backward difference to approximate $u'(1)$, i.e.,

$$u'(0) \approx D_h^+ u(0) = \frac{u(h) - u(0)}{h}, \quad u'(1) \approx D_h^- u(1) = \frac{u(1) - u(1-h)}{h}.$$

This suggests the following approximation of the boundary conditions:

$$\frac{u_0 - u_1}{h} = -g_0, \quad \frac{u_{N+1} - u_N}{h} = g_1.$$

The reason for the minus sign in the first relation is that the Neumann boundary condition at $x = 0$ is more naturally written as $-u'(0) = -g_0$, in terms of integration by parts and other considerations that we will see later. The discrete problem is thus equivalent to the linear system

$$B_h U_h = F_h, \tag{2.20}$$

with

$$B_h = \frac{1}{h^2} \begin{pmatrix} h & -h & 0 & \dots & \dots & 0 \\ -1 & 2 + h^2 c(x_1) & -1 & \ddots & & \vdots \\ 0 & -1 & 2 + h^2 c(x_2) & -1 & \ddots & \vdots \\ 0 & \ddots & \ddots & \ddots & \ddots & 0 \\ \vdots & & \ddots & -1 & 2 + h^2 c(x_N) & -1 \\ 0 & \dots & \dots & 0 & -h & h \end{pmatrix},$$

where the unknown U_h is the vector in $\mathbb{R}^{N+2}$ of components u_i, $i \in \{0, \dots, N+1\}$. Note that B_h is an $(N+2) \times (N+2)$ matrix. The right-hand side of the linear system is given by

$$F_h = \begin{pmatrix} -g_0 \\ f(x_1) \\ \vdots \\ f(x_N) \\ g_1 \end{pmatrix}.$$

We first remark that the matrix B_h is no longer symmetric due to the first and last lines. Before studying the convergence of the numerical method, let us make a few comments on the case $c = 0$. In this case, it is easily seen that the kernel of B_h is the one-dimensional space spanned by the vector $(1, 1, \ldots, 1)^T$, so that the matrix of the linear system (2.20) is not invertible. Therefore the numerical method does not work when $c = 0$, which is perfectly consistent with what happens at the continuous level. We will see in Proposition 2.7 that when there exists a positive constant c_0 such that $c \geq c_0$, the matrix B_h is invertible, hence the scheme is well defined.

Let us now study the convergence of the method when $c \geq c_0 > 0$, i.e., its consistency (Proposition 2.6 below) and its stability (Proposition 2.7).

Proposition 2.6 *Let us suppose that the solution u of problem (2.19) is C^4 on* $[0, 1]$. *Then the scheme (2.20) is consistent of order one.*

Proof Let $\varepsilon_h(u) = B_h S_h(u) - F_h$ be the truncation error, where S_h denotes here the sampling operator on all the grid points, and is thus $\mathbb{R}^{N+2}$-valued. Using the first equation in (2.19) and (2.3), we get for any $i \in \{1, \ldots, N\}$,

$$(\varepsilon_h(u))_i = \frac{-u(x_{i+1}) + 2u(x_i) - u(x_{i-1})}{h^2} + c(x_i)u(x_i) - f(x_i) = -\frac{h^2}{12}u^{(4)}(\xi_i),$$

for some ξ_i between x_{i-1} and x_{i+1}. For $i = 0$, using the left boundary condition in (2.19) and (2.1), we obtain

$$(\varepsilon_h(u))_0 = \frac{u(x_0) - u(x_1)}{h} + g_0 = \frac{u(0) - u(h)}{h} + u'(0) = -\frac{h}{2}u''(\xi_0),$$

for some ξ_0 between 0 and h. Similarly

$$(\varepsilon_h(u))_{N+1} = \frac{-u(x_N) + u(x_{N+1})}{h} - g_1 = \frac{u(1) - u(1-h)}{h} - u'(1) = -\frac{h}{2}u''(\xi_{N+1}),$$

for some ξ_{N+1} between $1 - h$ and 1. Let us set

$$M = \max\Bigl(\frac{1}{12}\max_{y\in[0,1]}|u^{(4)}(y)|, \frac{1}{2}\max_{y\in[0,1]}|u''(y)|\Bigr).$$

Since $h \leq 1$, we have $h^2 \leq h$ and therefore, for any $i \in \{0, \ldots, N+1\}$, we see that $|(\varepsilon_h(u))_i| \leq Mh$. In other words,

$$\|\varepsilon_h(u)\|_\infty \le Mh, \tag{2.21}$$

which shows that the scheme (2.20) is consistent of order one. □

Remark 2.9 Note that the terms coming from the approximation of the boundary conditions are dominant with respect to the second derivative approximation in the truncation error. □

We can show the following stability result.

Proposition 2.7 *Let us suppose that there exists a constant $c_0 > 0$ such that $c \ge c_0$. Then, the matrix B_h is inverse nonnegative, thus invertible and we have the following estimate:*

$$|||(B_h)^{-1}|||_\infty \le C, \tag{2.22}$$

where C is a constant which does not depend on h.

Proof We decompose the matrix as follows

$$B_h = B_h^0 + \begin{pmatrix} 0 & \cdots & \cdots & \cdots & \cdots & 0 \\ 0 & c(x_1)-c_0 & 0 & \cdots & \cdots & 0 \\ \vdots & \ddots & c(x_2)-c_0 & \ddots & & \vdots \\ \vdots & & \ddots & \ddots & \ddots & \vdots \\ 0 & \cdots & \cdots & 0 & c(x_N)-c_0 & 0 \\ 0 & \cdots & \cdots & \cdots & \cdots & 0 \end{pmatrix},$$

where

$$B_h^0 = \frac{1}{h^2}\begin{pmatrix} h & -h & 0 & \cdots & \cdots & 0 \\ -1 & 2+c_0h^2 & -1 & \ddots & & \vdots \\ 0 & \ddots & \ddots & \ddots & \ddots & \vdots \\ \vdots & \ddots & \ddots & \ddots & \ddots & 0 \\ \vdots & & \ddots & -1 & 2+c_0h^2 & -1 \\ 0 & \cdots & \cdots & 0 & -h & h \end{pmatrix}.$$

We first show that B_h is inverse nonnegative using the characterization of inverse nonnegative matrices given in Lemma 2.2. Let X be a vector in $\mathbb{R}^{N+2}$ such that $B_hX \ge 0$. We have to show that $X \ge 0$ and this is equivalent to showing that $X_{i_0} \ge 0$, where i_0 is the smallest index such that $X_{i_0} \le X_i$, for all $i \in \{0, \dots, N+1\}$. Let us first consider the case $i_0 = 0$. As $(B_hX)_0 \ge 0$, we have $X_0 \ge X_1$. On the other hand, by definition of i_0, we also have $X_0 \le X_1$, so that $X_0 = X_1$. Now, as $(B_hX)_1 \ge 0$, it follows that

$$-X_0 + (2 + c(x_1)h^2)X_1 - X_2 = (X_0 - X_2) + c(x_1)h^2X_0 \ge 0.$$

Now $X_0 - X_2 \leq 0$ and $c(x_1) \geq c_0 > 0$, therefore $X_0 \geq 0$, which is the expected result.

Let us now consider the case $i_0 \in \{1, \dots, N\}$. The condition $(B_h X)_{i_0} \geq 0$ reads

$$-X_{i_0-1} + (2 + c(x_{i_0})h^2)X_{i_0} - X_{i_0+1} = (X_{i_0} - X_{i_0-1}) + (X_{i_0} - X_{i_0+1}) + c(x_{i_0})h^2 X_{i_0} \geq 0.$$

We have $X_{i_0} - X_{i_0-1} \leq 0$ and $X_{i_0} - X_{i_0+1} \leq 0$, by definition of i_0. Since $c(x_{i_0}) \geq c_0 > 0$, this implies that $X_{i_0} \geq 0$.

Finally, the case $i_0 = N + 1$ cannot happen. Indeed, $X_{N+1} \geq X_N$ contradicts the definition of i_0. We have thus proven that B_h is inverse nonnegative. Since B_h^0 is a particular case of B_h, B_h^0 is also inverse nonnegative.

Let us now prove estimate (2.22). We proceed as in Proposition 2.5. As before, $B_h^{-1} - (B_h^0)^{-1} = B_h^{-1}[B_h^0 - B_h](B_h^0)^{-1}$. Since $c \geq c_0$, the matrix $B_h^0 - B_h$ is diagonal with nonpositive entries and $B_h^{-1} \geq 0$ and $(B_h^0)^{-1} \geq 0$ as we have just seen. Consequently, $B_h^{-1} - (B_h^0)^{-1} \leq 0$. It follows that $\|B_h^{-1}\|_\infty \leq \|(B_h^0)^{-1}\|_\infty$.

In order to obtain estimate (2.22), it is thus sufficient to show that

$$\|(B_h^0)^{-1}\|_\infty = \|(B_h^0)^{-1}E\|_\infty \leq C,$$

where C is a constant which does not depend on h and E the vector in $\mathbb{R}^{N+2}$ with all components equal to 1. We notice that $(B_h^0)^{-1}E$ is precisely the discrete solution U_h of system (2.20) in the particular case $c = c_0$, $f = 1$ and $-g_0 = g_1 = 1$. In other words, it is the discrete solution associated with the following boundary value problem:

$$\begin{cases} -\bar{u}''(x) + c_0\bar{u}(x) = 1, & x \in]0, 1[, \\ \bar{u}'(0) = -1, \quad \bar{u}'(1) = 1. \end{cases}$$

A simple computation shows that $\bar{u}$ has the following expression:

$$\bar{u}(x) = \frac{1}{\sqrt{c_0}} \frac{\cosh\big(\sqrt{c_0}(x - \frac{1}{2})\big)}{\sinh\big(\frac{\sqrt{c_0}}{2}\big)} + \frac{1}{c_0}.$$

This function is of class C^∞, hence C^4. By definition of the truncation error $\varepsilon_h(\bar{u}) = B_h^0 S_h(\bar{u}) - E$, we have

$$(B_h^0)^{-1}E = S_h(\bar{u}) - (B_h^0)^{-1}\varepsilon_h(\bar{u}).$$

Consequently,

$$\|(B_h^0)^{-1}\|_\infty = \|(B_h^0)^{-1}E\|_\infty \leq \|S_h(\bar{u})\|_\infty + \|(B_h^0)^{-1}\|_\infty \|\varepsilon_h(\bar{u})\|_\infty.$$

Now $\bar{u}$ is C^4, hence by Proposition 2.6, $\|\varepsilon_h(\bar{u})\|_\infty \leq Ch$ where C does not depend on h. Therefore, for h sufficiently small, this quantity can be bounded from above by $\frac{1}{2}$. It follows that

$$\|(B_h^0)^{-1}\|_\infty \leq 2\|S_h(\bar{u})\|_\infty \leq 2\Big(\frac{1}{\sqrt{c_0}}\coth\Big(\frac{\sqrt{c_0}}{2}\Big)+\frac{1}{c_0}\Big),$$

due to the expression of $\bar{u}$. The values of h that are not sufficiently small in the above sense are only finite in number, thus this completes the proof. □

Note that the above estimate gets worse and worse as $c_0 \to 0$, which is consistent with what happens when $c_0 = 0$ and B_h^0 is no longer invertible. We deduce from Proposition 2.6 (consistency) and Proposition 2.7 (stability) the following final convergence result:

Corollary 2.2 *Assume that there exists a constant $c_0 > 0$ such that $c \geq c_0$ and that the solution u of problem (2.19) is C^4 on $[0, 1]$. Then the scheme (2.20) is convergent of order one. More precisely, we have*

$$\|U_h - S_h(u)\|_\infty \leq Ch, \tag{2.23}$$

where C is a constant that does not depend on h.

Proof The proof is the same as for the Dirichlet case. We reproduce it for completeness. By definition of the scheme and of the truncation error, we have

$$B_h U_h = F_h \quad \text{and} \quad B_h S_h(u) = F_h + \varepsilon_h(u),$$

from which we deduce that

$$B_h(U_h - S_h(u)) = -\varepsilon_h(u).$$

Since B_h is invertible, this implies that the error is given by

$$U_h - S_h(u) = -(B_h)^{-1}\varepsilon_h(u),$$

so that we obtain

$$\|U_h - S_h(u)\|_\infty \leq \|(B_h)^{-1}\|_\infty \, \|\varepsilon_h(u)\|_\infty.$$

Estimate (2.23) then follows from (2.21) and (2.22). □

Remark 2.10 The final error estimate is only of order one, due to the decentered scheme chosen for the approximation of the Neumann boundary condition and in spite of the order two approximation of the second derivative inside the domain.

It could be thought that the order one error remains somehow concentrated near the boundary and that the approximation is better inside. This is not the case. The lower order approximation of the boundary condition "pollutes" the discrete solution everywhere, as we will see on a numerical example below. □

We can in fact improve the accuracy by choosing a central scheme for the approximation of the Neumann boundary condition, instead of the decentered approximations used before. In order to do that, we first add two fictitious grid points $x_{-1} = -h$ and $x_{N+2} = 1 + h$, which are outside the interval $[0, 1]$, and we use the following approximations of the first derivative at both ends:

$$u'(0) \approx \frac{u(h) - u(-h)}{2h}, \quad u'(1) \approx \frac{u(1+h) - u(1-h)}{2h},$$

assuming u has been adequately extended outside of $[0, 1]$, somehow.[2] We also add two fictitious discrete unknowns, denoted by u_{-1} and u_{N+2}. In order to have as many unknowns as equations, we extend equation (2.19) up to the boundary of the domain, i.e., we discretize it at each grid point x_i, internal or not, which finally gives the scheme

$$\begin{cases} -\dfrac{u_{i+1} - 2u_i + u_{i-1}}{h^2} + c(x_i)u_i = f(x_i), \quad i \in \{0, \dots, N+1\}, \\ \dfrac{u_{-1} - u_1}{2h} = -g_0, \quad \dfrac{u_{N+2} - u_N}{2h} = g_1. \end{cases}$$

It can be shown that the truncation error of this new scheme is of order two.

In order to illustrate the compared performance of the previous schemes, we plot the results of a few Scilab computations. We consider the case $c(x) = 4$, $f(x) = 1$ and $-g_0 = g_1 = 1$, for which we have an exact solution in closed form as seen before. First we compute the results of the first order scheme and plot them in Fig. 2.4.

We plot the discrete solutions, again interpolated for easier visualization, for $N = 8, 12, 16, 20$, and 24. Clearly, the convergence is fairly slow, and we can see that the order 1 error is uniformly distributed over the whole domain, even though it is only due to the approximation of the boundary conditions at the ends of the interval, and the truncation error inside is of order 2.

Next, we plot the second order scheme for the same data (we do not plot the fictitious values).

In this case, the marks for the discrete values are virtually on the graph of the exact solution, thus indicating a much faster convergence. It should be noted that at first glance, the matrices of each method do not look that different from each other. The results are nonetheless dramatically different, see Fig. 2.5.

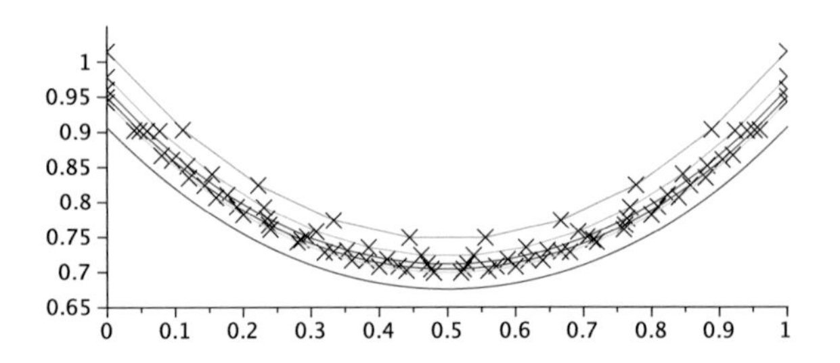

Fig. 2.4 First order scheme with × marks. Exact solution in *solid line* with no marks

[2] We skip the details.

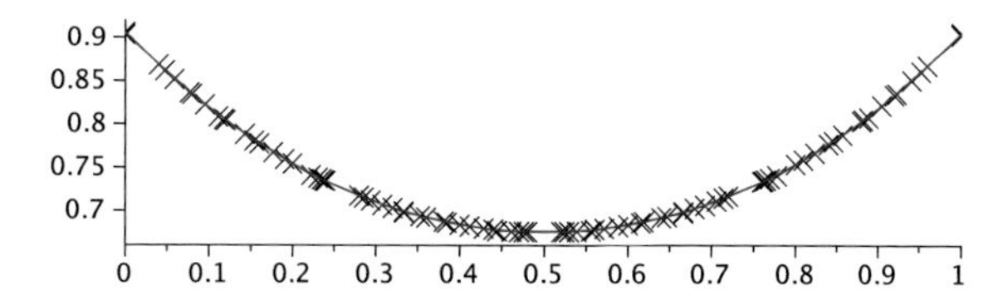

Fig. 2.5 Second order scheme with $\times$ marks. Exact solution in *solid line* with no marks

Remark 2.11 Let us make a final remark concerning the resolution of the discrete problem (2.20) in practice. This must be performed using a numerical method for linear systems implemented in software. Of course, such numerical methods work in floating point arithmetic, hence are subject to round-off errors. This thus raises the important question of controlling the effect of such errors on the computed solution.

Let us consider a $N \times N$ linear system $AX = F$ where A is invertible. We wish to measure the error δX produced on the solution X by an error δF on the right-hand side F, i.e., $A(X + \delta x) = F + \delta F$. Subtracting the two equations, we obtain $A\delta X = \delta F$, from which we deduce that $\|\delta X\| \leq |||A^{-1}||| \, \|\delta F\|$ (where $\|\cdot\|$ is a given norm on $\mathbb{R}^N$ and $|||\cdot|||$ the associated matrix norm). On the other hand, we also have $\|F\| \leq |||A||| \, \|X\|$, from which we deduce that $\frac{1}{\|X\|} \leq \frac{|||A|||}{\|F\|}$ when $F \neq 0$ and thus $X \neq 0$. Multiplying the two estimates together yields

$$\frac{\|\delta X\|}{\|X\|} \leq |||A||| \, |||A^{-1}||| \, \frac{\|\delta F\|}{\|F\|}. \tag{2.24}$$

The number $\mathrm{cond}\, A = |||A||| \, |||A^{-1}|||$ is called the *condition number* of A (with respect to the norm $\|\cdot\|$) [6, 18, 53]. It is always larger than 1 since $1 = |||I||| \leq |||A||| \, |||A^{-1}|||$ by submultiplicativity of an induced matrix norm. If it is roughly speaking small, we say that the matrix is *well-conditioned*. On the contrary, if it is large compared to 1, we say that the matrix is *ill-conditioned*. In the latter case, even a small relative error on the right-hand side may induce a large relative error on the solution, which may render the numerical computation meaningless. This is not the case if the matrix is well-conditioned, on account of estimate (2.24). Of course, different numerical methods have different abilities to handle ill-conditioned matrices. Some will fail on some ill-conditioned matrices, whereas others will succeed on the same matrices. The latter concept is not entirely well defined.

Let us go back to the first order scheme for the Neumann problem. We have written it in a "natural" form that allows for consistency and stability to be established. There are however infinitely many other different matrix forms, with the same solutions but different condition numbers.

In fact, we can for instance write this system in the following equivalent form:

$$B'_h U_h = F'_h, \tag{2.25}$$

with

$$B'_h = \frac{1}{h^2}\begin{pmatrix} 1 & -1 & 0 & \dots & \dots & 0 \\ -1 & 2+h^2c(x_1) & -1 & \ddots & & \vdots \\ 0 & -1 & 2+h^2c(x_2) & -1 & \ddots & \vdots \\ 0 & \ddots & \ddots & \ddots & \ddots & 0 \\ \vdots & & \ddots & -1 & 2+h^2c(x_N) & -1 \\ 0 & \dots & \dots & 0 & -1 & 1 \end{pmatrix},$$

and

$$F'_h = \begin{pmatrix} \frac{-g_0}{h} \\ f(x_1) \\ \vdots \\ f(x_N) \\ \frac{g_1}{h} \end{pmatrix},$$

where both the first and the last equations of system (2.20) have been divided by h. Even though the two systems (2.20) and (2.25) are strictly equivalent in terms of linear algebra, the new form (2.25) is not suitable for the computation of the truncation error. It is however better suited for numerical resolution because the matrix B'_h has a condition number that is smaller than that of matrix B_h.

We first plot on Fig. 2.6 the two condition numbers for the $\|\cdot\|_\infty$ norm, as a function of N in the case $c(x) = 1$. Next, Fig. 2.7, we plot their ratio, still as a function of N. For

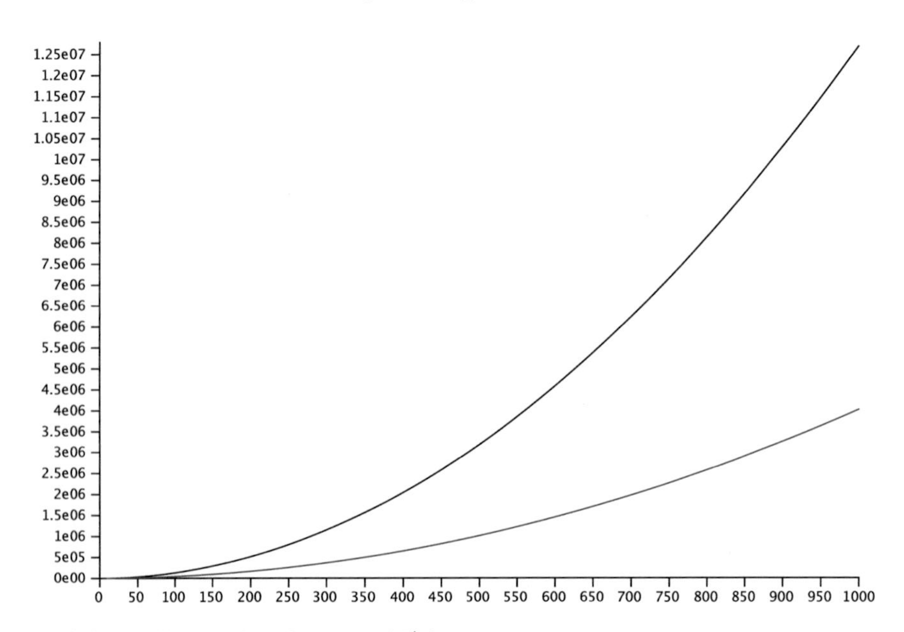

Fig. 2.6 Condition number of B_h *top* and B'_h *bottom*

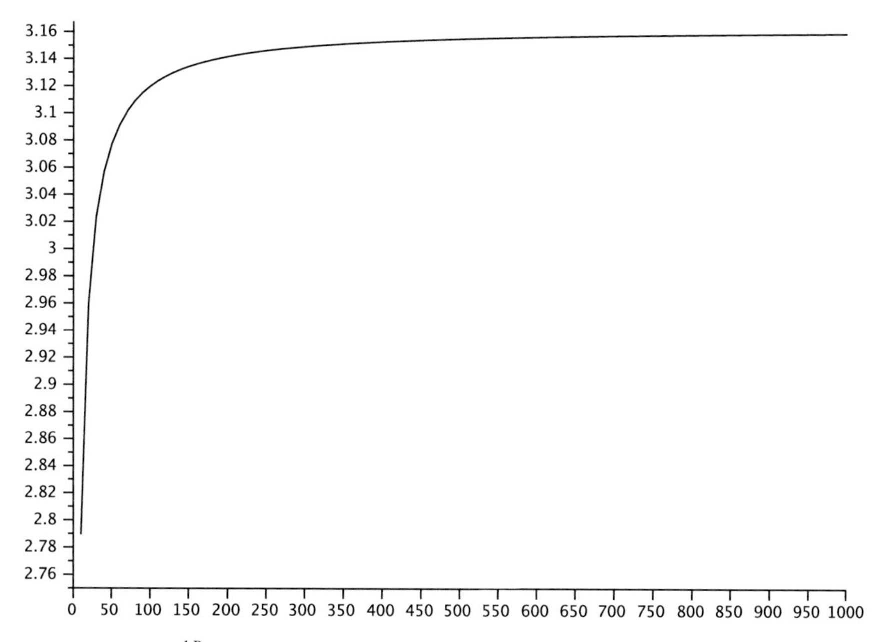

Fig. 2.7 Ratio $\frac{\operatorname{cond} B_h}{\operatorname{cond} B'_h}$

N large, both condition numbers are quite large, but there is a ratio of approximately 3 between the two, which is not spectacular, albeit appreciable. The general problem of finding a linear system equivalent to a given linear system but with a better condition number is called *preconditioning*. □

2.5 The Two-Dimensional Case

The finite difference method also applies to higher-dimensional elliptic problems, with some limitations. We describe here the approximation of a simple two-dimensional problem, and the generalization to three-dimensional and higher is easy to imagine.

Let us thus consider the homogeneous Dirichlet problem in $\Omega =]0,1[\times]0,1[$

$$\begin{cases} -\Delta u(x) = f(x) \text{ in } \Omega, \\ u(x) = 0 \text{ on } \Gamma, \end{cases} \tag{2.26}$$

where f is a given continuous function on $\bar{\Omega}$ and $\Gamma = \partial\Omega$. We will see in Chap. 4 that this problem has a unique solution. In general, there is no closed form solution, therefore our goal here is again to define a finite difference approximation for it. For

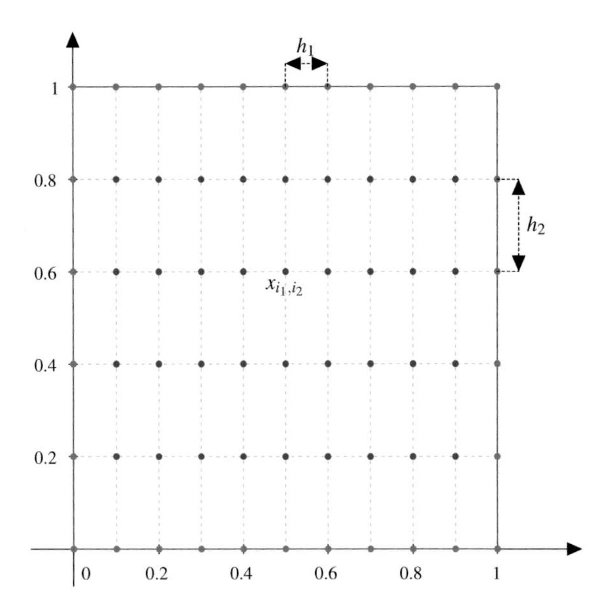

Fig. 2.8 A 2d grid

that purpose, we start by placing a uniform grid in $\bar{\Omega}$ as follows. Let us be given N_1 and N_2 two positive integers. We let $h_1 = 1/(N_1 + 1)$ be the grid step in direction x_1 and $h_2 = 1/(N_2 + 1)$ be the grid step in direction x_2. The grid points will then be $x_{i_1,i_2} = (i_1 h_1, i_2 h_2)$ for $i_1 \in \{0, \ldots, N_1 + 1\}$ and $i_2 \in \{0, \ldots, N_2 + 1\}$. The boundary grid points correspond to $i_1 \in \{0, N_1 + 1\}$ or $i_2 \in \{0, N_2 + 1\}$, and the internal grid points to $1 \leq i_1 \leq N_1$ and $1 \leq i_2 \leq N_2$, see Fig. 2.8.

Following the same idea as in the one-dimensional case, we want to compute an approximation u_{i_1,i_2} of $u(x_{i_1,i_2})$ for each $i_1 \in \{0, \ldots, N_1+1\}$ and $i_2 \in \{0, \ldots, N_2+1\}$. We naturally enforce the exact Dirichlet boundary condition on the boundary grid points, i.e., we set $u_{i_1,i_2} = 0$ if $i_1 \in \{0, N_1 + 1\}$ or if $i_2 \in \{0, N_2 + 1\}$.

We then need to approximate the Laplacian of u at internal grid points. Since partial derivatives are nothing more than usual one-dimensional derivatives with all the other variables frozen, we use the now familiar three point scheme for that purpose. Namely, we approximate

$$\frac{\partial^2 u}{\partial x_1^2}(x_1, \cdot) \approx \frac{u(x_1 + h_1, \cdot) - 2u(x_1, \cdot) + u(x_1 - h_1, \cdot)}{h_1^2}$$

and

$$\frac{\partial^2 u}{\partial x_2^2}(\cdot, x_2) \approx \frac{u(\cdot, x_2 + h_2) - 2u(\cdot, x_2) + u(\cdot, x_2 - h_2)}{h_2^2}.$$

Note that the variable x_2 plays no role in the first approximation and likewise for x_1 in the second approximation. Taking into account that $\Delta u = \frac{\partial^2 u}{\partial x_1^2} + \frac{\partial^2 u}{\partial x_2^2}$, this leads to the following scheme:

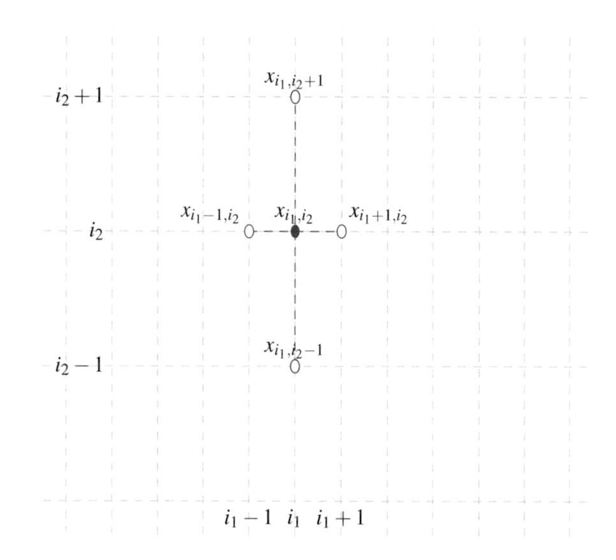

Fig. 2.9 The five point stencil for the Laplacian

$$\begin{cases} -\dfrac{u_{i_1+1,i_2}-2u_{i_1,i_2}+u_{i_1-1,i_2}}{h_1^2}-\dfrac{u_{i_1,i_2+1}-2u_{i_1,i_2}+u_{i_1,i_2-1}}{h_2^2} \\ \qquad = f(x_{i_1,i_2}), \text{ for } i_1\in\{1,\dots,N_1\}, i_2\in\{1,\dots,N_2\}, \\ u_{i_1,i_2} = 0, \text{ for } i_1\in\{0,N_1+1\} \text{ or } i_2\in\{0,N_2+1\}. \end{cases}$$

This scheme is called the *five point scheme for the Laplacian*. It is a central scheme. Indeed, in order to evaluate the discrete approximation of Δu at point x_{i_1,i_2}, we use the approximate values of u at the five grid points centered around x_{i_1,i_2}, namely the point x_{i_1,i_2} itself and its four neighboring points x_{i_1,i_2-1}, x_{i_1,i_2+1}, x_{i_1-1,i_2} and x_{i_1+1,i_2}. These five points constitute the *stencil* of the scheme, see Fig. 2.9.

We need to reformulate the above scheme as a linear system. In order to do that, we first have to number the unknowns. For example, we can decide to number them in the following way $u_{1,1},\cdots,u_{N_1,1},u_{1,2},\cdots,u_{N_1,2},\cdots,u_{N_1,1},\cdots,u_{N_1,N_2}$, i.e., line by line. Other numberings are possible, for instance column by column.

With this choice of numbering, the scheme has the following equivalent vector form

$$C_{h_1,h_2}U_{h_1,h_2} = F_{h_1,h_2}, \tag{2.27}$$

where the $N_1N_2\times N_1N_2$ matrix C_{h_1,h_2} has a block structure. More precisely, the matrix C_{h_1,h_2} is composed of N_2^2 blocks, each one of size $N_1\times N_1$ and tridiagonal,

$$C_{h_1,h_2} = \begin{pmatrix} A & -b_2 I & 0 & \dots & \dots & \dots & 0 \\ -b_2 I & A & -b_2 I & \ddots & & & \vdots \\ 0 & -b_2 I & A & -b_2 I & \ddots & & \vdots \\ \vdots & \ddots & \ddots & \ddots & \ddots & \ddots & \vdots \\ \vdots & & \ddots & -b_2 I & A & -b_2 I & 0 \\ \vdots & & & \ddots & -b_2 I & A & -b_2 I \\ 0 & \dots & \dots & \dots & 0 & -b_2 I & A \end{pmatrix},$$

where A is the $N_1 \times N_1$ matrix defined by

$$A = \begin{pmatrix} a & -b_1 & 0 & \dots & \dots & 0 \\ -b_1 & a & -b_1 & \ddots & & \vdots \\ 0 & \ddots & \ddots & \ddots & \ddots & \vdots \\ \vdots & \ddots & \ddots & \ddots & \ddots & 0 \\ \vdots & & \ddots & -b_1 & a & -b_1, \\ 0 & \dots & \dots & 0 & -b_1 & a \end{pmatrix},$$

and I is the $N_1 \times N_1$ identity matrix, where we have set

$$b_1 = \frac{1}{h_1^2}, \quad b_2 = \frac{1}{h_2^2}, \quad \text{and} \quad a = 2(b_1 + b_2)$$

for simplicity. Thus the matrix C_{h_1,h_2} is block tridiagonal. There are no more than five nonzero elements on each line (or each column) in the matrix C_{h_1,h_2}. Moreover, the matrix C_{h_1,h_2} is symmetric.

The unknown U_{h_1,h_2} and the right-hand side F_{h_1,h_2} have the same block structure given by

$$U_{h_1,h_2} = \begin{pmatrix} U^1 \\ \vdots \\ U^{i_2} \\ \vdots \\ U^{N_2} \end{pmatrix}, \text{ where } U^{i_2} = \begin{pmatrix} u_{1,i_2} \\ \vdots \\ u_{i_1,i_2} \\ \vdots \\ u_{N_1,i_2} \end{pmatrix} \in \mathbb{R}^{N_1},$$

and

$$F_{h_1,h_2} = \begin{pmatrix} F^1 \\ \vdots \\ F^{i_2} \\ \vdots \\ F^{N_2} \end{pmatrix}, \text{ where } F^{i_2} = \begin{pmatrix} f(h_1, i_2h_2) \\ \vdots \\ f(i_1h_1, i_2h_2) \\ \vdots \\ f(N_1h_1, i_2h_2) \end{pmatrix} \in \mathbb{R}^{N_1}.$$

We first note that the method is well-posed.

Proposition 2.8 *The matrix C_{h_1,h_2} is positive definite, hence invertible.*

Proof Same as in Proposition 2.1. □

Let us now study the convergence of the method. We start with the consistency. We first need to adapt the definition of the grid sampling operator to the present context. Here, the operator has values in $\mathbb{R}^{N_1N_2}$ and is simply defined by $S_{h_1,h_2}(v)_{i_1,i_2} = v(x_{i_1,i_2})$, for any $v \in C^0(\bar{\Omega})$.

Proposition 2.9 *Let us suppose that the solution u of problem (2.26) is C^4 on $\bar{\Omega}$. Then the scheme (2.27) is consistent of order two.*

Proof Let us denote by $\varepsilon_{h_1,h_2}(u) = C_{h_1,h_2}S_{h_1,h_2}(u) - F_{h_1,h_2}$ the truncation error. Using the same numbering as before, it has the block structure

$$\varepsilon_{h_1,h_2}(u) = \begin{pmatrix} \varepsilon_{h_1,h_2}(u)^1 \\ \vdots \\ \varepsilon_{h_1,h_2}(u)^{i_2} \\ \vdots \\ \varepsilon_{h_1,h_2}(u)^{N_2} \end{pmatrix}, \text{ where } \varepsilon_{h_1,h_2}(u)^{i_2} = \begin{pmatrix} \varepsilon_{h_1,h_2}(u)_{1,i_2} \\ \vdots \\ \varepsilon_{h_1,h_2}(u)_{i_1,i_2} \\ \vdots \\ \varepsilon_{h_1,h_2}(u)_{N_1,i_2} \end{pmatrix} \in \mathbb{R}^{N_1}.$$

We assume that u is C^4 on $\bar{\Omega}$. We have

$$\varepsilon_{h_1,h_2}(u)_{i_1,i_2} = -\frac{u(x_{i_1+1,i_2}) - 2u(x_{i_1,i_2}) + u(x_{i_1-1,i_2})}{h^2} \\ -\frac{u(x_{i_1,i_2+1}) - 2u(x_{i_1,i_2}) + u(x_{i_1,i_2-1})}{h^2} - f(x_{i_1,i_2}).$$

We remark that in each differential quotient, one variable is fixed (either the first one or the second one), so that we can still use the Taylor–Lagrange expansion (2.3). Thanks to the first equation in (2.26), we then obtain

$$(\varepsilon_{h_1,h_2}(u))_{i_1,i_2} = -\frac{1}{12}\Big[h_1^2\frac{\partial^4 u}{\partial x_1^4}(\xi_{i_1,i_2}, i_2h_2) + h_2^2\frac{\partial^4 u}{\partial x_2^4}(i_1h_1, \xi'_{i_1,i_2})\Big],$$

for some $\xi_{i_1,i_2} \in](i_1-1)h_1, (i_1+1)h_1[$ and $\xi'_{i_1,i_2} \in](i_2-1)h_2, (i_2+1)h_2[$. Setting $h = \max(h_1, h_2)$, we deduce that

$$\|\varepsilon_{h_1,h_2}(u)\|_\infty \leq Mh^2,$$

where

$$M = \frac{1}{12}\left(\max_{y\in\bar{\Omega}}\left|\frac{\partial^4 u}{\partial x_1^4}(y)\right| + \max_{y\in\bar{\Omega}}\left|\frac{\partial^4 u}{\partial x_2^4}(y)\right|\right). \tag{2.28}$$

This shows that the scheme (2.27) is consistent of order two. □

Let us now study the stability of the method.

Proposition 2.10 *We have the following estimate, for all h_1, h_2,*

$$|||(C_{h_1,h_2})^{-1}|||_\infty \leq \frac{1}{8}. \tag{2.29}$$

Proof Let Ω_{h_1,h_2} be the set of the indices of the grid points which are inside Ω, Γ_{h_1,h_2} the set of the indices of the grid points which are on Γ and $\bar{\Omega}_{h_1,h_2} = \Omega_{h_1,h_2} \cup \Gamma_{h_1,h_2} = \{0,\ldots,N_1+1\} \times \{0,\ldots,N_2+1\}$ the set of all grid indices. We consider the operator $D\colon \mathbb{R}^{\bar{\Omega}_{h_1,h_2}} \to \mathbb{R}^{\Omega_{h_1,h_2}}$ defined by

$$(D\widetilde{Z})_{i_1,i_2} = a\widetilde{Z}_{i_1,i_2} - b_1(\widetilde{Z}_{i_1-1,i_2} + \widetilde{Z}_{i_1+1,i_2}) - b_2(\widetilde{Z}_{i_1,i_2-1} + \widetilde{Z}_{i_1,i_2+1}).$$

The operator D is just the discrete five point Laplacian without boundary conditions. We first establish a discrete maximum principle for D. More precisely, we claim that if $D\widetilde{Z} \leq 0$, then

$$\max_{\bar{\Omega}_{h_1,h_2}} \widetilde{Z}_{i_1,i_2} = \max_{\Gamma_{h_1,h_2}} \widetilde{Z}_{i_1,i_2}. \tag{2.30}$$

Indeed, if the maximum in question is attained on Γ_{h_1,h_2}, there is nothing to prove. So let $(m_1, m_2) \in \Omega_{h_1,h_2}$ be an index such that $\widetilde{Z}_{m_1,m_2} \geq \widetilde{Z}_{i_1,i_2}$, for all $(i_1, i_2) \in \bar{\Omega}_{h_1,h_2}$. Since $(D\widetilde{Z})_{m_1,m_2} \leq 0$ and $a = 2(b_1 + b_2)$, we have

$$a\widetilde{Z}_{m_1,m_2} \leq b_1(\widetilde{Z}_{m_1+1,m_2} + \widetilde{Z}_{m_1-1,m_2}) + b_2(\widetilde{Z}_{m_1,m_2+1} + \widetilde{Z}_{m_1,m_2-1}) \leq a\widetilde{Z}_{m_1,m_2}.$$

Since $b_1 > 0$ and $b_2 > 0$, it follows that

$$\widetilde{Z}_{m_1,m_2} = \widetilde{Z}_{m_1+1,m_2} = \widetilde{Z}_{m_1-1,m_2} = \widetilde{Z}_{m_1,m_2+1} = \widetilde{Z}_{m_1,m_2-1}.$$

In other words, the maximum is also attained for the neighboring indices (m_1+1, m_2), $(m_1 - 1, m_2)$, $(m_1, m_2 + 1)$ and $(m_1, m_2 - 1)$. By an immediate induction, we have $\widetilde{Z}_{m_1,m_2} = \widetilde{Z}_{m_1-1,m_2} = \cdots = \widetilde{Z}_{0,m_2}$. This means that the maximum is in fact attained on Γ_{h_1,h_2} as well and we are done with the claim.

Let us now establish estimate (2.29). Let X be an arbitrary vector in $\mathbb{R}^{\Omega_{h_1,h_2}}$. We define a vector $\widetilde{X} \in \mathbb{R}^{\bar{\Omega}_{h_1,h_2}}$ by $\widetilde{X}_{i_1,i_2} = X_{i_1,i_2}$ for $(i_1, i_2) \in \Omega_{h_1,h_2}$ and $\widetilde{X}_{i_1,i_2} = 0$ for $(i_1, i_2) \in \Gamma_{h_1,h_2}$. With this definition, we see that $(D\widetilde{X})_{i_1,i_2} = (C_{h_1,h_2}X)_{i_1,i_2}$, for all $(i_1, i_2) \in \Omega_{h_1,h_2}$.

Let now $\widetilde{Y} \in \mathbb{R}^{\bar{\Omega}_{h_1,h_2}}$ be defined by

$$\widetilde{Y}_{i_1,i_2} = \frac{1}{4}\Big[\Big(i_1 - \frac{N_1+1}{2}\Big)^2 h_1^2 + \Big(i_2 - \frac{N_2+1}{2}\Big)^2 h_2^2\Big].$$

By direct inspection, we see that $(D\widetilde{Y})_{i_1,i_2} = -1$ for all $(i_1, i_2) \in \Omega_{h_1,h_2}$.

We next choose $s = \pm 1$ in such a way that

$$\max_{(i_1,i_2)\in\Omega_{h_1,h_2}} |\widetilde{X}_{i_1,i_2}| = |\widetilde{X}_{n_1,n_2}| = s\widetilde{X}_{n_1,n_2},$$

for some $(n_1, n_2) \in \Omega_{h_1,h_2}$. This way, we have

$$\max_{(i_1,i_2)\in\Omega_{h_1,h_2}} |\widetilde{X}_{i_1,i_2}| \le \max_{(i_1,i_2)\in\Omega_{h_1,h_2}} (s\widetilde{X}_{i_1,i_2}).$$

Finally, let $\widetilde{Z} = s\widetilde{X} + \|C_{h_1,h_2}X\|_\infty \widetilde{Y}$. We note that

$$\|C_{h_1,h_2}X\|_\infty = \max_{(i_1,i_2)\in\Omega_{h_1,h_2}} |(D\widetilde{X})_{i_1,i_2}|.$$

Therefore, we have

$$(D\widetilde{Z})_{i_1,i_2} = s(D\widetilde{X})_{i_1,i_2} - \|C_{h_1,h_2}X\|_\infty \le 0,$$

for all $(i_1, i_2) \in \Omega_{h_1,h_2}$. Applying estimate (2.30) to $\widetilde{Z}$, we obtain

$$\begin{aligned}
\max_{(i_1,i_2)\in\Omega_{h_1,h_2}} (s\widetilde{X}_{i_1,i_2}) &\le \max_{(i_1,i_2)\in\Omega_{h_1,h_2}} \widetilde{Z}_{i_1,i_2} \\
&\le \max_{(i_1,i_2)\in\Gamma_{h_1,h_2}} \widetilde{Z}_{i_1,i_2} \\
&\le \max_{(i_1,i_2)\in\Gamma_{h_1,h_2}} (s\widetilde{X}_{i_1,i_2}) + \frac{\|C_{h_1,h_2}X\|_\infty}{8},
\end{aligned}$$

since $0 \le \widetilde{Y}_{i_1,i_2} \le \frac{1}{8}$, for all $(i_1, i_2) \in \bar{\Omega}_{h_1,h_2}$. Now $\widetilde{X}_{i_1,i_2} = 0$ on Γ_{h_1,h_2}, therefore

$$\|X\|_\infty = \max_{(i_1,i_2)\in\Omega_{h_1,h_2}} |\widetilde{X}_{i_1,i_2}| \le \frac{\|C_{h_1,h_2}X\|_\infty}{8},$$

which completes the proof. □

Remark 2.12 The proof above shows, with only minor modifications, that the matrix C_{h_1,h_2} is inverse nonnegative.

The proof also shows that any $\widetilde{Z}$ such that $D\widetilde{Z} \le 0$ and that attains its maximum in Ω_{h_1,h_2} is in fact constant.

As opposed to the one-dimensional case, we do not have a closed form formula for the solution of $-\Delta\bar{u} = 1$ in Ω, $\bar{u} = 0$ on Γ, at our disposal ($\bar{u}$ can however be expressed with Fourier series or approximated using finite differences, see below), hence no closed form expression for its maximum value. This explains the roundabout way of introducing $\widetilde{Y}$ to play the same role, without boundary conditions.

As a consequence, the resulting stability estimate is not optimal, contrarily to the one-dimensional case. Numerical evidence indicates that $|||(C_{h_1,h_2})^{-1}|||_\infty \approx 0.07356$ pretty much independently of h_1, h_2 in the cases we computed. □

We deduce from Proposition 2.9 (consistency) and Proposition 2.10 (stability) the final convergence result.

Corollary 2.3 *Let us suppose that the solution u of problem (2.26) is C^4 on $\bar{\Omega}$. Then the scheme (2.27) is convergent of order two. More precisely, we have*

$$\|U_{h_1,h_2} - S_{h_1,h_2}(u)\|_\infty \leq \frac{M}{8} h^2, \tag{2.31}$$

where M is given by (2.28) and $h = \max(h_1, h_2)$.

Proof Exactly the same as in Corollary 2.2. □

Remark 2.13 On the surface, it looks like the one-dimensional error estimate (2.13) and the two-dimensional error estimate (2.31) are basically the same. This is not actually so. Indeed, an important consideration in numerical methods is that of their cost. In effect, it is only possible to meaningfully compare two methods if they apply to data of the same size.

For example, let us assume that Gaussian elimination, see [6, 53, 71], is used to solve linear systems (we ignore the fact that more efficient methods may exist that are better adapted to these particular matrices). In the one dimensional case, this would lead to a compute time $T_{1d} = O(N^3) = O(h^{-3})$, whereas in the two-dimensional case, we would be looking at a compute time $T_{2d} = O(N^6) = O(h^{-6})$ since the system to be solved is a $N^2 \times N^2$ system.[3] Thus the time required to achieve a given error estimate in two dimensions is roughly the square of the time needed to achieve the same error estimate in one dimension.

This is a general fact: Computations become exponentially costlier and costlier as the dimension of the problem grows. This is called the *curse of dimensionality*. □

Let us illustrate the previous considerations in Fig. 2.10 with a plot of the finite difference solution of the Poisson problem $-\Delta\bar{u} = 1$ in Ω, $\bar{u} = 0$ on Γ, in the unit square. This plot can be visualized in 3D by squinting toward the middle of the page.

We have described the finite difference method for the Poisson equation in the unit square. It immediately generalizes to the case of a rectangle. The case of a more general two-dimensional domain Ω is more complicated.

[3] In practice, the cases considered here are all very small, and are solved almost instantaneously on a personal computer, but the remark applies to more computationally challenging problems.

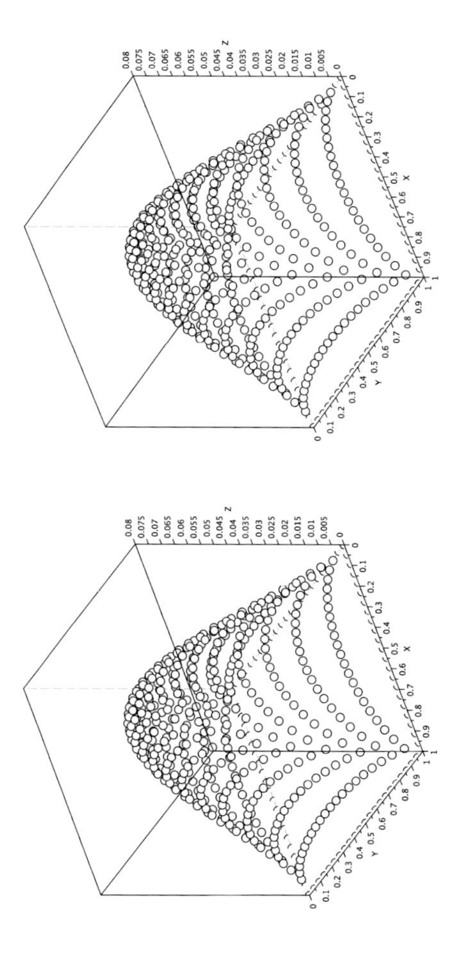

Fig. 2.10 Cross-eyed autostereogram for 3d visualization of the finite difference solution for $N = 21$

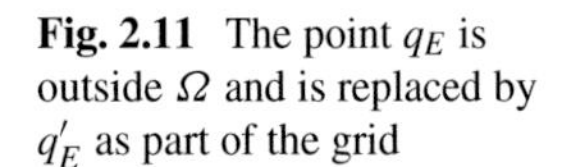
Fig. 2.11 The point q_E is outside Ω and is replaced by q'_E as part of the grid

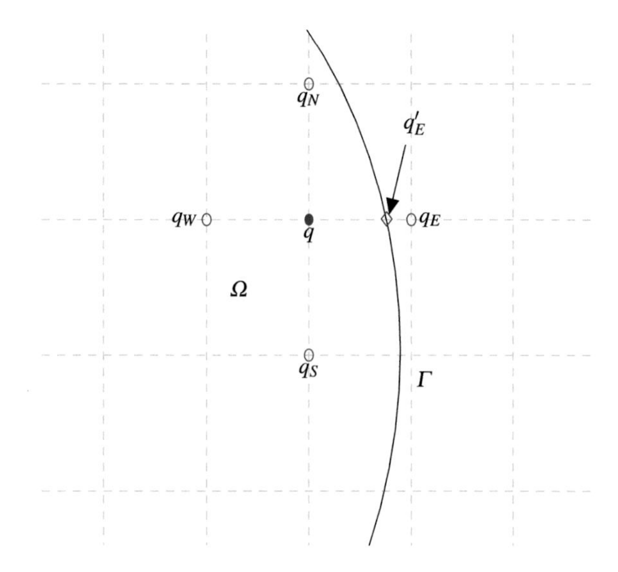

Let us consider the following non homogeneous Dirichlet problem: Find $u\colon \bar{\Omega} \to \mathbb{R}$ solution of

$$\begin{cases} -\Delta u(x) = f(x), & x \in \Omega, \\ u(x) = g(x), & x \in \Gamma, \end{cases}$$

where g is the given Dirichlet data on the boundary Γ of Ω.

For simplicity, we assume that $h_1 = h_2 = h$ and we consider the lattice $h\mathbb{Z}^2$ in the plane. The straight lines parallel to the axes and going through lattice points are called grid lines. In order to construct a finite difference grid for the problem, we first retain the lattice points that belong to Ω. The difficulty is that in order to impose the Dirichlet boundary condition, we need points on the boundary, whereas lattice points have no reason to fall on the boundary. To remedy this situation, the idea is to replace the lattice points that are closest to Ω outside of Ω by the intersection of grid lines with Γ. The finite difference scheme must be modified accordingly in the vicinity of such points.

To fix ideas, let us consider one possible configuration in Fig. 2.11. The grid point q is such that its lattice neighbors q_S, q_W and q_N are inside Ω. We keep them in the grid. On the contrary, the point q_E is outside Ω and closest to Γ. We replace it by point q'_E of intersection of the grid line with Γ. Its distance to point q is $h' < h$, as can be seen on the figure.

In the case of Fig. 2.11, we now look for an approximation of $\Delta u(q)$ that uses points q, q_S, q_W, q_N and q'_E. We just have to modify the approximation of the second order derivative with respect to variable x_1 (for the other variable, we can use the usual three point scheme using q, q_S and q_N which are inside Ω). Let us explain how this works. The trick consists in finding three coefficients α, β and γ such that

$$\alpha u(q_W) + \beta u(q) + \gamma u(q'_E) = \frac{\partial^2 u}{\partial x_1^2}(q) + O(h).$$

Using as usual Taylor–Lagrange expansions of $u(q_W)$ and of $u(q'_E)$ in a neighborhood of q, we get the following system:

$$\begin{aligned}\alpha + \beta + \gamma &= 0,\\ -\alpha h + \gamma h' &= 0,\\ \alpha\frac{h^2}{2} + \gamma\frac{h'^2}{2} &= 1,\end{aligned}$$

which admits a unique solution given by

$$\alpha = \frac{2}{h(h+h')}, \quad \beta = -\frac{2}{hh'}, \quad \gamma = \frac{2}{h'(h+h')}.$$

We can check that the truncation error is only of order one due to the fact that $h' \neq h$, hence a lesser quality of approximation compared with the rectangular case. Let us remark that the matrix of the linear system we have to solve is no longer symmetric.

2.6 Concluding Remarks

To conclude, we see that the finite difference method is not easily implemented in arbitrary domains in dimensions higher than 1 or for different boundary conditions. This is a serious limitation of the method for applications. It works well on domains with simple geometry, for example domains which are unions of rectangles, with Dirichlet boundary conditions. However, in the general case, i.e., for arbitrary domains and other boundary conditions, the finite element method that we will see in two dimensions in Chap. 6 will be generally preferred.

We will return to the finite difference method in Chap. 8 for the heat equation in one dimension of space and Chap. 9 for the wave equation also in one dimension of space.

The finite difference method for one-dimensional elliptic problems is mentioned in many references. The reader is referred to [18, 46, 47, 64, 70, 75] for example, for various extensions and points of view.

Chapter 3
A Review of Analysis

In order to go beyond the somewhat naive existence theory and finite difference method of approximation of elliptic boundary value problems seen in Chaps. 1 and 2, we need to develop a more sophisticated point of view. This requires in turn some elements of analysis pertaining to function spaces in several variables, starting with some abstract Hilbert space theory. This is the main object of this chapter.

As already mentioned in the preface, this chapter can be read quickly at first, for readers who are not too interested in the mathematical details and constructions therein. A summary of the important results needed for the subsequent chapters is thus provided at the end of the chapter.

3.1 Basic Hilbert Space Theory

Let us quickly review basic Hilbert space theory from the abstract viewpoint. Let H be a real Hilbert space, i.e., a real vector space endowed with a scalar product $(\cdot|\cdot)_H$ and associated norm $\|\cdot\|_H$ which is complete for this norm. The Cauchy–Schwarz inequality is really a hilbertian property.

Theorem 3.1 *For all $u, v \in H$, we have*

$$|(u|v)_H| \leq \|u\|_H \|v\|_H.$$

One of the most basic results in Hilbert space theory is the orthogonal projection theorem, see [9, 14, 32, 51] for a proof.

Theorem 3.2 *Let C be a non empty, convex, closed subset of H. For all $x \in H$, there exists a unique $p_C(x) \in C$ such that*

$$\|x - p_C(x)\|_H = \inf_{y \in C} \|x - y\|_H.$$

H. Le Dret and B. Lucquin, *Partial Differential Equations: Modeling, Analysis and Numerical Approximation*, International Series of Numerical Mathematics 168, DOI 10.1007/978-3-319-27067-8_3

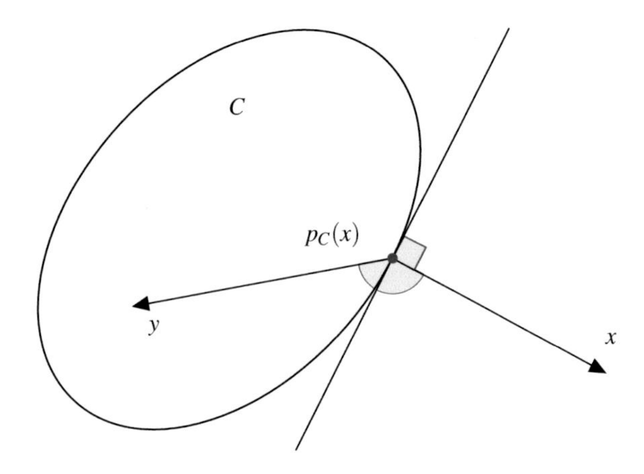

Fig. 3.1 The orthogonal projection on a closed convex subset C

The vector $p_C(x)$ *is called the* orthogonal projection of x on C. *It is also characterized by the inequality*

$$\forall y \in C, \quad (x - p_C(x)|y - p_C(x))_H \leq 0.$$

In addition, if the convex set is a closed vector subspace E *of* H, *then* p_E *is a continuous linear mapping from* H *to* E *which is also characterized by the equality*

$$\forall y \in E, \quad (x - p_E(x)|y)_H = 0.$$

The orthogonal projection of x on C is thus the element of C closest to x and the angle between $x - p_C(x)$ and $y - p_C(x)$ is larger than $\frac{\pi}{2}$, see Fig. 3.1. In particular, if $x \in C$, then $p_C(x) = x$. An important consequence of the last characterization in the case of a closed vector subspace E is that we can write $H = E \oplus E^\perp$ with continuous orthogonal projections on each factor. Indeed, we have $x = p_E(x) + (x - p_E(x))$ with $p_E(x) \in E$ by construction and $x - p_E(x) \in E^\perp$ by the second characterization. Hence $H = E + E^\perp$. To show that the sum is direct, it suffices to note that $E \cap E^\perp = \{0\}$ which is obvious since $x \in E \cap E^\perp$ implies $0 = (x|x)_H = \|x\|_H^2$, see Fig. 3.2.

Another important consequence is a characterization of dense subspaces.

Lemma 3.1 *A vector subspace* E *of* H *is dense in* H *if and only if* $E^\perp = \{0\}$.

Proof For any vector subspace, it is always true that $E^\perp = (\bar{E})^\perp$. Let E be a dense subspace, i.e., $\bar{E} = H$. Then, of course $E^\perp = H^\perp = \{0\}$. Conversely, if $E^\perp = \{0\}$, this implies that $(\bar{E})^\perp = \{0\}$ and since $H = (\bar{E})^\perp \oplus \bar{E}$, it follows that $\bar{E} = H$, and E is dense in H. □

The Riesz theorem provides a canonical way of identifying a Hilbert space and its dual.

Theorem 3.3 (Riesz) *Let* H *be a Hilbert space and* ℓ *an element of its dual* H'. *There exists a unique* $u \in H$ *such that*

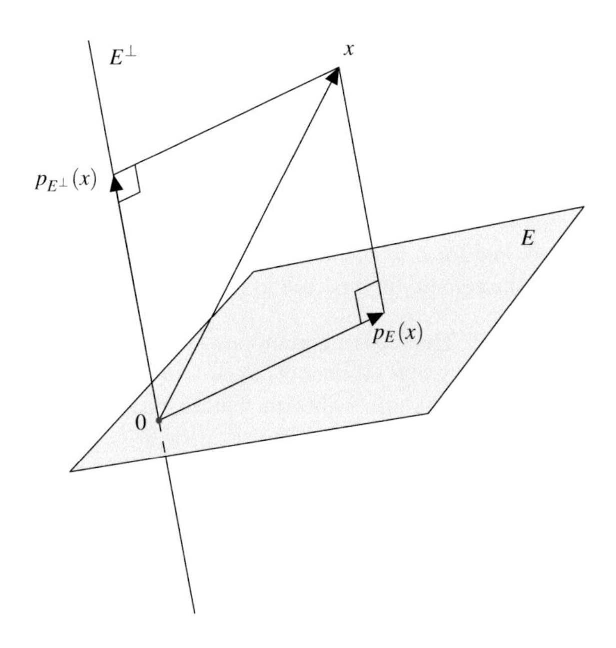

Fig. 3.2 The orthogonal projection on a closed vector subspace E

$$\forall v \in H, \quad \ell(v) = (u|v)_H.$$

Moreover

$$\|\ell\|_{H'} = \|u\|_H$$

and the linear mapping $\sigma : H' \to H, \ell \mapsto u$, *is an isometry.*

Proof For the uniqueness, assume u_1 and u_2 are two solutions, then for all $v \in H$, $(v|u_1 - u_2)_H = 0$. This is true in particular for $v = u_1 - u_2$, hence $u_1 = u_2$.

If $\ell = 0$, then we set $u = 0$ to be the unique u in question. Let $\ell \neq 0$. It is thus a nonzero continuous linear form, hence its kernel $\ker \ell$ is a closed hyperplane of H. Let us choose $u_0 \in (\ker \ell)^{\perp}$ with $\|u_0\|_H = 1$ (this is possible since $\ker \ell$ is not dense). Since $u_0 \notin \ker \ell$, we have $\ell(u_0) \neq 0$ and for all $v \in H$, we can set $w = v - \frac{\ell(v)}{\ell(u_0)} u_0$ so that

$$\ell(w) = \ell\Big(v - \frac{\ell(v)}{\ell(u_0)} u_0\Big) = \ell(v) - \frac{\ell(v)}{\ell(u_0)} \ell(u_0) = 0,$$

and $w \in \ker \ell$. Now writing $v = \frac{\ell(v)}{\ell(u_0)} u_0 + w$ and setting $u = \ell(u_0) u_0 \in (\ker \ell)^{\perp}$, we obtain

$$(v|u)_H = \Big(\frac{\ell(v)}{\ell(u_0)} u_0 \Big| u\Big)_H + (w|u)_H = \ell(v)(u_0|u_0)_H = \ell(v),$$

hence the existence of u.

The mapping σ is thus well defined and obviously linear. Finally, for the isometry, we have on the one hand

$$\|\ell\|_{H'} = \sup_{\|v\|_H \le 1} |\ell(v)| = \sup_{\|v\|_H \le 1} |(v|u)_H| \le \|v\|_H \|u\|_H \le \|u\|_H,$$

by the Cauchy–Schwarz inequality. On the other hand, equality trivially holds for $\ell = 0$, and for $\ell \neq 0$, choosing $v = \frac{u}{\|u\|_H}$ yields $(v|u)_H = \|u\|_H$ with $\|v\|_H = 1$, hence the equality in this case too. □

Remark 3.1 The Riesz theorem shows that the dual of a Hilbert space is also a Hilbert space for the scalar product $(\ell_1|\ell_2)_{H'} = (\sigma\ell_1|\sigma\ell_2)_H$ which induces the dual norm. Indeed it is not a priori obvious that the dual norm is hilbertian. It is often used to identify H and H' via the isometry σ or σ^{-1}. This identification is not systematic however. For example, when we have two Hilbert spaces H and V such that $V \hookrightarrow H$ and V is dense in H, the usual identification is to let

$$V \hookrightarrow H = H' \hookrightarrow V'$$

using the Riesz theorem for H, which is called the pivot space, but not for V, see [15, 28]. □

We now turn to the study of more concrete function spaces.

3.2 A Few Basic Function Spaces

Let us rapidly review the most basic function spaces that we will need. All these function spaces are real valued. In the sequel, Ω denotes an open subset of $\mathbb{R}^d$, $d \ge 1$. The canonical scalar product of two vectors x and y in $\mathbb{R}^d$ will be denoted $x \cdot y = \sum_{i=1}^d x_i y_i$, where x_i (resp. y_i) are the components of x (resp. y). The associated Euclidean norm is $\|x\| = (x \cdot x)^{1/2}$. We use the multiindex notation for partial derivatives. Let $\alpha = (\alpha_1, \alpha_2, \dots, \alpha_d) \in \mathbb{N}^d$ be a multiindex. The integer $|\alpha| = \sum_{i=1}^d \alpha_i$ is called the *length* of α and we set

$$\partial^\alpha u = \frac{\partial^{|\alpha|} u}{\partial x_1^{\alpha_1} \partial x_2^{\alpha_2} \cdots \partial x_d^{\alpha_d}},$$

whenever the function u is $|\alpha|$-times differentiable and the partial derivatives commute. The space $C^0(\Omega)$ is the space of real-valued, continuous functions on Ω, and for all $k \in \mathbb{N}$, we define

$$C^k(\Omega) = \{u; \quad \text{for all} \quad \alpha \in \mathbb{N}^d, |\alpha| \le k, \partial^\alpha u \in C^0(\Omega)\}$$

to be the space of k-times continuously differentiable functions on Ω. The space of indefinitely differentiable functions on Ω is defined by

$$C^\infty(\Omega) = \bigcap_{k\in\mathbb{N}} C^k(\Omega).$$

We do not specify the natural topologies of these vector spaces as we will not need them. Beware however that these natural topologies are not normed.

The *support* of a function u, $\operatorname{supp} u$, is the complement of the largest open subset of Ω on which u vanishes. It is thus a closed subset of Ω. A subset K of Ω is compact if and only if it is a closed, bounded subset that "does not touch the boundary" in the sense that $d(K, \complement_{\mathbb{R}^d}\Omega) > 0$. There is a "security strip" between K and $\partial\Omega$. Functions with compact support play an important role and deserve a notation of their own:

$$C_c^k(\Omega) = \mathscr{D}^k(\Omega) = \{u \in C^k(\Omega); \operatorname{supp} u \text{ is compact}\}$$

and

$$C_c^\infty(\Omega) = \mathscr{D}(\Omega) = \bigcap_{k\in\mathbb{N}} \mathscr{D}^k(\Omega).$$

Again, these vector spaces are endowed with natural topologies that we will not describe. We will return to these spaces later when talking about distributions.

Let $\bar\Omega$ be the closure of Ω in $\mathbb{R}^d$. The space $C^0(\bar\Omega)$ is the space of continuous functions on $\bar\Omega$. If $\bar\Omega$ is *compact*, that is to say, if Ω is bounded, this space is normed by

$$\|u\|_{C^0(\bar\Omega)} = \sup_{x\in\bar\Omega} |u(x)| = \max_{x\in\bar\Omega} |u(x)|.$$

The convergence associated to this normed topology is just uniform convergence.

Likewise, we define $C^k(\bar\Omega)$ to be the space of functions in $C^k(\Omega)$, all the partial derivatives of which up to order k have a continuous extension to $\bar\Omega$. Keeping the same symbol for this extension, the natural norm of this space when Ω is bounded is

$$\|u\|_{C^k(\bar\Omega)} = \max_{|\alpha|\le k} \|\partial^\alpha u\|_{C^0(\bar\Omega)},$$

and the convergence of a sequence in this space is the uniform convergence of all partial derivatives up to order k. All these spaces are Banach spaces, i.e., they are complete for the metric defined by their norm. We also define

$$C^\infty(\bar\Omega) = \bigcap_{k\in\mathbb{N}} C^k(\bar\Omega),$$

a space which is endowed with a natural topology that is again not a normed space, even when $\bar\Omega$ is compact.

For $0 < \beta \leq 1$ and Ω bounded, we define the spaces of Hölder functions [35, 40] (Lipschitz for $\beta = 1$) by

$$C^{0,\beta}(\bar{\Omega}) = \left\{u \in C^0(\bar{\Omega});\ \sup_{\substack{x,y\in\bar{\Omega}\\ x\neq y}} \frac{|u(x) - u(y)|}{\|x - y\|^\beta} < +\infty\right\}$$

and

$$C^{k,\beta}(\bar{\Omega}) = \{u \in C^k(\bar{\Omega});\ \partial^\alpha u \in C^{0,\beta}(\bar{\Omega}) \quad \text{for all} \quad \alpha \in \mathbb{N}^d, |\alpha| = k\}.$$

When equipped with the norms

$$\|u\|_{C^{k,\beta}(\bar{\Omega})} = \|u\|_{C^k(\bar{\Omega})} + \max_{|\alpha|=k}\Bigg(\sup_{\substack{x,y\in\bar{\Omega}\\ x\neq y}} \frac{|\partial^\alpha u(x) - \partial^\alpha u(y)|}{\|x - y\|^\beta}\Bigg),$$

these spaces also are Banach spaces. There are continuous injections

$$C^{k,\beta}(\bar{\Omega}) \hookrightarrow C^{k,\beta'}(\bar{\Omega}) \hookrightarrow C^k(\bar{\Omega}) \hookrightarrow C^{k-1,\gamma}(\bar{\Omega})$$

which are compact for $\beta' < \beta$ and $\gamma < 1$ by Ascoli's theorem (the compactness of the first embedding requires some regularity on Ω, see Sect. 3.3). A linear mapping f from a normed space E to a normed space F is continuous if and only if there exists a constant C such that for all $x \in E$, $\|f(x)\|_F \leq C\|x\|_E$. In the continuous injections above, f is just the identity, i.e., $f(u) = u$. A mapping is compact if it transforms bounded sets into relatively compact sets.

The other major family of function spaces that will be useful to us is that of the Lebesgue spaces (see for example [2, 15, 20, 51]),

$$L^p(\Omega) = \left\{u \text{ measurable};\ \int_\Omega |u(x)|^p\,dx < +\infty\right\}$$

for $1 \leq p < +\infty$ and

$$L^\infty(\Omega) = \left\{u \text{ measurable};\ \operatorname*{ess\,sup}_\Omega |u| < +\infty\right\},$$

where

$$\operatorname*{ess\,sup}_\Omega |u| = \inf\{M;\ |u| \leq M \text{ almost everywhere on } \Omega\}.$$

Now in these definitions, u is not strictly speaking a function, but an equivalence class of functions that are equal almost everywhere with respect to the Lebesgue measure. However, in practice and outside of very specific circumstances, it is harmless to think of u as just a function and not as an equivalence class. We just need to keep this fact at the back of our mind, just in case.

When equipped with the norms

$$\|u\|_{L^p(\Omega)} = \Big(\int_\Omega |u(x)|^p\,dx\Big)^{\frac{1}{p}}$$

for $1 \le p < +\infty$ and

$$\|u\|_{L^\infty(\Omega)} = \operatorname*{ess\,sup}_\Omega |u|,$$

the Lebesgue spaces are Banach spaces. For $p = 2$, the space $L^2(\Omega)$ is a Hilbert space for the scalar product

$$(u|v)_{L^2(\Omega)} = \int_\Omega u(x)v(x)\,dx,$$

see Sect. 3.1 for general Hilbert space theory. Hölder's inequality reads

$$\int_\Omega |u(x)v(x)|\,dx \le \Big(\int_\Omega |u(x)|^p\,dx\Big)^{\frac{1}{p}}\Big(\int_\Omega |v(x)|^{p'}\,dx\Big)^{\frac{1}{p'}}$$

when $1 < p, p' < +\infty$ are conjugate exponents, $\frac{1}{p} + \frac{1}{p'} = 1$ and

$$\int_\Omega |u(x)v(x)|\,dx \le (\operatorname*{ess\,sup}_\Omega |u|)\int_\Omega |v(x)|\,dx,$$

(the integrals do not need to be finite). In particular, if $u \in L^p(\Omega)$ and $v \in L^{p'}(\Omega)$, then $uv \in L^1(\Omega)$ and

$$\Big|\int_\Omega u(x)v(x)\,dx\Big| \le \|uv\|_{L^1(\Omega)} \le \|u\|_{L^p(\Omega)}\|v\|_{L^{p'}(\Omega)}.$$

For $p = 2$, we get the Cauchy–Schwarz inequality, which is actually a Hilbert space property as we have seen before,

$$|(u|v)_{L^2(\Omega)}| \le \|u\|_{L^2(\Omega)}\|v\|_{L^2(\Omega)}.$$

When Ω is bounded, there are continuous injections $C^k(\bar{\Omega}) \hookrightarrow L^p(\Omega) \hookrightarrow L^q(\Omega)$ whenever $q \le p$.[1]

[1] This is doubly false if Ω is not bounded.

The Lebesgue spaces admit local versions

$$L^p_{\text{loc}}(\Omega) = \{u;\, u_{|K} \in L^p(K) \quad \text{for all compact} \quad K \subset \Omega\}.$$

These vector spaces have a natural topology which is not a normed topology.

Clearly, in view of Hölder's inequality, we have $L^p_{\text{loc}}(\Omega) \subset L^q_{\text{loc}}(\Omega)$ whenever $q \leq p$. In particular, the space $L^1_{\text{loc}}(\Omega)$ is the largest of all these spaces, and actually the largest of all function spaces introduced up to now, which are all continuously embedded in it.

The following result is of importance [2, 20, 51].

Proposition 3.1 *Let $u \in L^1_{\text{loc}}(\Omega)$ be such that $\int_\Omega u\varphi\, dx = 0$ for all $\varphi \in \mathscr{D}(\Omega)$. Then $u = 0$ almost everywhere.*

Proof Note first that since φ has support in a compact subset K of Ω, so does the product $u\varphi$. Since φ is bounded, it follows that $u\varphi \in L^1(K)$ and the integral is well-defined.

Let $x_0 \in \Omega$ and n be large enough so that $B\big(x_0, \frac{1}{n}\big) \subset \Omega$. It is possible to construct a sequence $\varphi_k \in \mathscr{D}(\Omega)$ such that $\operatorname{supp}\varphi_k \subset B\big(x_0, \frac{1}{n}\big)$ and for all $x \in B\big(x_0, \frac{1}{n}\big)$, $\varphi_k(x) \to 1$ (we leave the proof as an exercise). Consequently, by the Lebesgue dominated convergence theorem, we have

$$0 = \int_{B(x_0, \frac{1}{n})} u\varphi_k\, dx \underset{k\to+\infty}{\longrightarrow} \int_{B(x_0, \frac{1}{n})} u\, dx.$$

Hence, since

$$0 = \frac{1}{\operatorname{meas} B\big(x_0, \frac{1}{n}\big)} \int_{B(x_0, \frac{1}{n})} u\, dx \underset{n\to+\infty}{\longrightarrow} u(x_0)$$

for almost all x_0 by the Lebesgue points theorem, see [68], we obtain the result. □

Remark 3.2 Here we see at work the idea of testing a function u with a test-function φ in order to obtain information on u. The general concept behind it is that of *duality* and it will be used in much larger generality in the context of distributions and variational formulations that we will see later on. □

3.3 Regularity of Open Subsets of $\mathbb{R}^d$

The structure of the open subsets of $\mathbb{R}^d$ for the usual topology is very simple for $d = 1$, since every open set is a union of an at most countable family of disjoint open intervals. In particular, a one-dimensional connected open set is just an open interval. The situation is more complicated in higher dimensions.

People tend to think of a connected open set of $\mathbb{R}^d$ as a potato-shaped object drawn in $\mathbb{R}^2$. This geometrical intuition is basically correct as far as the open set itself is concerned. It is misleading when the boundary of the open set is involved. In fact,

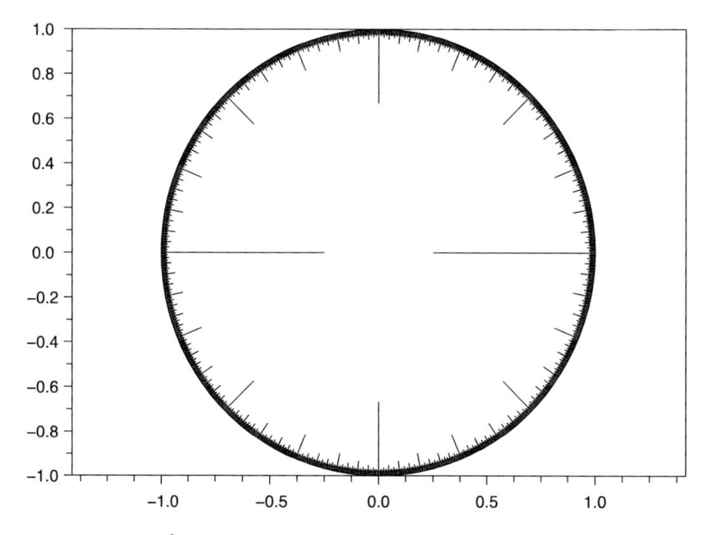

Fig. 3.3 An open set in $\mathbb{R}^2$ with a relatively wild boundary (imagine an infinity of little spikes pointing inward the disk)

Fig. 3.4 A zoom on the complement of the Mandelbrot set

the boundary of an open set in $\mathbb{R}^d$, $d > 1$, can be more or less regular, more or less smooth, as in Fig. 3.3.

There is worse: the Mandelbrot set is compact, its complement is open with a very convoluted boundary, see Fig. 3.4.

It is even possible to construct open sets in $\mathbb{R}^2$ (or in $\mathbb{R}^d$ for any d for that matter), the boundary of which has strictly positive Lebesgue measure, i.e., a strictly positive area! PDE problems are posed in open subsets of $\mathbb{R}^d$ and we often need a certain amount of regularity of the boundary of such open sets in order to deal with boundary conditions.

There are several ways of quantifying the regularity of an open set boundary, or in short the regularity of that open set. Let us give the definition that is the most adequate for our purposes here. Other definitions—equivalent or not—may be encountered in the literature [41, 44].

Definition 3.1 We say that a bounded open subset of $\mathbb{R}^d$ is Lipschitz *(resp.* of class $C^{k,\beta}$*)* if its boundary $\partial\Omega$ can be covered by a finite number of open hypercubes C_j, $j = 1, \ldots, m$, each with an attached system of orthonormal Cartesian coordinates, $y^j = (y_1^j, y_2^j, \ldots, y_d^j)$, in such a way that

$$C_j = \{y \in \mathbb{R}^d;\ |y_i^j| < a_j \text{ for } i = 1, \ldots, d\},$$

and there exists Lipschitz functions *(resp.* of class $C^{k,\beta}$*)* $\varphi_j : \mathbb{R}^{d-1} \to \mathbb{R}$ such that

$$\Omega \cap C_j = \{y \in C_j;\ y_d^j < \varphi_j((y^j)')\},$$

using the notation $(y^j)' = (y_1^j, y_2^j, \ldots, y_{d-1}^j) \in \mathbb{R}^{d-1}$.

The meaning of Definition 3.1 is that locally in C_j, Ω consists of those points located strictly below the graph of φ_j, in other words, the *hypograph* of φ_j, see Figs. 3.5 and 3.7. In particular, such an open set is situated on just one side of its boundary, which consists of pieces of graphs, since

$$\partial\Omega \cap C_j = \{y \in C_j;\ y_d^j = \varphi_j((y^j)')\}.$$

Remark 3.3 It is fairly clear that a bounded polygon is a Lipschitz open set in dimension 2. None of the wild examples of Figs. 3.3 and 3.4 is of class $C^{0,\beta}$.

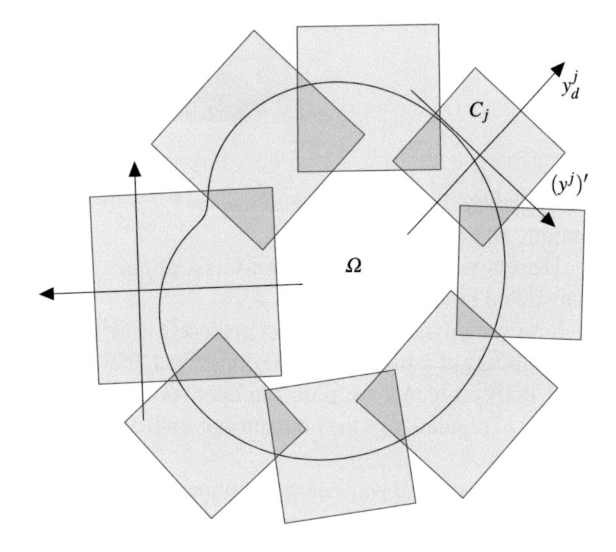

Fig. 3.5 Covering the boundary with hypercubes

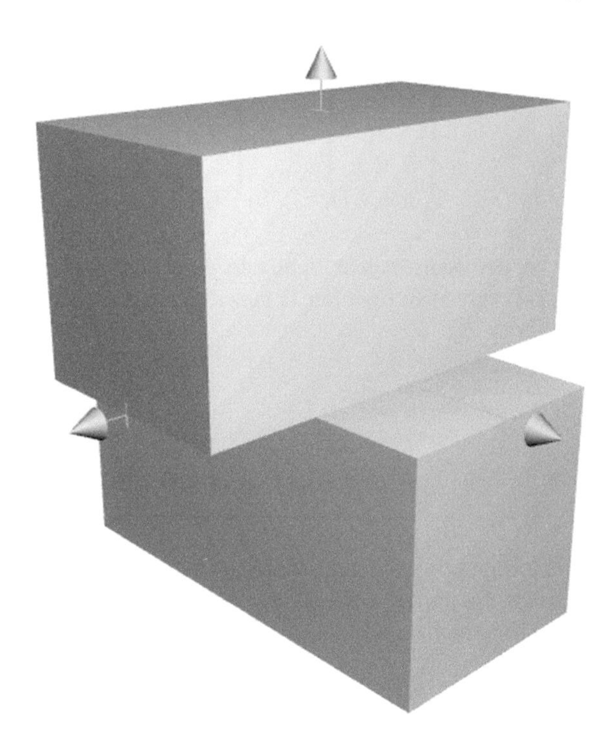

Fig. 3.6 Simple, however not Lipschitz

On the other hand, there are also perfectly nice open sets that are not Lipschitz in the previous sense. We give an example in Fig. 3.6, obtained by gluing together two parallelepipeds one on top of the other, adding the open square of contact. It is impossible to describe the resulting set as a hypograph at each vertex of that square. The open set is nonetheless perfectly tame, it is a polyhedron. □

The boundary of a Lipschitz open set, and a fortiori that of an open set of class $C^{k,\alpha}$, $k \geq 1$, possesses a certain number of useful properties.

Proposition 3.2 *Let Ω be a Lipschitz open set. There exists a normal unit exterior vector n, defined almost everywhere on $\partial\Omega$.*

Normal means orthogonal to the boundary, exterior means that it points toward the complement of Ω. We will go back to the meaning of almost everywhere later.

Proof Let us work in C_j and drop all j indices and exponents to simplify notation. We will admit *Rademacher's theorem*, a nontrivial result that says that a Lipschitz function on $\mathbb{R}^{d-1}$ is differentiable in the classical sense, almost everywhere with respect to the Lebesgue measure in $\mathbb{R}^{d-1}$, see [37].

Let y' be a point of differentiability of φ. At this point, the differentiability implies that the graph of φ has a tangent hyperplane generated by the $d-1$ vectors

$$a_1 = \begin{pmatrix} 1 \\ 0 \\ \vdots \\ 0 \\ \partial_1\varphi(y') \end{pmatrix}, a_2 = \begin{pmatrix} 0 \\ 1 \\ \vdots \\ 0 \\ \partial_2\varphi(y') \end{pmatrix}, \cdots, a_{d-1} = \begin{pmatrix} 0 \\ 0 \\ \vdots \\ 1 \\ \partial_{d-1}\varphi(y') \end{pmatrix},$$

(for brevity we use here a slightly different notation for partial derivatives, $\partial_i\varphi = \frac{\partial\varphi}{\partial y_i}$). The orthogonal straight line is generated by the vector

$$N = \begin{pmatrix} -\partial_1\varphi(y') \\ -\partial_2\varphi(y') \\ \vdots \\ -\partial_{d-1}\varphi(y') \\ 1 \end{pmatrix}.$$

which is clearly orthogonal to all a_i. To conclude, we just need to normalize it and notice that it points outwards due to the strictly positive last component and Ω lying under the graph,

$$n = \frac{1}{\sqrt{1 + \|\nabla\varphi(y')\|^2}} \begin{pmatrix} -\partial_1\varphi(y') \\ -\partial_2\varphi(y') \\ \vdots \\ -\partial_{d-1}\varphi(y') \\ 1 \end{pmatrix},$$

with $\|\nabla\varphi(y')\|^2 = \sum_{i=1}^{d-1}(\partial_i\varphi(y'))^2$. □

Remark 3.4 It should be noted that the normal vector n is an object of purely geometric nature that does not depend on the particular system of coordinates used to compute it. In particular, if we take another admissible covering of the boundary, the same formulas apply and compute the same vector in different coordinate systems. This geometrically obvious remark can also be checked by direct computation in two different coordinate systems, see Fig. 3.7.

The "almost everywhere" is meant in the sense of the space $\mathbb{R}^{d-1}$ associated with a local coordinate system. We give it an intrinsic meaning just below. □

If Ω is a Lipschitz subset of $\mathbb{R}^d$, there is a natural measure on $\partial\Omega$ that is inherited in a sense from the Lebesgue measure in $\mathbb{R}^d$, that we will call the *boundary measure*. We will not go into all the detail but give a few ideas on how this measure can be computed.

Let $A \subset \partial\Omega$ be a Borel subset of $\partial\Omega$. Since the open sets C_j cover the boundary, we can partition A with Borel sets $A_j \subset C_j$. Let Π_j be the orthogonal projection from C_j onto $\mathbb{R}^{d-1}$ according to the coordinate system associated with C_j. The restriction of the projection to the graph of φ_j is a homeomorphism, therefore $\Pi_j(A_j)$ is a Borel subset of $\mathbb{R}^{d-1}$.

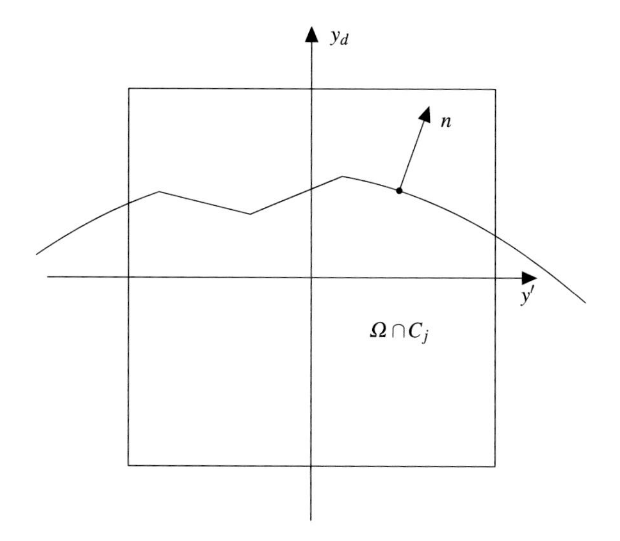

Fig. 3.7 Local aspect of the boundary of a Lipschitz open set and the normal unit exterior vector

We set

$$\mathscr{H}_{d-1}(A_j) = \int_{\Pi_j(A_j)} \sqrt{1 + \|\nabla\varphi_j((y^j)')\|^2}\, d(y^j)' \quad \text{and} \quad \mathscr{H}_{d-1}(A) = \sum_{j=1}^{m} \mathscr{H}_{d-1}(A_j).$$

It can be checked, although it is quite tedious, that this formula does not depend on the covering and coordinates chosen to compute it, and that it defines a Borel measure on $\partial\Omega$.

In the case when $\nabla\varphi_j$ is constant, that is to say if the graph is portion of a hyperplane, it is also easy to check that the formula above gives the $(d-1)$-dimensional Lebesgue measure on the hyperplane, using the same unit of length as in $\mathbb{R}^d$. In this sense, the boundary measure is inherited from $\mathbb{R}^d$.

The notation $\mathscr{H}_{d-1}$ alludes to the $(d-1)$-Hausdorff measure, a much more general and complicated object that coincides here with our hand-crafted measure. Let us develop an example.

In the example of Fig. 3.8, Ω is the unit square. We have figured just one square of the boundary covering, with the attached coordinate system. The part A of $\partial\Omega$ included in it is drawn with a thicker line. It is clearly described by the function $\varphi(y_1) = -|y_1|$, with $y_1 \in \left]-\frac{3}{2\sqrt{2}}, \frac{3}{2\sqrt{2}}\right[$. We have $\varphi'(y_1) = 1$ for $y_1 < 0$ and $\varphi'(y_1) = -1$ for $y_1 > 0$. Thus

$$\mathscr{H}_1(A) = \int_{-\frac{3}{2\sqrt{2}}}^{\frac{3}{2\sqrt{2}}} \sqrt{1+1}\, dy_1 = 3,$$

which is equal to the length of the thicker line.

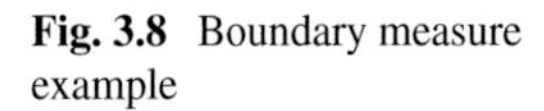

Fig. 3.8 Boundary measure example

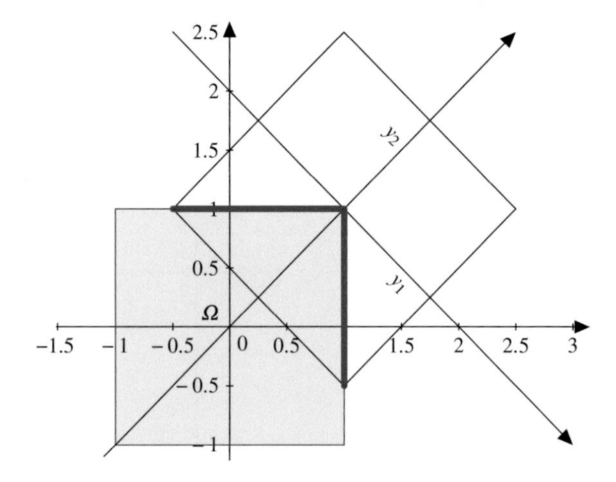

More generally, for $d = 2$, the boundary of Ω consists of curves and if these curves are regular, we recognize the length of the parametric curve $y_1 \mapsto (y_1, \varphi(y_1))$. The same interpretation holds for $d = 3$ with the area of a parametric surface.

It is now clear that the normal vector is defined almost everywhere with respect to the boundary measure. In addition, we can now define $L^p(\partial\Omega)$ spaces and compute all sorts of integrals on the boundary, using this measure. In order to have a more economical notation, we will write it $d\Gamma$ in the integrals. Thus, if g is a function on the boundary with support in C_j, we have

$$\int_{\partial\Omega} g\,d\Gamma = \int_{\Pi_j(C_j)} g(\Pi_j^{-1}(y'))\sqrt{1 + \|\nabla\varphi_j((y^j)')\|^2}\,d(y^j)'.$$

The formula is extended to all functions without condition of support by a partition of unity, see below.

3.4 Partitions of Unity

Partitions of unity [2] are a basic tool that is used in many contexts whenever the need arises to localize a function. In what follows, Ω will be a bounded open subset of $\mathbb{R}^d$ with a finite covering C_j, $j = 0, \ldots, m$, of its boundary $\partial\Omega$.[2]

Let us first state a few facts about convolution [15, 51]. Given two functions f and g in $L^1(\mathbb{R}^d)$, we define

[2]This particular assumption is only because this is the context in which we will use partitions of unity here. It should be clear from the proof, that the result extends to more general covers.

$$f \star g(x) = \int_{\mathbb{R}^d} f(x-y)g(y)\,dy.$$

The function $f \star g$, which is called the convolution of f and g, is well defined and belongs to $L^1(\mathbb{R}^d)$. If g is in addition of class C^∞ with integrable derivatives, so is $f \star g$, with

$$\partial^\alpha (f \star g) = f \star \partial^\alpha g$$

for all multiindices α. If we take a function ρ with support in the unit ball and integral equal to 1 and let $\rho_n(x) = n^d \rho(nx)$, so that ρ_n has support in the ball of radius $\frac{1}{n}$ and integral also equal to 1, then we have

$$f \star \rho_n \to f \text{ in } L^p(\mathbb{R}^d) \text{ when } n \to +\infty$$

as soon as $f \in L^p(\mathbb{R}^d)$, $1 \le p < +\infty$ [72]. If ρ is in addition of class C^∞, the sequence ρ_n is called a mollifying sequence or a sequence of mollifiers [2]. Indeed, the functions $f \star \rho_n$ are C^∞ approximates of f in the L^p-norm.

Proposition 3.3 *Let C_0 be an open set such that $\bar{C}_0 \subset \Omega$ and $\Omega \subset \cup_{j=0}^m C_j$. There exist $m+1$ functions $\psi_j \colon \mathbb{R}^d \to [0,1]$ of class C^∞ such that* $\operatorname{supp} \psi_j \subset \bar{C}_j$ *and $\sum_{j=0}^m \psi_j = 1$ in Ω.*

Proof Recall first that for any closed set A, the function

$$x \mapsto d(x, A) = \inf_{y \in A} \|x - y\|$$

is a continuous function from $\mathbb{R}^d$ into $\mathbb{R}_+$ that vanishes exactly on A.

We can choose $\eta > 0$ small enough so that:

1. The sets $C_j^\eta = \{x \in C_j; d(x, \partial C_j) > \eta\}$ still form an open cover of Ω in the sense that $\Omega \subset \cup_{j=0}^m C_j^\eta$.
2. We can take an open set C_{m+1}^η such that $\bar{C}_{m+1}^\eta \subset \mathbb{R}^d \setminus \Omega$ in order to cover the whole of $\mathbb{R}^d = \cup_{j=0}^{m+1} C_j^\eta$, and such that $d(\bar{C}_{m+1}^\eta, \bar{\Omega}) > \eta$).

This is possible by compactness of $\bar{\Omega}$ but we omit the (tedious) details.

The functions

$$\psi_j^\eta(x) = \frac{d(x, \mathbb{R}^d \setminus C_j^\eta)}{\sum_{k=0}^{m+1} d(x, \mathbb{R}^d \setminus C_k^\eta)} \tag{3.1}$$

are continuous on $\mathbb{R}^d$, indeed the denominator never vanishes because of the covering property. They are $[0,1]$-valued and ψ_j^η has support C_j^η. Finally, it is clear that $\sum_{j=0}^{m+1} \psi_j(x) = 1$ on $\mathbb{R}^d$, with ψ_{m+1}^η identically zero on the set $\{x; d(x, \bar{\Omega}) \le \eta\}$ which contains Ω.

This family of functions has all the desired properties except that the functions are not smooth. We thus use the convolution with a mollifier ρ_η with support in the ball $B(0, \eta)$. We have

$$1 = 1 \star \rho_\eta = \left(\sum_{j=0}^{m+1} \psi_j^\eta\right) \star \rho_\eta = \sum_{j=0}^{m+1} (\psi_j^\eta \star \rho_\eta).$$

Each function $\psi_j = \psi_j^\eta \star \rho_\eta$ has support in C_j for $j = 0, \ldots, m+1$, with $\psi_{m+1}^\eta \star \rho_\eta = 0$ on Ω (this is the reason why we shrank all open sets by η in the beginning since the convolution spreads supports by an amount η), and is of class C^∞. □

Let us give an example in dimension 1, see Figs. 3.9, 3.10, and 3.11, without the final smoothing step. We take $\Omega =]0, 1[$, $C_0 =]\frac{1}{8}, \frac{7}{8}[$, $C_1 =]-\frac{1}{4}, \frac{1}{4}[$, $C_2 =]\frac{3}{4}, \frac{5}{4}[$, $C_3 =]-\frac{1}{8}, \frac{3}{8}[$ and $C_4 =]-\infty, 0[\cup]1, +\infty[$. All functions can be computed explicitly. Thus, denoting $\xi_j(x) = d(x, \mathbb{R} \setminus C_j)$, we have

$$\xi_0(x) = \min\left\{\left(x - \frac{1}{8}\right)_+, \left(\frac{7}{8} - x\right)_+\right\},$$

and so on.

We note that the set C_3 is unnecessary to have a covering of Ω. We just added it to have a nicer picture. If we had not added it, the partition of unity would have been piecewise affine and it is a mistake to think the partitions of unity derived from formula (3.1) are always piecewise affine!

Let us also illustrate an example in dimension 2 (Figs. 3.12, 3.13, 3.14, 3.15, and 3.16), Ω is the unit disk covered by three squares of side 2.5, centered at 1, j and j^2 (identifying $\mathbb{R}^2$ and $\mathbb{C}$) and rotated so as to form a covering of the boundary as required. There is no C_0, since the three squares already cover Ω.

Corollary 3.1 *Let Ω be a bounded open set in $\mathbb{R}^d$ and u be a function on Ω. Let C_j, $j = 0, \ldots, m$, be an open cover as in Proposition 3.3. Then we can write $u = \sum_{j=0}^m u_j$ with* $\operatorname{supp} u_j \subset C_j$ *and u_j has the same smoothness or integrability as u.*

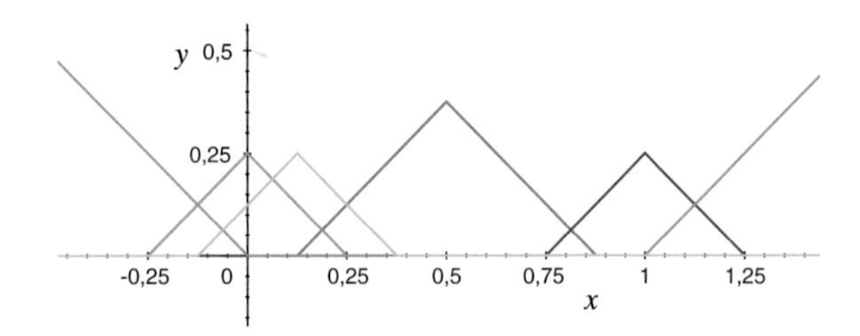

Fig. 3.9 The five functions ξ_j, $j = 0, \ldots, 4$

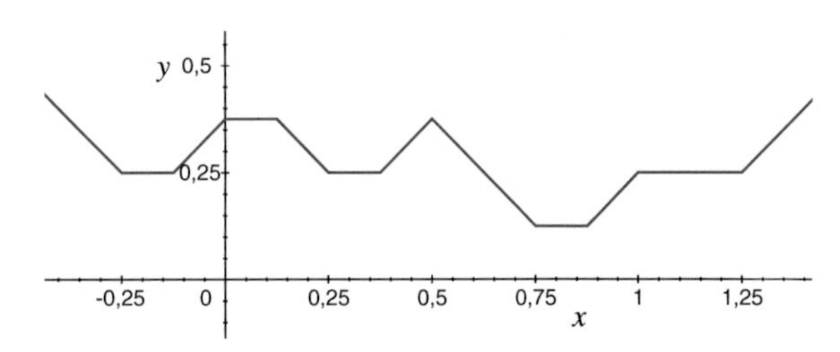

Fig. 3.10 Their sum, i.e., the denominator of (3.1), which never vanishes

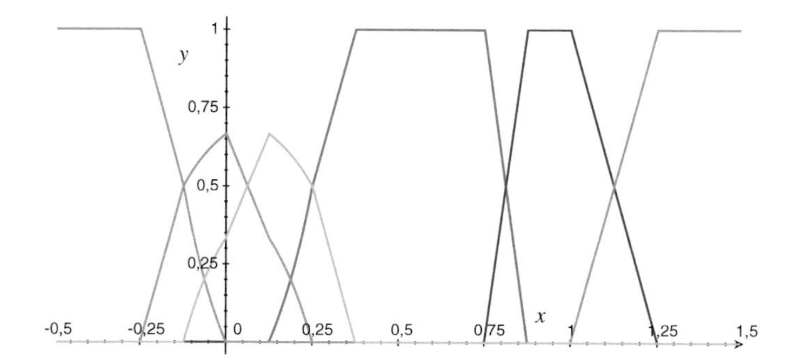

Fig. 3.11 The partition of unity ψ_j, $j = 0, \ldots, 4$

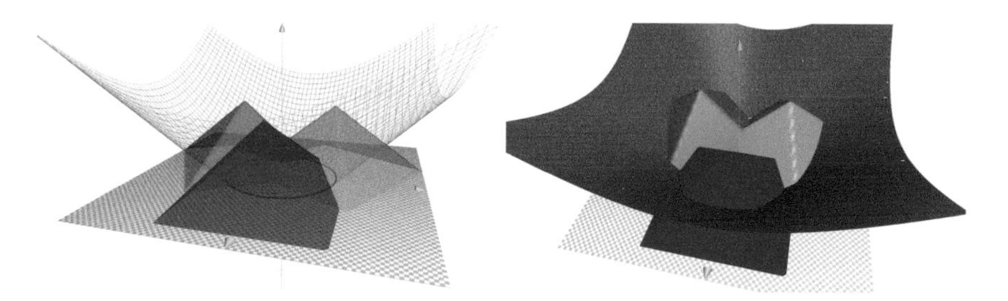

Fig. 3.12 The four functions ξ_j

Fig. 3.13 Their sum

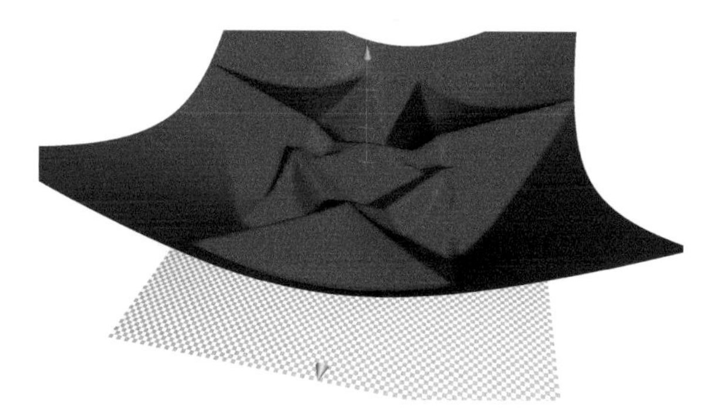

Proof We use the partition of unity ψ_j. Since $1 = \sum_{j=0}^{m} \psi_j$ on Ω, we can write

$$u = u \times 1 = u \sum_{j=0}^{m} \psi_j = \sum_{j=0}^{m} u\psi_j$$

and set $u_j = u\psi_j$. As supp $\psi_j \subset C_j$, it follows that u_j vanishes outside of C_j, and since ψ_j is C^∞ and between 0 and 1, u_j is as differentiable or as integrable as u already is. $\square$

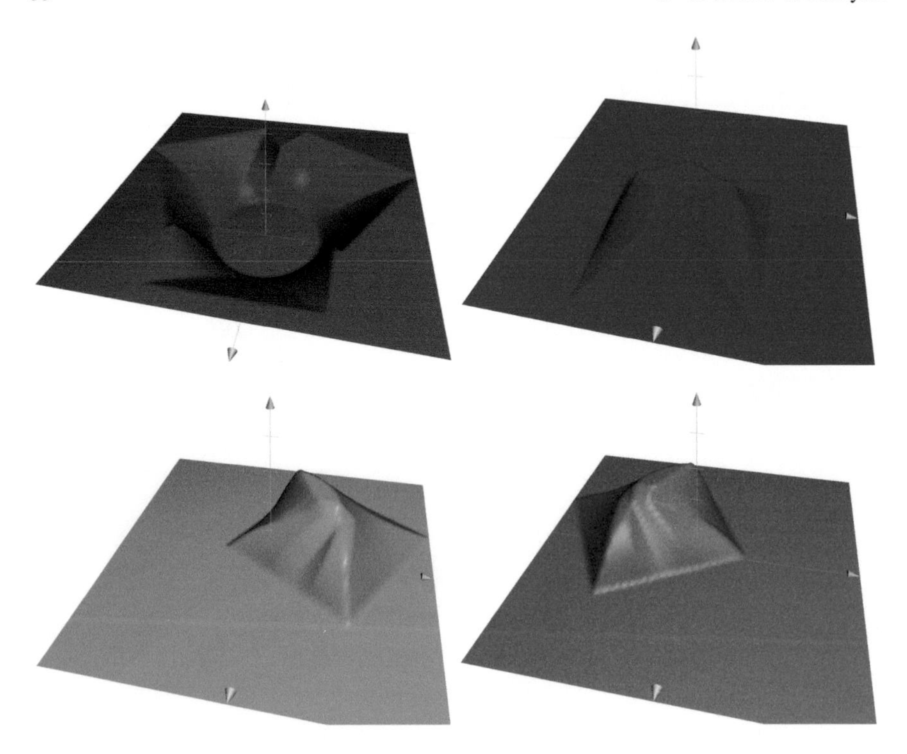

Fig. 3.14 The four functions ψ_j, drawn separately

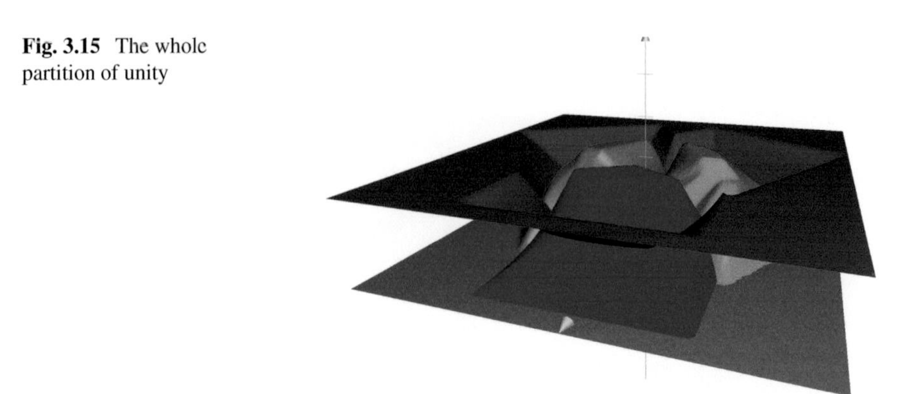

Fig. 3.15 The whole partition of unity

Fig. 3.16 Checking that $\psi_1 + \psi_2 + \psi_3 = 1$ on Ω

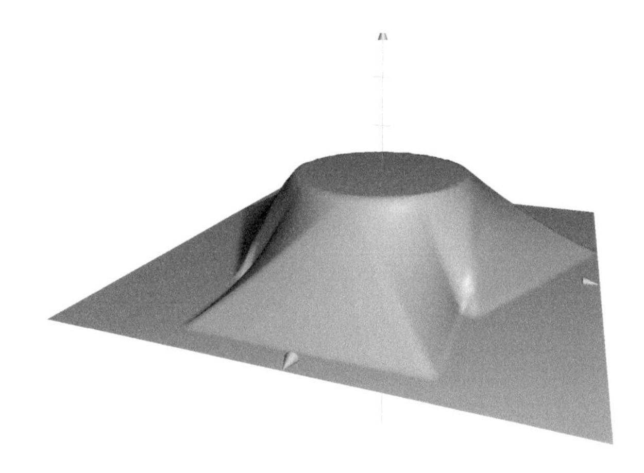

It is in this sense that partitions of unity are used to localize a function u. Such a function is decomposed into a sum of functions, each with support in a given open set of a covering. It is often easier to work with the localized parts u_j than with the function u itself. A prime example is integration by parts in the next section.

3.5 Integration by Parts in Dimension *d* and Applications

Integration by parts in $\mathbb{R}^d$ is a basic formula, that is not often entirely proved. In what follows, Ω will be an at least Lipschitz open subset of $\mathbb{R}^d$. The most crucial integration by parts formula, from which all the others follow, is given in the next Theorem.

Theorem 3.4 *Let Ω be a Lipschitz open set in $\mathbb{R}^d$ and $u \in C^1(\bar{\Omega})$. Then we have*

$$\int_\Omega \frac{\partial u}{\partial x_i}\,dx = \int_{\partial\Omega} u n_i\,d\Gamma, \tag{3.2}$$

where n_i is the ith *component of the normal unit exterior vector.*

Proof We will only write the proof in dimension $d = 2$, which is not a real restriction as the general case $d \geq 2$ follows from exactly the same arguments, and only in the case when Ω is of class C^1. This is a real restriction: there are additional technical difficulties in the Lipschitz case due to only almost everywhere differentiability.

We start with the partition of unity associated with the given covering C_j of the boundary completed by an open set C_0 to cover the interior. We have $u = \sum_{j=0}^m u_j$ with $u_j = u\psi_j$, and each u_j belongs to $C^1(\bar{\Omega})$ and has support in $\bar{C}_j$. Consequently, since formula (3.2) is linear with respect to u, it is sufficient to prove it for each u_j.

Let us start with the case $j = 0$. In this case, u_0 is compactly supported in Ω since $\bar{C}_0 \subset \Omega$. In particular, it vanishes on $\partial\Omega$, so that $\int_{\partial\Omega} u_0 n_i\,d\Gamma = 0$.

We extend u_0 by 0 to $\mathbb{R}^2$, thus yielding a $C^1(\mathbb{R}^2)$ function $\tilde{u}_0$. Since Ω is bounded, we choose a square that contains it, $\Omega \subset Q =]-M, M[^2$, for some M. Letting $i' = 1$ if $i = 2$, $i' = 2$ if $i = 1$, we obtain

$$\int_\Omega \frac{\partial u_0}{\partial x_i}\,dx = \int_Q \frac{\partial \tilde{u}_0}{\partial x_i}\,dx = \int_{-M}^{M}\Big(\int_{-M}^{M} \frac{\partial \tilde{u}_0}{\partial x_i}\,dx_i\Big)dx_{i'} = \int_{-M}^{M} [\tilde{u}_0]_{x_i=-M}^{x_i=M} dx_{i'} = 0,$$

by Fubini's theorem and the fact that $\tilde{u}_0 = 0$ sur ∂Q. Formula (3.2) is thus established for u_0.

The case $j > 0$ is a little more complicated. To simplify the notation, we omit all j indices and exponents. We thus have a function u with support in $\bar{C}$. In particular, $u = 0$ on $\partial C \cap \bar{\Omega}$. We have

$$\Omega \cap C = \{y \in C;\ y_2 < \varphi(y_1)\},$$

see also Fig. 3.7. We first establish formula (3.2) in the (y_1, y_2) coordinate system in which $C =]-a, a[^2$ for some a. We let $n_{y,i}$, $i = 1, 2$, denote the components of the normal vector in this coordinate system. There are two different computations depending on the coordinate under consideration.

Case $i = 1$. We first use Fubini's theorem

$$\int_\Omega \frac{\partial u}{\partial y_1}\,dy = \int_{-a}^{a}\Big(\int_{-a}^{\varphi(y_1)} \frac{\partial u}{\partial y_1}(y_1, y_2)\,dy_2\Big)\,dy_1,$$

see again Fig. 3.7. Now it is well-known from elementary calculus that

$$\frac{d}{dy_1}\Big(\int_{-a}^{\varphi(y_1)} u(y_1, y_2)\,dy_2\Big) = \int_{-a}^{\varphi(y_1)} \frac{\partial u}{\partial y_1}(y_1, y_2)\,dy_2 + u(y_1, \varphi(y_1))\varphi'(y_1),$$

(this is where the fact that φ is C^1 intervenes and where it would be a little harder to have φ only Lipschitz). Consequently,

$$\int_\Omega \frac{\partial u}{\partial y_1}\,dy = \int_{-a}^{a} \frac{d}{dy_1}\Big(\int_{-a}^{\varphi(y_1)} u(y_1, y_2)\,dy_2\Big)\,dy_1 - \int_{-a}^{a} u(y_1, \varphi(y_1))\varphi'(y_1)\,dy_1.$$

In the first integral, we integrate a derivative, so that

$$\int_{-a}^{a} \frac{d}{dy_1}\Big(\int_{-a}^{\varphi(y_1)} u(y_1, y_2)\,dy_2\Big)\,dy_1$$
$$= \int_{-a}^{\varphi(a)} u(a, y_2)\,dy_2 - \int_{-a}^{\varphi(-a)} u(-a, y_2)\,dy_2 = 0,$$

since $u = 0$ on ∂C. We thus see that

$$\begin{aligned}\int_\Omega \frac{\partial u}{\partial y_1}\,dy &= \int_{-a}^{a} u(y_1,\varphi(y_1))\frac{-\varphi'(y_1)}{\sqrt{1+\varphi'(y_1)^2}}\sqrt{1+\varphi'(y_1)^2}\,dy_1\\ &= \int_{-a}^{a} u(y_1,\varphi(y_1))n_{y,1}(y_1)\sqrt{1+\varphi'(y_1)^2}\,dy_1\\ &= \int_{C\cap\partial\Omega} un_{y,1}\,d\Gamma,\end{aligned}$$

by the formulas established in Sect. 3.3 for the normal vector components and the boundary measure. Hence formula (3.2) in this case.

Case $i = 2$. We start again with Fubini's theorem

$$\begin{aligned}\int_\Omega \frac{\partial u}{\partial y_2}\,dy &= \int_{-a}^{a}\Big(\int_{-a}^{\varphi(y_1)} \frac{\partial u}{\partial y_2}(y_1,y_2)\,dy_2\Big)\,dy_1\\ &= \int_{-a}^{a} u(y_1,\varphi(y_1))\,dy_1\\ &= \int_{-a}^{a} u(y_1,\varphi(y_1))\frac{1}{\sqrt{1+\varphi'(y_1)^2}}\sqrt{1+\varphi'(y_1)^2}\,dy_1\\ &= \int_{C\cap\partial\Omega} un_{y,2}\,d\Gamma,\end{aligned}$$

since $u(y_1,-a) = 0$. This proves the integration by parts formula in the (y_1, y_2) system attached to the cube C covering a part of the boundary.

We need to go back to the original coordinate system (x_1, x_2). Let us write the change of coordinates formulas

$$\begin{pmatrix}y_1\\y_2\end{pmatrix} = R\begin{pmatrix}x_1-c_1\\x_2-c_2\end{pmatrix} \text{ or again } \begin{pmatrix}x_1\\x_2\end{pmatrix} = R^T\begin{pmatrix}y_1\\y_2\end{pmatrix} + \begin{pmatrix}c_1\\c_2\end{pmatrix},$$

where R is an orthogonal matrix and (c_1, c_2) are the (x_1, x_2) coordinates of the center of C. Similarly, if v_x and v_y denote the column-vectors of the components of the same vector $v \in \mathbb{R}^2$ in each of the coordinate systems, we have

$$v_y = Rv_x \Longleftrightarrow v_x = R^T v_y.$$

This is true in particular for the normal vecteur n, $n_x = R^T n_y$. Let us note $\nabla_x u$ and $\nabla_y u$ the components of the gradient of u in the two coordinate systems, we see by the chain rule that

$$(\nabla_x u)_i = \frac{\partial u}{\partial x_i} = \sum_{j=1}^{2}\frac{\partial u}{\partial y_j}\frac{\partial y_j}{\partial x_i} = \sum_{j=1}^{2} R_{ji}\frac{\partial u}{\partial y_j} = (R^T\nabla_y u)_i,$$

hence the result for the localized part of the function, by linearity of the integrals, and then for the whole function again by linearity of the integrals. □

Once the basic formula is available, many other formulas are easily derived, that bear various names in the literature. We give below a short selection of such formulas that will be useful in the sequel. We do not specify the regularity of the functions below, it is understood that they are sufficiently differentiable for all derivatives and integrals to make sense.

Corollary 3.2 *Let Ω be a Lipschitz open set in $\mathbb{R}^d$. We have*
(i) Integration by parts strictly speaking

$$\int_\Omega \frac{\partial u}{\partial x_i} v\,dx = -\int_\Omega u\frac{\partial v}{\partial x_i}\,dx + \int_{\partial\Omega} uvn_i\,d\Gamma, \tag{3.3}$$

(ii) Green's formula

$$\int_\Omega (\Delta u)v\,dx = -\int_\Omega \nabla u\cdot\nabla v\,dx + \int_{\partial\Omega}\frac{\partial u}{\partial n}v\,d\Gamma, \tag{3.4}$$

where $\frac{\partial u}{\partial n} = \nabla u\cdot n = \sum_{i=1}^d \frac{\partial u}{\partial x_i}n_i$ denotes the normal derivative *of u on $\partial\Omega$.*
(iii) A slightly more symmetrical version of Green's formula

$$\int_\Omega (\Delta u)v\,dx = \int_\Omega u(\Delta v)\,dx + \int_{\partial\Omega}\Big(\frac{\partial u}{\partial n}v - u\frac{\partial v}{\partial n}\Big)\,d\Gamma, \tag{3.5}$$

(iv) Stokes formula

$$\int_\Omega \operatorname{div} U\,dx = \int_{\partial\Omega} U\cdot n\,d\Gamma, \tag{3.6}$$

where $U\colon\Omega\to\mathbb{R}^d$ is a vector field, its divergence is $\operatorname{div} U = \sum_{i=1}^d \frac{\partial U_i}{\partial x_i}$ and $U\cdot n = \sum_{i=1}^d U_in_i$ is the flux of the vector field through the boundary of Ω.

Proof For (i), we apply the basic formula (3.2) to the product uv, and so on. □

3.6 Distributions

In this section, Ω is an arbitrary open subset of $\mathbb{R}^d$.

It turns out that functions that are differentiable in the classical sense are not sufficient to work with PDEs. A more general concept is needed, which is called *distributions* [72, 73]. As we will see, distributions are a lot more general than functions. They can always be differentiated indefinitely, even when they correspond to functions that are not differentiable in the classical sense, and their derivatives are distributions. This is why distributional solutions to linear PDEs of arbitrary order make sense (with technical conditions on their coefficients). We will also use distributions to define an important class of function spaces for PDEs, the Sobolev spaces.

Let us first go back to the space $\mathscr{D}(\Omega)$ of indefinitely differentiable functions with compact support encountered in Sect. 3.2. It is trivial, but crucial for the sequel, that the space $\mathscr{D}(\Omega)$ is stable by differentiation of arbitrary order, i.e., if $\varphi \in \mathscr{D}(\Omega)$ then $\partial^\alpha \varphi \in \mathscr{D}(\Omega)$ for any multiindex α.

The function $\varphi(x) = e^{\frac{1}{\|x\|^2-1}}$ for $\|x\| < 1$, $\varphi(x) = 0$ otherwise, belongs to $\mathscr{D}(\mathbb{R}^d)$. It can be scaled to define a mollifier and thus construct infinitely many functions in $\mathscr{D}(\mathbb{R}^d)$ by convolution. Let us notice that if Ω contains the unit closed ball, then the restriction of φ to Ω belongs to $\mathscr{D}(\Omega)$. Conversely, it is quite clear that given a function in $\mathscr{D}(\Omega)$ for any Ω, by extending it by 0 to $\mathbb{R}^d \setminus \Omega$ we obtain a function in $\mathscr{D}(\mathbb{R}^d)$. In this sense, it can be said that a function of $\mathscr{D}(\Omega)$ vanishes on the boundary of Ω, even though it is *a priori* only defined on Ω.

As mentioned before, the vector space $\mathscr{D}(\Omega)$ has a natural topology that is a little difficult to understand (technically, it is an LF-space, a strict inductive limit of a sequence of Fréchet spaces and it is not metrizable, see [13]) and it is not very useful to master the details of this topology for the applications we have in mind. So we will just skip it.

The convergence of a sequence in $\mathscr{D}(\Omega)$ for its natural topology is given by the following proposition, which we admit, see [72, 73, 78, 80].

Proposition 3.4 *A sequence $\varphi_n \in \mathscr{D}(\Omega)$ converges to $\varphi \in \mathscr{D}(\Omega)$ in the sense of $\mathscr{D}(\Omega)$ if and only if*

(i) There exists a compact subset K of Ω such that $\operatorname{supp} \varphi_n \subset K$ *for all n.*

(ii) For all $\alpha \in \mathbb{N}^d$, $\partial^\alpha \varphi_n \to \partial^\alpha \varphi$ uniformly.

It follows clearly from the above proposition that if $\varphi_n \to \varphi$ in the sense of $\mathscr{D}(\Omega)$, then $\partial^\alpha \varphi_n \to \partial^\alpha \varphi$ in the sense of $\mathscr{D}(\Omega)$ for all $\alpha \in \mathbb{N}^d$. In addition, it is easy to see that $\mathscr{D}(\Omega) \subset L^p(\Omega)$ for all $p \in [1, +\infty]$ and that if $\varphi_n \to \varphi$ in the sense of $\mathscr{D}(\Omega)$, then $\varphi_n \to \varphi$ in $L^p(\Omega)$.

When a real (or complex) vector space is equipped with a topology that makes its vector space operations continuous, that is when it is a *topological vector space*, it makes sense to look at the vector space of continuous linear forms, that is the space of real (or complex) valued linear mappings that are continuous for the aforementioned topology. This space is called the *topological dual*, or in short dual space.

Definition 3.2 The *space of distributions on Ω*, $\mathscr{D}'(\Omega)$, is the dual of the space $\mathscr{D}(\Omega)$.

We will indifferently use the notations $T(\varphi) = \langle T, \varphi \rangle$ to denote the value of a distribution T on a test-function φ and the duality pairing between the two. Of course, since $\mathscr{D}'(\Omega)$ is a vector space, we can add distributions and multiply them by a scalar in the obvious way.

Now, not knowing the topology of $\mathscr{D}(\Omega)$ makes it a little difficult to decide which linear forms on $\mathscr{D}(\Omega)$ are continuous and which are not. Fortunately, even though the topology in question is not metrizable, the usual sequential criterion happens to still work in this particular case. We admit the following proposition, see [72, 73, 78, 80].

Proposition 3.5 *A linear form T on $\mathscr{D}(\Omega)$ is a distribution if and only if we have $T(\varphi_n) \to 0$ for all sequences $\varphi_n \in \mathscr{D}(\Omega)$ such that $\varphi_n \to 0$ in the sense of $\mathscr{D}(\Omega)$.*

Remark 3.5 Let us note that the property $T(\varphi_n) \to 0$ for all sequences $\varphi_n \in \mathscr{D}(\Omega)$ such that $\varphi_n \to 0$ immediately implies that $T(\varphi_n) \to T(\varphi)$ for all sequences φ_n such that $\varphi_n \to \varphi$ in the sense of $\mathscr{D}(\Omega)$, by linearity, hence the sequential continuity of the linear form T. The difficulty is that in a non metrizable topological space, sequential continuity does not imply continuity in general, even though, in this particular case, it does. □

Let us now see in which sense distributions generalize the usual notion of function.

Proposition 3.6 *For all $f \in L^1_{\text{loc}}(\Omega)$ there exists a distribution $\imath(f)$ on Ω defined by the formula*

$$\langle \imath(f), \varphi \rangle = \int_\Omega f\varphi\, dx$$

for all $\varphi \in \mathscr{D}(\Omega)$. The mapping $\imath\colon L^1_{\text{loc}}(\Omega) \to \mathscr{D}'(\Omega)$ is one-to-one.

Proof That the integral is well-defined has already been seen. It clearly defines a linear form on $\mathscr{D}(\Omega)$ by the linearity of the integral. What remains to be established for $\imath(f)$ to be a distribution, is its continuity. Let us thus be given a sequence $\varphi_n \to 0$ in the sense of $\mathscr{D}(\Omega)$, and K the associated compact set. We have

$$|\langle \imath(f), \varphi_n \rangle| = \left| \int_\Omega f\varphi_n\, dx \right| = \left| \int_K f\varphi_n\, dx \right|$$
$$\leq \int_K |f||\varphi_n|\, dx \leq \max_K |\varphi_n| \int_K |f|\, dx \to 0$$

since $f_{|K} \in L^1(K)$ and φ_n tends to 0 uniformly on K.

Let us now show that the mapping $\imath$ is one-to-one. Since it is clearly linear, it suffices to show that its kernel is reduced to the zero vector. Let $f \in \ker \imath$, which means that $\imath(f)$ is the zero linear form, or in other words that $\int_\Omega f\varphi\, dx = 0$ for all φ in $\mathscr{D}(\Omega)$. By Proposition 3.1, it follows that we have $f = 0$, and the proof is complete. □

Remark 3.6 The characterization of convergence in $\mathscr{D}(\Omega)$ of Proposition 3.4 is implied by the topology of $\mathscr{D}(\Omega)$. In the proof of Proposition 3.6, we can see the importance of having a fixed compact K containing all the supports. If it was not the case, the final estimate would break down. □

Remark 3.7 The mapping $\imath$ is not only one-to-one, it is also continuous (for the topology of $\mathscr{D}'(\Omega)$ as topological dual of $\mathscr{D}(\Omega)$, which we also keep shrouded in mystery). The mapping $\imath$ thus provides a faithful representation of one type of objects, L^1_{loc} functions, as objects of a completely different nature, distributions. It is so faithful that in day-to-day practice, we say that an L^1_{loc} function f *is* a distribution and dispense with the notation $\imath$ altogether, that is we just write $\langle f, \varphi \rangle$ for the duality bracket.

Conversely, when a distribution T belongs to the image of $\imath$, that is to say when there exists f in L^1_{loc} such that $\langle T, \varphi \rangle = \int_\Omega f\varphi\,dx$ for all φ in $\mathscr{D}(\Omega)$, we just say that T *is* a function and we write $T = f$. Beware however that most distributions are not functions and that the notation $\int_\Omega T\varphi\,dx$ is *unacceptable* for any distribution that is not in the range of the mapping $\imath$.

Proposition 3.6 is all the more important as it shows that the elements of all the function spaces introduced up to now actually are distributions, since $L^1_{\text{loc}}(\Omega)$ is the largest of all such spaces. □

Proposition 3.6 gives us our first examples of distributions. There are however many others which are not functions. Let us describe a couple of them.

We choose a point $a \in \Omega$ and define

$$\langle \delta_a, \varphi \rangle = \varphi(a)$$

for all $\varphi \in \mathscr{D}(\Omega)$. This is clearly a linear form on $\mathscr{D}(\Omega)$ and we just need to check its continuity. Let us thus be given again a sequence $\varphi_n \to 0$ in the sense of $\mathscr{D}(\Omega)$. In particular, there exists a compact subset K on which it converges uniformly to 0, and out of which is identically 0. Therefore, the sequence converges uniformly on Ω, hence pointwise. Thus $\varphi_n(a) \to 0$ and we are done: $\delta_a \in \mathscr{D}'(\Omega)$. This distribution is called the *Dirac mass* or *Dirac distribution* at point a. When $a = 0$, it is often simply denoted δ. The Dirac mass does not belong to the image of $\imath$, i.e., loosely speaking, it is not a function.

Let us quickly show this in the case $d = 1$. Assume that there exists a function $f \in L^1_{\text{loc}}(\mathbb{R})$ such that, for all $\varphi \in \mathscr{D}(\mathbb{R})$, we have

$$\int_{\mathbb{R}} f(x)\varphi(x)\,dx = \varphi(0).$$

Letting $g(x) = xf(x)$, we have $g \in L^1_{\text{loc}}(\mathbb{R})$, and we see that for all $\varphi \in \mathscr{D}(\mathbb{R})$,

$$\int_{\mathbb{R}} g(x)\varphi(x)\,dx = \int_{\mathbb{R}} xf(x)\varphi(x)\,dx = \int_{\mathbb{R}} f(x)(x\varphi(x))\,dx = 0$$

since $x \mapsto x\varphi(x)$ belongs to $\mathscr{D}(\mathbb{R})$ and vanishes at $x = 0$. This implies that $g = 0$ by Proposition 3.1, and therefore $f = 0$, which is a contradiction since $\delta \neq 0$.

Let us give a second example with $\Omega = \mathbb{R}$. The function $x \mapsto 1/x$ almost everywhere is not in $L^1_{\text{loc}}(\mathbb{R})$ because it is not integrable in a neighborhood of 0. Therefore, it cannot be identified with a distribution as in Proposition 3.6, which is rather unfortunate for such a simple function and a concept claiming to widely generalize functions. The distribution defined by

$$\Big\langle \mathrm{vp}\,\frac{1}{x}, \varphi\Big\rangle = \lim_{\varepsilon\to 0^+}\Big(\int_{-\infty}^{-\varepsilon}\frac{\varphi(x)}{x}\,dx + \int_{\varepsilon}^{+\infty}\frac{\varphi(x)}{x}\,dx\Big)$$

is called the *principal value of* $1/x$ and replaces the function $x \mapsto 1/x$ for all intents and purposes (exercise: show that it is a distribution). It is however *not* a function.

We have hinted at a topology on the space of distributions. Here again, it is not too important to know the details of this topology. The convergence of sequences is more than enough and is surprisingly simple. We admit the following result, see [72, 73, 78, 80], which can actually be taken as a definition.

Proposition 3.7 *A sequence $T_n \in \mathscr{D}'(\Omega)$ converges to $T \in \mathscr{D}'(\Omega)$ in the sense of $\mathscr{D}'(\Omega)$ if and only if $\langle T_n, \varphi\rangle \to \langle T, \varphi\rangle$ for all $\varphi \in \mathscr{D}(\Omega)$.*

Since distributions are linear forms on the space $\mathscr{D}(\Omega)$, we see that convergence in the sense of distributions is actually nothing but simple or pointwise convergence on $\mathscr{D}(\Omega)$. This makes it very easy to handle (and unfortunately, very easy to abuse. Remember, it is not magic!).

This notion of convergence agrees with all previous notions defined on smaller function spaces. In particular, we have

Proposition 3.8 *Let $u_n \to u$ in $L^p(\Omega)$ for some $p \in [1, +\infty]$. Then $u_n \to u$ in the sense of $\mathscr{D}'(\Omega)$.*

Proof For all $\varphi \in \mathscr{D}(\Omega)$, we have

$$|\langle u_n, \varphi\rangle - \langle u, \varphi\rangle| \le \int_\Omega |u_n - u||\varphi|\,dx \le \|u_n - u\|_{L^p(\Omega)}\|\varphi\|_{L^{p'}(\Omega)} \to 0$$

by Hölder's inequality. □

We have said earlier that distributions can be differentiated indefinitely, however in a specific sense.

Definition 3.3 Let T be a distribution on Ω. The formula

$$\langle S, \varphi\rangle = -\Big\langle T, \frac{\partial\varphi}{\partial x_i}\Big\rangle, \tag{3.7}$$

for all $\varphi \in \mathscr{D}(\Omega)$, defines a distribution S, which is called the (distributional) partial derivative of T with respect to x_i and is denoted $\frac{\partial T}{\partial x_i}$.

Proof This definition needs a proof. Formula (3.7) clearly defines a linear form on $\mathscr{D}(\Omega)$. Let us see that it is continuous. Let us be given a sequence $\varphi_n \to 0$ in $\mathscr{D}(\Omega)$. It is apparent that $\frac{\partial\varphi_n}{\partial x_i} \to 0$ in $\mathscr{D}(\Omega)$. Indeed, the support condition is the same, since the support of the partial derivative of a function is included in the support of this function, and the uniform convergence of all derivatives trivially holds true as all derivatives of $\frac{\partial \varphi_n}{\partial x_i}$ are derivatives of φ_n. Therefore,

$$\langle S, \varphi_n \rangle = -\Big\langle T, \frac{\partial \varphi_n}{\partial x_i} \Big\rangle \to 0$$

for all sequences $\varphi_n \to 0$ in $\mathscr{D}(\Omega)$. □

For example, the derivative of the Dirac mass δ in dimension one is the distribution

$$\langle \delta', \varphi \rangle = -\langle \delta, \varphi' \rangle = -\varphi'(0),$$

for all $\varphi \in \mathscr{D}(\mathbb{R})$.

The reason why it is reasonable to call this new distribution a partial derivative is in the next proposition.

Proposition 3.9 *Let u be a function in $C^1(\Omega)$. Then its distributional partial derivatives coincide with its classical partial derivatives.*

Proof Let $\varphi \in \mathscr{D}(\Omega)$. The support K of $u\varphi$ is bounded and we can include it in a hypercube C. We define v on C by $v(x) = u(x)\varphi(x)$ if $x \in K$, $v(x) = 0$ otherwise. It is easy to check that $v \in C^1(\bar{C})$ and $v = 0$ on ∂C. Since C is a Lipschitz open set, we can apply the integration by parts formula (3.3)[3] and obtain

$$\int_K \frac{\partial v}{\partial x_i}\,dx = \int_C \frac{\partial v}{\partial x_i}\,dx = 0.$$

Now, on K, we have $\frac{\partial v}{\partial x_i} = u\frac{\partial \varphi}{\partial x_i} + \frac{\partial u}{\partial x_i}\varphi$, so that

$$\int_\Omega \frac{\partial u}{\partial x_i}\varphi\,dx = \int_K \frac{\partial u}{\partial x_i}\varphi\,dx = -\int_K u\frac{\partial \varphi}{\partial x_i}\,dx = -\int_\Omega u\frac{\partial \varphi}{\partial x_i}\,dx,$$

since all intervening integrands are zero outside of K, which completes the proof by using Proposition 3.6. □

Remark 3.8 Be careful that the same result is false for functions that are only almost everywhere differentiable. Let us show an example, which is also a showcase example of how to compute a distributional derivative. Let H be the *Heaviside function* defined on $\mathbb{R}$ by $H(x) = 0$ for $x \le 0$, $H(x) = 1$ for $x > 0$. This function is classically differentiable with zero derivative for $x \neq 0$ and has a discontinuity of the first kind at $x = 0$. It is also in $L^\infty(\mathbb{R})$, hence in $L^1_{\text{loc}}(\mathbb{R})$, hence a distribution. Let us compute its distributional derivative. Take $\varphi \in \mathscr{D}(\mathbb{R})$ and let $R > 0$ be such that $\operatorname{supp}\varphi \subset [-R, R]$. We have

---

[3] Note that there is no regularity or boundedness hypothesis made on Ω itself.

$$\langle H', \varphi \rangle = -\langle H, \varphi' \rangle = -\int_0^R \varphi'(s)\, ds = \varphi(0) - \varphi(R) = \varphi(0),$$

since φ and φ' vanish for $x \geq R$. Therefore we see that

$$H' = \delta$$

even though the almost everywhere classical derivative of H is 0. This is an example of a function that is not differentiable in the classical sense, but that is also a distribution, hence has a distributional derivative and this derivative is not a function. The example also shows that H is a distributional primitive of the Dirac mass. □

Once a distribution is known to have partial derivatives of order one which are again distributions, it is obvious that the same operation can be repeated indefinitely and we have, for any distribution T and any multiindex α,

$$\langle \partial^\alpha T, \varphi \rangle = (-1)^{|\alpha|} \langle T, \partial^\alpha \varphi \rangle$$

for all $\varphi \in \mathscr{D}(\Omega)$, by induction on the length of α.

Differentiation in the sense of distributions is continuous, which is violently false in most function spaces. We just show here the sequential continuity, which is amply sufficient for the applications.

Proposition 3.10 *Let $T_n \to T$ in the sense of $\mathscr{D}'(\Omega)$. Then, for all multiindices α, we have $\partial^\alpha T_n \to \partial^\alpha T$ in the sense of $\mathscr{D}'(\Omega)$.*

Proof For all $\varphi \in \mathscr{D}(\Omega)$, we have

$$\langle \partial^\alpha T_n, \varphi \rangle = (-1)^{|\alpha|} \langle T_n, \partial^\alpha \varphi \rangle \to (-1)^{|\alpha|} \langle T, \partial^\alpha \varphi \rangle = \langle \partial^\alpha T, \varphi \rangle,$$

hence the result. □

This continuity provides another reason why the partial derivative terminology is adequate for distributions. Indeed, it can be shown that $C^\infty(\Omega)$ functions are dense in $\mathscr{D}'(\Omega)$. For any distribution T, there exists a sequence of indefinitely differentiable functions ψ_n that tends to T in the sense of distributions. Therefore, their distributional partial derivatives of arbitrary order, which coincide with their classical partial derivatives, also converge in the sense of distributions. So the distributional partial derivatives of a distribution appear as distributional limits of approximating classical partial derivatives.

Many other operations usually performed on functions can be extended to distributions using the same transposition trick as for partial derivatives. Let us just mention here the multiplication by a smooth function.

Definition 3.4 Let T be a distribution on Ω and $f \in C^\infty(\Omega)$. The formula

$$\langle fT, \varphi \rangle = \langle T, f\varphi \rangle,$$

for all $\varphi \in \mathscr{D}(\Omega)$, defines a distribution.

We leave the easy proof as an exercise. Of course, when $T \in L^1_{\text{loc}}(\Omega)$, fT coincides with the classical pointwise product and the mapping $T \mapsto fT$ is sequentially continuous on $\mathscr{D}'(\Omega)$. Note that it is not possible to define such a product in all generality by a function that is less smooth than C^∞. In particular, there is no product of two distributions with the reasonable properties to be expected from a product—a famous theorem by L. Schwartz that limits the usefulness of general distributions in dealing with nonlinear PDEs.

The partial derivatives of a distribution multiplied by a smooth function follow the classical Leibniz rule.

Proposition 3.11 *Let T be a distribution on Ω and $f \in C^\infty(\Omega)$. For all multiindices α such that $|\alpha| = 1$, we have*

$$\partial^\alpha (fT) = f\partial^\alpha T + \partial^\alpha f\, T. \tag{3.8}$$

Proof We just use the definitions. For all $\varphi \in \mathscr{D}(\Omega)$, we have

$$\begin{aligned}\langle \partial^\alpha (fT), \varphi\rangle &= -\langle fT, \partial^\alpha \varphi\rangle = -\langle T, f\partial^\alpha \varphi\rangle = -\langle T, \partial^\alpha (f\varphi)\rangle + \langle T, \partial^\alpha f\, \varphi\rangle \\ &= \langle \partial^\alpha T, f\varphi\rangle + \langle \partial^\alpha f\, T, \varphi\rangle = \langle f\partial^\alpha T, \varphi\rangle + \langle \partial^\alpha f\, T, \varphi\rangle \\ &= \langle f\partial^\alpha T + \partial^\alpha f\, T, \varphi\rangle\end{aligned}$$

by the Leibniz formula for smooth functions. □

We conclude this very brief review of distribution theory with the following result [2, 51].

Proposition 3.12 *Let Ω be a connected open set of $\mathbb{R}^d$ and T a distribution on Ω such that $\frac{\partial T}{\partial x_i} = 0$ for $i = 1, \dots, d$. Then, there exists a constant $c \in \mathbb{R}$ such that $T = c$.*

Proof We write the proof in the case $\Omega = \mathbb{R}^d$. The general case follows by a localization argument. First of all, we claim that if $\varphi \in \mathscr{D}(\mathbb{R}^d)$ is such that $\int_{\mathbb{R}^d} \varphi(x)\,dx = 0$, then there exists $\varphi_i \in \mathscr{D}(\mathbb{R}^d)$, $i = 1, \dots, d$, such that

$$\varphi = \sum_{i=1}^{d} \frac{\partial \varphi_i}{\partial x_i}. \tag{3.9}$$

The proof of the claim is by induction on the dimension d. For $d = 1$, the result holds true by taking $\varphi_1(x_1) = \int_{-\infty}^{x_1} \varphi(s)\,ds$, which is clearly C^∞. In addition, it is compactly supported. Indeed, let $a < b$ be such that $\operatorname{supp}\varphi \subset [a, b]$. If $x_1 < a$, we have $\varphi_1(x_1) = 0$ obviously, since the integrand vanishes. If $x_1 > b$, we have $0 = \int_{-\infty}^{+\infty} \varphi(s)\,ds = \varphi_1(x_1) + \int_{x_1}^{+\infty} \varphi(s)\,ds = \varphi_1(x_1)$ as well.

Assume now that decomposition (3.9) has been established for $d - 1$. Let φ be such that $\int_{\mathbb{R}^d} \varphi(x)\,dx = 0$. We set $\psi(x') = \int_{-\infty}^{+\infty} \varphi(x', s)\,ds$, so that $\psi \in \mathscr{D}(\mathbb{R}^{d-1})$. We have

$$\int_{\mathbb{R}^{d-1}} \psi(x')\,dx' = \int_{\mathbb{R}^d} \varphi(x)\,dx = 0$$

by Fubini's theorem.

Let now $a < b$ be such that $\operatorname{supp}\varphi \subset \mathbb{R}^{d-1} \times [a,b]$. We pick a function $\theta \in \mathscr{D}(\mathbb{R})$ such that $\operatorname{supp}\theta \subset [a,b]$ and $\int_{-\infty}^{+\infty} \theta(s)\,ds = 1$. Then we let

$$\varphi_d(x', x_d) = \int_{-\infty}^{x_d} \varphi(x', s)\,ds - \psi(x') \int_{-\infty}^{x_d} \theta(s)\,ds.$$

It is also clear that φ_d is C^∞. Let us check that φ_d is compactly supported. The variables x' pose no problem in this regard, so we just have to see what happens with respect to the variable x_d. For $x_d < a$, again obviously $\varphi_d(x', x_d) = 0$. For $x_d > b$, we have on the one hand $\int_{-\infty}^{x_d} \varphi(x', s)\,ds = \psi(x')$ and on the other hand $\int_{-\infty}^{x_d} \theta(s)\,ds = 1$, thus $\varphi_d(x', x_d) = 0$. This shows that $\varphi_d \in \mathscr{D}(\mathbb{R}^d)$. Now, by definition,

$$\frac{\partial \varphi_d}{\partial x_d}(x', x_d) = \varphi(x', x_d) - \psi(x')\theta(x_d),$$

so that

$$\varphi(x', x_d) = \psi(x')\theta(x_d) + \frac{\partial \varphi_d}{\partial x_d}(x', x_d).$$

The induction hypothesis applies to ψ, thus proving claim (3.9).

Let us now consider a distribution T whose derivatives vanish. Let us pick a function $\Theta \in \mathscr{D}(\mathbb{R}^d)$ such that $\int_{\mathbb{R}^d} \Theta(x)\,dx = 1$ and for all $\varphi \in \mathscr{D}(\mathbb{R}^d)$, let us set

$$\Phi(x) = \varphi(x) - \Big(\int_{\mathbb{R}^d} \varphi(y)\,dy\Big)\Theta(x).$$

Clearly, $\int_{\mathbb{R}^d} \Phi(x)\,dx = 0$, so we can apply decomposition (3.9) to Φ and write

$$\varphi - \Big(\int_{\mathbb{R}^d} \varphi(y)\,dy\Big)\Theta = \sum_{i=1}^{d} \frac{\partial \Phi_i}{\partial x_i},$$

so that

$$\varphi = \Big(\int_{\mathbb{R}^d} \varphi(y)\,dy\Big)\Theta + \sum_{i=1}^{d} \frac{\partial \Phi_i}{\partial x_i}.$$

It follows that

$$\begin{aligned}
\langle T, \varphi \rangle &= \Big(\int_{\mathbb{R}^d} \varphi(y)\,dy\Big)\langle T, \Theta \rangle + \sum_{i=1}^{d}\Big\langle T, \frac{\partial \Phi_i}{\partial x_i}\Big\rangle \\
&= \Big(\int_{\mathbb{R}^d} \varphi(y)\,dy\Big)\langle T, \Theta \rangle - \sum_{i=1}^{d}\Big\langle \frac{\partial T}{\partial x_i}, \Phi_i\Big\rangle \\
&= \int_{\mathbb{R}^d} c\varphi(y)\,dy,
\end{aligned}$$

where we have set $c = \langle T, \Theta \rangle$. This shows that T is identified with the $L^1_{\text{loc}}(\mathbb{R}^d)$ function $y \mapsto c$, which happens to be a constant function. □

This proposition states that distributions behave the same as functions when their gradient vanishes. There is nothing exotic added in this respect when generalizing from functions to distributions. In particular, such a T is a function in the sense of Proposition 3.6 which is equal to the constant c almost everywhere.

3.7 Sobolev Spaces

In this section, Ω is an arbitrary open subset of $\mathbb{R}^d$, unless otherwise specified. We now introduce and briefly study an important class of function spaces for PDEs, the Sobolev spaces. As we have seen, every function in $L^p(\Omega)$ is actually a distribution, therefore it has distributional partial derivatives. In general, these derivatives are not functions, of course. There are however some functions whose distributional derivative also are functions, even though they may not be differentiable in the classical sense. These are the functions we are going to be interested in.

Definition 3.5 Let $m \in \mathbb{N}$ and $p \in [1, +\infty]$. We define the Sobolev space

$$W^{m,p}(\Omega) = \{u \in L^p(\Omega); \partial^\alpha u \in L^p(\Omega) \quad \text{for all} \quad \alpha \quad \text{such that} \quad |\alpha| \le m\}.$$

When $p = 2$, we use the notation $W^{m,2}(\Omega) = H^m(\Omega)$.

Note the special case $m = 0$, where $W^{0,p}(\Omega) = L^p(\Omega)$ and $H^0(\Omega) = L^2(\Omega)$. So the notation is hardly ever used for $m = 0$. In this book, we will mainly use the $H^m(\Omega)$ spaces, with special emphasis on $H^1(\Omega)$. The natural Sobolev norms are as follows

$$\|u\|_{W^{m,p}(\Omega)} = \Big(\sum_{|\alpha| \le m} \|\partial^\alpha u\|^p_{L^p(\Omega)}\Big)^{\frac{1}{p}}$$

for $p < +\infty$ and

$$\|u\|_{W^{m,\infty}(\Omega)} = \max_{|\alpha|\le m} \|\partial^\alpha u\|_{L^\infty(\Omega)}.$$

In particular, for $p = 2$, we have

$$\|u\|_{H^m(\Omega)} = \Big(\sum_{|\alpha|\le m} \|\partial^\alpha u\|^2_{L^2(\Omega)}\Big)^{\frac{1}{2}}.$$

This latter norm is clearly a prehilbertian norm associated with the scalar product

$$(u|v)_{H^m(\Omega)} = \sum_{|\alpha|\le m} (\partial^\alpha u|\partial^\alpha u)_{L^2(\Omega)}.$$

The notations $\|u\|_{m,p}$ and $\|u\|_m$ for the $W^{m,p}$ and H^m norms are also encountered in the literature if the context is clear.

Remark 3.9 It follows from the definition that $W^{m+1,p}(\Omega) \subset W^{m,p}(\Omega)$ for all m, p. Moreover, if Ω is bounded $W^{m,p}(\Omega) \subset W^{m,q}(\Omega)$ whenever $q \le p$. Also if Ω is bounded, we have $C^m(\bar{\Omega}) \subset W^{m,p}(\Omega)$. If Ω is Lipschitz, we have in addition that $C^m(\bar{\Omega})$ is dense in $W^{m,p}(\Omega)$ (we will prove it later on for $m = 1$, $p = 2$). We also refer for example to [2, 25, 35, 40, 61] for other density results in $W^{m,p}(\Omega)$. Of course, there are functions in $W^{m,p}(\Omega)$ that are not of class C^m. For example, the function $x \mapsto x_+ = \max(x, 0)$ is in $H^1(]-1, 1[)$ since $(x_+)' = H_{|]-1,1[}$ in the sense of $\mathscr{D}'$ (exercise), but it is not differentiable in the classical sense at $x = 0$.

Similarly, there are functions in L^p that are not in $W^{1,p}$, such as the Heaviside function H whose derivative is δ, which is not a function. □

Theorem 3.5 *The spaces $W^{m,p}(\Omega)$ are Banach spaces. In particular, the spaces $H^m(\Omega)$ are Hilbert spaces.*

Proof We need to show that $W^{m,p}(\Omega)$ is complete for its norm. Let us thus be given a Cauchy sequence $(u_n)_{n\in\mathbb{N}}$ in $W^{m,p}(\Omega)$. In view of the definition of the norm, it follows that for each multiindex α, $|\alpha| \le m$, the sequence of partial derivatives $\partial^\alpha u_n$ is a Cauchy sequence in $L^p(\Omega)$. We know that $L^p(\Omega)$ is complete, therefore there exists $g_\alpha \in L^p(\Omega)$ such that $\partial^\alpha u_n \to g_\alpha$ in $L^p(\Omega)$. By Proposition 3.8, it follows that $\partial^\alpha u_n \to g_\alpha$ in the sense of $\mathscr{D}'(\Omega)$. Now by Proposition 3.10, we also know that $\partial^\alpha u_n \to \partial^\alpha u$ in the sense of $\mathscr{D}'(\Omega)$, where $u = g_{(0,0,\dots,0)}$ is the limit of the sequence in $L^p(\Omega)$. Therefore $\partial^\alpha u = g_\alpha \in L^p(\Omega)$ since the space of distributions is separated and thus a converging sequence can only have one limit. This shows that u belongs to $W^{m,p}(\Omega)$ on the one hand, and that $u_n \to u$ in $W^{m,p}(\Omega)$ since

$$\|u_n - u\|^p_{W^{m,p}(\Omega)} = \sum_{|\alpha|\le m} \|\partial^\alpha u_n - \partial^\alpha u\|^p_{L^p(\Omega)} = \sum_{|\alpha|\le m} \|\partial^\alpha u_n - g_\alpha\|^p_{L^p(\Omega)} \to 0,$$

for $p < +\infty$ and the same for $p = +\infty$. Therefore $W^{m,p}(\Omega)$ is complete, and so is the proof. □

From now on, we will mostly consider the case $p = 2$. Let us introduce an important subset of $H^m(\Omega)$.

Definition 3.6 The closure of $\mathscr{D}(\Omega)$ in $H^m(\Omega)$ is denoted $H_0^m(\Omega)$.

In other words, $H_0^m(\Omega)$ consists exactly of those functions u of $H^m(\Omega)$ which can be approximated in the sense of $H^m(\Omega)$ by indefinitely differentiable functions with compact support, i.e., such that there exists a sequence $\varphi_n \in \mathscr{D}(\Omega)$ with $\|\varphi_n - u\|_{H^m(\Omega)} \to 0$. By definition, it is a closed vector subspace of $H^m(\Omega)$ and thus a Hilbert space for the scalar product of $H^m(\Omega)$.

The following is a very important result. We introduce the semi-norm

$$|u|_{H^m(\Omega)} = \Big(\sum_{|\alpha|=m} \|\partial^\alpha u\|^2_{L^2(\Omega)}\Big)^{\frac{1}{2}}.$$

This semi-norm just retains the partial derivatives of the highest order compared with the norm.

Theorem 3.6 (Poincaré's inequality) *Let Ω be a bounded open subset of $\mathbb{R}^d$. There exists a constant C which only depends on Ω such that for all $u \in H_0^1(\Omega)$,*

$$\|u\|_{L^2(\Omega)} \le C|u|_{H^1(\Omega)}.$$

Proof Since Ω is assumed to be bounded, it is included in a strip[4] that we may assume to be of the form

$$\Omega \subset S_{a,b} = \{(x', x_d); x' \in \mathbb{R}^{d-1}, a < x_d < b\}$$

for some a and b, without loss of generality, see Fig. 3.17.

We argue by density. First let $\varphi \in \mathscr{D}(\Omega)$. We extend it by 0 to the whole of $\mathbb{R}^d$ and still call the extension φ. Let $\alpha_d = (0, 0, \dots, 0, 1)$ so that $\partial^{\alpha_d}\varphi = \frac{\partial \varphi}{\partial x_d}$. Since $\varphi(x', a) = 0$ for all $x' \in \mathbb{R}^{d-1}$ and φ is C^1 with respect to x_d, we can write

$$\varphi(x', x_d) = \int_a^{x_d} \partial^{\alpha_d}\varphi(x', s)\,ds$$

for all (x', x_d). In particular, for $a \le x_d \le b$, we obtain

$$\varphi(x', x_d)^2 \le (x_d - a)\int_a^{x_d} \big(\partial^{\alpha_d}\varphi(x', s)\big)^2\,ds \le (b-a)\int_a^{b} \big(\partial^{\alpha_d}\varphi(x', s)\big)^2\,ds$$

by the Cauchy–Schwarz inequality. We integrate the above inequality with respect to x'

---

[4]It is enough for Poincaré's inequality to be valid that Ω be included in such a strip although not necessarily bounded.

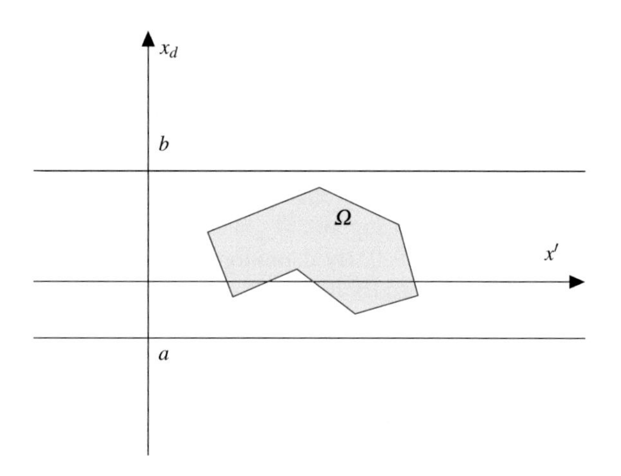

Fig. 3.17 The open set Ω included in a strip

$$\int_{\mathbb{R}^{d-1}} \varphi(x', x_d)^2 \, dx' \leq (b-a) \int_{S_{a,b}} \left(\partial^{\alpha_d} \varphi(x)\right)^2 dx$$

by Fubini's theorem. Now, because the support of φ is included in $\Omega \subset S_{a,b}$, it follows that

$$\int_{S_{a,b}} \left(\partial^{\alpha_d} \varphi(x)\right)^2 dx = \|\partial^{\alpha_d} \varphi\|^2_{L^2(\Omega)}.$$

We integrate again with respect to x_d between a and b and obtain

$$\|\varphi\|^2_{L^2(\Omega)} \leq (b-a)^2 \|\partial^{\alpha_d} \varphi\|^2_{L^2(\Omega)},$$

for the same reasons (Fubini and support of φ). Now by definition of the semi-norm, it follows that

$$\|\partial^{\alpha_d} \varphi\|^2_{L^2(\Omega)} \leq \sum_{|\alpha|=1} \|\partial^{\alpha} \varphi\|^2_{L^2(\Omega)} = |\varphi|^2_{H^1(\Omega)},$$

hence Poincaré's inequality for a function $\varphi \in \mathscr{D}(\Omega)$ with constant $C = (b-a)$.

We complete the proof by a density argument. Let $u \in H^1_0(\Omega)$. By definition of $H^1_0(\Omega)$ as the closure of $\mathscr{D}(\Omega)$ in $H^1(\Omega)$, there exists a sequence $\varphi_n \in \mathscr{D}(\Omega)$ such that $\varphi_n \to u$ in $H^1(\Omega)$. Inspection of the definition of the H^1 norm reveals that this is equivalent to $\varphi_n \to u$ in $L^2(\Omega)$ and $\partial^{\alpha}\varphi_n \to \partial^{\alpha} u$ for all $|\alpha| = 1$ also in $L^2(\Omega)$. Since all the L^2 norms then converge, we obtain in the limit

$$\|u\|_{L^2(\Omega)} \leq (b-a)|u|_{H^1(\Omega)},$$

which is Poincaré's inequality on $H^1_0(\Omega)$. □

Remark 3.10 Poincaré's inequality shows that $H^1_0(\Omega)$ is a strict subspace of $H^1(\Omega)$ when Ω is bounded. Indeed, the constant function $u = 1$ is in $H^1(\Omega)$ but does not satisfy the inequality, since all its partial derivatives vanish. It follows that it is impossible to approximate a non zero constant by a sequence in $\mathscr{D}(\Omega)$ in the norm of $H^1(\Omega)$. □

From now on, we will use the gradient notation ∇u to denote the vector of all first order distributional partial derivatives of u. We already used the simplified notation $\partial_i = \frac{\partial}{\partial x_i}$ for individual first order derivatives instead of the multiindex notation. Similarly, we note $\partial_{ij} = \frac{\partial^2}{\partial x_i \partial x_j}$ for second order derivatives. When $u \in H^1(\Omega)$, then we have $\nabla u \in L^2(\Omega; \mathbb{R}^d)$. Poincaré's inequality has an important corollary.

Corollary 3.3 *Let Ω be a bounded subset of $\mathbb{R}^d$. The H^1 semi-norm $|\cdot|_{H^1(\Omega)}$ is a norm on $H_0^1(\Omega)$ that is equivalent to the H^1 norm. It is also a Hilbertian norm associated with the scalar product*

$$(u|v)_{H_0^1(\Omega)} = \int_\Omega \nabla u \cdot \nabla v \, dx.$$

Proof First of all, it is clear that $|u|_{H^1(\Omega)} \le \|u\|_{H^1(\Omega)}$ for all $u \in H^1(\Omega)$, hence all $u \in H_0^1(\Omega)$, since the norm squared is the semi-norm squared plus the L^2 norm squared.

The converse inequality follows from Poincaré's inequality. Indeed, for $u \in H_0^1(\Omega)$, we have $\|u\|_{L^2(\Omega)} \le C|u|_{H^1(\Omega)}$. Therefore

$$\|u\|_{H^1(\Omega)} = (\|u\|^2_{L^2(\Omega)} + |u|^2_{H^1(\Omega)})^{\frac{1}{2}} \le (C^2+1)^{\frac{1}{2}} |u|_{H^1(\Omega)}.$$

This shows both that the semi-norm is a norm on $H_0^1(\Omega)$ and that it is equivalent to the full H^1 norm on $H_0^1(\Omega)$. This also shows that the bilinear form above is positive definite, hence a scalar product. □

Remark 3.11 The fact that the two norms are equivalent implies that $H_0^1(\Omega)$ is also complete for the semi-norm. Hence, it is also a Hilbert space for the scalar product corresponding to the semi-norm. Beware however that this is a different Hilbert structure from the one obtained by restricting the H^1 scalar product to H_0^1. Indeed, we now have two different notions of orthogonality, and (at least) two different ways of identifying the dual of H_0^1, see Theorem 3.3. □

Remark 3.12 The above results generalize to $H_0^m(\Omega)$ on which the semi-norm $|\cdot|_{H^m(\Omega)}$ is equivalent to the full H^m norm. They also generalize to the spaces $W_0^{m,p}(\Omega)$ defined in a obvious way. □

Let us give yet another way of identifying the dual of $H_0^1(\Omega)$ as a subspace of the space of distributions, see [2, 25, 35].

Definition 3.7 Let

$$H^{-1}(\Omega) = \{T \in \mathscr{D}'(\Omega); \exists C, \forall \varphi \in \mathscr{D}(\Omega), |\langle T, \varphi\rangle| \le C|\varphi|_{H^1(\Omega)}\}, \tag{3.10}$$

equipped with the norm

$$\|T\|_{H^{-1}(\Omega)} = \inf\{C \text{ appearing in formula (3.10)}\} = \sup_{\substack{\varphi\in\mathscr{D}(\Omega)\\ \varphi\neq 0}} \frac{|\langle T,\varphi\rangle|}{|\varphi|_{H^1(\Omega)}}.$$

Then $H^{-1}(\Omega)$ is isometrically isomorphic to $(H_0^1(\Omega))'$.

Proof Since $\mathscr{D}(\Omega) \subset H_0^1(\Omega)$ by definition, any linear form ℓ on $H_0^1(\Omega)$ defines a linear form on $\mathscr{D}(\Omega)$ by restriction. Moreover, if $\varphi_n \to 0$ in $\mathscr{D}(\Omega)$, we obviously have $\varphi_n \to 0$ in $H_0^1(\Omega)$ as well. Hence, if ℓ is continuous, that is $\ell \in (H_0^1(\Omega))'$, its restriction to $\mathscr{D}(\Omega)$ is a distribution $T \in \mathscr{D}'(\Omega)$. This distribution clearly belongs to $H^{-1}(\Omega)$.

Conversely, let us be given an element T of $H^{-1}(\Omega)$. By definition, it is a linear form defined on a dense subspace of $H_0^1(\Omega)$ and continuous with respect to the H_0^1-norm. Therefore, it extends to an element ℓ of the dual space $(H_0^1(\Omega))'$, with the same norm. □

Remark 3.13 The identification of the dual of $H_0^1(\Omega)$ with $H^{-1}(\Omega)$ is the one that leads to the pivot space inclusions of Remark 3.1, $V \hookrightarrow H \hookrightarrow V'$, in the case of $H = L^2(\Omega)$ and $V = H_0^1(\Omega)$. Indeed, the scalar product used in the identification of the pivot space with its dual is equal to the duality bracket of an L^2 function seen as a distribution and a $\mathscr{D}$ test-function, when the second argument in the scalar product is such a test-function, i.e., if $\varphi \in \mathscr{D}(\Omega)$ and $f \in L^2(\Omega)$, we have

$$\langle f,\varphi\rangle = \int_\Omega f\varphi\,dx.$$

The two scalar products that $H_0^1(\Omega)$ comes equipped with do not have this property. Consequently, an identification of $H_0^1(\Omega)$ with its dual using either one of the latter scalar products, even though it is legitimate, does not use the same duality as the one used to identify a function with a distribution. □

In order to explain the -1 exponent in the notation, we note the following.

Proposition 3.13 *Let $f \in L^2(\Omega)$, then $\partial_i f \in H^{-1}(\Omega)$ and $\langle \partial_i f,\varphi\rangle = -\int_\Omega f\partial_i\varphi\,dx$ for all $\varphi \in \mathscr{D}(\Omega)$.*

Proof By definition of distributional derivatives,

$$\langle \partial_i f,\varphi\rangle = -\langle f,\partial_i\varphi\rangle = -\int_\Omega f\partial_i\varphi\,dx,$$

since f is locally integrable. Thus

$$|\langle \partial_i f,\varphi\rangle| \le \|f\|_{L^2(\Omega)}\|\partial_i\varphi\|_{L^2(\Omega)} \le \|f\|_{L^2(\Omega)}|\varphi|_{H_0^1(\Omega)},$$

by the Cauchy–Schwarz inequality, hence $\partial_i f \in H^{-1}(\Omega)$ with $\|\partial_i f\|_{H^{-1}(\Omega)} \le \|f\|_{L^2(\Omega)}$. □

Remark 3.14 This shows that the operator ∂_i is linear continuous from $L^2(\Omega)$ $(= H^0(\Omega))$ into $H^{-1}(\Omega)$, just as it is linear continuous from $H^1(\Omega)$ into $L^2(\Omega)$. Each time, the exponent in the notation gets decremented by 1 as one derivative is lost. □

Remark 3.15 When Ω is regular, a distribution in $H^{-1}(\Omega)$ whose first order partial derivatives are all in $H^{-1}(\Omega)$ is in fact a function in $L^2(\Omega)$. The latter result is known as Lions's lemma. □

The above bracket formula is also valid for all $v \in H_0^1(\Omega)$, in the sense that

$$\langle \partial_i f, v\rangle_{H^{-1}(\Omega),H_0^1(\Omega)} = \int_\Omega f \partial_i v\, dx,$$

by density. In the same vein, we have

Corollary 3.4 *The operator $-\Delta$ is linear continuous from $H_0^1(\Omega)$ into $H^{-1}(\Omega)$ and for all $u, v \in H_0^1(\Omega)$, we have*

$$\langle -\Delta u, v\rangle_{H^{-1}(\Omega),H_0^1(\Omega)} = \int_\Omega \nabla u \cdot \nabla v\, dx.$$

The dual of $H_0^m(\Omega)$ is likewise identified with a subspace $H^{-m}(\Omega)$ of $\mathscr{D}'(\Omega)$.

3.8 Properties of Sobolev Spaces in One Dimension

The one-dimensional case is simple and useful to get acquainted with the properties of Sobolev spaces in general. For simplicity, we mostly consider $H^1(\Omega)$ where $\Omega =]a, b[$ is a bounded open interval of $\mathbb{R}$. Let us admit a density result that we will prove later in arbitrary dimension.

Proposition 3.14 *The space $C^1([a, b])$ is dense in $H^1(]a, b[)$.*

The density above is meant in the sense that the equivalence classes of elements of $C^1([a, b])$ are dense in $H^1(]a, b[)$. We have already seen examples of functions in dimension one that are H^1 but not C^1. All one-dimensional H^1 functions however are continuous, in the sense that each equivalence class contains one continuous representative. There is even a more precise embedding.

Theorem 3.7 *We have that $H^1(]a, b[) \hookrightarrow C^{0,1/2}([a, b])$.*

Recall that the hooked arrow means that there is an injection between the two spaces and that this injection is continuous.

Proof Let us be given $v \in H^1(]a, b[)$. The distributional derivative v' is in $L^2(a, b)$ hence is integrable on $[a, b]$. For all $x \in [a, b]$, we thus define

$$w(x) = \int_a^x v'(t)\,dt.$$

For all $x, y \in [a, b]$, we can consequently write (with the convention $\int_y^x g\,dt = -\int_x^y g\,dt$)

$$w(y) - w(x) = \int_x^y v'(t)\,dt.$$

Squaring this relation, we see that

$$(w(y) - w(x))^2 = \Big(\int_x^y v'(t)\,dt\Big)^2 \leq |y - x| \int_a^b (v'(t))^2\,dt \leq |y - x| |v|^2_{H^1(]a,b[)},$$

by the Cauchy–Schwarz inequality. Therefore, for all $x \neq y$, we obtain

$$\frac{|w(y) - w(x)|}{|y - x|^{1/2}} \leq |v|_{H^1(]a,b[)}. \tag{3.11}$$

It follows from this that w is Hölder continuous of exponent $\frac{1}{2}$ on $[a, b]$. Therefore, w is a distribution on $]a, b[$. Let us compute its derivative. For all $\varphi \in \mathscr{D}(]a, b[)$, we have

$$\begin{aligned}
\langle w', \varphi \rangle = -\langle w, \varphi' \rangle &= -\int_a^b w(x)\varphi'(x)\,dx \\
&= -\int_a^b \Big(\int_a^x v'(t)\,dt\Big)\varphi'(x)\,dx = -\int_a^b \Big(\int_t^b \varphi'(x)\,dx\Big)v'(t)\,dt \\
&= -\int_a^b (\varphi(b) - \varphi(t))v'(t)\,dt = \int_a^b \varphi(t)v'(t)\,dt = \langle v', \varphi \rangle,
\end{aligned}$$

since $\varphi(b) = 0$. The integral interchange is justified by Fubini's theorem. Therefore, we have shown that $w' = v'$, from which it follows that there exists a constant c such that $v = w + c$ by Proposition 3.12. Consequently, v has a representative that belongs to $C^{0,1/2}([a, b])$, namely $w + c$, and we can write

$$v(x) = v(y) + \int_y^x v'(t)\,dt, \tag{3.12}$$

for all[5] x, y in $[a, b]$ by using the definition of w. Squaring this relation and using the Cauchy–Schwarz inequality again, we obtain that

$$v(x)^2 \leq 2v(y)^2 + 2(b-a) \int_a^b v'(t)^2\,dt,$$

which we integrate with respect to y to obtain

$$(b-a)v(x)^2 \leq 2\|v\|^2_{L^2(a,b)} + 2(b-a)^2\|v'\|^2_{L^2(a,b)} \leq 2\max(1, (b-a)^2)\|v\|^2_{H^1(]a,b[)}.$$

As this holds true for all x in $[a, b]$, it follows that

$$\|v\|_{C^0([a,b])} \leq \sqrt{2\max\Big(\frac{1}{b-a}, b-a\Big)}\|v\|_{H^1(]a,b[)}. \tag{3.13}$$

Putting estimates (3.11) and (3.13) together, we obtain the announced continuous embedding. □

That all H^1 functions are continuous is specific to dimension one, as we will see later.

Note also that not all $C^{0,1/2}$ functions belong to H^1 (consider $x \mapsto \sqrt{x}$ on $]0, 1[$). The injection above is nonetheless optimal since for each $\beta > 1/2$, there is an H^1 function that is not $C^{0,\beta}$ (consider $x \mapsto x^{\frac{2\beta+1}{4}}$ on $]0, 1[$).

An important feature of the one-dimensional case is that pointwise values of a H^1 function are unambiguously defined as the pointwise value of its continuous representative. Moreover, such pointwise values depend continuously on the function in the H^1 norm by estimate (3.13). This is in particular true of the endpoint values at a and b, which can be surprising because the Sobolev space definition is based on the open set $]a, b[$, extremities excluded.

Corollary 3.5 *The linear mapping $H^1(]a, b[) \to \mathbb{R}^2$, $u \mapsto (u(a), u(b))$ is continuous.*

Proof Obviously $\max(|u(a)|, |u(b)|) \leq \|u\|_{C^{0,1/2}([a,b])} \leq C\|u\|_{H^1(]a,b[)}$. □

Remark 3.16 This is the one-dimensional version of the *trace theorem* that we will prove in all dimensions later on. The linear mapping in question is called the *trace mapping*. The result also shows that Dirichlet boundary conditions make sense for functions of $H^1(]a, b[)$, a fact that was not evident from the start.

Because of the continuity of the trace, it is clear that $H^1_0(]a, b[)$ is included in the kernel of the trace $\{u \in H^1(]a, b[); u(a) = u(b) = 0\}$. It suffices to take a sequence φ_n of $\mathscr{D}(]a, b[)$ that tends to u in $H^1(]a, b[)$. Actually, the reverse inclusion holds true so that

---

[5] And not only almost everywhere, since we are now talking about the continuous representative of v.

$$H_0^1(]a,b[) = \{u \in H^1(]a,b[);\, u(a) = u(b) = 0\},$$

see Proposition 3.16 in any dimension. The space $H_0^1(]a,b[)$ is thus adequate for homogeneous Dirichlet conditions for second order boundary value problems. □

Remark 3.17 We also have $H^m(]a,b[) \hookrightarrow C^{m-1,1/2}([a,b])$, the trace on $H^m(]a,b[)$ is $u \mapsto (u(a), u'(a), \dots, u^{(m-1)}(a), u(b), u'(b), \dots, u^{(m-1)}(b))$ and $H_0^m(]a,b[)$ is the set of u such that $u(a) = u'(a) = \cdots = u^{(m-1)}(a) = u(b) = u'(b) = \cdots = u^{(m-1)}(b) = 0$. Similar results can be written for the $W^{m,p}(]a,b[)$ spaces, not with the same Hölder exponent though (exercise). □

To conclude the one-dimensional case, let us mention the Rellich compact embedding theorem.

Theorem 3.8 *The injection $H^1(]a,b[) \to L^2(a,b)$, $u \mapsto u$ is compact.*

Proof A mapping is compact if it transforms bounded sets into relatively compact sets. Here, it is enough to take the unit ball of $H^1(]a,b[)$ by linearity. By estimates (3.11) and (3.13), this is a bounded subset of $C^{0,1/2}([a,b])$. Bounded sets of $C^{0,1/2}([a,b])$ are equicontinuous, therefore relatively compact in $C^0([a,b])$ by Ascoli's theorem [7, 32, 33, 69]. Finally the embedding $C^0([a,b]) \hookrightarrow L^2(a,b)$ is continuous, thus transforms relatively compact sets into relatively compact sets. □

Remark 3.18 The Rellich theorem is true in arbitrary dimension d, i.e., the embedding $H^1(\Omega) \to L^2(\Omega)$ is compact, provided that Ω is bounded and sufficiently regular, for example Lipschitz, see [15, 35, 66]. □

3.9 Density of Smooth Functions and Trace in Dimension d

We have seen that Sobolev functions in dimension one are continuous. This is no longer true in dimensions 2 and higher. We will concentrate on the space $H^1(\Omega)$. Note however that functions in $W^{1,p}(\Omega)$, where Ω is an open subset of $\mathbb{R}^d$, are continuous for $p > d$, this is known as Morrey's theorem, [15, 35]. See also [2, 58] for functions in $W^{m,p}$ for $mp > d$.

Let D be the unit disk in $\mathbb{R}^2$. It can be checked (exercise) that the function $u : x \mapsto \ln(|\ln(\|x\|/e)|)$ is in $H_0^1(D)$. This function tends to $+\infty$ at the origin, thus there is no continuous function in its equivalence class, see Fig. 3.18.

Now we can do much worse! We extend u by 0 to $\mathbb{R}^2$, which still is a function in $H^1(\mathbb{R}^2)$. Next, let $(x_i)_{i\in\mathbb{N}}$ be countable, dense set of points in $\mathbb{R}^2$. Then the function $v(x) = \sum_{i=0}^{+\infty} 2^{-i} u(x - x_i)$ is in $H^1(\mathbb{R}^2)$, since $\|u(\cdot - x_i)\|_{H^1(\mathbb{R}^2)} = \|u\|_{H^1(\mathbb{R}^2)}$ and we have a normally convergent series, but this function tends to $+\infty$ at all points x_i, which are dense. Therefore, v is not locally bounded: there is no open set on which it is bounded. This sounds pretty bad, even though it is a perfectly legitimate, although hard to mentally picture, function of $H^1(\mathbb{R}^2)$, see Fig. 3.19.

Fig. 3.18 A discontinuous H^1-function

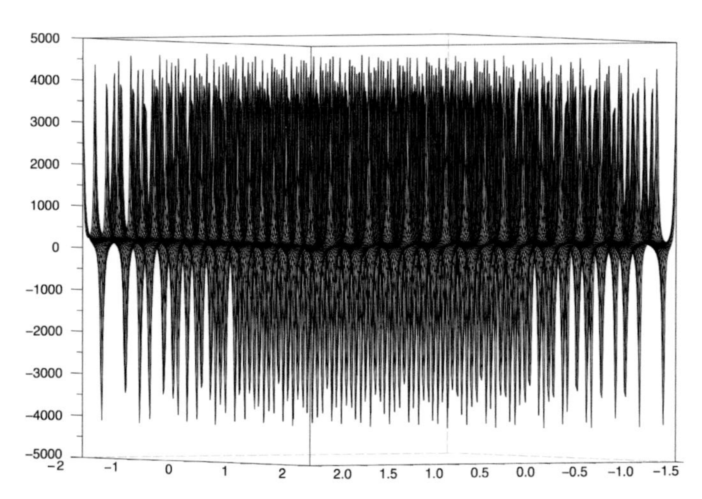

Fig. 3.19 An attempt to draw a very bad H^1-function (graphics cheat: the spikes should be thinner, (infinitely) higher and (infinitely) denser)

In higher dimensions, we can picture such singularities occurring on a dense set of curves or submanifolds of dimension $d-2$. In view of this state of things, ascribing some kind of boundary value to a H^1 function that would be a reasonably defined continuous extension from the values taken in Ω seems difficult. In PDE problems, we nonetheless need boundary values, to write Dirichlet conditions for example.

The definition of a good boundary value for H^1 functions is by means of a mapping called the *trace mapping*. This mapping is defined by density of smooth functions, so let us deal with that first. Besides, as should already be quite clear, density arguments are very useful in Sobolev spaces [2, 25, 35, 40, 61].

Theorem 3.9 *Let Ω be a Lipschitz open subset of $\mathbb{R}^d$. Then the space $C^1(\bar{\Omega})$ is dense in $H^1(\Omega)$.*

Proof We use the same partition of unity as before. It suffices to construct a $C^1(\bar{\Omega})$ approximation for each part $u_j = \psi_j u$ of u. Indeed, we have $u_j \in H^1(\Omega)$ for all j by Proposition 3.11.

We start with the case $j = 0$. Since u_0 is compactly supported in Ω, its extension by 0 to the whole of $\mathbb{R}^d$ belongs to $H^1(\mathbb{R}^d)$ as is easily checked. Let us take a mollifier ρ, that is to say a C^∞ function with compact support in the unit ball B and such that $\int_B \rho(y)\,dy = 1$ as in Sect. 3.4.

For all integers $n \geq 1$, we set $\rho_n(y) = n^d \rho(ny)$ and $u_{0,n} = \rho_n \star u_0$, where the star denotes the convolution as usual. By the general properties of convolution, $u_{0,n}$ is compactly supported in Ω for n sufficiently large, and we have $u_{0,n} \in C^\infty(\mathbb{R}^d) \cap L^2(\mathbb{R}^d)$ and $u_{0,n} \to u_0$ in $L^2(\mathbb{R}^d)$ when $n \to +\infty$. Moreover, since $\partial_i u_{0,n} = \rho_n \star \partial_i u_0$, the same argument shows that $\partial_i u_{0,n} \to \partial_i u_0$ in $L^2(\mathbb{R}^d)$, whence $u_{0,n} \to u_0$ in $H^1(\mathbb{R}^d)$ when $n \to +\infty$. This settles the case $j = 0$ because the restriction of $u_{0,n}$ to $\bar{\Omega}$ is of class C^1 (in fact, it is even compactly supported as soon as n is large enough) and the H^1 norm on Ω is smaller than the H^1 norm on $\mathbb{R}^d$.

The regularity of Ω comes into play for $j > 0$, in the hypercubes C_j that cover the boundary. We drop again all subscripts or superscripts j for brevity. The difficulty compared with $j = 0$ is that we cannot extend u by 0 to $\mathbb{R}^d$ and remain in $H^1(\mathbb{R}^d)$. For example, it is easy to see that the function equal to 1 in Ω and 0 outside is not in $H^1(\mathbb{R}^d)$. We will use a two step process, first a translation, then a convolution.

Let $n \in \mathbb{N}^*$. We set $u_n(y) = u(y', y_d - 1/n)$, which is a function defined on the translated set $\Omega_n = \{y \in \mathbb{R}^d; (y', y_d - 1/n) \in \Omega \cap C\}$, see Fig. 3.20. We extend u_n by 0 to $\mathbb{R}^d$ and let $\widetilde{u}_n$ denote this extension. Since u is compactly supported in C, and the translation shifts it upwards, the restriction of $\widetilde{u}_n$ to $\Omega \cap C$ is still in $H^1(\Omega \cap C)$ for n large enough.

It can be shown[6] that the translation is continuous on $L^2(\mathbb{R}^d)$ in the sense that $\widetilde{u}_n \to \widetilde{u}$ in $L^2(\mathbb{R}^d)$ when $n \to +\infty$, thus by restriction we have $\widetilde{u}_{n|\Omega\cap C} \to u_{|\Omega\cap C}$ in $L^2(\Omega \cap C)$. Computing the partial derivatives in the sense of distributions shows that $\partial_i(\widetilde{u}_{n|\Omega\cap C}) = (\partial_i u)_{n|\Omega\cap C}$ using the same notation for the translation. Therefore, we have the same convergence for the partial derivatives, which shows that $\widetilde{u}_{n|\Omega\cap C} \to$

[6]The fairly easy proof uses the density of continuous, compactly supported functions in $L^2(\mathbb{R}^d)$.

$u_{|\Omega\cap C}$ in $H^1(\Omega\cap C)$ when $n\to+\infty$. To conclude, we just need to approximate $\widetilde{u}_{n|\Omega\cap C}$ for any given n by a $C^1(\bar{\Omega})$ function, and use a double limit argument.

We now use the convolution by a mollifier again and set $\widetilde{u}_{n,p}=\widetilde{u}_n\star\rho_p$. By construction, $\widetilde{u}_{n,p}$ is of class C^∞ on $\mathbb{R}^d$ and $\widetilde{u}_{n,p}\to\widetilde{u}_n$ in $L^2(\mathbb{R}^d)$ when $p\to+\infty$. Now for the subtle point. We do not have L^2 convergence of the gradients, because in general $\partial_i\widetilde{u}_n$ is not a function, let alone in $L^2(\mathbb{R}^d)$. Take for example $\varphi=0$ and $u=1$, then $\partial_d\widetilde{u}_n$ is a Dirac mass on the hyperplane $y_d=\frac{1}{n}$, *cf.* the one-dimensional case. However, since we have shifted the discontinuity outside of Ω by the translation, there is hope that the restrictions to Ω still converge.

To see that this is the case, we let $\widetilde{\partial_i u_n}$ denote the extension of $\partial_i u_n$ to $\mathbb{R}^d$ by 0. We have $\widetilde{\partial_i u_n}\in L^2(\mathbb{R}^d)$ and $\widetilde{\partial_i u_n}\star\rho_p\to\widetilde{\partial_i u_n}$ in $L^2(\mathbb{R}^d)$ when $p\to+\infty$ by the properties of convolution again. Of course, as already noted, $\widetilde{\partial_i u_n}\neq\partial_i\widetilde{u}_n$ so that $\widetilde{\partial_i u_n}\star\rho_p\neq\partial_i\widetilde{u}_{n,p}$. We will show that, for p large enough, we nonetheless have $(\widetilde{\partial_i u_n}\star\rho_p)_{|\Omega\cap C}=(\partial_i\widetilde{u}_{n,p})_{|\Omega\cap C}$. As we have just seen that $\widetilde{\partial_i u_n}\star\rho_p$ converges in $L^2(\mathbb{R}^d)$, this will lead to the conclusion that $(\partial_i\widetilde{u}_{n,p})_{|\Omega\cap C}\to\partial_i u_n$ in $L^2(\Omega\cap C)$, hence $\widetilde{u}_{n,p|\Omega\cap C}\to u_{n|\Omega\cap C}$ in $H^1(\Omega\cap C)$ when $p\to+\infty$. Since $\widetilde{u}_{n,p|\overline{\Omega\cap C}}\in C^1(\overline{\Omega\cap C})$, we will have our approximation.

To show that the restrictions are equal, we go back to the convolution formula

$$\widetilde{\partial_i u_n}\star\rho_p(x)=\int_{\mathbb{R}^d}\rho_p(x-y)\widetilde{\partial_i u_n}(y)\,dy=\int_{B(x,1/p)}\rho_p(x-y)\widetilde{\partial_i u_n}(y)\,dy$$

since ρ has support in the unit ball. Now, if for all $x\in\Omega\cap C$, we had $B\big(x,\frac{1}{p}\big)\subset\Omega_n$, then the only values of $\widetilde{\partial_i u_n}$ in the integral would coincide with those of $\partial_i\widetilde{u}_n$, see Fig. 3.20. Hence the equality of the restrictions since $\partial_i\widetilde{u}_{n,p}=\rho_p\star\partial_i\widetilde{u}_n$.

We are thus down to a geometry question, where the regularity of Ω intervenes (at last). We need to estimate the distance between the graph of φ, denoted G, and the same graph translated upwards by $\frac{1}{n}$, denoted G_n. Let L be the Lipschitz constant of φ and take two points $x\in G$ and $y\in G_n$. We have

$$\|y-x\|^2=\|y'-x'\|^2+\Big(\varphi(y')-\varphi(x')+\frac{1}{n}\Big)^2,$$

using the prime notation to denote the projection on $\mathbb{R}^{d-1}$ as usual. Now

$$\varphi(y')-\varphi(x')+\frac{1}{n}\geq\frac{1}{n}-L\|y'-x'\|,$$

therefore if $\|y'-x'\|\leq\frac{1}{2nL}$, then $\|y-x\|\geq\frac{1}{2n}$. On the other hand, if $\|y'-x'\|\geq\frac{1}{2nL}$, it follows trivially that $\|y-x\|\geq\frac{1}{2nL}$. We thus see that

$$\|y-x\|\geq\min\Big(\frac{1}{2n},\frac{1}{2nL}\Big).$$

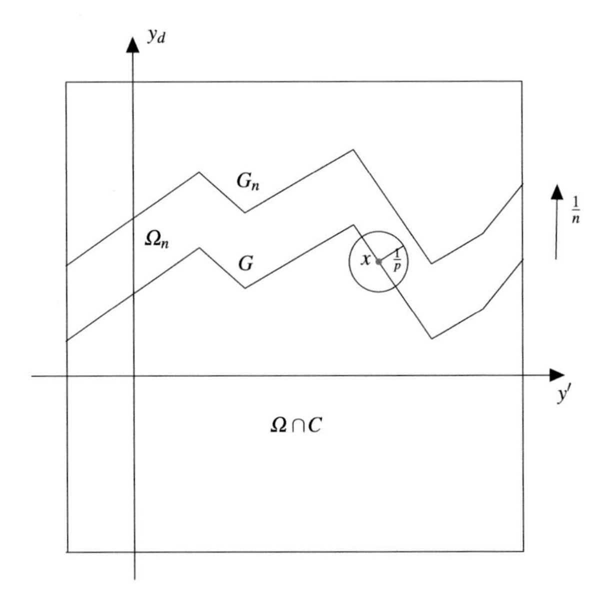

Fig. 3.20 The translated open set Ω_n and the ball of radius $\frac{1}{p}$ used to compute the convolution at point x. For x in $\bar{\Omega} \cap C$, the ball remains included in Ω_n uniformly for $p \to +\infty$, n fixed

If we choose

$$p > \frac{1}{\min\left(\frac{1}{2n}, \frac{1}{2nL}\right)},$$

then $B\left(x, \frac{1}{p}\right) \subset \Omega_n$, hence the final result. □

Remark 3.19 There is a slight cheat in the geometric part of the above proof, in that we have ignored what happens on the lateral sides of C. Indeed, there is actually no problem, since u vanishes there. □

Remark 3.20 Warning: there are open sets less regular than Lipschitz on which not only does the above proof not work, but the density result is false (exercise, find a simple example). It is however always true that $C^1(\Omega) \cap H^1(\Omega)$ is dense in $H^1(\Omega)$, a weaker result (this is the Meyers–Serrin theorem, see [1]) which is sometimes sufficient, although not here for the existence of the trace mapping. □

Once the density of $C^1(\bar{\Omega})$ is established, we can prove the trace theorem.

Theorem 3.10 *Let Ω be a Lipschitz open subset of $\mathbb{R}^d$. There exists a unique continuous linear mapping $\gamma_0 \colon H^1(\Omega) \to L^2(\partial\Omega)$ such that for all $u \in C^1(\bar{\Omega})$, we have*

$$\gamma_0(u) = u_{|\partial\Omega}.$$

In particular, there exists a constant C_{γ_0} such that, for all $u \in H^1(\Omega)$,

$$\|\gamma_0(u)\|_{L^2(\partial\Omega)} \le C_{\gamma_0} \|u\|_{H^1(\Omega)}.$$

In other words, the trace is the unique reasonable way of defining a boundary value for $H^1(\Omega)$ functions, as the continuous extension of the restriction to the boundary for functions for which this restriction makes sense unambiguously, i.e., functions in $C^1(\bar{\Omega})$.

Proof We write the proof only in dimension $d = 2$, but the general case is strictly identical, up to heavier notation.

Let $u \in C^1(\bar{\Omega})$. By partition of unity, we consider $u\psi$ which is supported in one of the $C_j = C$ for $j = 1, \dots, m$. Let $G = \partial\Omega \cap C$ be the part of the boundary included in C. By definition of the boundary measure, we have

$$\begin{aligned}\|u\psi\|^2_{L^2(G)} &= \int_{-a}^{a} (u\psi)(y_1, \varphi(y_1))^2 \sqrt{1+\varphi'(y_1)^2}\,dy_1 \\ &= \int_{-a}^{a}\Big(\int_{-a}^{\varphi(y_1)} \frac{\partial(u\psi)}{\partial y_2}(y_1, y_2)\,dy_2\Big)^2 \sqrt{1+\varphi'(y_1)^2}\,dy_1,\end{aligned}$$

since $\psi(y_1, -a) = 0$. By the Cauchy–Schwarz inequality, we have

$$\Big(\int_{-a}^{\varphi(y_1)} \frac{\partial(u\psi)}{\partial y_2}(y_1, y_2)\,dy_2\Big)^2 \le |\varphi(y_1) + a| \int_{-a}^{\varphi(y_1)} \Big(\frac{\partial(u\psi)}{\partial y_2}(y_1, y_2)\Big)^2\,dy_2,$$

with $|\varphi(y_1) + a| \le 2a$. Let us set $M = \max_{[-a,a]} \sqrt{1+\varphi'(y_1)^2}$. We obtain

$$\|u\psi\|^2_{L^2(G)} \le 2aM \int_{-a}^{a}\int_{-a}^{\varphi(y_1)} \Big(\frac{\partial(u\psi)}{\partial y_2}\Big)^2\,dy_2dy_1 = 2aM \int_{\Omega\cap C} \Big(\frac{\partial(u\psi)}{\partial y_2}\Big)^2\,dx.$$

Now $\frac{\partial(u\psi)}{\partial y_2} = \psi\frac{\partial u}{\partial y_2} + u\frac{\partial \psi}{\partial y_2}$, so that

$$\begin{aligned}\|u\psi\|^2_{L^2(G)} &\le 4aM\Big(\int_{\Omega\cap C} \psi^2\Big(\frac{\partial u}{\partial y_2}\Big)^2\,dx + \int_{\Omega\cap C} u^2\Big(\frac{\partial \psi}{\partial y_2}\Big)^2\,dx\Big) \\ &\le 4aM\Big[\Big\|\frac{\partial u}{\partial y_2}\Big\|^2_{L^2(\Omega\cap C)} + \max_{\Omega\cap C}\Big(\frac{\partial \psi}{\partial y_2}\Big)^2 \|u\|^2_{L^2(\Omega\cap C)}\Big] \\ &\le K^2\|u\|^2_{H^1(\Omega)},\end{aligned}$$

(remembering that ψ is $[0, 1]$-valued) where $K = K_j$ is a constant which depends on j.

We put all the estimates together by partition of unity and the triangle inequality:

$$\|u\|_{L^2(\partial\Omega)} \le \sum_{j=0}^{m} \|u\psi_j\|_{L^2(G_j)} \le (m+1)K\|u\|_{H^1(\Omega)},$$

for all $u \in C^1(\bar{\Omega})$ where $K = \max_{j=1,\dots,m} K_j$. The linear mapping $u \mapsto u_{|\partial\Omega}$ defined on $C^1(\bar{\Omega})$ is thus continuous in the $H^1(\Omega)$ and $L^2(\partial\Omega)$ norms. Since $C^1(\bar{\Omega})$ is dense in $H^1(\Omega)$, the mapping has a unique continuous extension to $H^1(\Omega)$ with values in $L^2(\partial\Omega)$, which is called the trace mapping γ_0. □

Remark 3.21 Note that if $u \in H^1(\Omega) \cap C^0(\bar{\Omega})$ the trace $\gamma_0(u)$ is equal to the restriction of the function to $\partial\Omega$, $u_{|\partial\Omega}$, even if u is not in $C^1(\bar{\Omega})$. □

Remark 3.22 Again, there are open sets less regular than Lipschitz on which no trace mapping can be defined. □

We now are in a position to extend the integration by parts formula(s) to elements of Sobolev spaces.

Theorem 3.11 *Let Ω be a Lipschitz open set and $u, v \in H^1(\Omega)$. Then we have*

$$\int_\Omega \frac{\partial u}{\partial x_i} v\,dx = -\int_\Omega u \frac{\partial v}{\partial x_i}\,dx + \int_{\partial\Omega} \gamma_0(u)\gamma_0(v) n_i\,d\Gamma, \tag{3.14}$$

where n_i is the ith component of the normal unit exterior vector n.

Proof We argue by density. We already know that formula (3.14) holds true on $C^1(\bar{\Omega})$, see Corollary 3.2. Let u_n, v_n be sequences in $C^1(\bar{\Omega})$ such that $u_n \to u$ and $v_n \to v$ in $H^1(\Omega)$ when $n \to +\infty$. This means that $u_n \to u$, $v_n \to v$, $\partial_i u_n \to \partial_i u$ and $\partial_i v_n \to \partial_i u$ in $L^2(\Omega)$. Therefore, $\partial_i u_n v_n \to \partial_i u v$ and $u_n \partial_i v_n \to u \partial_i v$ in $L^1(\Omega)$ and the left-hand side integral and the first integral in the right-hand side pass to the limit. Secondly, we have $\gamma_0(u_n) \to \gamma_0(u)$ and $\gamma_0(v_n) \to \gamma_0(v)$ in $L^2(\partial\Omega)$ since the trace mapping is continuous, hence $\gamma_0(u_n)\gamma_0(v_n) \to \gamma_0(u)\gamma_0(v)$ in $L^1(\partial\Omega)$ and the second integral in the right-hand side also passes to the limit. □

The various corollaries of the integration by parts formula also hold true, provided all the integrals make sense. For instance, for all $u \in H^1(\Omega)$,

$$\int_\Omega \frac{\partial u}{\partial x_i}\,dx = \int_{\partial\Omega} \gamma_0(u) n_i\,d\Gamma,$$

as is seen from taking $v = 1$.

The formulas that entail second derivatives should be applied to H^2 functions. Such functions u are in $H^1(\Omega)$, thus have a trace $\gamma_0(u)$ and they also have a second trace $\gamma_1(u)$, called the normal trace, that plays the role of the normal derivative for a regular function. Indeed, $\partial_i u \in H^1(\Omega)$ therefore $\gamma_1(u) = \sum_{i=1}^d \gamma_0(\partial_i u) n_i$ is well defined and continuous from $H^2(\Omega)$ into $L^2(\partial\Omega)$, because the functions n_i are in $L^\infty(\partial\Omega)$. Furthermore, if $u \in C^2(\bar{\Omega})$, then $\gamma_1(u) = \frac{\partial u}{\partial n}$. We thus establish the following result (Green's formula for Sobolev spaces).

Proposition 3.15 *Let Ω be a Lipschitz open set. For all $u \in H^2(\Omega)$ and all $v \in H^1(\Omega)$, we have*

$$\int_\Omega (\Delta u) v\,dx = -\int_\Omega \nabla u \cdot \nabla v\,dx + \int_{\partial\Omega} \gamma_1(u)\gamma_0(v)\,d\Gamma. \tag{3.15}$$

Proof The proof is by density of $C^2(\bar{\Omega})$ in $H^2(\Omega)$ and of $C^1(\bar{\Omega})$ in $H^1(\Omega)$, starting from formula (3.4). □

Proposition 3.16 *Let Ω be a Lipschitz open set. Then we have*

$$H_0^1(\Omega) = \ker \gamma_0.$$

Proof One inclusion is easy. The space $H_0^1(\Omega)$ is by definition the closure of the space $\mathscr{D}(\Omega)$ in $H^1(\Omega)$. Thus, if $u \in H_0^1(\Omega)$, then there exist $\varphi_n \in \mathscr{D}(\Omega)$ such that $\varphi_n \to u$ in $H^1(\Omega)$. It is clear by definition of the trace mapping that $\gamma_0(\varphi_n) = 0$, thus $u \in \ker \gamma_0$ by continuity of the trace, or in other words $H_0^1(\Omega) \subset \ker \gamma_0$.

We just give the idea for the reverse inclusion [25, 35, 66]. Take a zero trace function u, use a partition of unity adapted to the boundary, extend all the parts to the whole of $\mathbb{R}^d$ by 0 (the integration by parts formula (3.14) shows that the extension remains in $H^1(\mathbb{R}^d)$ this time), translate each function downwards[7] by a small amount in its cube then perform the convolution step which provides a compactly supported, C^∞ approximation. We leave the details to the reader, all the necessary technical elements have already been introduced previously. □

Remark 3.23 By Remark 3.21, if $u \in H^1(\Omega) \cap C^0(\bar{\Omega})$ is such that $u_{|\partial\Omega} = 0$, then $u \in H_0^1(\Omega)$. □

Remark 3.24 Similar arguments show that $H_0^2(\Omega) = \ker \gamma_0 \cap \ker \gamma_1$ and so on. Everything we have said in terms of traces can also naturally be done in the spaces $W^{m,p}(\Omega)$. □

It should be noted that the trace mapping γ_0 is not onto.

Proposition 3.17 *The image space of the trace mapping,* im γ_0, *is a strict, dense subspace of $L^2(\partial\Omega)$. This space is denoted $H^{1/2}(\partial\Omega)$. The norm*

$$\|g\|_{H^{1/2}(\partial\Omega)} = \inf_{\substack{v \in H^1(\Omega) \\ \gamma_0(v)=g}} \|v\|_{H^1(\Omega)}$$

makes $H^{1/2}(\partial\Omega)$ into a Hilbert space.

There are other equivalent norms on $H^{1/2}(\partial\Omega)$. We do not pursue the study of the trace space $H^{1/2}(\partial\Omega)$ here, see for example [25, 58, 61].

3.10 A Summary of Important Results

Let us now give a quick review of the results of this chapter that are essential for the following chapters, i.e., everything that concerns variational formulations and variational approximation methods.

[7] Instead of upwards for the density result.

The elementary properties of Hilbert spaces of Sect. 3.1 are useful to understand the abstract variational problems considered in the next chapter.

The Green's formulas of Corollary 3.2 in the classical case p. 90, and of Proposition 3.15 in the Sobolev case p. 114, are essential to establish the variational formulation of elliptic boundary value problems in more than one dimension of space.

There is no real need for a deep understanding of distribution theory, see [72, 73]. However, the characterization of distributions of Proposition 3.5, p. 92, the definition of distributional derivatives of Definition 3.3, p. 94, and the characterization of the convergence in the sense of distributions of Proposition 3.7, p. 94 are always good to know.

The hilbertian Sobolev spaces $H^1(\Omega)$ and $H_0^1(\Omega)$ introduced in Definitions 3.5, p. 99 and 3.6, p. 101, are also essential as the basic function spaces in which variational problems are set.

We will have to use Poincaré's inequality, Theorem 3.6, p. 101, and its consequences, for example Corollary 3.3, p. 103, for homogeneous Dirichlet problems.

The existence and properties of the trace mapping, Theorem 3.10, p. 112 and Proposition 3.16, p. 115, are also key for what follows.

Chapter 4
The Variational Formulation of Elliptic PDEs

We now begin the theoretical study of elliptic boundary value problems in a context that is more general than the one-dimensional model problem treated in Chap. 1. We will focus on one approach, which is called the variational approach. There are other ways of solving elliptic problems, such as working with Green functions as seen in Chap. 2. The variational approach is quite simple and well suited for a whole class of approximation methods, as we will see later.

4.1 Model Boundary Value Problems

Let us start with a few more model problems. The simplest of all is a slight generalization of the Poisson equation with a homogeneous Dirichlet boundary condition. Let us thus be given an open Lipschitz subset Ω of $\mathbb{R}^d$, a function $c \in L^\infty(\Omega)$ and another function $f \in L^2(\Omega)$. We are looking for a function $u\colon \bar{\Omega} \to \mathbb{R}$ such that

$$\begin{cases} -\Delta u + cu = f \text{ in } \Omega, \\ \qquad\quad u = 0 \text{ on } \partial\Omega. \end{cases} \tag{4.1}$$

We are going to transform the boundary value problem (4.1) into an entirely different kind of problem that is amenable to an existence and uniqueness theory, as well as the definition of approximation methods.

Proposition 4.1 *Assume that $u \in H^2(\Omega)$ solves the PDE in problem (4.1), i.e., the first equation in (4.1). Then, for all $v \in H_0^1(\Omega)$, we have*

$$\int_\Omega \nabla u \cdot \nabla v \, dx + \int_\Omega cuv \, dx = \int_\Omega fv \, dx. \tag{4.2}$$

H. Le Dret and B. Lucquin, *Partial Differential Equations: Modeling, Analysis and Numerical Approximation*, International Series of Numerical Mathematics 168, DOI 10.1007/978-3-319-27067-8_4

Proof We take an arbitrary $v \in H_0^1(\Omega)$, multiply the equation by v, which yields

$$-(\Delta u)v + cuv = fv,$$

and then integrate the result over Ω. Indeed, every term is integrable. First of all, $u \in H^2(\Omega)$ hence $\Delta u \in L^2(\Omega)$, and $v \in L^2(\Omega)$ imply $(\Delta u)v \in L^1(\Omega)$. Moreover, $c \in L^\infty(\Omega)$, $u \in L^2(\Omega)$ and $v \in L^2(\Omega)$ imply $cuv \in L^1(\Omega)$. Finally, $f \in L^2(\Omega)$ implies $fv \in L^1(\Omega)$. We thus obtain

$$-\int_\Omega (\Delta u)v\,dx + \int_\Omega cuv\,dx = \int_\Omega fv\,dx.$$

We now use Green's formula (3.15), according to which

$$\int_\Omega (\Delta u)v\,dx = -\int_\Omega \nabla u \cdot \nabla v\,dx + \int_{\partial\Omega} \gamma_1(u)\gamma_0(v)\,d\Gamma,$$

and we conclude since $v \in H_0^1(\Omega)$ is equivalent to $\gamma_0(v) = 0$. □

Concerning the second equation in (4.1), i.e., the boundary condition, we have to interpret it in the sense of traces in the Sobolev context. In fact, as we have seen in the previous chapter, the reasonable way to impose the Dirichlet boundary condition is to require that $\gamma_0(u) = 0$, or in other words, that $u \in H_0^1(\Omega)$. The conjunction of (4.2) with the requirement that $u \in H_0^1(\Omega)$ is called the *variational formulation* of problem (4.1). The functions v are called *test-functions*.

Let us rewrite the variational formulation in a standard, abstract form. We let $V = H_0^1(\Omega)$, it is a Hilbert space. Then we have a bilinear form on $V \times V$

$$a(u, v) = \int_\Omega (\nabla u \cdot \nabla v + cuv)\,dx$$

and a linear form on V

$$\ell(v) = \int_\Omega fv\,dx.$$

The variational formulation then reads: Find $u \in V$ such that

$$\forall v \in V, \quad a(u, v) = \ell(v), \tag{4.3}$$

and we have shown that a solution of the boundary value problem with the additional regularity $u \in H^2(\Omega)$ is a solution of the variational problem (4.3).

Now what about the reverse implication? Does a solution of the variational problem solve the boundary value problem? The answer is basically yes, the two problems are equivalent.

Proposition 4.2 *Assume that $u \in H_0^1(\Omega)$ solves the variational problem (4.3). Then we have*

$$-\Delta u + cu = f \text{ in the sense of } \mathscr{D}'(\Omega).$$

Moreover $\Delta u \in L^2(\Omega)$ and the PDE is also satisfied almost everywhere on Ω.

Proof First of all, note that the variational formulation (4.2) makes sense for $u \in H_0^1(\Omega)$. We have $\mathscr{D}(\Omega) \subset H_0^1(\Omega)$, therefore we can take $v = \varphi \in \mathscr{D}(\Omega)$ as test-function in (4.3). Let us examine each term separately.

For the first term, we have

$$\begin{aligned}\int_\Omega \nabla u \cdot \nabla \varphi \, dx &= \int_\Omega \Big(\sum_{i=1}^d \partial_i u \partial_i \varphi\Big)\, dx = \sum_{i=1}^d \Big(\int_\Omega \partial_i u \partial_i \varphi \, dx\Big) \\ &= \sum_{i=1}^d \langle \partial_i u, \partial_i \varphi \rangle = \sum_{i=1}^d -\langle \partial_{ii} u, \varphi \rangle = -\Big\langle \sum_{i=1}^d \partial_{ii} u, \varphi \Big\rangle = -\langle \Delta u, \varphi \rangle,\end{aligned}$$

by definition of distributional derivatives. Similarly

$$\int_\Omega cu\varphi \, dx = \langle cu, \varphi \rangle \text{ and } \int_\Omega f\varphi \, dx = \langle f, \varphi \rangle.$$

Therefore, we have for all $\varphi \in \mathscr{D}(\Omega)$

$$\langle -\Delta u + cu - f, \varphi \rangle = 0$$

or

$$-\Delta u + cu - f = 0 \text{ in the sense of } \mathscr{D}'(\Omega)$$

and the PDE is satisfied in the sense of distributions. The Dirichlet boundary condition is also satisfied by the simple fact that $u \in H_0^1(\Omega)$, hence the boundary value problem is solved.

To conclude, we note that $\Delta u = cu - f \in L^2(\Omega)$. This implies that the distribution Δu is an L^2-function and thus that the PDE is satisfied almost everywhere in Ω. □

Remark 4.1 Note that the condition $\gamma_0(u) = 0$ also means in a sense that u vanishes almost everywhere on the boundary $\partial\Omega$. □

Remark 4.2 The two problems are thus equivalent, except for the fact that we have assumed $u \in H^2(\Omega)$ in one direction, and only recuperated $\Delta u \in L^2(\Omega)$ in the other.[1] Actually, the assumption $u \in H^2(\Omega)$ is somewhat artificial and made only to make

[1]The Laplacian is a specific linear combination of some of the second order derivatives. So it being in L^2 is a priori less than all individual second order derivatives, even those not appearing in the Laplacian, being in L^2, except when $d = 1$.

use of Green's formula (3.15). It is possible to dispense with it with a little more work, but that would take us too far.

It should be noted in any case, that if $u \in H_0^1(\Omega)$, $\Delta u \in L^2(\Omega)$ and Ω is for example of class C^2, then $u \in H^2(\Omega)$. This is very profound result in *elliptic regularity theory*, far beyond the scope of these notes (see [15, 25, 35] for example). We will come back to this point at the end of the chapter. It is trivial in dimension one though.

Of course, so far we have no indication that either problem has a solution. The fact is that the variational formulation is significantly easier to treat, once the right point of view is found. And the right point of view is an abstract point of view, as is often the case, more of this in Sect. 4.2. □

Before we start delving in the abstract, let us give a couple more model problems of a different kind. First is the higher dimensional analogue of the *Neumann boundary condition* already seen in one dimension in Chap. 2:

$$\begin{cases} -\Delta u + cu = f \text{ in } \Omega, \\ \dfrac{\partial u}{\partial n} = g \text{ on } \partial\Omega. \end{cases} \tag{4.4}$$

When $g = 0$, it is naturally called a homogeneous Neumann boundary condition. In terms of modeling, the Neumann condition is a flux condition. For instance, in the heat equilibrium interpretation, the condition corresponds to an imposed heat flux through the boundary, as opposed to the Dirichlet condition which imposes a given temperature on the boundary. The case $g = 0$ corresponds to perfect thermal insulation: no heat is allowed to enter or leave Ω.

Let us derive the variational formulation informally. Assume first that $u \in H^2(\Omega)$, take $v \in H^1(\Omega)$, multiply, integrate and use Green's formula to obtain

$$\forall v \in H^1(\Omega), \quad \int_\Omega (\nabla u \cdot \nabla v + cuv)\, dx = \int_\Omega fv\, dx + \int_{\partial\Omega} g\gamma_0(v)\, d\Gamma.$$

Note the different test-function space and the additional boundary term in the right-hand side.

The converse is more interesting. Let $u \in H^2(\Omega)$ be a solution of the above variational problem. Taking first $v = \varphi \in \mathscr{D}(\Omega)$, we obtain

$$-\Delta u + cu = f \text{ in the sense of } \mathscr{D}'(\Omega)$$

exactly as in the Dirichlet case. Of course, a test-function with compact support does not see what happens on the boundary, and no information on the Neumann condition is recovered. Thus, in a second step, we take v arbitrary in $H^1(\Omega)$. By Green's formula again, we have

$$\int_\Omega \nabla u \cdot \nabla v\, dx = -\int_\Omega (\Delta u)v\, dx + \int_{\partial\Omega} \gamma_1(u)\gamma_0(v)\, d\Gamma.$$

Recall that the normal trace $\gamma_1(u)$ plays the role of the normal derivative. Since u is a solution of the variational problem, it follows that

$$\int_\Omega (-\Delta u + cu)v\,dx + \int_{\partial\Omega} \gamma_1(u)\gamma_0(v)\,d\Gamma = \int_\Omega fv\,dx + \int_{\partial\Omega} g\gamma_0(v)\,d\Gamma.$$

But we already know that $\int_\Omega(-\Delta u + cu)v\,dx = \int_\Omega fv\,dx$ by the previous step, hence we are left with

$$\int_{\partial\Omega} \gamma_1(u)\gamma_0(v)\,d\Gamma = \int_{\partial\Omega} g\gamma_0(v)\,d\Gamma,$$

for all $v \in H^1(\Omega)$. For simplicity, we assume here that $g \in H^{1/2}(\partial\Omega)$, the image of the trace γ_0, see Proposition 3.17 of Chap. 3, and that Ω is smooth. Since $u \in H^2(\Omega)$, it follows that $\gamma_1(u) = \sum_{i=1}^d \gamma_0(\partial_i u)n_i \in H^{1/2}(\partial\Omega)$. Therefore, there exists $v \in H^1(\Omega)$ such that $\gamma_0(v) = \gamma_1(u) - g$. With this choice of v, we obtain

$$\int_{\partial\Omega} (\gamma_1(u) - g)^2\,d\Gamma = 0,$$

hence $\gamma_1(u) = g$, which is the Neumann condition. The last hypotheses ($u \in H^2(\Omega)$ and $g \in H^{1/2}(\partial\Omega)$) are made for brevity only. They are not at all necessary to conclude.

Another problem of interest is the *non homogeneous Dirichlet problem.*

$$\begin{cases} -\Delta u + cu = f \text{ in } \Omega, \\ \qquad\quad\; u = g \text{ on } \partial\Omega, \end{cases}$$

with $g \in H^{3/2}(\partial\Omega)$.[2] This problem is reduced to the homogeneous problem by taking a function $G \in H^2(\Omega)$ such that $\gamma_0(G) = g$ and setting $U = u - G$. Then clearly $U \in H_0^1(\Omega)$ and $-\Delta U + cU = -\Delta u + cu + \Delta G - cG = f + \Delta G - cG$. Then we just write the variational formulation of the homogeneous problem for U with right-hand side $F = f + \Delta G - cG \in L^2(\Omega)$. Note that it is also possible to solve the problem under the more natural assumption $g \in H^{1/2}(\partial\Omega)$.

The Dirichlet and Neumann conditions can be mixed together, but *not at the same place* on the boundary, yielding the so-called *mixed problem*. More precisely, let Γ_1 and Γ_2 be two subsets of $\partial\Omega$ such that $\Gamma_1 \cap \Gamma_2 = \emptyset$, $\bar{\Gamma}_1 \cup \bar{\Gamma}_2 = \partial\Omega$. Then the mixed problem reads

$$\begin{cases} -\Delta u + cu = f \text{ in } \Omega, \\ \qquad\quad\; u = g_1 \text{ on } \Gamma_1, \\ \qquad\quad \dfrac{\partial u}{\partial n} = g_2 \text{ on } \Gamma_2. \end{cases} \tag{4.5}$$

[2]The space $H^{3/2}(\partial\Omega)$ is the space of traces of $H^2(\Omega)$ functions.

The variational formulation for the mixed problem (in the case $g_1 = 0$ for brevity, if not follow the above route) is to let $V = \{v \in H^1(\Omega); \gamma_0(v) = 0 \text{ on } \Gamma_1\}$ and

$$\forall v \in V, \quad \int_\Omega (\nabla u \cdot \nabla v + cuv)\, dx = \int_\Omega f v\, dx + \int_{\Gamma_2} g_2 \gamma_0(v)\, d\Gamma,$$

with $u \in V$. Note that the mixed problem reduces to the Neumann problem when meas $(\Gamma_1) \neq 0$ and to the Dirichlet problem when meas $(\Gamma_2) \neq 0$.

Remark 4.3 An important rule of thumb to be remembered from the above examples is that (homogeneous) Dirichlet conditions are taken into account in the test-function space, whereas Neumann boundary conditions are taken into account in the linear form via boundary integrals and a larger test-function space.

4.2 Abstract Variational Problems

We now describe the general abstract framework for all variational problems. We have just seen that some boundary value problems can be recast in the following form. We are given a Hilbert space V (in the examples we have seen before $H^1_0(\Omega)$ or $H^1(\Omega)$), a bilinear form a on $V \times V$ and a linear form ℓ on V. The solution of the boundary value problem is then a solution of problem (4.3). At this point, we completely abstract the boundary value problem aspect.

Definition 4.1 An abstract variational problem consists in finding $u \in V$ such that

$$\forall v \in V, \quad a(u, v) = \ell(v), \tag{4.6}$$

where V is a Hilbert space, a is a bilinear form on $V \times V$ and ℓ is a linear form on V.

The basic tool for solving abstract variational problems is the Lax–Milgram theorem [55]. This theorem is important, not because it is in any way difficult, which it is not, but because it has a very wide range of applicability as we will see later.

Theorem 4.1 (Lax–Milgram) *Let V be a Hilbert space, a be a bilinear form and ℓ be a linear form. Assume that*

(i) *The bilinear form a is continuous, i.e., there exists a constant M such that $|a(u, v)| \leq M\|u\|_V\|v\|_V$ for all $u, v \in V$,*

(ii) *The bilinear form a is V*-elliptic[3], *i.e., there exists a constant $\alpha > 0$ such that $a(v, v) \geq \alpha\|v\|_V^2$ for all $v \in V$,*

(iii) *The linear form ℓ is continuous, i.e., there exists a constant C such that $|\ell(v)| \leq C\|v\|_V$ for all $v \in V$.*

---

[3]This condition is also sometimes called coerciveness.

Under the above assumptions, there exists a unique $u \in V$ that solves the abstract variational problem (4.6).

Proof Let us start with the uniqueness. Let u_1 and u_2 be two solutions of problem (4.6). Since a is linear with respect to its first argument, it follows that $a(u_1 - u_2, v) = 0$ for all $v \in V$. In particular, for $v = u_1 - u_2$, we obtain

$$0 = a(u_1 - u_2, u_1 - u_2) \geq \alpha \|u_1 - u_2\|_V^2,$$

so that $\|u_1 - u_2\|_V = 0$ since $\alpha > 0$.

We next prove the existence of a solution. We first note that for all $u \in V$, the mapping $v \mapsto a(u, v)$ is linear (by bilinearity of a) and continuous (by i) continuity of a). Therefore, there exists a unique element Au of V' such that $a(u, v) = \langle Au, v\rangle_{V',V}$. Moreover, the bilinearity of a shows that the mapping $A\colon V \to V'$ thus defined is linear. It is also continuous since for all $v \in V$ with $\|v\|_V \leq 1$,

$$|\langle Au, v\rangle_{V',V}| = |a(u, v)| \leq M\|u\|_V\|v\|_V \leq M\|u\|_V$$

so that

$$\|Au\|_{V'} = \sup_{\|v\|_V \leq 1} |\langle Au, v\rangle_{V',V}| \leq M\|u\|_V.$$

We rewrite the variational problem as: Find $u \in V$ such that

$$\forall v \in V, \quad \langle Au - \ell, v\rangle_{V',V} = 0$$

or

$$Au = \ell,$$

and this is where the continuity of ℓ is used.

Thus, proving the existence is equivalent to showing that the mapping A is onto.[4] We do this in two independent steps: we show that $\operatorname{im} A$ is closed on the one hand and that it is dense on the other hand.[5] For the closedness of the image, we use assumption (ii) of V-ellipticity. Let ℓ_n be a sequence in $\operatorname{im} A$ such that $\ell_n \to \ell$ in V'. We want to show that $\ell \in \operatorname{im} A$, which will imply that $\operatorname{im} A$ is closed. The sequence ℓ_n is a Cauchy sequence in V', and for all n, there exists $u_n \in V$ such that $Au_n = \ell_n$. By V-ellipticity,

$$\begin{aligned}\|u_n - u_m\|_V^2 &\leq \frac{1}{\alpha} a(u_n - u_m, u_n - u_m) = \frac{1}{\alpha}\langle Au_n - Au_m, u_n - u_m\rangle_{V',V} \\ &= \frac{1}{\alpha}\langle \ell_n - \ell_m, u_n - u_m\rangle_{V',V} \leq \frac{1}{\alpha}\|\ell_n - \ell_m\|_{V'}\|u_n - u_m\|_V,\end{aligned}$$

[4] Since we already know it is one-to-one, it will then be an isomorphism.

[5] This is a pretty common strategy, to be kept in mind.

by the definition of the dual norm. Therefore, if $\|u_n - u_m\|_V = 0$ we are happy, otherwise we divide by $\|u_n - u_m\|_V$ and in both cases

$$\|u_n - u_m\|_V \leq \frac{1}{\alpha}\|\ell_n - \ell_m\|_{V'},$$

so that u_n is a Cauchy sequence in V. Since V is complete, there exists $u \in V$ such that $u_n \to u$ in V. Since A is continuous, it follows that $\ell_n = Au_n \to Au$ in V'. Hence $\ell = Au \in \operatorname{im} A$.

To show the density, we show that $(\operatorname{im} A)^\perp = \{0\}$ (according to Lemma 3.1 of Chap. 3). We note that $(Au|\ell)_{V'} = (\sigma Au|\sigma\ell)_V = \langle Au, \sigma\ell\rangle_{V',V} = a(u, \sigma\ell)$, where σ is the Riesz isomorphism. Let $\ell \in (\operatorname{im} A)^\perp$. For all $u \in V$, we thus have $a(u, \sigma\ell) = 0$. In particular, for $u = \sigma\ell$, we obtain $0 = a(\sigma\ell, \sigma\ell) \geq \alpha\|\sigma\ell\|_V^2$ by V-ellipticity. Since $\alpha > 0$, it follows that $\ell = 0$. □

Remark 4.4 There is another classical proof of the Lax–Milgram theorem using the Banach fixed point theorem [15, 19, 25]. Note that if V is separable, there is yet another proof based on the Galerkin method which can be generalized to nonlinear variational problems [57]. Finally, there is a generalization of the Lax–Milgram theorem, known as the Stampacchia theorem, [15], which is used to solve variational inequalities, see for example [21].

Remark 4.5 It should be noted that the Lax–Milgram theorem is not a particular case of the Riesz theorem. It is actually more general, since it applies to bilinear forms that are not necessarily symmetric, and it implies the Riesz theorem when the bilinear form is just the scalar product.

Sometimes when the bilinear form a is symmetric, people think it advantageous to apply Riesz's theorem in place of the Lax–Milgram theorem. This is usually an illusion: indeed, if a new scalar product defined by the bilinear form is introduced, in order to apply Riesz's theorem, it is necessary to show that the space equipped with the new scalar product is still a Hilbert space, i.e., is complete. This is done by V-ellipticity, hence nothing is gained (although this is the part that people who think they are seeing a good deal usually forget). The continuity of the linear form for the new norm must also be checked, which amounts to having the bilinear form and linear form continuous for the original norm, again, no gain.

The only case when Riesz's theorem can be deemed advantageous over the Lax–Milgram theorem, is when both above facts to be checked are already known. An example is the bilinear form $a(u, v) = \int_\Omega \nabla u \cdot \nabla v\, dx$ on $V = H_0^1(\Omega)$. □

Remark 4.6 In the case of complex Hilbert spaces and complex-valued variational problems, the Lax–Milgram theorem still holds true for a bilinear or sesquilinear form. The V-ellipticity assumption can even be relaxed to only involve the real part of a, i.e., $\operatorname{Re}(a(u, u)) \geq \alpha\|u\|^2$ (or the imaginary part), which is rather useful as the imaginary part can then be pretty arbitrary (exercise). □

Remark 4.7 Let us emphasize again that the Lax–Milgram theorem only gives sufficient conditions for existence and uniqueness of the solution of an abstract variational problem. More specifically, V-ellipticity is not necessary. Indeed, when V is finite dimensional, V-ellipticity is just the positive definiteness of the operator A (identifying V and V' without second thoughts). Obviously, there are more isomorphisms in $\mathscr{L}(V)$ than just positive definite linear mappings.

Replacing V-ellipticity with the following *inf-sup condition*:

$$\inf_{u \in V \setminus \{0\}} \sup_{v \in V \setminus \{0\}} \frac{a(u, v)}{\|u\|_V \|v\|_V} > 0,$$

combined with the requirement that if v is such that $a(u, v) = 0$ for all $u \in V$, then $v = 0$, we obtain a set of necessary and sufficient conditions, as is easily seen along the same lines as before [7, 41]. Both conditions are clearly implied by V-ellipticity. As a rule, elliptic problems are usually amenable to the Lax–Milgram theorem. □

The linear form in the right-hand side of a variational problem should be thought of as data. In this respect, the solution depends continuously on the data.

Proposition 4.3 *The mapping $V' \to V$, $\ell \mapsto u$ defined by the Lax–Milgram theorem is linear and continuous.*

Proof The operator A is linear and invertible, therefore so is A^{-1}. The continuity of A^{-1} stems from Banach's theorem, see [15]. We actually have a more precise result since

$$\alpha \|u\|_V^2 \le a(u, u) = \ell(u) \le \|\ell\|_{V'} \|u\|_V,$$

hence

$$\|u\|_V \le \frac{1}{\alpha} \|\ell\|_{V'},$$

which shows that the continuity constant of A^{-1} is smaller than the inverse of the V-ellipticity constant of a. □

Proposition 4.4 *Let the hypotheses of the Lax–Milgram theorem be satisfied. Assume in addition that the bilinear form a is symmetric. Then the solution u of the variational problem (4.6) is also the unique solution of the minimization problem:*

$$J(u) = \inf_{v \in V} J(v) \quad \textit{with} \quad J(v) = \frac{1}{2} a(v, v) - \ell(v).$$

Proof Let u be the Lax–Milgram solution. For all $v \in V$, we let $w = v - u$ and

$$\begin{aligned} J(v) = J(u + w) &= \frac{1}{2} a(u, u) + \frac{1}{2} a(u, w) + \frac{1}{2} a(w, u) + \frac{1}{2} a(w, w) - \ell(u) - \ell(w) \\ &= J(u) + a(u, w) - \ell(w) + \frac{1}{2} a(w, w) \\ &\ge J(u), \end{aligned}$$

since $a(w, w) \geq 0$. Make note of where the symmetry is used. Hence, u minimizes J on V.

Conversely, assume that u minimizes J on V. Then, for all $\lambda > 0$ and all $v \in V$, we have $J(u + \lambda v) \geq J(u)$. Expanding the left-hand side, we get

$$\frac{1}{2}a(u, u) + \lambda a(u, v) + \frac{\lambda^2}{2}a(v, v) - \ell(u) - \lambda\ell(v) \geq J(u)$$

so that dividing by λ

$$a(u, v) - l(v) + \frac{\lambda}{2}a(v, v) \geq 0.$$

We then let $\lambda \to 0$, hence

$$a(u, v) - l(v) \geq 0,$$

and finally change v in $-v$ to obtain

$$a(u, v) - l(v) = 0,$$

for all $v \in V$. □

Remark 4.8 Taking $\lambda > 0$, dividing by λ and then letting $\lambda \to 0$ is quite clever, and known as Minty's trick. □

Remark 4.9 When the bilinear form a is not symmetric, we can still define the functional J in the same fashion as before and try to minimize it. It is clear from the above proof that the minimizing element u does not solve the variational problem associated with a but the variational problem associated with the symmetric part of a. Note that u exists because we can apply the Lax–Milgram theorem to the symmetric part of the bilinear form a. Of course, when both variational problems are translated into PDEs, we get entirely different equations. □

4.3 Application to the Model Problems, and More

Here again, Ω is a Lipschitz open subset of $\mathbb{R}^d$. We now apply the previous abstract results to concrete examples. We start with the first model problem (4.1).

Proposition 4.5 *Let $f \in L^2(\Omega)$, $c \in L^\infty(\Omega)$. Assume that $c \geq 0$. Then the problem: Find $u \in V = H_0^1(\Omega)$ such that*

$$\forall v \in V, \quad \int_\Omega (\nabla u \cdot \nabla v + cuv)\, dx = \int_\Omega fv\, dx,$$

has one and only one solution.

Proof We already know that V is a Hilbert space, for both scalar products that we defined earlier. Of course

$$a(u, v) = \int_\Omega (\nabla u \cdot \nabla v + cuv)\, dx$$

clearly defines a bilinear form on $V \times V$ and

$$\ell(v) = \int_\Omega f v\, dx$$

a linear form on V. Hence, we have an abstract variational problem. Let us try and apply the Lax–Milgram theorem. We need to check the theorem hypotheses. For definiteness, we choose to work with the full H^1 norm.

First of all, for all $(u, v) \in V \times V$,

$$\begin{aligned}
|a(u, v)| &= \left| \int_\Omega (\nabla u \cdot \nabla v + cuv)\, dx \right| \\
&\le \int_\Omega |\nabla u \cdot \nabla v + cuv|\, dx \\
&\le \int_\Omega |\nabla u \cdot \nabla v|\, dx + \int_\Omega |cuv|\, dx \\
&\le \|\nabla u\|_{L^2(\Omega)} \|\nabla v\|_{L^2(\Omega)} + \|c\|_{L^\infty(\Omega)} \|u\|_{L^2(\Omega)} \|v\|_{L^2(\Omega)} \\
&\le \max\big(1, \|c\|_{L^\infty(\Omega)}\big) \|u\|_{H^1(\Omega)} \|v\|_{H^1(\Omega)},
\end{aligned}$$

by the Cauchy–Schwarz inequality to go from the third line to the fourth line, and again the Cauchy–Schwarz inequality in $\mathbb{R}^2$ to go from the fourth line to the fifth line, hence the continuity of the bilinear form a.

Next is the V-ellipticity. For all $v \in V$, we have

$$a(v, v) = \int_\Omega (\|\nabla v\|^2 + cv^2)\, dx \ge \int_\Omega \|\nabla v\|^2\, dx \ge \alpha \|v\|^2_{H^1(\Omega)}$$

with $\alpha = (C^2 + 1)^{-1/2} > 0$ by Corollary 3.3 of Chap. 3, where C is the Poincaré inequality constant, and since $c \ge 0$.

Finally, we check the continuity of the linear form. For all $v \in V$,

$$|\ell(v)| = \left| \int_\Omega f v\, dx \right| \le \|f\|_{L^2(\Omega)} \|v\|_{L^2(\Omega)} \le \|f\|_{L^2(\Omega)} \|v\|_{H^1(\Omega)}$$

by the Cauchy–Schwarz inequality again.

All the hypotheses of the Lax–Milgram theorem are satisfied, therefore there is one and only one solution $u \in V$. $\square$

Remark 4.10 Now is a time to celebrate since we have successfully solved our first boundary value problem in arbitrary dimension. Indeed, we have already seen that any solution of the variational problem is a solution of the PDE in the distributional sense and in the L^2 sense. The solution u depends continuously in H^1 on f in L^2. Note that we have also solved the non homogeneous Dirichlet problem at the same time.

It is an instructive exercise to redo the proof using the H^1 semi-norm in place of the full norm. The same ingredients are used, but not at the same spots.

This is a case of a symmetric bilinear form, therefore the solution u also minimizes the so-called energy functional

$$J(v) = \frac{1}{2}\int_\Omega (\|\nabla v\|^2 + cv^2)\,dx - \int_\Omega fv\,dx$$

over V. □

Remark 4.11 It should be noted that the positivity condition $c \geq 0$ is by no means a necessary condition for existence and uniqueness via the Lax–Milgram theorem. With a little more work, it is not too hard to allow the function c to take some negative values. However, we have seen an example at the very beginning of Chap. 1 with a negative function c for which existence and uniqueness fails.

One should also be aware that there is an existence and uniqueness theory that goes beyond the Lax–Milgram theorem, which only gives a sufficient condition for existence and uniqueness. □

Let us now consider the non homogeneous Neumann problem (4.4). The hypotheses are slightly different.

Proposition 4.6 *Let $f \in L^2(\Omega)$, $c \in L^\infty(\Omega)$, $g \in L^2(\partial\Omega)$. Assume that there exists a constant $c_0 > 0$ such that $c \geq c_0$ almost everywhere. Then the problem: Find $u \in V = H^1(\Omega)$ such that*

$$\forall v \in V,\quad \int_\Omega (\nabla u \cdot \nabla v + cuv)\,dx = \int_\Omega fv\,dx + \int_{\partial\Omega} g\gamma_0(v)\,dx,$$

has one and only one solution.

Proof We have a different Hilbert space (but known to be Hilbert, nothing to check here), the same bilinear form and a different linear form

$$\ell(v) = \int_\Omega fv\,dx + \int_{\partial\Omega} g\gamma_0(v)\,dx.$$

We have already shown that the bilinear form is continuous in the H^1 norm.[6] The V-ellipticity is clear since, for all $v \in V$,

$$a(v,v) = \int_\Omega (\|\nabla v\|^2 + cv^2)\,dx \geq \int_\Omega \|\nabla v\|^2\,dx + c_0 \int_\Omega v^2\,dx \geq \min(1,c_0)\|v\|^2_{H^1(\Omega)},$$

with $\min(1,c_0) > 0$. The continuity of the linear form is also clear

$$|\ell(v)| \leq \|f\|_{L^2(\Omega)}\|v\|_{L^2(\Omega)} + \|g\|_{L^2(\partial\Omega)}\|\gamma_0(v)\|_{L^2(\partial\Omega)} \leq (\|f\|_{L^2(\Omega)} + C_{\gamma_0}\|g\|_{L^2(\partial\Omega)})\|v\|_{H^1(\Omega)}$$

by the Cauchy–Schwarz inequality, where C_{γ_0} is continuity constant of the trace mapping. □

The mixed problem (4.5) is a nice mixture of the Dirichlet and the Neumann problems.

Proposition 4.7 *Same hypotheses as in Proposition 4.6 and let Γ_1 and Γ_2 be two subsets of $\partial\Omega$ such that $\Gamma_1 \cap \Gamma_2 = \emptyset$, $\bar\Gamma_1 \cup \bar\Gamma_2 = \partial\Omega$ and meas$(\Gamma_1) \neq 0$. Then the problem: Find $u \in V = \{v \in H^1(\Omega); \gamma_0(v) = 0$ on $\Gamma_1\}$ such that*

$$\forall v \in V, \quad \int_\Omega (\nabla u \cdot \nabla v + cuv)\,dx = \int_\Omega fv\,dx + \int_{\Gamma_2} g\gamma_0(v)\,d\Gamma,$$

has one and only one solution.

Proof The only real difference with Proposition 4.6 lies with the space V, which we do not know yet to be a Hilbert space. It suffices to show that V is a closed subspace of $H^1(\Omega)$. Let v_n be a sequence in V such that $v_n \to v$ in $H^1(\Omega)$. By continuity of the trace mapping, we have $\gamma_0(v_n) \to \gamma_0(v)$ in $L^2(\partial\Omega)$. Therefore, there exists a subsequence $\gamma_0(v_{n_p})$ that converges to $\gamma_0(v)$ almost everywhere on $\partial\Omega$. Since $\gamma_0(v_n) = 0$ almost everywhere on Γ_1, it follows that $\gamma_0(v) = 0$ almost everywhere on Γ_1, hence $v \in V$, which is thus closed. □

Proposition 4.7 also holds true under the less demanding hypothesis $c \geq 0$, using a different argument for V-ellipticity.

A natural question arises about the Neumann problem for $c = 0$, see Chap. 2, Sect. 2.4 in one dimension. Now, this is an entirely different problem from the previous ones. First we have to find the variational formulation of the boundary value problem and show that it is equivalent to the boundary value problem, then we have to apply the Lax–Milgram theorem.

Let us thus consider the Neumann problem

$$\begin{cases} -\Delta u = f \text{ in } \Omega, \\ \frac{\partial u}{\partial n} = g \text{ on } \partial\Omega, \end{cases} \tag{4.7}$$

[6] If we had worked with the semi-norm for the Dirichlet problem, we would have had to do the continuity all over again here...

in a Lipschitz open set Ω in $\mathbb{R}^d$. We see right away that things are going to be different since we do not have uniqueness here. Indeed, if u is a solution, then $u + s$ is also a solution for any constant s. Furthermore, by Green's formula (3.15) with $v = 1$, it follows that if there is a solution, then, necessarily

$$\int_\Omega f\,dx + \int_{\partial\Omega} g\,d\Gamma = 0. \tag{4.8}$$

If the data f, g does not satisfy the compatibility condition (4.8), there is thus no solution. The two remarks, non uniqueness and non existence, are actually dual to each other.

There are several ways of going around both problems, thus several variational formulations.[7] We choose to set

$$V = \Big\{v \in H^1(\Omega); \int_\Omega v\,dx = 0\Big\}. \tag{4.9}$$

This is well defined, since Ω is bounded and we thus have $H^1(\Omega) \subset L^2(\Omega) \subset L^1(\Omega)$. Note that V is the L^2-orthogonal in $H^1(\Omega)$, which also happens to be the H^1-orthogonal in this case, to the one-dimensional space of constant functions. Note that these functions are precisely the cause of non uniqueness.

Lemma 4.1 *The space V is a Hilbert space for the scalar product of $H^1(\Omega)$.*

Proof It suffices to show that V is closed. Let v_n be a sequence in V such that $v_n \to v$ in $H^1(\Omega)$. Of course, $v_n \to v$ in $L^2(\Omega)$ and by the Cauchy–Schwarz inequality, $v_n \to v$ in $L^1(\Omega)$. Therefore

$$0 = \int_\Omega v_n\,dx \to \int_\Omega v\,dx,$$

and $v \in V$. □

We introduce the bilinear form a defined on $V \times V$ by $a(u, v) = \int_\Omega \nabla u \cdot \nabla v\,dx$ and the linear form ℓ defined on V by $\ell(v) = \int_\Omega f v\,dx + \int_{\partial\Omega} g\gamma_0(v)\,d\Gamma$.

Proposition 4.8 *Assume that $f \in L^2(\Omega)$, $g \in L^2(\partial\Omega)$ satisfy the compatibility condition (4.8). Then, any solution $u \in H^2(\Omega)$ of the Neumann problem (4.7) is a solution of the variational problem defined by the triple (V, a, ℓ). Conversely, any solution $u \in H^2(\Omega)$ of the variational problem is a solution of problem (4.7).*

Proof Multiplying the PDE by $v \in V$ and using Green's formula, we easily see that if $u \in H^2(\Omega)$ solves problem (4.7), then we have for all $v \in V$, $a(u, v) = \ell(v)$.

Conversely, let us be given a function $u \in V \cap H^2(\Omega)$ such that for all $v \in V$, $a(u, v) = \ell(v)$. We would like to proceed as before and take $v \in \mathcal{D}(\Omega)$ to derive the PDE. This does not work here because $\mathcal{D}(\Omega) \not\subset V$. For all $\varphi \in \mathcal{D}(\Omega)$, we set

[7] We have always said *the* variational formulation, but there is no evidence that it is unique in general.

$$\psi = \varphi - \frac{1}{\text{meas}\,\Omega}\int_\Omega \varphi(x)\,dx,$$

so that $\psi \in V$ and we can use ψ as a test-function. Now φ and ψ differ by a constant $k = \frac{1}{\text{meas}\,\Omega}\int_\Omega \varphi(x)\,dx$, therefore $\nabla\psi = \nabla\varphi$. We thus obtain,

$$\begin{aligned}\int_\Omega \nabla u\cdot\nabla\varphi\,dx &= \int_\Omega \nabla u\cdot\nabla\psi\,dx = \int_\Omega f\psi\,dx + \int_{\partial\Omega} g\psi\,d\Gamma\\ &= \int_\Omega f(\varphi+k)\,dx + \int_{\partial\Omega} g(\varphi+k)\,d\Gamma\\ &= \int_\Omega f\varphi\,dx + k\Big(\int_\Omega f\,dx + \int_{\partial\Omega} g\,d\Gamma\Big) = \int_\Omega f\varphi\,dx,\end{aligned}$$

since φ vanishes on $\partial\Omega$ and f, g satisfy condition (4.8). So we can deduce right away that $-\Delta u = f$ in the sense of distributions, and since $f \in L^2(\Omega)$ in the sense of $L^2(\Omega)$ as well.

We next pick an arbitrary $v \in V$ and apply Green's formula again. This yields

$$\int_\Omega fv\,dx + \int_{\partial\Omega} g\gamma_0(v)\,d\Gamma = -\int_\Omega (\Delta u)v\,dx + \int_{\partial\Omega}\gamma_1(u)\gamma_0(v)\,d\Gamma.$$

Hence, taking into account that $-\Delta u = f$, we obtain

$$\int_{\partial\Omega}(g-\gamma_1(u))\gamma_0(v)\,d\Gamma = 0$$

for all $v \in V$. Now it is clear that $\gamma_0(V) = H^{1/2}(\partial\Omega)$. Indeed, let us pick a $\theta \in \mathscr{D}(\Omega)$ such that $\int_\Omega \theta\,dx = 1$. Then, for all $w \in H^1(\Omega)$, $v = w - \big(\int_\Omega w\,dx\big)\theta \in V$ and $\gamma_0(v) = \gamma_0(w)$. Therefore, there are enough test-functions in V to conclude that $\gamma_1(u) = g$, since $H^{1/2}(\partial\Omega)$ is dense in $L^2(\partial\Omega)$. □

Remark 4.12 It is possible to establish a variational formulation of the Neumann problem without the artificial hypothesis $u \in H^2(\Omega)$. □

To apply the Lax–Milgram theorem, we need a new inequality.

Theorem 4.2 (Poincaré–Wirtinger inequality) *Assume that Ω is connected. There exists a constant C such that, for all $v \in H^1(\Omega)$,*

$$\Big\| v - \frac{1}{\text{meas}\,\Omega}\int_\Omega v\,dx\Big\|_{L^2(\Omega)} \le C\|\nabla v\|_{L^2(\Omega)}. \tag{4.10}$$

Proof We use a contradiction argument. Assume that there is no such constant C. For all $n \in \mathbb{N}^*$, we can thus find $v_n \in H^1(\Omega)$ such that

$$\Big\| v_n - \frac{1}{\text{meas}\,\Omega}\int_\Omega v_n\,dx\Big\|_{L^2(\Omega)} > n\|\nabla v_n\|_{L^2(\Omega)}.$$

In particular, the left-hand side is strictly positive. We let

$$w_n = \frac{v_n - \frac{1}{\operatorname{meas}\Omega}\int_\Omega v_n\,dx}{\left\|v_n - \frac{1}{\operatorname{meas}\Omega}\int_\Omega v_n\,dx\right\|_{L^2(\Omega)}}.$$

By construction, we have $\|w_n\|_{L^2(\Omega)} = 1$ and w_n belongs to the L^2-orthogonal of the one-dimensional space of constant functions, which is closed in $L^2(\Omega)$.

Moreover, $\nabla w_n = \nabla v_n$ and we have

$$\|\nabla w_n\|_{L^2(\Omega)} < \frac{1}{n} \to 0 \text{ when } n \to +\infty.$$

In particular, the sequence w_n is bounded in $H^1(\Omega)$. By the Rellich theorem, see Remark 3.18 of Chap. 3, it is relatively compact in $L^2(\Omega)$. We may thus find a subsequence w_{n_p} and an element $w \in L^2(\Omega)$ such that $w_{n_p} \to w$ strongly in $L^2(\Omega)$ when $p \to +\infty$.

On the one hand we have $\|w\|_{L^2(\Omega)} = 1$ and w also belongs to the L^2-orthogonal of the space of constant functions.

On the other hand, $\nabla w_{n_p} \to 0$ strongly in $L^2(\Omega)$, hence in the sense of distributions, so that $\nabla w = 0$. As Ω is connected, this implies that w belongs to the space of constant functions.

We thus see that w belongs to the intersection of one subspace and its orthogonal, so that $w = 0$. This contradicts $\|w\|_{L^2(\Omega)} = 1$. □

Remark 4.13 Even though there is a certain formal similarity with the Poincaré inequality, there are major differences. In particular, the Poincaré–Wirtinger inequality fails for open sets that are not regular enough whereas no regularity is needed for the Poincaré inequality. Note that the contradiction argument above is not constructive. It gives no indication about the actual value of C, as opposed to the proof of the Poincaré inequality given earlier. □

Proposition 4.9 *Assume that Ω is connected, $f \in L^2(\Omega)$ and $g \in L^2(\partial\Omega)$. Then the problem: Find $u \in V$, V given by (4.9), such that*

$$\forall v \in V, \quad \int_\Omega \nabla u \cdot \nabla v\,dx = \int_\Omega f v\,dx + \int_{\partial\Omega} g\gamma_0(v)\,d\Gamma,$$

has one and only one solution.

Proof We have already shown that V is a Hilbert space for the H^1 scalar product. The continuity of both bilinear and linear forms have also already been proved. Only the V-ellipticity remains.

For all $v \in V$, we have $\int_\Omega v\,dx = 0$, hence by the Poincaré–Wirtinger inequality (4.10),

$$\|v\|^2_{H^1(\Omega)} = \|v\|^2_{L^2(\Omega)} + \|\nabla v\|^2_{L^2(\Omega)} \leq (C^2 + 1)\|\nabla v\|^2_{L^2(\Omega)}.$$

Therefore,

$$a(v, v) = \|\nabla v\|^2_{L^2(\Omega)} \geq \alpha \|v\|^2_{H^1(\Omega)}$$

with $\alpha = \frac{1}{(C^2+1)} > 0$. □

Remark 4.14 The compatibility condition (4.8) plays no role in the application of the Lax–Milgram theorem. So exercise: What happens when it is not satisfied? What exactly are we solving then? □

Remark 4.15 Since the space V is a hyperplane of H^1 that is L^2 orthogonal to the constants, it follows that the general solution of the Neumann problem is of the form $v + s$, where $v \in V$ is the unique solution of the variational problem above and $s \in \mathbb{R}$ is arbitrary. □

We now introduce a new kind of boundary condition, the Fourier condition (also called the Robin condition or the third boundary condition). The boundary value problem reads

$$\begin{cases} -\Delta u + cu = f \text{ in } \Omega, \\ bu + \dfrac{\partial u}{\partial n} = g \text{ on } \partial\Omega, \end{cases} \tag{4.11}$$

where b and c are given functions. When $b = 0$, we recognize the Neumann problem (and, in a sense, when $b = +\infty$ the Dirichlet problem). This condition is called after Fourier who introduced it in the context of the heat equation, see Chap. 1, Sect. 1.7. In the heat interpretation, $\frac{\partial u}{\partial n}$ represents the heat flux through the boundary. Let us assume that we are modeling a situation in which the boundary is actually a very thin wall that insulates Ω from the outside where the temperature is $0°$. If $g = 0$, the Fourier condition states that $\frac{\partial u}{\partial n} = -bu$, that is to say that the heat flux passing through the wall is proportional to the temperature difference between the inside and the outside. For this interpretation to be physically reasonable, it is clearly necessary that $b \geq 0$, i.e., the heat flows inwards when the outside is warmer than the inside and conversely. It thus to be expected that the sign of b will play a role.

We follow the same pattern as before: First find a variational formulation for the boundary value problem (4.11), second apply the Lax–Milgram theorem to prove existence and uniqueness. We introduce the triple

$$V = H^1(\Omega),$$

$$a(u, v) = \int_\Omega (\nabla u \cdot \nabla v + cuv)\, dx + \int_{\partial\Omega} b\gamma_0(u)\gamma_0(v)\, d\Gamma,$$

$$\ell(v) = \int_\Omega f v\, dx + \int_{\partial\Omega} g\gamma_0(v)\, d\Gamma.$$

Proposition 4.10 *Assume that we have $f \in L^2(\Omega)$, $g \in L^2(\partial\Omega)$, $c \in L^\infty(\Omega)$ and $b \in L^\infty(\partial\Omega)$. Then, any solution $u \in H^2(\Omega)$ of the Fourier problem (4.11) is a solution of the variational problem defined by the triple (V, a, ℓ). Conversely, any solution $u \in H^2(\Omega)$ of the variational problem is a solution of problem (4.11).*

Proof As always, we multiply the PDE by $v \in V$ and use Green's formula,

$$\begin{aligned}\int_\Omega (\nabla u\cdot\nabla v + cuv)\,dx &= \int_\Omega fv\,dx + \int_{\partial\Omega}\gamma_1(u)\gamma_0(v)\,d\Gamma\\ &= \int_\Omega fv\,dx + \int_{\partial\Omega}(g - b\gamma_0(u))\gamma_0(v)\,d\Gamma,\end{aligned}$$

hence

$$\int_\Omega (\nabla u\cdot\nabla v + cuv)\,dx + \int_{\partial\Omega} b\gamma_0(u)\gamma_0(v)\,d\Gamma = \int_\Omega fv\,dx + \int_{\partial\Omega} g\gamma_0(v)\,d\Gamma, \tag{4.12}$$

for all $v \in V$.

Conversely, let us be given a solution $u \in H^2(\Omega)$ of the variational problem (4.12). Taking first $v = \varphi \in \mathscr{D}(\Omega)$, all the boundary integrals vanish and we obtain $-\Delta u + cu = f$ exactly as before. Taking then $v \in H^1(\Omega)$ arbitrary, using Green's formula and the PDE just obtained, we get

$$\int_{\partial\Omega}\gamma_1(u)\gamma_0(v)\,d\Gamma + \int_{\partial\Omega} b\gamma_0(u)\gamma_0(v)\,d\Gamma = \int_{\partial\Omega} g\gamma_0(v)\,d\Gamma,$$

so that

$$\int_{\partial\Omega}(\gamma_1(u) + b\gamma_0(u) - g)\gamma_0(v)\,d\Gamma = 0,$$

for all $v \in V = H^1(\Omega)$, hence the Fourier boundary condition. □

Remark 4.16 A natural question to ask is why not keep the term $\gamma_1(u)$ in the bilinear form? The answer is that, while it is true that $\gamma_1(u)$ exists when $u \in H^2(\Omega)$ is a solution of either the boundary value problem or the variational problem, it does not exist for a general $v \in H^1(\Omega)$, hence cannot appear in a bilinear form that is defined on $H^1(\Omega) \times H^1(\Omega)$. Besides, how would b appear otherwise? □

Let us give a first existence and uniqueness result.

Proposition 4.11 *Let $f \in L^2(\Omega)$, $g \in L^2(\partial\Omega)$, $c \in L^\infty(\Omega)$ and $b \in L^\infty(\partial\Omega)$. Assume that $c \geq c_0 > 0$ for some constant c_0 and that $\|b_-\|_{L^\infty(\partial\Omega)} < \frac{\min(1,c_0)}{C_{\gamma_0}^2}$, where C_{γ_0} is the continuity constant of the trace mapping. Then the problem: Find $u \in V = H^1(\Omega)$ such that*

$$\forall v \in V, \int_\Omega (\nabla u\cdot\nabla v + cuv)\,dx + \int_{\partial\Omega} b\gamma_0(u)\gamma_0(v)\,d\Gamma = \int_\Omega fv\,dx + \int_{\partial\Omega} g\gamma_0(v)\,d\Gamma,$$

has one and only one solution.

Here $b_- = -\min(0, b)$ denotes the negative part of b.

Proof We check the hypotheses of the Lax–Milgram theorem. We already know that V is a Hilbert space. The continuity of the bilinear form a has also already been checked, except for the boundary integral terms

$$\left|\int_{\partial\Omega} b\gamma_0(u)\gamma_0(v)\,d\Gamma\right| \leq \|b\|_{L^\infty(\partial\Omega)}\|\gamma_0(u)\|_{L^2(\partial\Omega)}\|\gamma_0(v)\|_{L^2(\partial\Omega)}$$
$$\leq C_{\gamma_0}^2\|b\|_{L^\infty(\partial\Omega)}\|u\|_{H^1(\Omega)}\|v\|_{H^1(\Omega)}$$

for all u and v. The linear form is also known to be continuous. Let us check the V-ellipticity. Obviously $b \geq -b_-$, thus

$$\int_\Omega (\|\nabla v\|^2 + cv^2)\,dx + \int_{\partial\Omega} b\gamma_0(v)^2\,d\Gamma$$
$$\geq \min(1, c_0)\|v\|_{H^1(\Omega)}^2 - \|b_-\|_{L^\infty(\partial\Omega)}\|\gamma_0(v)\|_{L^2(\partial\Omega)}^2$$
$$\geq \big(\min(1, c_0) - C_{\gamma_0}^2\|b_-\|_{L^\infty(\partial\Omega)}\big)\|v\|_{H^1(\Omega)}^2,$$

hence the V-ellipticity. □

Remark 4.17 Under the previous hypotheses, we have existence and uniqueness via the Lax–Milgram theorem provided b is not too negative in some sense. □

All these hypotheses only give sufficient conditions. Let us give another set of such hypotheses.

Proposition 4.12 *Same hypotheses except that we assume that $c \geq 0$ and that $b \geq \mu > 0$ for some constant μ. Then the Fourier problem (4.11) has one and only one solution.*

Proof The only point to be established is V-ellipticity. We use a compactness argument by contradiction based on Rellich's theorem, see Remark 3.18 of Chap. 3 again. We have

$$\int_\Omega (\|\nabla v\|^2 + cv^2)\,dx + \int_{\partial\Omega} b\gamma_0(v)^2\,d\Gamma \geq \int_\Omega \|\nabla v\|^2\,dx + \mu\int_{\partial\Omega} \gamma_0(v)^2\,d\Gamma.$$

Let us assume for contradiction that there is no constant $\alpha > 0$ such that

$$\int_\Omega \|\nabla v\|^2\,dx + \mu\int_{\partial\Omega} \gamma_0(v)^2\,d\Gamma \geq \alpha\|v\|_{H^1(\Omega)}^2.$$

This implies that for all $n \in \mathbb{N}^*$, there exists $v_n \in H^1(\Omega)$ such that

$$\int_\Omega \|\nabla v_n\|^2\,dx + \mu\int_{\partial\Omega} \gamma_0(v_n)^2\,d\Gamma < \frac{1}{n}\|v_n\|_{H^1(\Omega)}^2.$$

We can assume without loss of generality that

$$\|v_n\|^2_{H^1(\Omega)} = 1, \tag{4.13}$$

and that we have

$$\int_\Omega \|\nabla v_n\|^2\,dx + \mu \int_{\partial\Omega} \gamma_0(v_n)^2\,d\Gamma \to 0. \tag{4.14}$$

Now v_n is bounded in $H^1(\Omega)$ by (4.13), thus relatively compact in $L^2(\Omega)$ by Rellich's theorem. We may extract a subsequence, still denoted v_n, and $v \in L^2(\Omega)$ such that $v_n \to v$ in $L^2(\Omega)$. By (4.14), $\|\nabla v_n\|_{L^2(\Omega)} \to 0$, therefore, since $\nabla v_n \to \nabla v$ in $\mathscr{D}'(\Omega)$, we have $\nabla v = 0$ and v is constant on each connected component of Ω. Therefore $v \in H^1(\Omega)$ and

$$\|v_n - v\|^2_{H^1(\Omega)} = \|\nabla v_n\|^2_{L^2(\Omega)} + \|v_n - v\|^2_{L^2(\Omega)} \to 0 \tag{4.15}$$

so that, by continuity of the trace mapping $\gamma_0(v_n) \to \gamma_0(v)$ in $L^2(\partial\Omega)$. By (4.14) again, we also have $\|\gamma_0(v_n)\|_{L^2(\partial\Omega)} \to 0$ since $\mu > 0$ and therefore $\gamma_0(v) = 0$. It follows that v being a constant with zero trace, it vanishes in each connected component, i.e., $v = 0$. We now realize that (4.13) and (4.15) contradict each other, therefore our premise that there exists no V-ellipticity constant α is false. □

Remark 4.18 As for the proof of the Poincaré–Wirtinger inequality, this is a typical compactness-contradiction argument: we can prove that the constant exists but we have no idea of its value. □

4.4 General Second Order Elliptic Problems

Up to now, the partial differential operator always was the Laplacian. Let us rapidly consider more general second order elliptic operators in a Lipschitz open subset Ω of $\mathbb{R}^d$. We are given a $d \times d$ matrix-valued function $A(x) = (a_{ij}(x))$ with $a_{ij} \in C^1(\bar{\Omega})$. Let $u \in C^2(\Omega)$ (we can lower this regularity considerably), then $A\nabla u$ is a vector field with components

$$(A\nabla u)_i = \sum_{j=1}^d a_{ij}\partial_j u$$

whose divergence is given by

$$\begin{aligned}\operatorname{div}(A\nabla u) &= \sum_{i=1}^d \partial_i (A\nabla u)_i \\ &= \sum_{i,j=1}^d a_{ij}\partial_{ij}u + \sum_{j=1}^d \Bigl(\sum_{i=1}^d \partial_i a_{ij}\Bigr)\partial_j u.\end{aligned}$$

The principal part of this operator $\sum_{i,j=1}^{d} a_{ij}\partial_{ij}$ is of the second order. We will consider the boundary value problem

$$\begin{cases} -\text{div}\,(A\nabla u) + cu = f \text{ in } \Omega, \\ u = h \text{ on } \Gamma_0, \\ bu + n\cdot A\nabla u = g \text{ on } \Gamma_1, \end{cases} \tag{4.16}$$

where c, b, f, g and h are given functions and Γ_0, Γ_1 a partition of $\partial\Omega$ as in the mixed problem. When $A = I$, we recognize $-\text{div}\,(A\nabla u) = -\Delta u$ and $n\cdot A\nabla u = \frac{\partial u}{\partial n}$, so that we are generalizing all the model problems seen up to now. First of all, we reduce the study to the case $h = 0$ by subtracting a function with the appropriate trace, as before.

Proposition 4.13 *Assume that $f \in L^2(\Omega)$, $g \in L^2(\Gamma_1)$, $c \in L^\infty(\Omega)$ and $b \in L^\infty(\Gamma_1)$. Then the triple*

$$V = \{v \in H^1(\Omega);\, \gamma_0(v) = 0 \text{ on } \Gamma_1\},$$

$$a(u,v) = \int_\Omega (A\nabla u\cdot\nabla v + cuv)\,dx + \int_{\Gamma_1} b\gamma_0(u)\gamma_0(v)\,d\Gamma,$$

$$\ell(v) = \int_\Omega fv\,dx + \int_{\Gamma_1} g\gamma_0(v)\,d\Gamma,$$

defines a variational formulation for problem (4.16), at least for $H^2(\Omega)$ solutions.

Proof The proof is routine, but we partially write it down for completeness. It is easy to check that $(A\nabla u)_i \in H^1(\Omega)$ for all i. We multiply the PDE by $v \in V$ and integrate by parts. This yields first

$$-\int_\Omega \Big(\sum_{i=1}^{d} \partial_i(A\nabla u)_i\Big)v\,dx + \int_\Omega cuv\,dx = \int_\Omega fv\,dx,$$

then

$$\int_\Omega \sum_{i=1}^{d}(A\nabla u)_i\partial_i v\,dx - \int_{\Gamma_1}\Big(\sum_{i=1}^{d}\gamma_0\big((A\nabla u)_i\big)n_i\Big)\gamma_0(v)\,d\Gamma + \int_\Omega cuv\,dx = \int_\Omega fv\,dx,$$

and finally

$$\int_\Omega (A\nabla u\cdot\nabla v + cuv)\,dx + \int_{\Gamma_1} b\gamma_0(u)\gamma_0(v)\,d\Gamma = \int_\Omega fv\,dx + \int_{\Gamma_1} g\gamma_0(v)\,d\Gamma.$$

We leave the converse argument to the reader. □

Proposition 4.14 *Let $f \in L^2(\Omega)$, $g \in L^2(\Gamma_1)$, $c \in L^\infty(\Omega)$ and $b \in L^\infty(\Gamma_1)$. We assume that the matrix A is* uniformly elliptic, *that is to say that there exists a constant $\alpha > 0$ such that*

$$\sum_{i,j=1}^{d} a_{ij}(x)\xi_i\xi_j \geq \alpha \|\xi\|^2$$

for all $x \in \bar{\Omega}$ and all $\xi \in \mathbb{R}^d$. We assume in addition that $c \geq c_0 > 0$ for some constant c_0 and that $b \geq 0$. Then the problem: Find $u \in V = \{v \in H^1(\Omega); \gamma_0(v) = 0 \text{ on } \Gamma_1\}$ such that

$$\forall v \in V, \int_\Omega (A\nabla u \cdot \nabla v + cuv)\,dx + \int_{\Gamma_1} b\gamma_0(u)\gamma_0(v)\,d\Gamma = \int_\Omega fv\,dx + \int_{\Gamma_1} g\gamma_0(v)\,d\Gamma,$$

has one and only one solution.

Proof That V is a Hilbert space and that ℓ is continuous are already known facts. The proof of the continuity of the bilinear form, which is implied by the boundedness of the matrix coefficients $a_{ij}(x)$, is left to the reader. The V-ellipticity is also quite obvious, since

$$\begin{aligned} a(v,v) &= \int_\Omega (A\nabla v \cdot \nabla v + cv^2)\,dx + \int_{\Gamma_1} b\gamma_0(v)^2\,d\Gamma \\ &\geq \alpha \int_\Omega \|\nabla v\|^2\,dx + c_0 \int_\Omega v^2\,dx \\ &\geq \min(\alpha, c_0)\|v\|^2_{H^1(\Omega)}, \end{aligned}$$

hence the existence, uniqueness and continuous dependence of the solution on the data by the Lax–Milgram theorem. □

Remark 4.19 When the matrix A is not symmetric, neither is the bilinear form a, even though the principal part of the operator is symmetric since $\sum_{i,j=1}^d a_{ij}\partial_{ij} = \sum_{i,j=1}^d \frac{a_{ij}+a_{ji}}{2}\partial_{ij}$ due to the fact that $\partial_{ij} = \partial_{ji}$. When A is symmetric, then so is the bilinear form and we have an equivalent minimization problem with

$$J(v) = \frac{1}{2}\Big[\int_\Omega (A\nabla v \cdot \nabla v + cv^2)\,dx + \int_{\Gamma_1} b\gamma_0(v)^2\,d\Gamma\Big] - \int_\Omega fv\,dx - \int_{\Gamma_1} g\gamma_0(v)\,d\Gamma,$$

to be minimized over V.

It is quite clear that we can reduce the regularity of A down to L^∞ without loosing the existence and uniqueness of the variational problem. The interpretation in terms of PDEs stops at the divergence form $-\text{div}\,(A\nabla u) + cu = f$ since we cannot develop the divergence using Leibniz formula in this case. Such lack of regularity of the coefficients is useful to model heterogeneous media. □

We now give another example of a non symmetric problem, the convection–diffusion problem. Let us be given a vector field σ. The convection–diffusion problem reads

$$\begin{cases} -\Delta u + \sigma \cdot \nabla u + cu = f \text{ in } \Omega, \\ u = 0 \text{ on } \partial\Omega. \end{cases} \tag{4.17}$$

We have a diffusion term $-\Delta u$ and a transport term $\sigma \cdot \nabla u$ in the same equation that compete with each other.

Proposition 4.15 *Assume that $f \in L^2(\Omega)$, $\sigma \in C^1(\bar{\Omega}; \mathbb{R}^d)$ and $c \in L^\infty(\Omega)$. Then the triple*

$$V = H_0^1(\Omega),$$
$$a(u, v) = \int_\Omega \big(\nabla u \cdot \nabla v + (\sigma \cdot \nabla u + cu)v\big)\, dx,$$
$$\ell(v) = \int_\Omega f v\, dx,$$

defines a variational formulation for problem (4.17).

Proof The proof follows the same lines as before and we leave it as an exercise. Note that the bilinear form a is not symmetric. □

Proposition 4.16 *Let $f \in L^2(\Omega)$, $\sigma \in C^1(\bar{\Omega}; \mathbb{R}^d)$ and $c \in L^\infty(\Omega)$. We assume that $c - \frac{1}{2} \operatorname{div} \sigma \geq 0$. Then the problem: Find $u \in V$ such that*

$$\forall v \in V, \int_\Omega \big(\nabla u \cdot \nabla v + (\sigma \cdot \nabla u + cu)v\big)\, dx = \int_\Omega f v\, dx,$$

has one and only one solution.

Proof We just prove the V-ellipticity. We have for all $v \in V$

$$a(v, v) = \int_\Omega \big(\|\nabla v\|^2 + cv^2 + (\sigma \cdot \nabla v)v\big)\, dx.$$

It can be checked that $\sigma_i v \in H^1(\Omega)$ and that the Leibniz formula (3.8) holds in this case for first derivatives. Let us integrate the last integral by parts

$$\int_\Omega (\sigma \cdot \nabla v)v\, dx = \int_\Omega \Big(\sum_{i=1}^d \sigma_i \partial_i v\Big) v\, dx$$
$$= -\int_\Omega \Big(\sum_{i=1}^d \partial_i(\sigma_i v)\Big) v\, dx = -\int_\Omega \Big(\sum_{i=1}^d \partial_i \sigma_i\Big) v^2\, dx - \int_\Omega \Big(\sum_{i=1}^d \sigma_i \partial_i v\Big) v\, dx$$
$$= -\int_\Omega \operatorname{div} \sigma\, v^2\, dx - \int_\Omega (\sigma \cdot \nabla v)v\, dx,$$

since all boundary terms vanish, so that

$$\int_\Omega (\sigma \cdot \nabla v) v\, dx = -\frac{1}{2}\int_\Omega \operatorname{div}\sigma v^2\, dx.$$

Therefore

$$a(v,v) = \int_\Omega \Big(\|\nabla v\|^2 + \Big(c - \frac{1}{2}\operatorname{div}\sigma\Big)v^2\Big)\, dx \geq |v|^2_{H^1(\Omega)},$$

hence the result by the equivalence of the H^1 semi-norm and the H^1 norm on $H^1_0(\Omega)$, see Corollary 3.3 of Chap. 3. ☐

Remark 4.20 We thus have existence and uniqueness if $c = 0$ and $\operatorname{div}\sigma = 0$. The case $\operatorname{div}\sigma = 0$ is interesting because if σ represents the velocity field of such a fluid as air or water, the divergence free condition is the expression of the incompressibility of the fluid. Under usual experimental conditions, both fluids are in fact considered to be incompressible. ☐

Let us now give a fourth order example, even though only second order problems were advertised in the section title. We consider a slight variant of the plate problem involving the bilaplacian with homogeneous Dirichlet boundary conditions

$$\begin{cases} \Delta^2 u + cu = f \text{ in } \Omega, \\ u = 0 \text{ on } \partial\Omega, \\ \dfrac{\partial u}{\partial n} = 0 \text{ on } \partial\Omega. \end{cases}$$

The derivation of a variational formulation is again fairly routine, but since this is our first (and only) fourth order problem, we give some detail. The variational space for this Dirichlet problem is $V = H^2_0(\Omega)$ which incorporates the two boundary conditions. Assume that $u \in H^4(\Omega) \cap H^2_0(\Omega)$. Then $\Delta u \in H^2(\Omega)$ and we can use Green's formula

$$\begin{aligned} \int_\Omega (\Delta^2 u) v\, dx &= \int_\Omega (\Delta(\Delta u)) v\, dx \\ &= \int_\Omega \Delta u\, \Delta v\, dx + \int_{\partial\Omega} (\gamma_0(v)\gamma_1(\Delta u) - \gamma_1(v)\gamma_0(\Delta u))\, d\Gamma \\ &= \int_\Omega \Delta u\, \Delta v\, dx, \end{aligned}$$

since $\gamma_0(v) = \gamma_1(v) = 0$ for all $v \in H^2_0(\Omega)$. So we have our variational formulation

$$\forall v \in V,\ \int_\Omega (\Delta u\, \Delta v + cuv)\, dx = \int_\Omega f v\, dx, \tag{4.18}$$

which is easily checked to give rise to a solution of the boundary value problem.

Let $\nabla^2 v$ denote the collection of all second order partial derivatives of v. We let

$$\|\nabla^2 v\|_{L^2(\Omega)}^2 = \sum_{1\le i,j\le d} \Big\| \frac{\partial^2 v}{\partial x_i \partial x_j} \Big\|_{L^2(\Omega)}^2.$$

We have

Lemma 4.2 *The semi-norm $\|\nabla^2 v\|_{L^2(\Omega)}$ is a norm on $H_0^2(\Omega)$ that is equivalent to the H^2 norm.*

Proof It is enough to establish a bound from below. Let $v \in H_0^2(\Omega)$. Then we have $\partial_i v \in H_0^1(\Omega)$ for all i. Therefore $\|\nabla(\partial_i v)\|_{L^2(\Omega)}^2 \ge C^2 \|\partial_i v\|_{H^1(\Omega)}^2$, as a consequence of Poincaré's inequality, and $C \le 1$. Now of course

$$\|\partial_i v\|_{H^1(\Omega)}^2 = \|\nabla(\partial_i v)\|_{L^2(\Omega)}^2 + \|\partial_i v\|_{L^2(\Omega)}^2,$$

so summing over i, we get

$$\begin{aligned}\|\nabla^2 v\|_{L^2(\Omega)}^2 = \sum_{i=1}^d \|\nabla(\partial_i v)\|_{L^2(\Omega)}^2 &\ge C^2\Big(\|\nabla^2 v\|_{L^2(\Omega)}^2 + |v|_{H^1(\Omega)}^2\Big)\\ &\ge C^2\|\nabla^2 v\|_{L^2(\Omega)}^2 + C^4\|v\|_{H^1(\Omega)}^2 \ge C^4\|v\|_{H^2(\Omega)}^2,\end{aligned}$$

since $v \in H_0^1(\Omega)$. □

Proposition 4.17 *Let $f \in L^2(\Omega)$ and $c \in L^\infty(\Omega)$. We assume that $c \ge 0$. Then problem (4.18) has one and only one solution.*

Proof We just prove the V-ellipticity. We have

$$a(v, v) \ge \int_\Omega (\Delta v)^2\,dx.$$

We argue by density. Let $\varphi \in \mathscr{D}(\Omega)$, since $\Delta\varphi = \sum_{i=1}^d \partial_{ii}\varphi$, we can write

$$\begin{aligned}\int_\Omega (\Delta\varphi)^2\,dx = \int_\Omega \Big(\sum_{i=1}^d \partial_{ii}\varphi\Big)\Big(\sum_{j=1}^d \partial_{jj}\varphi\Big)\,dx &= \sum_{i,j=1}^d \int_\Omega \partial_{ii}\varphi\,\partial_{jj}\varphi\,dx\\ &= -\sum_{i,j=1}^d \int_\Omega \partial_i\varphi\,\partial_{ijj}\varphi\,dx = \sum_{i,j=1}^d \int_\Omega \partial_{ij}\varphi\,\partial_{ij}\varphi\,dx\end{aligned}$$

with two successive integrations by parts, the first one with respect to x_i and the second one with respect to x_j. Hence, for all $\varphi \in \mathscr{D}(\Omega)$, we obtain

$$\int_\Omega (\Delta\varphi)^2\,dx = \sum_{i,j=1}^d \int_\Omega (\partial_{ij}\varphi)^2\,dx = \|\nabla^2\varphi\|^2_{L^2(\Omega)}. \tag{4.19}$$

Now, by definition, $H^2_0(\Omega)$ is the closure of $\mathscr{D}(\Omega)$ in $H^2(\Omega)$, thus for all $v \in H^2_0(\Omega)$, there exists a sequence $\varphi_n \in \mathscr{D}(\Omega)$ such that $\varphi_n \to v$ in $H^2(\Omega)$. Passing to the limit in the above equality, we thus get

$$\int_\Omega (\Delta v)^2\,dx = \|\nabla^2 v\|^2_{L^2(\Omega)},$$

since $\partial_{ij}\varphi_n \to \partial_{ij}v$ in $L^2(\Omega)$, hence the result by Lemma 4.2. □

Remark 4.21 Notice the trick used in the above proof. To establish an equality for H^2 functions, we need to use third derivatives, which make no sense as functions in this context. However, all formulas are valid for smooth functions, for which third derivatives are not a problem, and since in the end, the resulting equality (4.19) does not involve any derivatives of order higher than two, it extends to H^2 by density.

The formula is actually surprising, since Δv does not contain any derivative $\partial_{ij}v$ with $i \neq j$, and only the sum of all $\partial_{ii}v$ derivatives. Its L^2 norm squared is nonetheless equal to the sum of the L^2 norms squared of all individual second derivatives. This is related to elliptic regularity, which was mentioned in passing before. □

Remark 4.22 This is another symmetric problem, hence we have an equivalent energy minimization formulation with

$$J(v) = \frac{1}{2}\int_\Omega \big((\Delta v)^2 + cv^2\big)\,dx - \int_\Omega fv\,dx,$$

to be minimized on $H^2_0(\Omega)$. □

4.5 Concluding Remarks

To conclude this section, we discuss the general three point strategy for solving elliptic problems that was repeatedly applied here. First we establish a variational formulation: (homogeneous) Dirichlet boundary conditions are enforced by the test-function space, which is included in H^1 for second order problems; we multiply the PDE by a test-function—possibly assuming additional regularity on the solution—and use integration by parts or Green's formula to obtain the variational problem. The bilinear form must be well-defined on the test-function space.

The second point is to check that the variational formulation actually gives rise to a solution of the boundary value problem. This point is usually itself in two steps: first obtain the PDE in the sense of distributions by using test-functions in $\mathscr{D}$, second retrieve Neumann or Fourier boundary conditions by using the full test-function space. The first two points can appear somewhat formal because of the assumed regularity on the solution that is not always easily obtained in the end. This is not a real problem, since it is possible to write rigorous arguments, at the expense of more theory than we need here.

The final third point is to try and apply the Lax–Milgram theorem, by making precise regularity and possibly sign assumptions on the data and coefficients in order to ensure continuity of the linear and bilinear forms as well as V-ellipticity. Here we prove existence and uniqueness of the solution to the variational problem.

A question that can be asked is what is the relevance of such solutions to a boundary value problem, in which the partial derivatives are taken in a rather weak sense. This is where elliptic regularity theory comes into play. Using elliptic regularity, it is possible to show that the variational solution given by the Lax–Milgram theorem is indeed a classical solution, provided the coefficients, right-hand side, boundary of Ω and so on are smooth enough.

To be a little more precise on these regularity results, let us mention that they proceed in two steps

- Local regularity, which only depends on the regularity of the coefficients and right-hand side of the PDE. For example, for the model Dirichlet variational problem (4.3), with $c \in C^\infty(\Omega)$, we have for any integer $k \geq 0$

$$\text{if } f \in H^k_{\text{loc}}(\Omega), \text{ then } u \in H^{k+2}_{\text{loc}}(\Omega),$$

where

$$H^k_{\text{loc}}(\Omega) = \{f \in \mathscr{D}'(\Omega), \text{ such that for any } \varphi \in \mathscr{D}(\Omega),\ f\varphi \in H^k(\Omega)\}.$$

- Global regularity, i.e., up to the boundary Γ of Ω, which also depends on the regularity of the boundary and on the boundary conditions, see [3, 4, 15, 25, 35].

We have for example the following result, still for the model Dirichlet problem, see [44].

Proposition 4.18 *Let u be the solution of (4.3) with $f \in L^2(\Omega)$. If Γ is of class $C^{1,1}$, then $u \in H^2(\Omega)$.*

The same type of result holds for the Neumann problem of Proposition 4.6 with $g \in H^{1/2}(\Gamma)$, see [44]. We refer for example to [25] for a counterexample when Ω is not regular enough. □

In this chapter, we applied the variational method to solve elliptic boundary value problems from the theoretical point of view. We are now going to see that the variational approach is also very well suited to numerical approximation, in particular via the finite element method.

Chapter 5
Variational Approximation Methods for Elliptic PDEs

One of the virtues of the variational approach is that it leads naturally to a whole family of approximation methods. Let us emphasize again that the reason why approximation methods for PDEs are needed is that, even though we may be able to prove the existence of a solution, in general there is no closed form formula for it. In addition, such approximations must be effectively computable.

There are other approximation methods that are not variational, such as the finite difference method seen earlier, the finite volume method that we will see in Chap. 10, and yet many other methods that we will not consider in this book.

5.1 The General Abstract Variational Approximation Scheme

As we have seen, boundary value problems naturally take place in infinite dimensional vector spaces. An infinite dimensional space is too large to fit inside a computer, thus the main idea is to build finite dimensional approximations thereof. Any approximation method of this kind falls under the general heading of a *Galerkin method* [7, 11, 19, 28, 63, 65, 66]. Let us start with a few definitions that pertain to the variational case.

Definition 5.1 Let V be a Hilbert space and $(V_n)_{n\in\mathbb{N}}$ be a sequence of finite dimensional vector subspaces of V. We say that this sequence is a *conforming approximation sequence* if for all $u \in V$, there exists a sequence $(v_n)_{n\in\mathbb{N}}$ such that

$$v_n \in V_n \text{ and } \|u - v_n\|_V \to 0 \text{ when } n \to +\infty.$$

H. Le Dret and B. Lucquin, *Partial Differential Equations: Modeling, Analysis and Numerical Approximation*, International Series of Numerical Mathematics 168, DOI 10.1007/978-3-319-27067-8_5

Remark 5.1 Note that in general, we do not have $V_n \subset V_{n+1}$, i.e., the approximation spaces do not need to be nested. The conforming approximation condition implies that $\bigcup_{n\in\mathbb{N}} V_n$ is dense in V.

There are situations in which *non conforming approximations* are called for, that is to say $V_n \not\subset V$, see for example [11, 19, 45]. Of course, in this case $\|u - v_n\|_V$ does not make sense, and another definition is needed.

The traditional notation for an approximation sequence is V_h instead of V_n, where h is a discretization parameter that is assumed to belong to a sequence that tends to 0 instead of having $n \to +\infty$, the two being equivalent. We will from now on stick with the tradition. □

The main abstract result is the following, also known under the name of Céa's lemma, cf. [7, 9, 14].

Theorem 5.1 (Céa's Lemma) *Let V be a Hilbert space, a be a bilinear form and ℓ be a linear form satisfying the hypotheses of the Lax-Milgram theorem. Let V_h be a closed subspace of V. Then there exists a unique $u_h \in V_h$ such that*

$$\forall v_h \in V_h, \quad a(u_h, v_h) = \ell(v_h), \tag{5.1}$$

and we have

$$\|u - u_h\|_V \leq \frac{M}{\alpha} \inf_{v_h \in V_h} \|u - v_h\|_V = \frac{M}{\alpha} d(u, V_h),$$

where M is the continuity constant of a and α its V-ellipticity constant.

Proof Since V_h is closed, it is a Hilbert space for the restriction of the scalar product of V. The Lax-Milgram hypotheses for the variational problem on V_h are thus satisfied and the existence and uniqueness of u_h is assured.

Now we have $a(u, v) = \ell(v)$ for all $v \in V$, thus in particular for $v = w_h \in V_h$. On the other hand, we also have $a(u_h, w_h) = \ell(w_h)$, so that subtracting the two

$$0 = a(u, w_h) - a(u_h, w_h) = a(u - u_h, w_h) \tag{5.2}$$

for all $w_h \in V_h$. By V-ellipticity, for all $v_h \in V_h$,

$$\begin{aligned} \alpha\|u - u_h\|_V^2 &\leq a(u - u_h, u - u_h) \\ &\leq a(u - u_h, u - v_h) + a(u - u_h, v_h - u_h) \\ &= a(u - u_h, u - v_h) \\ &\leq M\|u - u_h\|_V \|u - v_h\|_V, \end{aligned}$$

since $w_h = v_h - u_h \in V_h$. The case $\|u - u_h\|_V = 0$ is ideal and nothing needs to be done. If the norm is non zero, we divide by it and obtain

$$\|u - u_h\|_V \leq \frac{M}{\alpha} \|u - v_h\|_V$$

for all $v_h \in V_h$, thus the theorem by taking the infimum of the right-hand side over V_h. □

Remark 5.2 A word of warning about the traditional notation $v_h \in V_h$. This traditional notation is rather unfortunate, since v_h is not a function of h, but an arbitrary element of V_h. The subscript h must thus not be understood as a regular subscript, but just as a typographical reminder that we are talking about an arbitrary element of V_h. On the other hand, the solution u_h can be considered as a function of h insofar as V_h can be considered as a function of h. □

We now apply Céa's lemma to the case of a conforming approximation sequence.

Corollary 5.1 *Let V_h be a conforming approximation sequence. Then the sequence $u_h \in V_h$ of approximated solutions converges to the solution u in V, with the* a priori error estimate

$$\|u - u_h\|_V \leq \frac{M}{\alpha} d(u, V_h) \to 0 \text{ when } h \to 0.$$

Proof Each subspace V_h is finite dimensional, hence closed. We thus apply Theorem 5.1 and obtain the convergence result since $d(u, V_h) \leq \|u - v_h\|_V$ where v_h is given by the definition of conforming approximation for this u. □

Remark 5.3 We also trivially have $\|u - u_h\|_V \geq d(u, V_h)$, thus the error estimate is optimal in terms of order of magnitude when $h \to 0$. Now, if the constant M/α is very large, then the numerical error can be large too with respect to $d(u, V_h)$.

In the case when a is symmetric, the constant in Céa's lemma can be improved to $\sqrt{M/\alpha}$. Indeed, in this case, a defines a second scalar product on V, for which V is also a Hilbert space. Equation (5.2) then says that $u - u_h$ is orthogonal to V_h for the new scalar product. Therefore, by the orthogonal projection Theorem 3.2 of Chap. 3, it minimizes the new distance to V_h and we thus have

$$\alpha\|u - u_h\|_V^2 \leq a(u - u_h, u - u_h) \leq a(u - v_h, u - v_h) \leq M\|u - v_h\|_V^2$$

for all $v_h \in V_h$. Taking the infimum in the right-hand side with respect to v_h yields the improved estimate. Of course, $M \geq \alpha$, so this is a real improvement of the constant.

An interesting feature of Céa's lemma is that it decomposes the error estimate into two basically independent parts: The constant M/α which only depends on the bilinear form, i.e., the PDE, and not on the approximation method, and $d(u, V_h)$ which depends mostly on the approximation properties of the space V_h. In practice, the second part will be estimated by constructing a linear operator $\Pi_h \colon V \to V_h$, writing that

$$d(u, V_h) \leq \|u - \Pi_h u\|_V \leq \|I - \Pi_h\| \|u\|_V$$

and estimating the term $\|I - \Pi_h\|$ which depends only on V_h. □

The approximation u_h lives in a finite dimensional space, therefore it is computable, at least in principle. Let us see how to proceed in practice.

Proposition 5.1 *Let $N_h = \dim V_h$ and $(w^1, w^2, \dots, w^{N_h})$ be a basis of V_h. We write $u_h = \sum_{j=1}^{N_h} u_{h,j} w^j$. We introduce an $N_h \times N_h$ matrix A defined by $A_{ij} = a(w^j, w^i)$ and two vectors $B \in \mathbb{R}^{N_h}$ by $B_i = \ell(w^i)$ and $X \in \mathbb{R}^{N_h}$ by $X_j = u_{h,j}$. Then the matrix A is invertible and we have $AX = B$. Conversely, the solution of this linear system is the vector of coordinates of u_h in the basis (w^i).*

Proof Let us take $v_h = w^i$ in the variational formulation of the finite dimensional problem. This yields

$$B_i = \ell(w^i) = a(u_h, w^i) = a\Big(\sum_{j=1}^{N_h} u_{h,j} w^j, w^i\Big) = \sum_{j=1}^{N_h} u_{h,j} a(w^j, w^i) = \sum_{j=1}^{N_h} A_{ij} X_j = (AX)_i$$

for all i. Hence $AX = B$.

Conversely, if $AX = B$, then by the above computation, $\ell(w^i) = a(\widetilde{u}_h, w^i)$ where $\widetilde{u}_h = \sum_{j=1}^{N_h} X_j w^j$. For all $v_h \in V_h$, we have $v_h = \sum_{i=1}^{N_h} v_{h,i} w^i$, therefore

$$\ell(v_h) = \sum_{i=1}^{N_h} v_{h,i} \ell(w^i) = \sum_{i=1}^{N_h} v_{h,i} a(\widetilde{u}_h, w^i) = a\Big(\widetilde{u}_h, \sum_{i=1}^{N_h} v_{h,i} w^i\Big) = a(\widetilde{u}_h, v_h)$$

therefore, by the uniqueness of the Lax-Milgram solution, we have $\widetilde{u}_h = u_h$. Thus the variational problem and the linear system are equivalent. Since the variational problem has one and only one solution for any ℓ, it follows that A is invertible. □

Remark 5.4 The problem of computing the finite dimensional approximation u_h is thus reduced to that of computing the matrix A and the right-hand side B once a basis of V_h is chosen, which is called *assembling the system*, and then of solving the linear system $AX = B$. In practical applications, N_h is typically large, ranging from the thousands to the millions or billions. This is a whole other subject with many facets: matrix conditioning, efficient algorithms for large linear systems, high performance computing. We will not touch on this.

It is important not to lose sight of the fact that the size of the matrix A and of the right-hand side B depend on h, via N_h, even though the notation fails to make this dependence apparent. In particular, when $h \to 0$, we have $N_h \to +\infty$.

Do not forget the exchange of indices $A_{ij} = a(w^j, w^i)$ and *not* $A_{ij} = a(w^i, w^j)$! Of course if a is symmetric, then the matrix A is symmetric. It is also positive, definite, with $a(v_h, v_h) = Y^T AY$ where Y is the vector of coordinates of $v_h \in V_h$ in the basis (w^i). □

We now introduce the main example of variational approximation method, the finite element method (FEM). There are other variational approximation methods, such as the spectral method, see [10, 11, 17, 43, 50, 63, 65] as well as such extensions as discontinuous Galerkin methods, see [14, 31, 52], which are non conforming methods, among many others.

In the rest of this chapter, we consider the finite element method in the one-dimensional case. The two-dimensional case will be the subject of the next chapter.

5.2 The Finite Element Method in Dimension One

Let $\Omega =]a, b[$ and consider the model problem

$$\begin{cases} -u'' + cu = f \text{ in } \Omega, \\ u(a) = u(b) = 0. \end{cases} \tag{5.3}$$

When $f \in L^2(a, b)$, $c \in L^\infty(a, b)$ and $c \geq 0$, we know that this problem has one and only one solution by using the variational formulation $V = H_0^1(]a, b[)$, $a(u, v) = \int_\Omega (u'v' + cuv)\,dx$ and $\ell(v) = \int_\Omega fv\,dx$.

The idea of the FEM is to take approximation spaces V_h composed of functions that are piecewise polynomial of low degree, with lots of pieces. In one dimension, we have $H^1(]a, b[) \subset C^0([a, b])$, thus for the approximation to be conforming, we need to impose $V_h \subset C^0([a, b])$ as well.

The FEM is based on the notion of *mesh*. In one dimension, a mesh is just a subdivision of $]a, b[$ into a finite number of subintervals. Each one of the small intervals is called an *element*. We will only consider uniform meshes for simplicity. Nonuniform meshes pose no additional conceptual difficulty as should become clear. Let N be a positive integer. We set $h = \frac{b-a}{N+1}$, which is called the *mesh size*, and let $x_i = a + ih$, $i = 0, \ldots, N+1$, be the *nodes* of the mesh.

We thus have $N + 1$ subintervals $[x_i, x_{i+1}]$ of length h, N interior nodes x_i, $i = 1, \ldots, N$, and 2 boundary nodes x_0 and x_{N+1}. Even though Fig. 5.1 is virtually identical to Fig. 2.1 of Chap. 2 depicting a one-dimensional grid, the two concepts of mesh and grid are really different. In a sense, in a grid only the grid points count and nothing in between, whereas in a mesh both the elements and nodes are of importance. Of course, the visual difference between grids and meshes is more apparent in dimensions 2 and higher.

We now define

$$V_h = \{v_h \in C^0([a, b]);\ v_{h|[x_i, x_{i+1}]} \text{ is affine for } i = 0, \ldots, N,\ v_h(a) = v_h(b) = 0\}.$$

Note that here, the subscript h in V_h is actually the same h as the mesh size. Since $h \to 0$ when $N \to +\infty$, we thus have a sequence of spaces. We first need to verify that these spaces are subspaces of V.

Proposition 5.2 *We have* $V_h \subset H_0^1(]a, b[)$.

Proof First of all, since $V_h \subset C^0([a, b])$ with $[a, b]$ compact, we have $V_h \subset L^2(a, b)$. Let us compute the distributional derivative of an element v_h of V_h. Since $v_{h|[x_i, x_{i+1}]}$

Fig. 5.1 A uniform 1d mesh

is an affine function, we can write $v_h(x) = \lambda_i x + \mu_i$ for $x \in [x_i, x_{i+1}]$, where λ_i, μ_i are constants that depend on the subinterval. For all $\varphi \in \mathscr{D}(]a, b[)$, we have

$$\begin{aligned}\langle v_h', \varphi\rangle = -\langle v_h, \varphi'\rangle &= -\int_a^b v_h(x)\varphi'(x)\,dx\\ &= -\sum_{i=0}^N \int_{x_i}^{x_{i+1}} v_h(x)\varphi'(x)\,dx\\ &= -\sum_{i=0}^N \int_{x_i}^{x_{i+1}} (\lambda_i x + \mu_i)\varphi'(x)\,dx.\end{aligned}$$

Now we can classically integrate each element integral by parts,

$$\begin{aligned}-\int_{x_i}^{x_{i+1}} (\lambda_i x + \mu_i)\varphi'(x)\,dx &= \int_{x_i}^{x_{i+1}} \lambda_i\varphi(x)\,dx - [v_h(x)\varphi(x)]_{x_i}^{x_{i+1}}\\ &= \int_{x_i}^{x_{i+1}} \lambda_i\varphi(x)\,dx - v_h(x_{i+1})\varphi(x_{i+1}) + v_h(x_i)\varphi(x_i),\end{aligned}$$

since v_h is continuous on $[a, b]$ its right and left limits at x_i and x_{i+1} respectively are just its value at these points.

Now, if we let

$$g(x) = \sum_{i=0}^N \lambda_i \mathbf{1}_{]x_i, x_{i+1}[}(x),$$

then obviously g is a piecewise constant function, hence is bounded, and thus in $L^2(a, b)$ and

$$\sum_{i=0}^N \int_{x_i}^{x_{i+1}} \lambda_i\varphi(x)\,dx = \int_a^b g(x)\varphi(x)\,dx.$$

On the other hand,

$$\begin{aligned}-\sum_{i=0}^N [v_h(x)\varphi(x)]_{x_i}^{x_{i+1}} = &- v_h(x_1)\varphi(x_1) + v_h(x_0)\varphi(x_0)\\ &- v_h(x_2)\varphi(x_2) + v_h(x_1)\varphi(x_1) - \cdots\\ &\cdots - v_h(x_{N+1})\varphi(x_{N+1}) + v_h(x_N)\varphi(x_N) = 0,\end{aligned}$$

since all terms involving interior nodes appear twice with opposite signs, and $\varphi(x_0) = \varphi(x_{N+1}) = 0$ since φ has compact support. Finally, we see that

$$\langle v_h', \varphi\rangle = \int_a^b g(x)\varphi(x)\,dx = \langle g, \varphi\rangle,$$

with $g \in L^2(a, b)$ which shows that $v_h \in H^1(]a, b[)$ and $v_h' = g$. Now all elements of V_h also satisfy $v_h(a) = v_h(b) = 0$ so that $v_h \in H_0^1(]a, b[)$. □

It is fairly clear that the space V_h is finite dimensional, since any of its elements is determined by a finite number of constants λ_i and μ_i. The space V_h is therefore closed, and the general abstract principle applies, i.e., there exists a unique $u_h \in V_h$ such that $a(u_h, v_h) = \ell(v_h)$ for all $v_h \in V_h$, and we have Céa's lemma error estimate. Let us see how this estimate can be exploited to quantify the convergence rate. Let us start with a general purpose lemma concerning V_h.

Lemma 5.1 *There exists a unique continuous linear mapping $\Pi_h : H_0^1(]a, b[) \to V_h$, called the V_h-*interpolation operator *such that for all v in $H_0^1(]a, b[)$, $v(x_i) = \Pi_h v(x_i)$ for $i = 0, \dots, N + 1$.*

Proof First of all, we note that $H_0^1(]a, b[) \hookrightarrow C^0([a, b])$, therefore the nodal values $v(x_i)$ are unambiguously defined and $v(x_0) = v(x_{N+1}) = 0$.

Now an affine function on $[x_i, x_{i+1}]$ is uniquely determined by its values at x_i and x_{i+1}. Thus, the relations $v(x_i) = \Pi_h v(x_i)$ for $i = 0, \dots, N + 1$ define a unique piecewise affine function on the mesh, that is continuous and vanishes at both ends, thus belongs to V_h. Let $\Pi_h v$ be this function. Clearly, the mapping $v \mapsto \Pi_h v$ is linear from $H_0^1(]a, b[)$ into V_h. Finally we infer from the fact that the values taken by an affine function on an interval lie between the values at the endpoints, that $\max_{x \in [x_i, x_{i+1}]} |\Pi_h v(x)| = \max\big(|v(x_i)|, |v(x_{i+1})|\big)$, and therefore

$$
\begin{aligned}
\|\Pi_h v\|_{C^0([a,b])} &= \max_{i=0,\dots,N} \max_{x \in [x_i, x_{i+1}]} |\Pi_h v(x)| \\
&= \max_{i=0,\dots,N} \max\big(|v(x_i)|, |v(x_{i+1})|\big) \\
&\le \max_{x \in [a,b]} |v(x)| = \|v\|_{C^0([a,b])} \le C\|v\|_{H^1(]a,b[)},
\end{aligned}
$$

by Theorem 3.7 of Chap. 3. Consequently, the V_h-interpolation operator is continuous. □

Remark 5.5 A picture is in order here. As can be seen on Fig. 5.2, $\Pi_h v$, which we call the *V_h-interpolate of v*, is the unique element of V_h that coincides with v at all nodes of the mesh. □

Theorem 5.2 *Assume c and f are continuous on $[a, b]$. Then the solution u of problem* (5.3) *is of class $C^2([a, b])$ and there exists a constant C independent of u such that*

$$
\|u - u_h\|_V \le Ch \max_{x \in [a,b]} |u''(x)|. \tag{5.4}
$$

Proof If f and c are continuous, since u is also continuous by Theorem 3.7 of Chap. 3, then $u'' = cu - f$ is continuous on $[a, b]$ and $u \in C^2([a, b])$. By Céa's lemma (viz. Theorem 5.1), we have

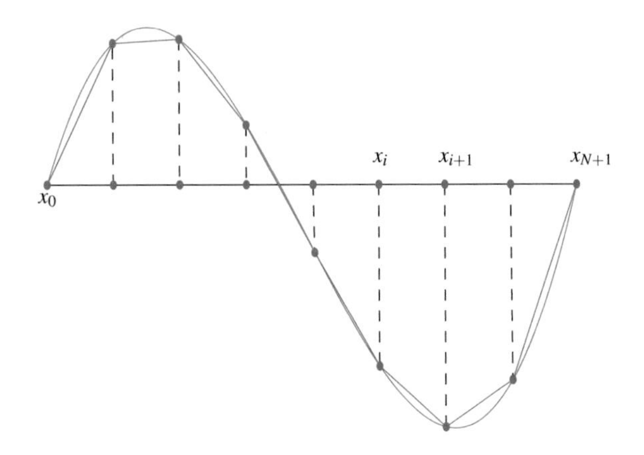

Fig. 5.2 The V_h-interpolate $\Pi_h v$ of a function v

$$\|u - u_h\|_V \leq \frac{M}{\alpha} \inf_{v_h \in V_h} \|u - v_h\|_V.$$

We choose $v_h = \Pi_h u$. It follows that

$$\|u - u_h\|_V \leq \frac{M}{\alpha} \|u - \Pi_h u\|_V,$$

and we are left with estimating the rightmost norm.

Let us take the H^1 semi-norm as a norm on V (this makes for simpler computations). We have

$$\|u - \Pi_h u\|_V^2 = \int_a^b ((u - \Pi_h u)'(x))^2\, dx = \sum_{i=0}^{N} \int_{x_i}^{x_{i+1}} (u' - (\Pi_h u)'(x))^2\, dx.$$

Let us consider the function $w = u - \Pi_h u$ on $[x_i, x_{i+1}]$. By definition of V_h-interpolation, we have $w(x_i) = w(x_{i+1}) = 0$. Since w is C^1 on $[x_i, x_{i+1}]$, Rolle's theorem applies and there exists $c \in]x_i, x_{i+1}[$ such that $w'(c) = 0$. Now, w is also of class C^2 on $[x_i, x_{i+1}]$ so that

$$w'(x) = \int_c^x w''(t)\, dt = \int_c^x u''(t)\, dt$$

for all $x \in [x_i, x_{i+1}]$, since $\Pi_h u$ is affine there, thus its second derivative vanishes. It follows from this equality that

$$|w'(x)| \leq \int_c^x |u''(t)|\, dt \leq \int_{x_i}^{x_{i+1}} |u''(t)|\, dt \leq h \max_{t \in [x_i, x_{i+1}]} |u''(t)| \leq h \max_{x \in [a,b]} |u''(x)|,$$

for all $x \in [x_i, x_{i+1}]$. Squaring and integrating, we thus see that

$$\int_{x_i}^{x_{i+1}} ((u - (\Pi_h u))'(x))^2 \, dx = \int_{x_i}^{x_{i+1}} (w'(x))^2 \, dx \leq h^2 (x_{i+1} - x_i) \max_{x \in [a,b]} |u''(x)|^2.$$

Now we sum from $i = 0$ to N

$$\|u - \Pi_h u\|_V^2 \leq h^2 \Big(\sum_{i=0}^{N} (x_{i+1} - x_i) \Big) \max_{x \in [a,b]} |u''(x)|^2 = h^2 (b - a) \max_{x \in [a,b]} |u''(x)|^2.$$

Finally, we obtain

$$\|u - u_h\|_V \leq \Big(\frac{M}{\alpha} \sqrt{b - a} \Big) h \max_{x \in [a,b]} |u''(x)|,$$

which completes the proof. □

Remark 5.6 Note that we have not proved that the sequence V_h is a conforming approximation sequence in the sense of Definition 5.1. Rather, we have exploited Céa's error estimate directly, coupled with an additional regularity hypothesis, here that u be C^2 essentially, to obtain an explicit error estimate and a convergence order in $O(h)$ when $h \to 0$, see Fig. 5.3. Note that $u \in H^2(]a, b[)$ is sufficient to obtain the same error estimate, see Theorem 6.2 of Chap. 6 for a proof in the two-dimensional case. More precisely, the following inequality holds true

$$\|u - \Pi_h u\|_V \leq Ch \|u\|_{H^2(]a,b[)}, \tag{5.5}$$

which in turns implies the error estimate

$$\|u - u_h\|_V \leq Ch \|u\|_{H^2(]a,b[)}, \tag{5.6}$$

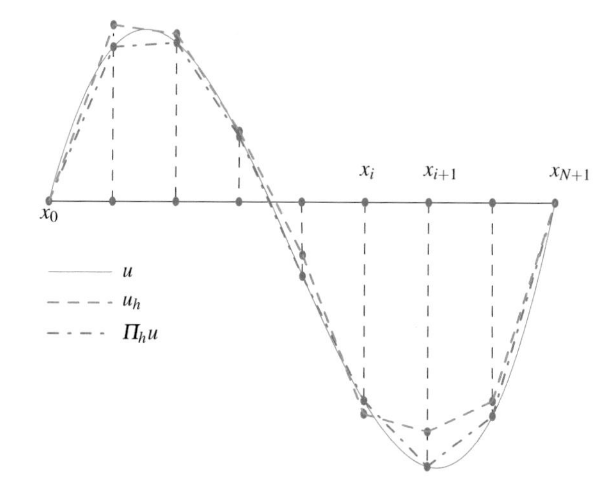

Fig. 5.3 A fictitious computation: the continuous solution u, the discrete solution u_h, and the V_h-interpolate $\Pi_h u$ of u used to control the error between the former two. Note that only u_h is effectively computable

due to Céa's lemma. This will be a general fact: additional regularity hypotheses on the solution will be needed for explicit error estimates. Such regularity can however often be deduced from elliptic regularity theory.

The sequence V_h is in fact a conforming approximation sequence, but this does not turn out to be too useful, as the convergence toward a generic element of H^1 can be much slower than $O(h)$. □

Let us now talk about the choice of a basis in V_h. Even though in principle, the resolution of the finite dimensional problem should not depend on the basis choice, in practice this is an extremely important issue since the choice of basis directly impacts the matrix A. A wrong choice of basis can lead to a linear problem that cannot be solved numerically (bad conditioning, see Remark 2.11 of Chap. 2, full matrix) in the sense that all theoretically convergent algorithms may fail or take too long or use up too much computer memory. Recall that for a basis (w^j), the matrix coefficients are given by

$$A_{ij} = a(w^j, w^i) = \int_a^b ((w^j)'(w^i)' + cw^j w^i)\, dx.$$

For numerical purposes, full matrices are to be avoided and sparse matrices preferred. Now, there is an easy way of making sure that $A_{ij} = 0$, given the above formula, and that is to arrange for the supports of w^i and w^j to have negligible intersection. So we want to find a basis for V_h for which the supports are as small as possible, in order to minimize the intersections. Now clearly, the support of any function of V_h is at least comprised of two elements. We thus define

Definition 5.2 For $i = 1, \dots, N$, let $w_h^i \in V_h$ be defined by $w_h^i(x_i) = 1$ and $w_h^i(x_j) = 0$ for $j = 0, \dots, N+1$, $j \neq i$. We call these functions the *hat functions* or *basis functions* for P_1 Lagrange interpolation, see Fig. 5.4.

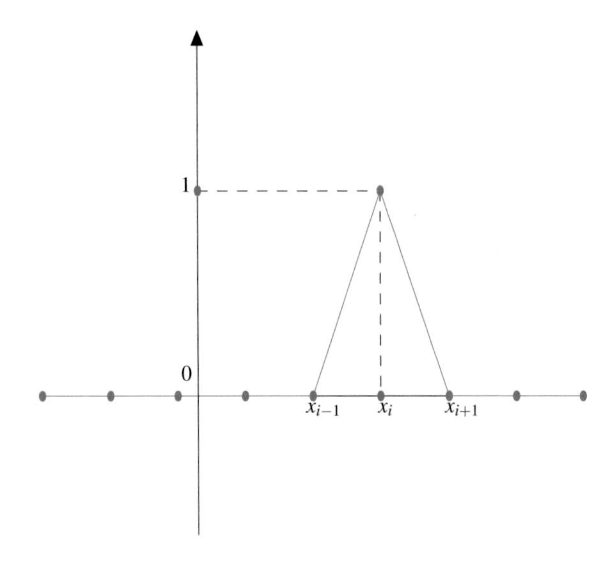

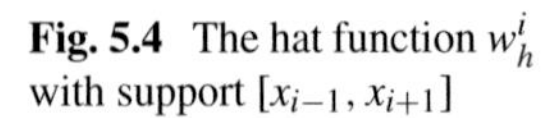
Fig. 5.4 The hat function w_h^i with support $[x_{i-1}, x_{i+1}]$

As we have said before, all functions in V_h are determined by their nodal values. In particular here, $w_h^i(a) = w_h^i(b) = 0$, since the endpoints correspond to $j = 0$ and $j = N + 1$, and $1 \le i \le N$. The term P_1 Lagrange interpolation stems from the fact that affine functions are polynomials of degree at most 1, hence P_1, and that these functions are also used for Lagrange interpolation in V_h, as we will see shortly.

Proposition 5.3 *The family* $(w_h^i)_{i=1,\dots,N}$ *is a basis of* V_h. *Thus* dim $V_h = N$. *Moreover, we have the* interpolation property

$$\forall v_h \in V_h, \quad v_h(x) = \sum_{i=1}^{N} v_h(x_i) w_h^i(x), \tag{5.7}$$

for all $x \in [a, b]$.

Proof We use the Kronecker delta symbol: $\delta_{ij} = 1$ if $i = j$, $\delta_{ij} = 0$ otherwise. The hat functions thus satisfy $w_h^i(x_j) = \delta_{ij}$ for $i = 1, \dots, N$ and $j = 0, \dots, N + 1$.

Let us first show that the family is linearly independent. Let λ_i be scalars such that

$$\sum_{i=1}^{N} \lambda_i w_h^i = 0.$$

Evaluating this zero function at point x_j yields

$$0 = \sum_{i=1}^{N} \lambda_i w_h^i(x_j) = \sum_{i=1}^{N} \lambda_i \delta_{ij} = \lambda_j$$

since in the last sum, the only nonzero term corresponds to $i = j$. Thus all coefficients vanish and the family is linearly independent.

Next we show that the family spans V_h. For all $v_h \in V_h$, we define

$$\widetilde{v}_h = \sum_{i=1}^{N} v_h(x_i) w_h^i \in V_h.$$

Now, of course $v_h - \widetilde{v}_h \in V_h$ and since $\widetilde{v}_h(x_j) = \sum_{i=1}^{N} v_h(x_i) w_h^i(x_j) = \sum_{i=1}^{N} v_h(x_i)$ $\delta_{ij} = v_h(x_j)$ (same computation as above), then $(v_h - \widetilde{v}_h)(x_j) = 0$ for all $j = 0, \dots, N + 1$. For each element $[x_j, x_{j+1}]$, we thus see that $v_h - \widetilde{v}_h$ is affine on the segment and vanishes at both endpoints, hence is identically zero on $[x_j, x_{j+1}]$. As this is true for all j, we have $v_h - \widetilde{v}_h = 0$ on $[a, b]$, that is to say $v_h = \widetilde{v}_h$, which shows both that the family is spanning and that we have formula (5.7).

The family $(w^i)_{i=1,\dots,N}$ is linearly independent and spanning, thus is a basis of V_h. It has N elements so that dim $V_h = N$. $\square$

Remark 5.7 The Lagrange interpolation property (5.7) is very important. It shows that with this specific choice of basis, the coordinates of a function v_h are precisely its nodal values $v_h(x_i)$. Hence solving the linear system $AX = B$ is going to directly provide the nodal values of the discrete solution u_h, without any post-processing. The linear forms $v_h \mapsto v_h(x_i)$, which belong to the dual V_h^* of V_h, are called the *degrees of freedom* in the FEM context. From the point of view of linear algebra, they are just the dual basis of the basis $(w^i)_{i=1,\dots,N}$. □

Corollary 5.2 *The $N \times N$ matrix A is tridiagonal in the hat functions basis.*

Proof Indeed, the support of w_h^i is $[x_{i-1}, x_{i+1}]$, therefore if $|i - j| \geq 2$, then $x_{i-1} \geq x_{j+1}$ or $x_{j-1} \geq x_{i+1}$ and the intersection of both supports is of zero measure, hence $A_{ij} = 0$. Thus, on any given line of the matrix A, we have at most three nonzero coefficients: $A_{i,i-1}$ corresponding to the subdiagonal, A_{ii} corresponding to the diagonal and $A_{i,i+1}$ corresponding to the superdiagonal. □

Of course, a tridiagonal matrix is the best kind of matrix that can be expected, apart from a diagonal matrix which cannot occur. This is because the problem is exceedingly simple. Actually, it is easy to compute all nonzero coefficients.

Proposition 5.4 *If $c = c_0$ and $f = f_0$ are constant, then we have*

$$A_{ii} = \frac{2}{h} + \frac{2h}{3}c_0, \quad A_{i,i-1} = A_{i,i+1} = -\frac{1}{h} + \frac{h}{6}c_0 \quad \textit{and} \quad B_i = hf_0.$$

Proof We start by noticing that $w_h^i(x) = w_h^1(x - x_i)$ (extending w_h^1 by zero outside of $[a, b]$), thus $A_{ii} = A_{11}$ and $A_{i,i-1} = A_{i,i+1} = A_{12}$.

It is easy to see that $w_h^1(x) = \frac{x}{h}$ on $[x_0, x_1]$ and $w_h^1(x) = 2 - \frac{x}{h}$ on $[x_1, x_2]$, 0 elsewhere. Therefore, $(w_h^1)'(x) = \frac{1}{h}$ on $[x_0, x_1]$, $(w_h^1)'(x) = -\frac{1}{h}$ on $[x_1, x_2]$, 0 elsewhere.

Thus

$$\begin{aligned}
A_{11} &= \int_{x_0}^{x_2} ((w_h^1)'(x)^2 + c_0 w_h^1(x)^2)\,dx \\
&= \int_{x_0}^{x_1} ((w_h^1)'(x)^2 + c_0 w_h^1(x)^2)\,dx + \int_{x_1}^{x_2} ((w_h^1)'(x)^2 + c_0 w_h^1(x)^2)\,dx \\
&= \frac{1}{h^2} \times h + \frac{c_0}{h^2}\int_{x_0}^{x_1} x^2\,dx + \frac{1}{h^2} \times h + \frac{c_0}{h^2}\int_{x_1}^{x_2} (2h - x)^2\,dx \\
&= \frac{2}{h} + \frac{2h}{3}c_0.
\end{aligned}$$

The intersection of the supports of w_h^1 and w_h^2 is $[x_1, x_2]$, hence

$$
\begin{aligned}
A_{12} &= \int_{x_1}^{x_2} ((w_h^1)'(x)(w_h^2)'(x) + c_0 w_h^1(x) w_h^2(x))\,dx \\
&= -\frac{1}{h^2} \times h + \frac{c_0}{h^2} \int_{x_1}^{x_2} (2h - x)(x - h)\,dx \\
&= -\frac{1}{h} + \frac{h}{6} c_0.
\end{aligned}
$$

We leave the last value to the reader. □

Remark 5.8 When c or f is not constant, the corresponding terms may not necessarily be exactly computable and it may be necessary to resort to numerical integration [23, 64, 71]. These terms however are corrections to the dominant terms $\frac{2}{h}$ and $-\frac{1}{h}$, so that it can be shown that choosing a sufficiently accurate numerical integration rule does not modify the final error estimate.

Numerical methods for linear systems are especially efficient in the case of a tridiagonal matrix. The LU factorization is very efficient, but other methods can be used such as the Cholesky factorization (the matrix is symmetric), the conjugate gradient method, and so on, see [6, 18, 59]. □

Let us now make a rapid comparison between the finite element method and the finite difference method seen in Chap. 2.

5.3 Comparison with the Finite Difference Method

We have shown that the finite element method is of order one in the case of the model one-dimensional example, cf. estimate (5.4), under the hypotheses c and f continuous. Under a slightly stronger hypothesis, namely c and f of class C^2, the finite difference error estimate (2.13) is of order two. It could thus be thought that the finite difference method is better than the finite element method.

It should however be noticed that these errors are not measured in the same norms. In particular, the finite difference error estimate does not involve the derivative of u, whereas the finite element does. If we do not take the derivative into account, the finite element method also yields an error of order two in the L^2 norm. This is called the Aubin-Nitsche duality trick that we now explain.

Proposition 5.5 *Assume that $c \in L^\infty(a, b)$ and $f \in L^2(a, b)$, then $u \in H^2(]a, b[)$ and we have*

$$
\|u - u_h\|_{L^2(a,b)} \leq Ch^2 \|u\|_{H^2(]a,b[)}. \tag{5.8}
$$

Proof We have seen that $u'' = cu - f$ in the sense of $\mathscr{D}'(]a, b[)$. Since $c \in L^\infty(a, b)$ and $u \in L^2(a, b)$, it follows that $u'' \in L^2(a, b)$, hence that $u \in H^2(]a, b[)$. Let us set $e_h = u - u_h$. We define the adjoint variational problem: Find $w \in V$ such that

$$\forall v \in V, \quad a(w, v) = \int_a^b e_h(x)v(x)\,dx. \tag{5.9}$$

Clearly, $w \in H^2(]a, b[)$ with $-w'' + cw = e_h$ almost everywhere in $]a, b[$. By Proposition 4.3 of Chap. 4, we thus have $\|w\|_V \leq C\|e_h\|_{L^2(a,b)}$. Therefore,

$$\|w''\|_{L^2(a,b)} \leq \|c\|_{L^\infty(a,b)}\|w\|_{L^2(a,b)} + \|e_h\|_{L^2(a,b)} \leq C\|e_h\|_{L^2(a,b)},$$

where C is a constant that changes from line to line. It follows that $\|w\|_{H^2(]a,b[)} \leq C\|e_h\|_{L^2(a,b)}$.[1]

We know that $a(e_h, v_h) = 0$ for all $v_h \in V_h$. Consequently

$$\begin{aligned}\|e_h\|^2_{L^2(a,b)} &= \int_a^b e_h(x)e_h(x)\,dx = a(w, e_h) = a(e_h, w)\\ &= a(e_h, w - \Pi_h w) \leq M\|e_h\|_V\|w - \Pi_h w\|_V,\end{aligned}$$

where we have used the dual problem (5.9), and the symmetry and continuity of the bilinear form a. By the error estimate (5.6), we have

$$\|e_h\|_V \leq Ch\|u\|_{H^2(]a,b[)},$$

on the one hand, and by the interpolation estimate (5.5), we have

$$\|w - \Pi_h w\|_V \leq Ch\|w\|_{H^2(]a,b[)} \leq Ch\|e_h\|_{L^2(a,b)},$$

on the other hand. Combining the above estimates, we obtain

$$\|e_h\|^2_{L^2(a,b)} \leq Ch^2\|u\|_{H^2(]a,b[)}\|e_h\|_{L^2(a,b)},$$

hence the result. □

It is interesting to make a numerical comparison of the finite element method with the finite difference method in the present context. When c and f are smooth, we know from the theory that both methods should give good results of about the same order, and this is confirmed by numerical experiments. We can make the situation a little more challenging for the finite difference method by taking nonsmooth data, which the finite element method should be able to handle correctly.

Consider thus the homogeneous Dirichlet problem $-u'' = f$ in $]0, 1[$, $u(0) = u(1) = 0$, with $f(x) = 20$ for $x \in]\frac{1}{2} - \frac{1}{40}, \frac{1}{2} + \frac{1}{40}[$, $f(x) = 0$ otherwise.[2] We plot in Fig. 5.5 the exact solution, the numerical finite element solution and the numerical finite difference solution.

---

[1] This is an elliptic regularity estimate.

[2] This right-hand side is an approximation of the Dirac mass $\delta_{\frac{1}{2}}$.

Estimate (5.4) is not valid since f is not continuous. However, this function is in $L^2(0, 1)$, and estimate (5.6) applies. Now it is not entirely clear whether or not the finite difference method converges in this case, but it is obviously having a significantly harder time than the finite element method.

In particular, we see on Fig. 5.5 that the convergence of the finite element method is very good, with an excellent agreement even for small values of N. On the other hand, the behavior of the finite difference approximation is curious. There is a marked difference between successive values of N, depending on parity. In any case, the numerical convergence of the finite difference approximation is here much slower than the convergence of the finite element method.

5.4 A Fourth Order Example

Let us now briefly consider the beam problem

$$\begin{cases} u^{(4)} + cu = f \text{ in } \Omega, \\ u(a) = u(b) = u'(a) = u'(b) = 0. \end{cases} \tag{5.10}$$

The variational formulation of this problem is set in $V = H_0^2(]a, b[)$ and since $H^2(]a, b[) \hookrightarrow C^1([a, b])$, the previous P_1 finite element method is not adapted (exercise, show that if v is piecewise affine, then $v'' = \sum_{i=1}^{N}(\lambda_i - \lambda_{i-1})\delta_{x_i} \notin L^2(a, b)$). We need higher degree polynomials to not only match the values, but also the derivatives at mesh nodes. In the following, P_k denotes the space of polynomials of degree at most k. We thus define

$$V_h = \{v_h \in C^1([a, b]);\ v_{h|[x_i, x_{i+1}]} \in P_3 \text{ for } i = 0, \dots, N,$$
$$v_h(a) = v_h(b) = v_h'(a) = v_h'(b) = 0\}.$$

A natural question is why not simply use P_2 polynomials. The reason is that even though in this case, the space V_h is not reduced to $\{0\}$, its description is very unwieldy and it is not clear that it can be used for approximation purposes. This is because the determination of a P_2 polynomial on an interval requires three interpolation data. It is thus possible to interpolate the value of the polynomial and of its derivative at one end of an interval, but then, there is only one interpolation value left at the other end, which makes it difficult to ensure that the piecewise P_2 function thus constructed is globally C^1 and satisfies the boundary conditions.

The above difficulty disappears for degrees $k \geq 3$ as we will see shortly.

Proposition 5.6 *We have $V_h \subset H_0^2(]a, b[)$.*

Proof Argue as in the proof of Proposition 5.2. □

We thus need C^1 functions, and in order to ensure the continuity and derivability at the endpoints of each element, we need to be able to specify both the value of

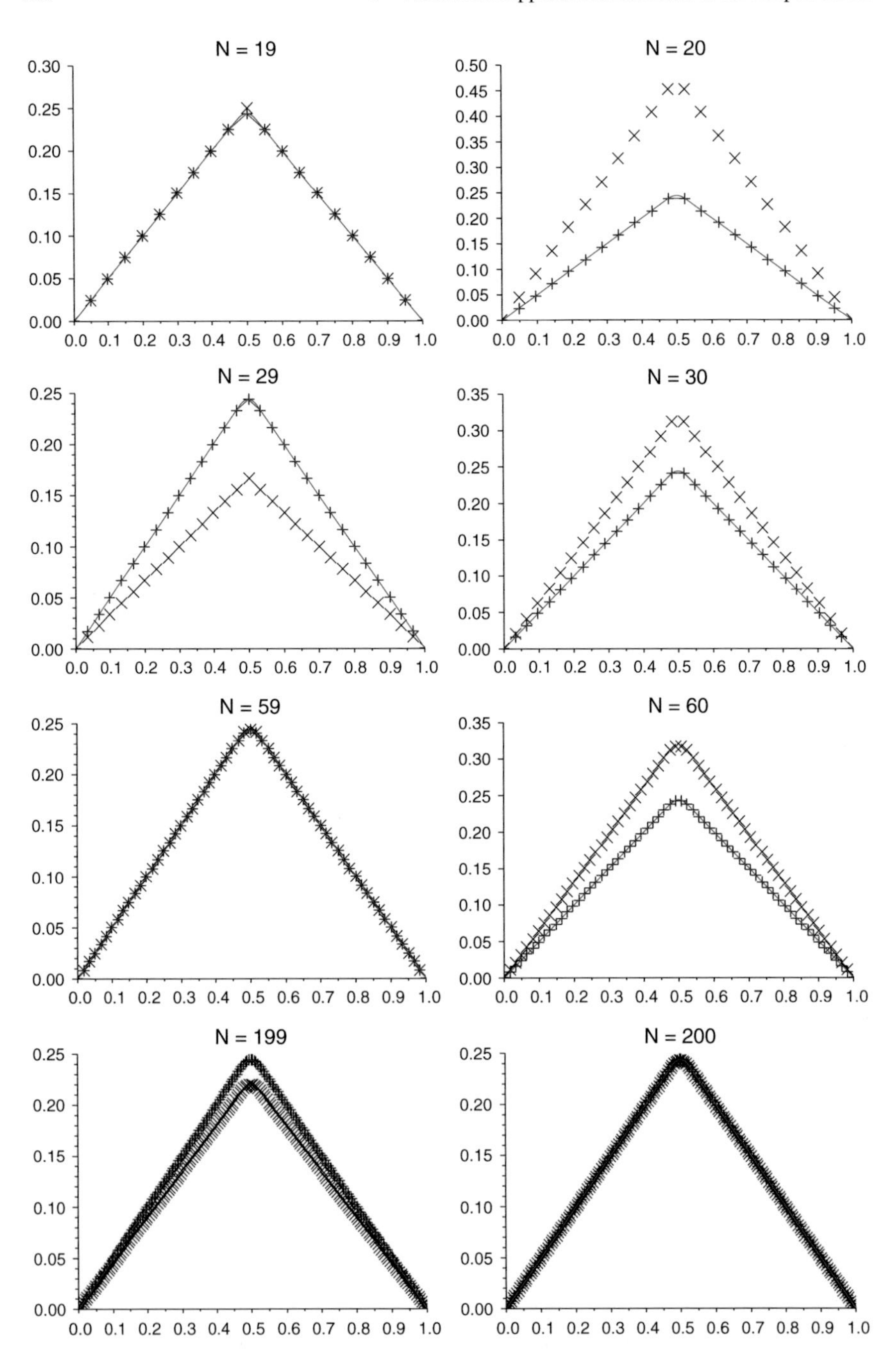

Fig. 5.5 Comparison finite difference (× *marks*), finite elements (+ *marks*) and exact solution (*solid curve*), for different values of N

the polynomial and of its derivative. The simplest way to achieve this is to use P_3 Hermite interpolation. Let us rapidly present this interpolation.

Proposition 5.7 *For all quadruplets* $(\alpha_0, \alpha_1, \beta_0, \beta_1)$ *of scalars, there exists a unique polynomial* $P \in P_3$ *such that*

$$P(0) = \alpha_0, \quad P(1) = \alpha_1, \quad P'(0) = \beta_0, \quad P'(1) = \beta_1.$$

This polynomial is given by

$$P = \alpha_0 p_0 + \alpha_1 p_1 + \beta_0 q_0 + \beta_1 q_1,$$

where the P_3 *Hermite basis polynomials* p_0, p_1, q_0 *and* q_1 *are given by*

$$p_0(x) = (1-x)^2(1+2x), p_1(x) = x^2(3-2x),$$
$$q_0(x) = x(1-x)^2, q_1(x) = x^2(x-1). \quad (5.11)$$

Proof The proof of Proposition 5.7 follows from a simple dimension argument: we show that the linear mapping $P_3 \to \mathbb{R}^4$, $P \mapsto (P(0), P(1), P'(0), P'(1))$ is an isomorphism. Since P_3 is four-dimensional, it suffices to show that its kernel is trivial. But a polynomial such that $P(0) = P(1) = P'(0) = P'(1) = 0$ has a double root at $x = 0$ and another double root at $x = 1$, hence a number of roots counting multiplicities of at least four. We know that a nonzero polynomial of degree at most three has at most three roots. Hence $P = 0$.

The inverse image of the canonical basis of $\mathbb{R}^4$ by the previous isomorphism forms a basis of P_3. Its elements are uniquely determined by the following interpolation values: $(\alpha_0, \alpha_1, \beta_0, \beta_1) = (1, 0, 0, 0)$ for p_0, $(\alpha_0, \alpha_1, \beta_0, \beta_1) = (0, 1, 0, 0)$ for p_1, $(\alpha_0, \alpha_1, \beta_0, \beta_1) = (0, 0, 1, 0)$ for q_0 and $(\alpha_0, \alpha_1, \beta_0, \beta_1) = (0, 0, 0, 1)$ for q_1. Therefore, any polynomial P of P_3 is uniquely written as

$$P = P(0)p_0 + P(1)p_1 + P'(0)q_0 + P'(1)q_1.$$

Formulas (5.11) can then be checked by hand, see Fig. 5.6. □

In FEM language, the linear forms $P \mapsto P(0)$, $P \mapsto P(1)$, $P \mapsto P'(0)$ and $P \mapsto P'(1)$ are the degrees of freedom of P_3 Hermite interpolation on the reference element $[0, 1]$.

Once we have Hermite interpolation on the reference element $[0, 1]$, we have Hermite interpolation on any element $[x_i, x_{i+1}]$ by a simple affine change of variables: $p_0\left(\frac{x-x_i}{h}\right)$, $p_1\left(\frac{x-x_i}{h}\right)$, $hq_0\left(\frac{x-x_i}{h}\right)$ and $hq_1\left(\frac{x-x_i}{h}\right)$.

Lemma 5.2 *There exists a unique continuous linear mapping* $\Pi_h \colon H_0^2(]a, b[) \to V_h$*, again called the* V_h-interpolation operator *such that for all* v *in* $H_0^2(]a, b[)$, $v(x_i) = \Pi_h v(x_i)$ *and* $v'(x_i) = (\Pi_h v)'(x_i)$ *for* $i = 0, \dots, N+1$.

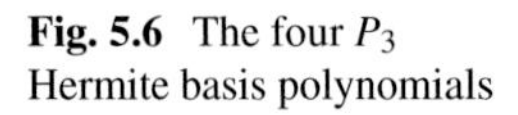
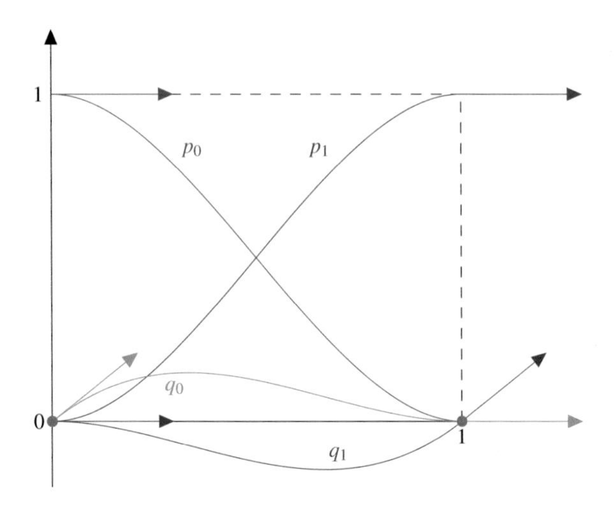

Fig. 5.6 The four P_3 Hermite basis polynomials

Proof First of all, we note that $H_0^2(]a,b[) \hookrightarrow C^1([a,b])$, therefore the nodal values $v(x_i)$ and $v'(x_i)$ are unambiguously defined.

Now a P_3 polynomial on $[x_i, x_{i+1}]$ is uniquely determined by its values and the values of its derivative at x_i and x_{i+1}, by Hermite interpolation. Thus, the relations $v(x_i) = \Pi_h v(x_i)$ and $v'(x_i) = (\Pi_h v)'(x_i)$ for $i = 0, \ldots, N+1$ define a unique piecewise P_3 function on the mesh, that is globally C^1 and vanishes at both ends together with its derivatives, thus belongs to V_h. Let $\Pi_h v$ be this function. Clearly, the mapping $v \mapsto \Pi_h v$ is linear from $H_0^1(]a,b[)$ into V_h. We leave the continuity to the reader. □

The above proof also shows that $V_h \neq \{0\}$. We can now show an error estimate, along the same lines as before.

Proposition 5.8 *Assume c and f are continuous on* $[a,b]$. *Then the solution u of problem* (5.10) *is of class* $C^4([a,b])$ *and there exists a constant C independent of u such that*

$$\|u - u_h\|_V \leq Ch^2 \max_{x\in[a,b]} |u^{(4)}(x)|.$$

Proof If f and c are continuous, since u is also continuous by Theorem 3.7 of Chap. 3, then $u^{(4)} = cu - f$ is continuous on $[a,b]$ and thus $u \in C^4([a,b])$. By Céa's lemma, i.e., Theorem 5.1, we have

$$\|u - u_h\|_V \leq \frac{M}{\alpha}\|u - \Pi_h u\|_V.$$

We use the H^2 semi-norm as a norm on V. We have

$$\|u - \Pi_h u\|_V^2 = \int_a^b ((u - \Pi_h u)''(x))^2\,dx = \sum_{i=0}^{N} \int_{x_i}^{x_{i+1}} (u''(x) - (\Pi_h u)''(x))^2\,dx.$$

Let us consider the function $w = u - \Pi_h u$ on $[x_i, x_{i+1}]$. By definition of V_h-interpolation, we have $w(x_i) = w(x_{i+1}) = 0$. Since w is C^1 on $[x_i, x_{i+1}]$, Rolle's theorem applies and there exists $c_1 \in]x_i, x_{i+1}[$ such that $w'(c_1) = 0$. Now, we also have $w'(x_i) = w'(x_{i+1}) = 0$ by V_h-interpolation, and w' is also of class C^1 so that Rolle applies again and there exists $c_2 < c_1 < c_3$ such that $w''(c_2) = w''(c_3) = 0$. We apply Rolle one last time since w'' is C^1 and obtain a point $c_4 \in [x_i, x_{i+1}]$ such that $w'''(c_4) = 0$. Consequently

$$w'''(x) = \int_{c_4}^{x} w^{(4)}(t)\,dt = \int_{c_4}^{x} u^{(4)}(t)\,dt$$

for all $x \in [x_i, x_{i+1}]$, since $\Pi_h u$ is of degree at most three there, thus its fourth derivative vanishes. It follows from this equality that

$$|w'''(x)| \leq \int_{c_4}^{x} |u^{(4)}(t)|\,dt \leq \int_{x_i}^{x_{i+1}} |u^{(4)}(t)|\,dt \leq h \max_{x\in[a,b]} |u^{(4)}(x)|,$$

for all $x \in [x_i, x_{i+1}]$. We also have

$$w''(x) = \int_{c_2}^{x} w'''(t)\,dt,$$

so that substituting the previous estimate yields

$$|w''(x)| \leq h^2 \max_{x\in[a,b]} |u^{(4)}(x)|.$$

Squaring and integrating, we thus see that

$$\int_{x_i}^{x_{i+1}} (u''(x) - (\Pi_h u)''(x))^2\,dx = \int_{x_i}^{x_{i+1}} (w''(x))^2\,dx \leq h^4 (x_{i+1} - x_i) \max_{x\in[a,b]} |u^{(4)}(x)|^2.$$

Now we sum from $i = 0$ to N

$$\|u - \Pi_h u\|_V^2 \leq h^4 (b-a) \max_{x\in[a,b]} |u^{(4)}(x)|^2,$$

which completes the proof. □

Remark 5.9 Under regularity hypotheses, we thus have convergence of the P_3 Hermite FEM based on the smallness of the interpolation error. It should be noted that this kind of proof relying on Rolle's theorem is not very natural in a Sobolev space context. There are better proofs using Hilbertian arguments. □

Let us say a few words about bases and matrices. It is apparent that the operator Π_h only uses the nodal values of the function and its derivatives. Hence, any set of interpolation data with N elements for the values and N elements for the derivative values gives rise to one and only one element of V_h. We thus define

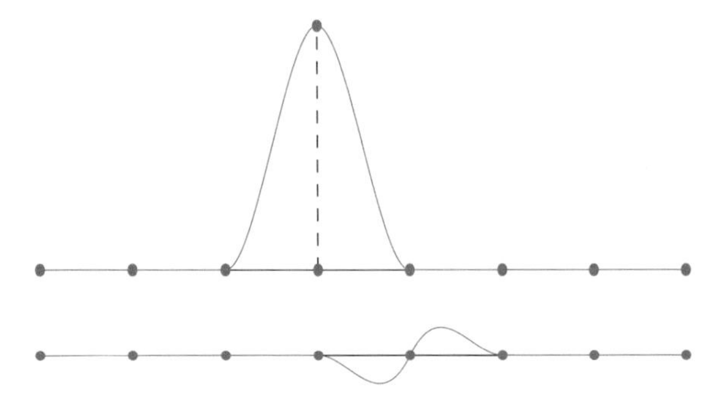

Fig. 5.7 A w_h^i basis function

Fig. 5.8 A z_h^i basis function

Definition 5.3 For $i = 1, \dots, N$, let $w_h^i \in V_h$ be defined by

$$w_h^i(x_j) = \delta_{ij} \text{ and } (w_h^i)'(x_j) = 0,$$

and $z_h^i \in V_h$ be defined by

$$z_h^i(x_j) = 0 \text{ and } (z_h^i)'(x_j) = \delta_{ij},$$

for $j = 0, \dots, N+1$. We call these functions the *basis functions* for P_3 Hermite interpolation on the mesh.

The function w_h^i is thus equal to 1 at x_i and zero at all other nodes, with zero derivatives at all nodes, whereas the function z_h^i has derivative 1 at x_i and zero at all other nodes, with zero values at all nodes, see Figs. 5.7 and 5.8. Clearly they are constructed by pairing together the Hermite basis interpolation polynomials in each element $[x_{i-1}, x_i]$ and $[x_i, x_{i+1}]$, which are also called *shape functions* in the FEM context.

Proposition 5.9 *The family* $(w_h^i, z_h^i)_{i=1,\dots,N}$ *is a basis of* V_h. *Thus* $\dim V_h = 2N$. *Moreover, we have the* interpolation property

$$\forall v_h \in V_h, \quad v_h(x) = \sum_{i=1}^{N} v_h(x_i) w_h^i(x) + \sum_{i=1}^{N} v_h'(x_i) z_h^i(x).$$

Proof Similar to the proof of Proposition 5.3 but using P_3 Hermite interpolation in each element. □

The supports of the basis functions are again $[x_{i-1}, x_{i+1}]$, thus we can expect lots of zero coefficients in the matrix. We do not write the detail here. Let us just mention that there is an issue of numbering. In the P_1 Lagrange case, there was a natural numbering of basis functions, which was that of the nodes. Here we have several choices, leading to different matrices. If we choose to number the basis elements as $(w_h^1, w_h^2, \dots, w_h^N, z_h^1, z_h^2, \dots, z_h^N)$, then the $2N \times 2N$ matrix A is comprised of four

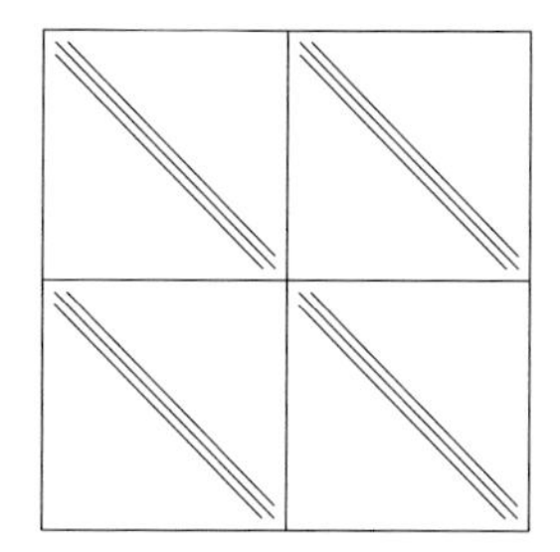
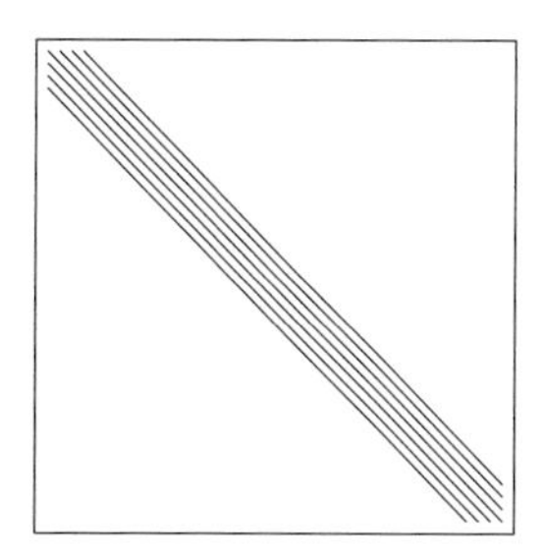

Fig. 5.9 Matrix structure. *Left* block tridiagonal, *right* interlaced

$N \times N$ blocks, and each one of the blocks is tridiagonal. If on the other hand, we interlace the basis functions like $(w_h^1, z_h^1, w_h^2, z_h^2, \dots, w_h^N, z_h^N)$, we obtain a matrix whose nonzero coefficients are grouped around the diagonal, this is called a *band matrix*. More precisely, each row of A has at most six nonzero coefficients resulting in seven nonzero diagonal rows: the diagonal plus three above the diagonal and three under the diagonal, see Fig. 5.9.

5.5 Neumann and Fourier Conditions

Let us briefly indicate how to deal with Neumann and Fourier conditions for the model second order problem. There are several changes: the test-function space must not enforce boundary conditions, i.e., $V = H^1(]a, b[)$, additional terms come up in the right hand-side for both problems, and there is an additional term in the bilinear form for the Fourier condition. Let us just consider the case $c \geq c_0 > 0$. We thus let

$$V_h = \{v_h \in C^0([a, b]); v_{h|[x_i, x_{i+1}]} \in P_1\}.$$

Compared to the previous version of V_h, we have added two degrees of freedom $v_h \mapsto v(a)$ and $v_h \mapsto v(b)$, hence $\dim V_h = N + 2$. We must accordingly complete the basis by adding two more basis functions w_h^0 and w_h^{N+1} defined by $w_h^0(x_j) = \delta_{0j}$ and $w_h^{N+1}(x_j) = \delta_{N+1,j}$ for all $j \in \{0, \dots, N+1\}$, see Fig. 5.10.

The variational formulation for the Fourier problem (replacing b by d in the Fourier condition to avoid a conflict in notation with the boundary b) is

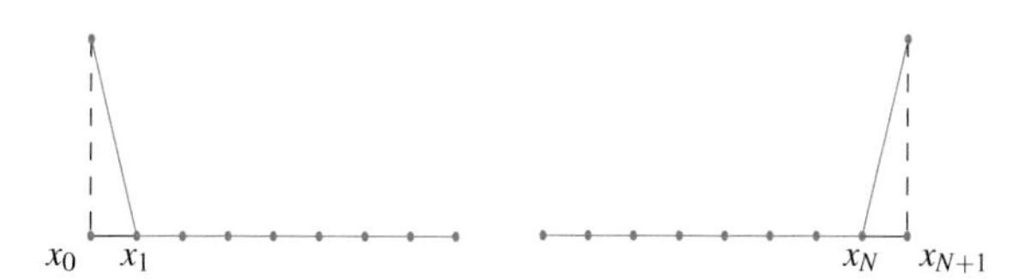

Fig. 5.10 The two additional basis functions w_h^0 *left* and w_h^{N+1} *right*

$$\int_a^b (u'v' + cuv)\,dx + (duv)(a) + (duv)(b) = \int_a^b fv\,dx + (gv)(a) + (gv)(b).$$

We just set $d = 0$ for the Neumann problem. In matrix terms, the $(N+2) \times (N+2)$ matrix A is still symmetric tridiagonal, and we have for c and d constants,

$$A_{00} = A_{N+1,N+1} = \frac{1}{h} + \frac{h}{3}c + d$$

and

$$A_{01} = A_{10} = A_{N,N+1} = A_{N+1,N} = -\frac{1}{h} + \frac{h}{6}c.$$

Of course

$$A_{0i} = A_{i0} = A_{j,N+1} = A_{N+1,j} = 0$$

for $i \geq 2$ and $j \leq N-1$. The other coefficients are unchanged. The right-hand side has two additional components $B_0 = \frac{hf}{2} + g(a)$ and $B_{N+1} = \frac{hf}{2} + g(b)$.

Similar changes must be made to treat a fourth order Neumann problem.

In the next chapter, we consider the finite element method in two dimensions. The abstract framework does not change, but the algebraic and geometric aspects are considerably more elaborate.

Chapter 6
The Finite Element Method in Dimension Two

It should already be clear that there is no difference between elliptic problems in one dimension and elliptic problems in several dimensions from the variational viewpoint. The same goes for the abstract part of variational approximations. The difference lies in the description of the finite dimensional approximation spaces. The FEM in any dimension of space is based on the same principle as in one dimension, that is to say, we consider spaces of piecewise polynomials of low degree, with lots of pieces for accuracy. Now things are right away quite different, and actually considerably more complicated, since polynomials have several variables, and open sets are much more varied than in dimension one. For simplicity, we limit ourselves to the two-dimensional case.

6.1 Meshes in 2d

Let Ω be an open subset of $\mathbb{R}^2$. The idea is to cover $\bar{\Omega}$ with a finite number of closed sets T_k of simple shape, $\bar{\Omega} = \bigcup_{k=1}^{N_{\mathscr{T}}} T_k$, with $\mathscr{T} = \{T_k\}_{1\leq k\leq N_{\mathscr{T}}}$ and $N_{\mathscr{T}} = \text{card}\,\mathscr{T}$. This decomposition will be used to decompose integrals over Ω into sums of integrals over the T_k, thus we impose that $\text{meas}\,(T_k \cap T_{k'}) = 0$ for $k \neq k'$. We will only consider two cases:

- The T_k are closed rectangles and $\bar{\Omega}$ is a union of rectangles. For definiteness, the sides of the rectangles will be parallel to the coordinate axes, without loss of generality, see Fig. 6.1.
- The T_k are closed triangles and $\bar{\Omega}$ is a polygon, see Fig. 6.2.

Such a structure will be called a *triangulation* (even in the case of rectangles …) or *mesh* on Ω. The T_k are called the elements, their sides are called the edges, and

H. Le Dret and B. Lucquin, *Partial Differential Equations: Modeling, Analysis and Numerical Approximation*, International Series of Numerical Mathematics 168, DOI 10.1007/978-3-319-27067-8_6

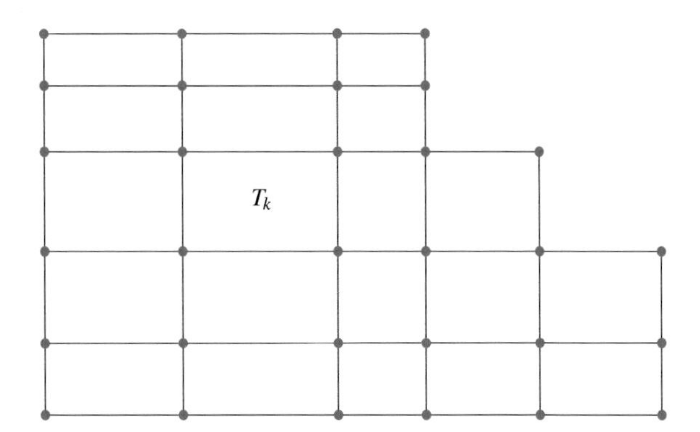

Fig. 6.1 A rectangular mesh

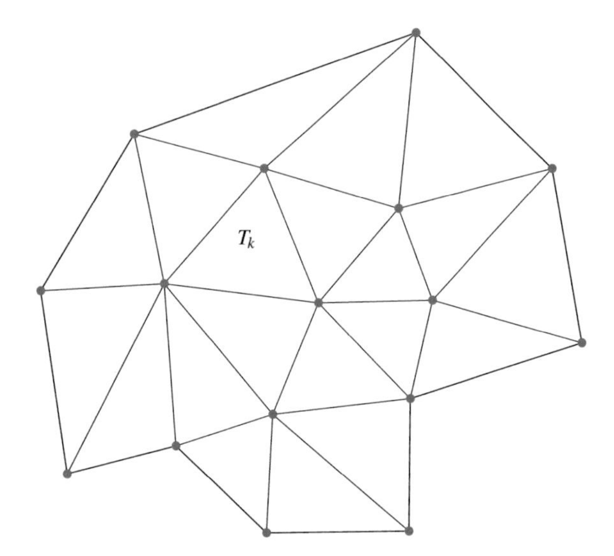

Fig. 6.2 A triangular mesh

their vertices are called mesh nodes.[1] The fact that $\bar{\Omega}$ must be a union of rectangles or more generally a polygon can appear to be unduly restrictive. There are however ways of going around this restriction and to cover very general domains, see [19, 21] for example.

In practice, meshes in a domain are not given but must be constructed by computer. This is a subject all by itself called *automatic mesh generation*. We will not pursue this subject here, but refer the reader to [39] for example. See Fig. 6.3 for a real-life example of three-dimensional mesh.

From now on, $\bar{\Omega}$ will always be a polygon in $\mathbb{R}^2$.

Given a mesh $\mathscr{T} = \{T_k\}_{1\leq k\leq N_{\mathscr{T}}}$, we let $h(T_k) = \operatorname{diam} T_k = \sup_{x,y\in T_k} \|x - y\|$ and

$$h = \max_{T_k\in\mathscr{T}} h(T_k).$$

[1] There are often additional mesh nodes, as we will see later.

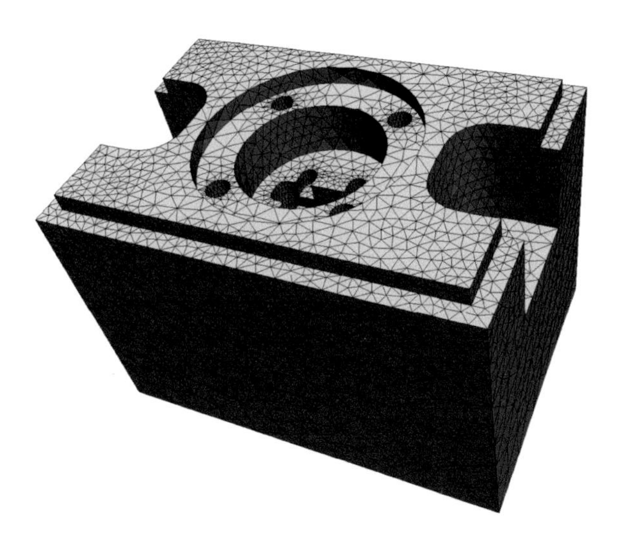

Fig. 6.3 A real life mesh in 3d. The elements are tetrahedra that fill the volume, we just see triangular faces of those tetrahedra that touch the boundary. We will not talk about 3d problems in this book, even though most real life problems occur in 3d. The conceptual difference between 3d and 2d is much less marked than between 2d and 1d

The scalar h is called the mesh size. The approximation spaces will thus be of the form

$$V_h = \{v \in V;\, v_{|T_k} \text{ is a low degree polynomial}\},$$

to be made more precise later. This is again a case of bad traditional notation, since V_h does not depend solely on h, but on the whole mesh, of which h is but one characteristic length. Accordingly, when we say $h \to 0$, this means that we are given a sequence $(\mathscr{T}_n)_{n\in\mathbb{N}}$ of meshes whose mesh size tends to 0 when $n \to +\infty$, and we will use the classical notation $\mathscr{T}_h$ instead. Naturally in practice, computer calculations are executed on one or a small number of meshes. The convergence $h \to 0$ is only for theoretical purposes.

In order to be of use, a triangulation must satisfy a certain number of properties.

Definition 6.1 A mesh is said to be *admissible* if

(i) For all $k \neq k'$, $T_k \cap T_{k'}$ is either empty, or consists of exactly one node or of one entire edge.
(ii) No T_k is of zero measure.

Condition (ii) means that no triangle or rectangle is degenerated, that is to say that no element has all its vertices on a straight line. Condition (i) is easier to understand in terms of which situations it precludes. For instance, any one of the three cases shown on Fig. 6.4 is forbidden.

For any triangle T, let $\rho(T)$ be the diameter of the inscribed circle (the center of the inscribed circle is called the incenter and is located at the intersection of the three internal angle bissectors, see Fig. 6.5).

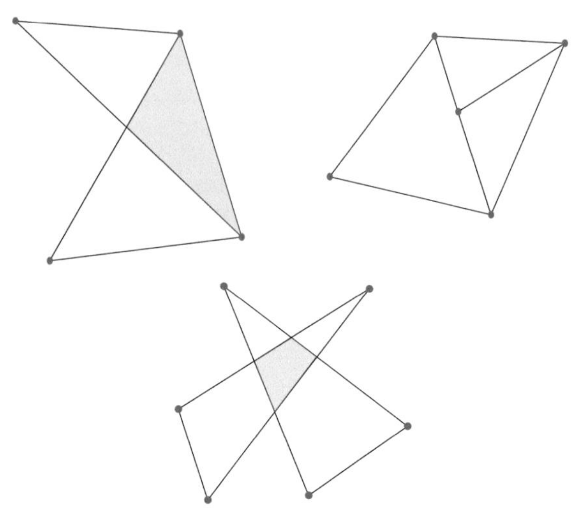

Fig. 6.4 Forbidden meshes according to rule (i)

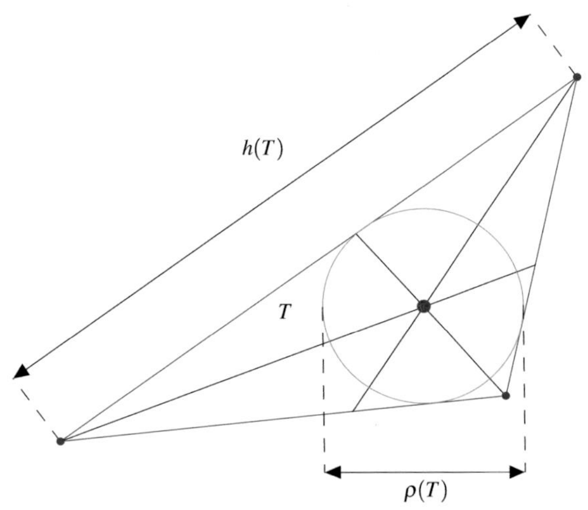

Fig. 6.5 Triangle incircle and diameter

Definition 6.2 Let $\mathscr{T}_h$ be a sequence of triangular meshes whose mesh size h tends to 0. We say that the sequence is a *regular family* if there exists a constant $C > 0$ such that for all h,

$$\max_{T \in \mathscr{T}_h} \frac{h(T)}{\rho(T)} \leq C.$$

For a sequence of meshes, not to be regular means that there are smaller and smaller triangles that become arbitrarily flat. Of course, the definition needs an infinite sequence of meshes to make sense. A similar condition for rectangular meshes is that the ratio of the longer side by the smaller side of each rectangle remains bounded from above. This property is needed for convergence results, as we will see in the proof of Lemma 6.4, p. 181. Note moreover that computations on a mesh such that $\max_{T\in\mathscr{T}_h}\frac{h(T)}{\rho(T)}$ is large may run into numerical difficulties, so the regularity condition is also relevant from the practical point of view.

Let us now give a general purpose proposition on piecewise regular functions on a mesh.

Proposition 6.1 *Let $\mathscr{T}$ be an admissible mesh on Ω. Define*

$$X_h = \{v \in C^0(\bar{\Omega});\ v_{|T_k} \in C^1(T_k) \textit{ for all } T_k \in \mathscr{T}\}.$$

Then we have $X_h \subset H^1(\Omega)$ and $\partial_i v = \sum_{k=1}^{N_{\mathscr{T}}} \partial_i(v_{|T_k})\mathbf{1}_{T_k}$ for all $v \in X_h$.

Proof Let $v \in X_h$. Clearly, $v \in L^2(\Omega)$ and we just need to compute its partial derivatives in the sense of distributions. Let us thus take an arbitrary function $\varphi \in \mathscr{D}(\Omega)$. We have

$$\langle \partial_i v, \varphi\rangle = -\langle v, \partial_i\varphi\rangle = -\int_\Omega v\partial_i\varphi\,dx = -\sum_{k=1}^{N_{\mathscr{T}}}\int_{T_k} v\partial_i\varphi\,dx.$$

Now v is C^1 on each T_k, which is a closed triangle or rectangle, therefore we can use the integration by parts formula to obtain

$$\begin{aligned}
-\int_{T_k} v\partial_i\varphi\,dx &= \int_{T_k}\partial_i(v_{|T_k})\varphi\,dx - \int_{\partial T_k} vn_{k,i}\varphi\,d\Gamma\\
&= \int_\Omega \partial_i(v_{|T_k})\mathbf{1}_{T_k}\varphi\,dx - \int_{\partial T_k} vn_{k,i}\varphi\,d\Gamma,
\end{aligned}$$

where n_k denotes the unit exterior normal vector to ∂T_k. Note that since $v \in C^0(\bar{\Omega})$ there is no need to take the restriction of v to T_k in the boundary term. Summing on all triangles or rectangles, we obtain

$$\langle \partial_i v, \varphi\rangle = \int_\Omega\Big(\sum_{k=1}^{N_{\mathscr{T}}}\partial_i(v_{|T_k})\mathbf{1}_{T_k}\Big)\varphi\,dx - \sum_{k=1}^{N_{\mathscr{T}}}\int_{\partial T_k} vn_{k,i}\varphi\,d\Gamma.$$

Now, each ∂T_k is composed of three or four edges and there are two cases:

1. Either the edge in question is included in $\partial\Omega$ and in this case $\varphi = 0$, hence the corresponding integral vanishes.
2. Or the edge is included in Ω (except possibly for one node) and in this case, by condition (i) of mesh admissibility, see Definition 6.1, this edge is the intersection of exactly two elements T_k and $T_{k'}$. The two integrals corresponding to this edge cancel out each other, since $v\varphi$ is continuous, it takes the same value on $\partial T_k \cap \partial T_{k'}$ as seen from either side, and $n_k = -n_{k'}$, see Fig. 6.6.

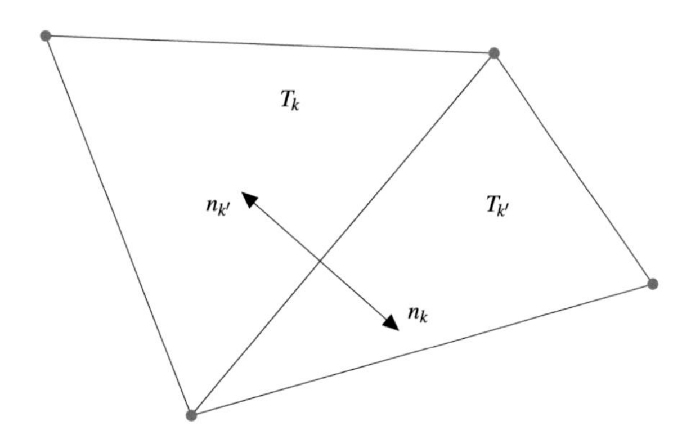

Fig. 6.6 Pairwise cancellation of edge integrals

Finally, we see that

$$\sum_{k=1}^{N_{\mathcal{T}}} \int_{\partial T_k} v n_{k,i} \varphi \, d\Gamma = 0$$

and since the function $\sum_{k=1}^{N_{\mathcal{T}}} \partial_i (v_{|T_k}) \mathbf{1}_{T_k}$ is bounded, it is also in $L^2(\Omega)$. □

Remark 6.1 The above proof shows that in fact $X_h \subset W^{1,\infty}(\Omega)$. □

6.2 Rectangular Q_1 Finite Elements

We start over with the model problem

$$\begin{cases} -\Delta u + cu = f \text{ in } \Omega, \\ \qquad\quad u = 0 \text{ on } \partial\Omega, \end{cases} \tag{6.1}$$

with $f \in L^2(\Omega)$, $c \in L^\infty(\Omega)$, $c \geq 0$ and $\Omega =]0,1[\times]0,1[$. The variational formulation is of course of the general form (4.6) with $V = H_0^1(\Omega)$, $a(u,v) = \int_\Omega (\nabla u \cdot \nabla v + cuv)\, dx$ and $\ell(v) = \int_\Omega f v \, dx$.

Let us be given two positive integers N_1 and N_2 and let $h_1 = \frac{1}{N_1+1}$ and $h_2 = \frac{1}{N_2+1}$. We define a rectangular mesh on Ω by setting

$$R_k = \{(x_1, x_2);\, i_1 h_1 \leq x_1 \leq (i_1+1)h_1,\, i_2 h_2 \leq x_2 \leq (i_2+1)h_2, \\ i_1 = 0, \ldots, N_1,\, i_2 = 0, \ldots, N_2\}.$$

The elements are rectangles with sides parallel to the coordinate axes and of lengths h_1 and h_2, see Fig. 6.7. There are $N_{\mathcal{T}} = (N_1+1)(N_2+1)$ elements. In the above formula, we have $k = 1, \ldots, N_{\mathcal{T}}$. As we will see later on, the actual numbering of the rectangles, i.e., the function $(i_1, i_2) \mapsto k$, is largely irrelevant. The mesh size is

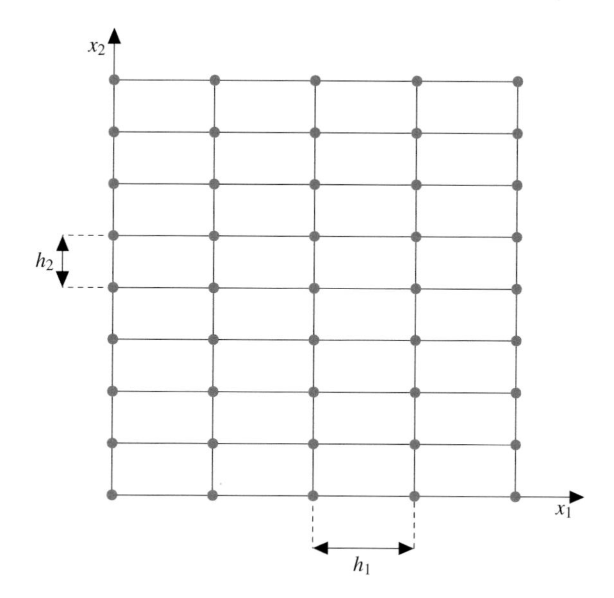

Fig. 6.7 A rectangular mesh on $\Omega =]0, 1[^2$, 32 elements, 45 nodes

$h = \sqrt{h_1^2 + h_2^2}$. Actually, since $\max(h_1, h_2) \leq h \leq \sqrt{2}\max(h_1, h_2)$, we may as well take $h = \max(h_1, h_2)$. The inscribed circle has diameter $\min(h_1, h_2)$, so the regularity requirement for a family of such meshes would be that $\frac{\max(h_1,h_2)}{\min(h_1,h_2)} \leq C$, or roughly speaking that N_1 and N_2 be of the same order of magnitude.

The mesh nodes are the points (i_1h_1, i_2h_2), $i_1 = 0, \dots, N_1 + 1, \ , i_2 = 0, \dots, N_2 + 1$. There is a total of $N_{\text{tot}} = (N_1 + 2)(N_2 + 2)$ nodes, including $2(N_1 + 1) + 2(N_2 + 1) = 2(N_1 + N_2) + 4 = N_{\text{bdy}}$ boundary nodes located on $\partial\Omega$ and $N_{\text{int}} = N_1N_2$ interior nodes located in Ω. Of course, $N_{\text{tot}} = N_{\text{int}} + N_{\text{bdy}}$. We will talk about numbering issues later (numbering of nodes, numbering of elements).

Let us now talk about the discrete approximation space. We first state a few facts about the algebra of polynomials in several variables. First of all, there are several notions of degree for such polynomials. The *total degree* of a nonzero monomial in two variables $ax_1^nx_2^m$ is $n + m$ (and the obvious generalization for more variables, that we will not use here). The total degree of a polynomial is the maximum total degree of its monomials. The *partial degree* of the same monomial is $\max(n, m)$. The partial degree of a polynomial is the maximum partial degree of its monomials. Since we are working on an infinite number field, $\mathbb{R}$, we can identify polynomials and polynomial functions on an open set of $\mathbb{R}^2$. We will perform this identification freely.

There are two families of spaces of polynomials that will be of interest to us.

Definition 6.3 For each $k \in \mathbb{N}$, we denote by P_k the space of polynomials of total degree less than or equal to k and by Q_k the space of polynomials of partial degree less than or equal to k.

Both spaces obviously are vector spaces. It is an easy exercise in algebra to establish that $\dim P_k = \frac{(k+1)(k+2)}{2}$ and $\dim Q_k = (k+1)^2$.

Since the total degree of a polynomial is always larger than its partial degree, it follows that $P_k \subset Q_k$. Moreover, as the Q_k monomial $x_1^k x_2^k$ is clearly of the highest possible total degree, we also have $Q_k \subset P_{2k}$. The only value of k for which these spaces coincide is thus $k = 0$, with only constant polynomials. The space P_1 is the space of affine functions

$$P_1 = \{p; p(x) = a_0 + a_1x_1 + a_2x_2, a_i \in \mathbb{R}\}$$

and the space Q_1 is described in terms of its canonical basis

$$Q_1 = \{p; p(x) = a_0 + a_1x_1 + a_2x_2 + a_3x_1x_2, a_i \in \mathbb{R}\}.$$

We can now introduce the corresponding approximation spaces. We start with a version without boundary conditions

$$W_h = \{v_h \in C^0(\bar{\Omega}); \forall R_k \in \mathscr{T}, v_{h|R_k} \in Q_1\},$$

and the subspace thereof that includes homogeneous Dirichlet conditions

$$V_h = \{v_h \in W_h; v_h = 0 \text{ on } \partial\Omega\},$$

see Remark 3.23 of Chap. 3.

The space W_h thus consists of globally continuous functions the restriction of which to each element coincides with one Q_1 polynomial per element. It is the same idea as in dimension 1. Since Q_1 polynomials are of course of class C^1, Proposition 6.1 immediately implies

Proposition 6.2 *We have $W_h \subset H^1(\Omega)$ and $V_h \subset H_0^1(\Omega)$.*

We now establish interpolation results for Q_1 polynomials and piecewise Q_1 functions. We start with a uniqueness result.

Proposition 6.3 *A function of W_h is uniquely determined by its values at the nodes of the mesh.*

Proof A function v_h in W_h is uniquely determined by the values it takes in each rectangular element, that is to say by the $N_{\mathscr{T}}$ polynomials in Q_1 that correspond to each element. It is thus sufficient to argue element by element. Let R be such an element and $S^i = (x_1^i, x_2^i)$ be its four vertices numbered counterclockwise starting from the lower left corner, see Fig. 6.8.

We have $h_1 = x_1^i - x_1^1$ for $i = 2, 3$ and $h_2 = x_2^i - x_2^1$ for $i = 3, 4$. Since v_h is equal to a Q_1 polynomial in R, there exists four constants α_j, $j = 1, \dots, 4$ such that

$$v_h(x) = \alpha_1 + \alpha_2(x_1 - x_1^1) + \alpha_3(x_2 - x_2^1) + \alpha_4(x_1 - x_1^1)(x_2 - x_2^1).$$

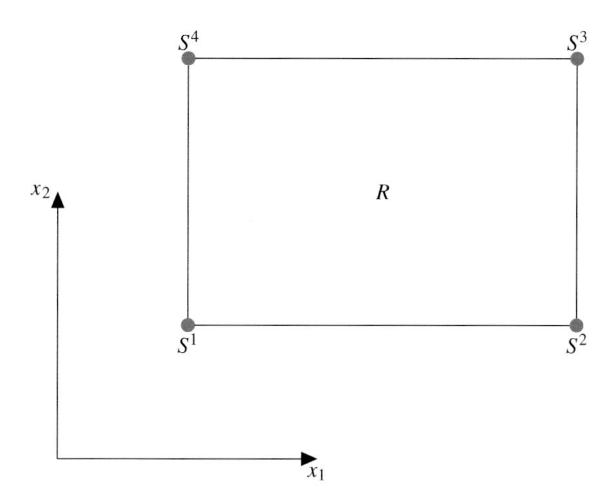

Fig. 6.8 A generic rectangle R in the mesh

Let us express the values of v_h at the four vertices.

$$
\begin{aligned}
v_h(S^1) &= \alpha_1 \\
v_h(S^2) &= \alpha_1 + \alpha_2(x_1^2 - x_1^1) + \alpha_3(x_2^2 - x_2^1) + \alpha_4(x_1^2 - x_1^1)(x_2^2 - x_2^1) \\
&= \alpha_1 + \alpha_2 h_1
\end{aligned}
$$

since $x_2^2 = x_2^1$,

$$
\begin{aligned}
v_h(S^4) &= \alpha_1 + \alpha_3 h_2 \\
v_h(S^3) &= \alpha_1 + \alpha_2 h_1 + \alpha_3 h_2 + \alpha_4 h_1 h_2.
\end{aligned}
$$

This is a 4×4 linear system in the four unknowns α_j which we can rewrite in matrix form

$$
\begin{pmatrix} 1 & 0 & 0 & 0 \\ 1 & h_1 & 0 & 0 \\ 1 & 0 & h_2 & 0 \\ 1 & h_1 & h_2 & h_1 h_2 \end{pmatrix}
\begin{pmatrix} \alpha_1 \\ \alpha_2 \\ \alpha_3 \\ \alpha_4 \end{pmatrix}
=
\begin{pmatrix} v_h(S^1) \\ v_h(S^2) \\ v_h(S^4) \\ v_h(S^3) \end{pmatrix}.
$$

The determinant of the triangular matrix above is $h_1^2 h_2^2 \neq 0$, hence the system has one and only one solution for any given vertex values for v_h. Therefore, we have the announced uniqueness. □

We also have an existence result.

Proposition 6.4 *For any set of values assigned to the nodes of the mesh, there exists one and only one element v_h of W_h that takes these values at the nodes.*

Proof The previous proof shows that four values for the four vertices of an element determine one and only one Q_1 polynomial that interpolates the values at the vertices inside the element. Therefore, if we are given a set of values for each node in the

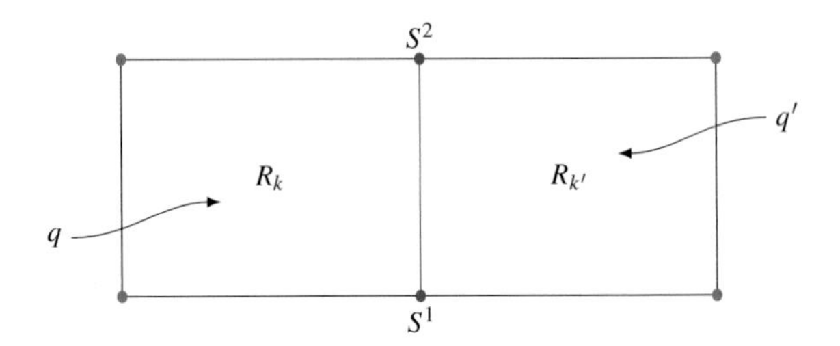

Fig. 6.9 Continuity across an internal edge $[S^1, S^2]$

mesh, this set determines one and only one Q_1 polynomial per element. The only thing to be checked is that these polynomials combine into a globally C^0 function. Indeed, discontinuities could arise at internal edges, those that are common to two elements. We have to show that this is not the case.

Let us thus consider the situation of Fig. 6.9, where the common edge between the two rectangles is parallel to the x_2 axis, without loss of generality.

We thus have two Q_1 polynomials q and q' such that $q(S^1) = q'(S^1)$ and $q(S^2) = q'(S^2)$. We can write

$$(q - q')(x) = \alpha_1 + \alpha_2(x_1 - x_1^1) + \alpha_3(x_2 - x_2^1) + \alpha_4(x_1 - x_1^1)(x_2 - x_2^1),$$

for some constants α_j, $j = 1, \dots, 4$.

Now, any point on the segment $[S^1, S^2]$ is such that $x_1 = x_1^1$. Therefore, on this segment, we have

$$(q - q')(x) = \alpha_1 + \alpha_3(x_2 - x_2^1).$$

Now, $(q - q')(S^1) = 0$ implies $\alpha_1 = 0$ and then $(q - q')(S^2) = 0$ implies $\alpha_3 = 0$. Consequently, $(q - q')_{|[S^1,S^2]}$ vanishes identically and thus, the function defined by $q(x)$ if $x \in R_k$, $q'(x)$ if $x \in R_{k'} \setminus R_k$ is continuous on $R_k \cup R_{k'}$. □

Remark 6.2 Notice that, in the above proof, the global continuity follows from the fact that Q_1 polynomials are affine on any segment that is parallel to the coordinate axes. If two such polynomials coincide at two points of such a segment, they then coincide on the whole straight line going through the two points. Of course, they are in general not affine on any segment that is not parallel to the coordinate axes. □

Corollary 6.1 *Let S^j, $j = 1, \dots, N_{\text{tot}}$, be a numbering of the mesh nodes. There exists a unique family $(w_h^i)_{i=1,\dots,N_{\text{tot}}}$ such that $w_h^i \in W_h$ and $w_h^i(S^j) = \delta_{ij}$. This family is a basis of W_h, which is of dimension N_{tot}, and for all $v_h \in W_h$, we have*

$$v_h = \sum_{i=1}^{N_{\text{tot}}} v_h(S^i) w_h^i. \tag{6.2}$$

Proof The existence and uniqueness of w_h^i follow readily from Propositions 6.3 and 6.4, since for all i, $\{\delta_{ij};\ 1 \le j \le N_{\text{tot}}\}$ is a set of values for all the nodes.

These propositions also show that the linear mapping $W_h \to \mathbb{R}^{N_{\text{tot}}}$, $v_h \mapsto (v_h(S^i))$ is an isomorphism, hence $\dim W_h = N_{\text{tot}}$. The family $(w_h^i)_{i=1,\dots,N_{\text{tot}}}$ is the inverse image of the canonical basis of $\mathbb{R}^{N_{\text{tot}}}$ by this isomorphism, therefore it is a basis of W_h. Finally, every element v_h of W_h is decomposed on this basis as $v_h = \sum_{i=1}^{N_{\text{tot}}} \lambda_i w_h^i$, so that taking $x = S^j$, we obtain

$$v_h(S^j) = \sum_{i=1}^{N_{\text{tot}}} \lambda_i w_h^i(S^j) = \sum_{i=1}^{N_{\text{tot}}} \lambda_i \delta_{ij} = \lambda_j$$

which establishes Eq. (6.2). □

We can now characterize the elements of V_h, i.e., those functions of W_h that vanish on $\partial\Omega$.

Corollary 6.2 *Assume, for convenience only, that the nodes S^j, $j = 1, \dots, N_{\text{int}}$ are the interior nodes. Then the family $(w_h^i)_{i=1,\dots,N_{\text{int}}}$ is a basis of V_h, and V_h is of dimension N_{int}.*

Proof If a function is in V_h, then $v_h(S^j) = 0$ for $j > N_{\text{int}}$. Therefore, we necessarily have

$$v_h = \sum_{i=1}^{N_{\text{int}}} v_h(S^i) w_h^i.$$

It remains to be seen that $w_h^i \in V_h$ for $i \leq N_{\text{int}}$. This is clear, since these functions vanish on all boundary nodes. Hence by the same token as before, they vanish on all the edges joining boundary nodes, and the whole boundary $\partial\Omega$ is composed of such edges. □

Remark 6.3 The functions w_h^i are called the basis functions for Q_1 Lagrange interpolation. The linear mappings $v_h \mapsto v_h(S^j)$ are again called the *degrees of freedom*.

It is easy to see that the support of w_h^i is composed of the four elements surrounding S^i when S^i is an interior node, see Fig. 6.14, two elements when it is a boundary node, but not a vertex of Ω, and just one element when it is one of the four vertices of Ω.

The graph of a basis function corresponding to an interior node over its support is made of four hyperbolic paraboloid pieces that look like a tent,[2] see Fig. 6.10. □

In Figs. 6.11 and 6.12, we show pictures of elements of V_h.

6.3 Convergence and Error Estimate for the Q_1 FEM

The approximation space V_h is finite-dimensional, therefore closed, hence Céa's lemma, i.e., Theorem 5.1 of Chap. 5, applies and we denote by u_h the solution of the discrete variational problem (5.1). We thus need to estimate such quantities as

[2] Which is why they are sometimes called tent-functions.

Fig. 6.10 Two views of a Q_1 basis function for an interior node

Fig. 6.11 The graph of a random element of V_h. The fact that functions in V_h are piecewise affine on segments parallel to the coordinate axes is apparent, see Sect. 6.4

Fig. 6.12 The graph of the V_h-interpolate of the function $(x_1, x_2) \mapsto \sin(\pi x_1)\sin(\pi x_2)$

$\|u - \Pi_h u\|_{H^1(\Omega)}$, where Π_h is some interpolation operator with values in V_h in order to obtain an error estimate and prove convergence. We now encounter a new difficulty, which is that H^1 functions are not continuous in two dimensions, therefore, the nodal values of u a priori do not make any sense and it is not possible to perform any kind of Lagrange interpolation on H^1.

We will thus need to make regularity hypotheses. We will admit the following particular case of the Sobolev embedding theorems, which is valid in dimension two, see for example [1].

Theorem 6.1 *There is a continuous embedding* $H^2(\Omega) \hookrightarrow C^0(\bar{\Omega})$.

With this theorem at hand, we can V_h-interpolate H^2 functions.

Let us thus be given a regular family of admissible meshes, that we index by $h = \max(h_1, h_2)$, regularity meaning here that there exists a constant C such that $\frac{\max(h_1,h_2)}{\min(h_1,h_2)} \leq C$. Let u be the solution of problem (6.1) in variational form and u_h its variational approximation on V_h. We will prove the following convergence and error estimate theorem.

Theorem 6.2 *There exists a constant C such that, if $u \in H^2(\Omega)$, we have*

$$\|u - u_h\|_{H^1(\Omega)} \leq Ch|u|_{H^2(\Omega)}.$$

The constant C is naturally not the same constant as a couple of lines higher. Actually, the proof of Theorem 6.2 will be broken into a series of lemmas, and constants C will come up that generally vary from line to line. This is what is called a generic constant.... The important thing is not their actual value, but that they do not depend on any of the other quantities that appear, in this specific case, h and u.

Let $\widehat{R} = [0, 1] \times [0, 1]$ be the *reference rectangle*[3] or reference element. Its four vertices $\widehat{S}^j$, $j = 1, \ldots, 4$, are $(0, 0)$, $(1, 0)$, $(1, 1)$ and $(0, 1)$. We let $\widehat{\Pi}$ denote the Q_1 interpolation operator on the four vertices $\widehat{S}^j$. The Q_1 Lagrange interpolation basis polynomials, or shape functions, on the reference rectangle are

$$\hat{p}_1(\hat{x}) = (1 - \hat{x}_1)(1 - \hat{x}_2),\ \hat{p}_2(\hat{x}) = \hat{x}_1(1 - \hat{x}_2),$$
$$\hat{p}_3(\hat{x}) = \hat{x}_1\hat{x}_2,\ \hat{p}_4(\hat{x}) = (1 - \hat{x}_1)\hat{x}_2, \tag{6.3}$$

as can be checked by hand. For all $\hat{v} \in C^0(\widehat{R})$, we thus have

$$\widehat{\Pi}\hat{v} = \sum_{j=1}^{4} \hat{v}(\widehat{S}^j)\hat{p}_j. \tag{6.4}$$

Let us begin our series of lemmas.

Lemma 6.1 *The operator $\widehat{\Pi}$ is continuous from $H^2(\widehat{R})$ to $H^1(\widehat{R})$.*

Proof We equip Q_1 with the $H^1(\widehat{R})$ norm (recall that all norms are equivalent on Q_1 since it is finite dimensional). By Theorem 6.1, we have for all $\hat{v} \in H^2(\widehat{R})$

$$\|\hat{v}\|_{C^0(\widehat{R})} \leq C\|\hat{v}\|_{H^2(\widehat{R})},$$

for some constant C. By formula (6.4), we have

$$\begin{aligned}\|\widehat{\Pi}\hat{v}\|_{H^1(\widehat{R})} &\leq \sum_{j=1}^{4} |\hat{v}(\widehat{S}^j)|\|\hat{p}_j\|_{H^1(\widehat{R})} \\ &\leq \Big(\sum_{j=1}^{4} \|\hat{p}_j\|_{H^1(\widehat{R})}\Big)\|\hat{v}\|_{C^0(\widehat{R})}\end{aligned}$$

[3] Ok, it's a square, and unluckily it happens to look a lot like Ω, although there is no conceptual connection between the two.

$$\leq C\Big(\sum_{j=1}^{4}\|\hat{p}_j\|_{H^1(\widehat{R})}\Big)\|\hat{v}\|_{H^2(\widehat{R})},$$

which completes the proof. □

Lemma 6.2 *There exists a constant C such that, for all $\hat{v} \in H^2(\widehat{R})$*

$$\|\hat{v} - \widehat{\Pi}\hat{v}\|_{H^1(\widehat{R})} \leq C\|\nabla^2\hat{v}\|_{L^2(\widehat{R})}.$$

Proof We note that $P_1 \subset Q_1$, thus for all $p \in P_1$, we have $\widehat{\Pi}p = p$. Therefore

$$\|\hat{v} - \widehat{\Pi}\hat{v}\|_{H^1(\widehat{R})} = \|\hat{v} - p - \widehat{\Pi}(\hat{v} - p)\|_{H^1(\widehat{R})} \leq \|I - \widehat{\Pi}\|_{\mathscr{L}(H^2;H^1)}\|\hat{v} - p\|_{H^2(\widehat{R})}$$

for all $\hat{v} \in H^2(\widehat{R}), p \in P_1$, by Lemma 6.1. Consequently

$$\|\hat{v} - \widehat{\Pi}\hat{v}\|_{H^1(\widehat{R})} \leq C \inf_{p\in P_1}\|\hat{v} - p\|_{H^2(\widehat{R})} = C\|\hat{v} - P\hat{v}\|_{H^2(\widehat{R})},$$

where P denotes the H^2 orthogonal projection onto P_1.

Let us now show that there is a constant C such that

$$\|\hat{v} - P\hat{v}\|_{H^2(\widehat{R})} \leq C\|\nabla^2\hat{v}\|_{L^2(\widehat{R})},$$

which will complete the proof of the Lemma. We argue by contradiction and assume there is no such constant C. In this case, as in the proof of the Poincaré–Wirtinger inequality, i.e., Theorem 4.2 of Chap. 4, there exists a sequence $\hat{v}_n \in H^2(\widehat{R})$ such that

$$\|\hat{v}_n - P\hat{v}_n\|_{H^2(\widehat{R})} = 1 \quad \text{and} \quad \|\nabla^2\hat{v}_n\|_{L^2(\widehat{R})} \to 0,$$

when $n \to +\infty$. Let us set $\hat{w}_n = \hat{v}_n - P\hat{v}_n$, which belongs to $P_1^\perp$. The second derivatives of a P_1 polynomial vanish, so that $\nabla^2\hat{w}_n = \nabla^2\hat{v}_n$. By Rellich's compact embedding theorem, see Remark 3.18 of Chap. 3, there exists a sequence, still denoted $\hat{w}_n$ and a $\hat{w} \in H^1(\widehat{R})$ such that $\hat{w}_n \to \hat{w}$ in $H^1(\widehat{R})$. Then, the condition $\|\nabla^2\hat{w}_n\|_{L^2(\widehat{R})} \to 0$ shows that $\hat{w}_n$ is a Cauchy sequence in $H^2(\widehat{R})$, which is complete. Hence, $\hat{w} \in H^2(\widehat{R})$ and $\hat{w}_n \to \hat{w}$ in $H^2(\widehat{R})$ as well. Now, the space $P_1^\perp$ is a H^2 orthogonal, hence is closed in $H^2(\widehat{R})$, from which it follows that $\hat{w} \in P_1^\perp$. On the other hand, we have $\nabla^2\hat{w} = 0$, so that $\hat{w} \in P_1$. Consequently, $\hat{w} = 0$ and $\hat{w}_n \to 0$ in $H^2(\widehat{R})$, which contradicts $\|\hat{w}_n\|_{H^2(\widehat{R})} = 1$. The proof is complete. □

Remark 6.4 The above proof does not really use Q_1-interpolation, but only P_1-interpolation. It would thus equally apply for triangular elements, which we will discuss later. The proof is non constructive in the sense that we do not have any idea of the actual value of the constant. □

We now perform a change of variable between the reference element $\widehat{R}$ and a generic element R_k of the mesh.

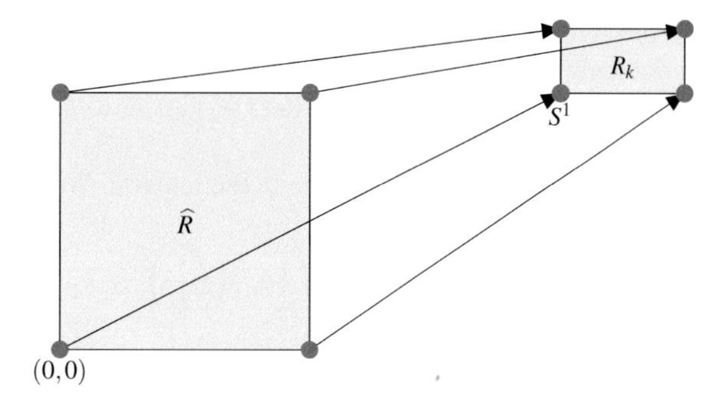

Fig. 6.13 The affine change of variable from the reference element to the generic element

Lemma 6.3 *Let R_k be an element of the mesh. There exists a unique affine bijective mapping F_k such that $F_k(\widehat{R}) = R_k$ and that maps the vertices counted counterclockwise from the lower left corner to their counterparts in R_k.*

Proof Consider Fig. 6.13.

In view of the figure, it is clearly enough to map the origin to point S^1 of coordinates $(x_1(R_k), x_2(R_k))$ and then to multiply abscissae by h_1 and ordinates by h_2. This yields

$$F_k(\hat{x}) = \begin{pmatrix} x_1(R_k) + h_1\hat{x}_1 \\ x_2(R_k) + h_2\hat{x}_2 \end{pmatrix}.$$

The inverse mapping is given by

$$F_k^{-1}(y) = \begin{pmatrix} \frac{y_1 - x_1(R_k)}{h_1} \\ \frac{y_2 - x_2(R_k)}{h_2} \end{pmatrix}.$$

It is also affine, naturally. □

Lemma 6.4 *There exists a constant C such that for all elements R_k and all $v \in H^1(R_k)$, setting $\hat{v}(\hat{x}) = v(F_k(\hat{x}))$, we have*

$$\int_{R_k} \|\nabla v\|^2 \, dx \leq C \int_{\widehat{R}} \|\nabla \hat{v}\|^2 \, d\hat{x}.$$

Proof This is brute force computation. We have $v(x) = \hat{v}(F_k^{-1}(x))$ thus

$$\begin{aligned} \frac{\partial v}{\partial x_i}(x) &= \sum_{j=1}^{2} \frac{\partial \hat{v}}{\partial \hat{x}_j}(F_k^{-1}(x)) \frac{\partial (F_k^{-1})_j}{\partial x_i}(x) \\ &= h_i^{-1} \frac{\partial \hat{v}}{\partial \hat{x}_i}(F_k^{-1}(x)), \end{aligned}$$

by the multidimensional chain rule. We also need the Jacobian of the change of variables $x = F_k(\hat{x})$

$$dx = |\det DF_k(\hat{x})|\, d\hat{x} = h_1 h_2\, d\hat{x}$$

to perform the change of variable in the integral. We obtain

$$\begin{aligned}\int_{R_k} \|\nabla v\|^2\, dx &= \int_{\widehat{R}} \Big[h_1^{-2}\Big(\frac{\partial \hat{v}}{\partial \hat{x}_1}\Big)^2 + h_2^{-2}\Big(\frac{\partial \hat{v}}{\partial \hat{x}_2}\Big)^2\Big] h_1 h_2\, d\hat{x} \\ &\leq (\min(h_1, h_2))^{-2} h_1 h_2 \int_{\widehat{R}} \|\nabla \hat{v}\|^2\, d\hat{x}.\end{aligned}$$

Now this is where the regularity of the mesh family intervenes. According to Definition 6.2, letting $h = \max(h_1, h_2)$ and $\rho = \min(h_1, h_2)$, we have $\frac{h}{\rho} \leq C$. Therefore $(\min(h_1, h_2))^{-2} h_1 h_2 \leq \frac{C^2}{h^2} h^2 = C^2$, and the proof is complete. □

We now are in a position to prove Theorem 6.2.

Proof of Theorem 6.2. We use the H^1 semi-norm. Let Π_h be the V_h-interpolation operator and let $v_k = (u - \Pi_h u)_{|R_k}$ and $u_k = u_{|R_k}$. It is important to note that

$$\widehat{\big(\Pi_h u_{|R_k}\big)} = \widehat{\Pi}\, \widehat{u_{|R_k}} = \widehat{\Pi} \widehat{u_k},$$

using the same hat notation as in Lemma 6.4 for the change of variables in functions. This is because an affine change of variables of the form of F_k maps Q_1 polynomials to Q_1 polynomials due to their special structure. Moreover, the two sides of the above equality satisfy the same interpolation conditions at the four vertices of the reference element, hence are equal everywhere. Therefore, we have

$$\widehat{v_k} = \widehat{u_k} - \widehat{\Pi} \widehat{u_k}.$$

We decompose the semi-norm squared as a sum over all elements

$$|u - \Pi_h u|^2_{H^1(\Omega)} = \sum_{k=1}^{N_{\mathcal{T}}} \int_{R_k} \|\nabla v_k\|^2\, dx \leq C \sum_{k=1}^{N_{\mathcal{T}}} \int_{\widehat{R}} \|\nabla \widehat{v_k}\|^2\, d\hat{x},$$

by Lemma 6.4.

By Lemma 6.2, we have

$$\begin{aligned}\int_{\widehat{R}} \|\nabla \widehat{v_k}\|^2\, d\hat{x} &\leq C \int_{\widehat{R}} \|\nabla^2 \widehat{u_k}\|^2\, d\hat{x} \\ &\leq C \sum_{i,j=1}^{2} \int_{\widehat{R}} \Big(\frac{\partial^2 \widehat{u_k}}{\partial \hat{x}_i \partial \hat{x}_j}\Big)^2\, d\hat{x}\end{aligned}$$

$$
\begin{aligned}
&= C \sum_{i,j=1}^{2} \int_{R_k} \left(h_i h_j \frac{\partial^2 u_k}{\partial x_i \partial x_j} \right)^2 \frac{1}{h_1 h_2} \, dx \\
&\leq C h^2 \int_{R_k} \|\nabla^2 u_k\|^2 \, dx
\end{aligned}
$$

by performing the reverse change of variables, and using the regularity of the mesh family again. It follows that

$$
|u - \Pi_h u|^2_{H^1(\Omega)} \leq C h^2 \sum_{k=1}^{N_{\mathcal{T}}} \int_{R_k} \|\nabla^2 u_k\|^2 \, dx = C h^2 \|\nabla^2 u\|^2_{L^2(\Omega)},
$$

and the proof is complete since the H^1 semi-norm is equivalent to the H^1 norm on $H^1_0(\Omega)$, see Corollary 3.3 of Chap. 3. □

Remark 6.5 Under the hypothesis $u \in H^2(\Omega)$, which is satisfied in this particular case, due to elliptic regularity in a convex polygon, we thus have convergence of the Q_1 FEM when $h \to 0$, and we have an error estimate with a constant C that depends neither on h nor on u. The drawback however is that the proof does not tell us how large this constant is, see Remark 6.4. □

6.4 Assembling the Matrix

Let us assume that a numbering of the interior nodes, and thus of the basis functions of V_h, has been chosen: S^j and w_h^j, $j = 1, \ldots, N_{\text{int}}$. We have, by Q_1 interpolation

$$
u_h = \sum_{j=1}^{N_{\text{int}}} u_h(S^j) w_h^j
$$

and the matrix A has coefficients

$$
A_{ij} = a(w_h^j, w_h^i) = \int_{\Omega} (\nabla w_h^j \cdot \nabla w_h^i + c w_h^j w_h^i) \, dx.
$$

If we set

$$
A_{ij}(R_k) = \int_{R_k} (\nabla w_h^j \cdot \nabla w_h^i + c w_h^j w_h^i) \, dx,
$$

we see that

$$
A_{ij} = \sum_{k=1}^{N_{\mathcal{T}}} A_{ij}(R_k),
$$

and the coefficients can thus be computed element-wise. The idea is that many of the numbers $A_{ij}(R_k)$ do not need to be computed, since it is known that they vanish as soon as the intersection of the supports of w_h^j and w_h^i does not meet R_k. This vastly reduces the computer load.

Likewise, the right-hand side of the linear system can be written as

$$B_i = \int_\Omega f w_h^i \, dx = \sum_{k=1}^{N_\mathcal{T}} \int_{R_k} f w_h^i \, dx = \sum_{k=1}^{N_\mathcal{T}} B_i(R_k),$$

with only four nonzero terms in the last sum.

Now the restriction of w_h^j to R_k is either zero, or one of the four Q_1 interpolation basis polynomials on R_k, which we denote p_i^k, $i = 1, \dots, 4$. Here again, the reference element $\widehat{R}$ can be used with profit to compute the coefficients of the matrix. The Q_1 Lagrange interpolation basis polynomials, or shape functions, on the reference rectangle are given by formula (6.3).

We have already noticed that $p_i^k(x) = \hat{p}_i(F_k^{-1}(x))$ because both sides are Q_1 and satisfy the same interpolation conditions at the vertices. Let us give an example of computation with $\hat{p}_3$. We thus have

$$p_3^k(x) = \hat{p}_3(F_k^{-1}(x)) = \Big(\frac{x_1 - x_1(R_k)}{h_1}\Big)\Big(\frac{x_2 - x_2(R_k)}{h_2}\Big).$$

Therefore

$$\|\nabla p_3^k(x)\|^2 = \frac{1}{h_1^2 h_2^2}\big((x_1 - x_1(R_k))^2 + (x_2 - x_2(R_k))^2\big),$$

and assuming S^i is the upper right corner of R_k, we obtain by computing the integrals on R_k

$$A_{ii}(R_k) = \frac{h_1 h_2}{3}\Big(\frac{1}{h_1^2} + \frac{1}{h_2^2}\Big) + c_0 \frac{h_1 h_2}{9}$$

in the case when $c = c_0$ is a constant. Now there are four such contributions to A_{ii} coming from the four rectangles that surround S^i (see Fig. 6.14), which are all equal, hence

$$A_{ii} = \frac{4h_1 h_2}{3}\Big(\frac{1}{h_1^2} + \frac{1}{h_2^2}\Big) + c_0 \frac{4h_1 h_2}{9}.$$

In the case when $h_1 = h_2 = h$, we thus obtain $A_{ii} = \frac{8}{3} + \frac{4}{9} c_0 h^2$.

The diagonal coefficients do not depend on the node numbering, but the off-diagonal ones do depend completely on it. So we have to talk about numbering, since in the 2d case, as opposed to the 1d case, no natural numbering appears at the onset.

We first note that there is a connection between Q_1 Lagrange approximation in two dimensions and P_1 Lagrange approximation in one dimension.

The four basis polynomials on $\widehat{R}$ are given by Eq. (6.3). In one dimension, the basis polynomials for P_1 Lagrange interpolation on [0, 1] are

$$\ell_1(x) = 1 - x, \quad \ell_2(x) = x.$$

Therefore, we see that

$$\hat{p}_1(x) = \ell_1(\hat{x}_1)\ell_1(\hat{x}_2), \hat{p}_2(x) = \ell_2(\hat{x}_1)\ell_1(\hat{x}_2),$$
$$\hat{p}_3(x) = \ell_2(\hat{x}_1)\ell_2(\hat{x}_2), \hat{p}_4(x) = \ell_1(\hat{x}_1)\ell_2(\hat{x}_2).$$

In this context, we introduce a useful notation. Let f and g be two functions in one variable. We define a function in two variables $f \otimes g$ by $f \otimes g(x_1, x_2) = f(x_1)g(x_2)$. This function is called the *tensor product* of f and g.[4] With this notation, we thus have $p_1 = \ell_1 \otimes \ell_1$ and so on.

This tensor product decomposition extends to the basis functions on Ω themselves. Let S^i be an interior node of coordinates (i_1h_1, i_2h_2) and R^i_k, $k = 1, \ldots, 4$, the four elements surrounding it.

By direct verification of the interpolation relations, we easily check that

$$w^i_h(x) = \begin{cases} \ell_2\big(\frac{x_1}{h_1} - (i_1 - 1)\big)\ell_2\big(\frac{x_2}{h_2} - (i_2 - 1)\big) & \text{in } R^i_1, \\ \ell_1\big(\frac{x_1}{h_1} - i_1\big)\ell_2\big(\frac{x_2}{h_2} - (i_2 - 1)\big) & \text{in } R^i_2, \\ \ell_1\big(\frac{x_1}{h_1} - i_1\big)\ell_1\big(\frac{x_2}{h_2} - i_2\big) & \text{in } R^i_3, \\ \ell_2\big(\frac{x_1}{h_1} - (i_1 - 1)\big)\ell_1\big(\frac{x_2}{h_2} - i_2\big) & \text{in } R^i_4, \\ 0 & \text{elsewhere.} \end{cases}$$

We remark that the support of w^i_h is the union of the four rectangles surrounding S^i, see Fig. 6.14. Therefore, if $w^{i_1}_{h_1}$ denotes the 1d hat function associated with node i_1h_1 of the 1d mesh of [0, 1] of mesh size h_1, and likewise for $w^{i_2}_{h_2}$, we see that

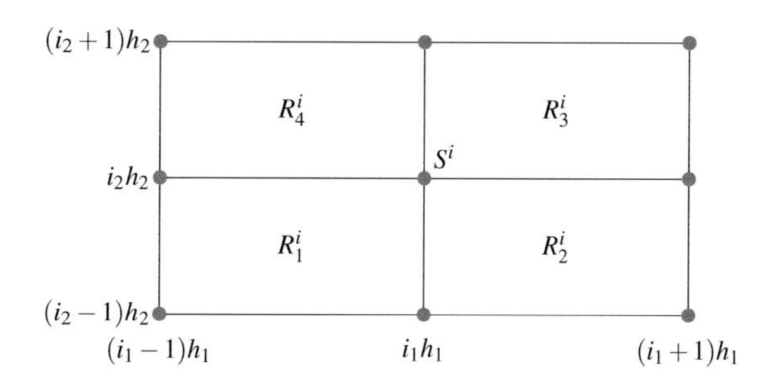

Fig. 6.14 The support of w^i_h

[4] Consider this to be just vocabulary. We do not need to know anything about tensor products in general.

$$w_h^i = w_{h_1}^{i_1} \otimes w_{h_2}^{i_2}.$$

In other words, the basis functions of the Q_1 FEM in 2d are nothing but the tensor products of the one-dimensional P_1 basis functions *cf.* also Fig. 6.14.

Let us use this tensor product decomposition to number the basis functions. The idea is to use the indices i_1 and i_2 to sweep the rows and then the columns of the mesh.[5] We thus define a mapping $\{1, 2, \dots, N_1\} \times \{1, 2, \dots, N_2\} \to \{1, 2, \dots, N_{\text{int}}\}$ by

$$(i_1, i_2) \mapsto i = i_1 + (i_2 - 1)N_1.$$

It is clearly a bijection (recall that $N_{\text{int}} = N_1 N_2$). To compute the inverse mapping, we note that $i_1 - 1$ is the remainder of the Euclidean division of $i - 1$ by N_1, thus

$$i_1 = i - \left\lfloor \frac{i-1}{N_1} \right\rfloor N_1, \quad i_2 = \left\lfloor \frac{i-1}{N_1} \right\rfloor + 1. \tag{6.5}$$

Now, the support of a tensor product is the Cartesian product of the supports. Thus

$$\operatorname{supp} w_h^i = \operatorname{supp} w_{h_1}^{i_1} \times \operatorname{supp} w_{h_2}^{i_2} = [(i_1 - 1)h_1, (i_1 + 1)h_1] \times [(i_2 - 1)h_2, (i_2 + 1)h_2].$$

If an index j in the numbering corresponds to a couple (j_1, j_2), we thus see that $A_{ij} \neq 0$ if and only if the supports have non negligible intersection, that is to say

$$A_{ij} \neq 0 \iff |j_1 - i_1| \leq 1 \text{ and } |j_2 - i_2| \leq 1.$$

In view of the numbering formulas above, saying that $|j_1 - i_1| \leq 1$ is equivalent to saying that $j - i = \alpha + kN_1$ with $\alpha = i_1 - j_1 = -1$, 0 or 1, and k an integer. Since we also have $i_2 = \frac{i - i_1}{N_1}$, it follows that $i_2 - j_2 = k = -1$, 0 or 1. Therefore, for a given i, that is a given row of A, there are at most nine values of j, that is nine columns, that contain a nonzero coefficient. Of course, not all rows contain nine nonzero coefficients. For example, the first row has four nonzero coefficients, the second row has six nonzero coefficients, and so on. Rows that correspond to index pairs (i_1, i_2) such that $2 \leq i_1, i_2 \leq N_1 - 1$ do have nine nonzero coefficients (they correspond to interior nodes with nine neighboring interior nodes, including themselves). Such a row looks like Fig. 6.15.

We see three tridiagonal $N_1 \times N_1$ blocks emerging, that are themselves arranged block tridiagonally. The whole $(N_1 N_2) \times (N_1 N_2)$ matrix is thus composed of N_2^2 blocks A^{kl}, $1 \leq k, l, \leq N_2$, of size $N_1 \times N_1$, that are either zero or tridiagonal. Indeed, if we define the $N_1 \times N_1$ matrix A^{kl} to be the block comprised of lines $(k - 1)N_1 + 1$ to kN_1 and columns $(l - 1)N_1 + 1$ to lN_1, then using the inverse numbering (6.5), we see that

$$A_{ij} = a(w_h^j, w_h^i) = a(w_{h_1}^{j_1} \otimes w_{h_2}^l, w_{h_1}^{i_1} \otimes w_{h_2}^k),$$

[5] Or the other way around. But let's stick to this one here.

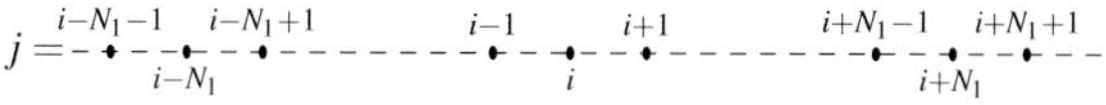

Fig. 6.15 A typical row in the matrix

$j =$ $i-N_1-1$, $i-N_1$, $i-N_1+1$, $i-1$, i, $i+1$, $i+N_1-1$, $i+N_1$, $i+N_1+1$

for all (i, j) in this block. Therefore we have $(A^{kl})_{i_1 j_1} = a(w_{h_1}^{j_1} \otimes w_{h_2}^{l}, w_{h_1}^{i_1} \otimes w_{h_2}^{k})$, thus $A^{kl} = 0$ as soon as $|k - l| \geq 2$ and is tridiagonal for $|k - l| \leq 1$, for reasons of supports. We thus have

$$A = \begin{pmatrix} A^{11} & A^{12} & 0 & \cdots & 0 \\ A^{21} & A^{22} & A^{23} & \cdots & 0 \\ 0 & A^{32} & A^{33} & \ddots & 0 \\ \vdots & \vdots & \ddots & \ddots & \vdots \\ 0 & \cdots & 0 & A^{N_2-1,N_2} & A^{N_2 N_2} \end{pmatrix},$$

where the block tridiagonal structure appears, see also Fig. 6.16.

The sweep columns then rows numbering scheme thus gives rise to a well-structured matrix for which there exist efficient numerical methods. It is instructive to see what kind of matrix would result from other numberings that could be considered just as natural, such as the numbering used to prove that $\mathbb{N}^2$ is countable (although limited to a square here): start from the lower left node, go east one node, then north west, then north, then south east, etc., see Fig. 6.17.

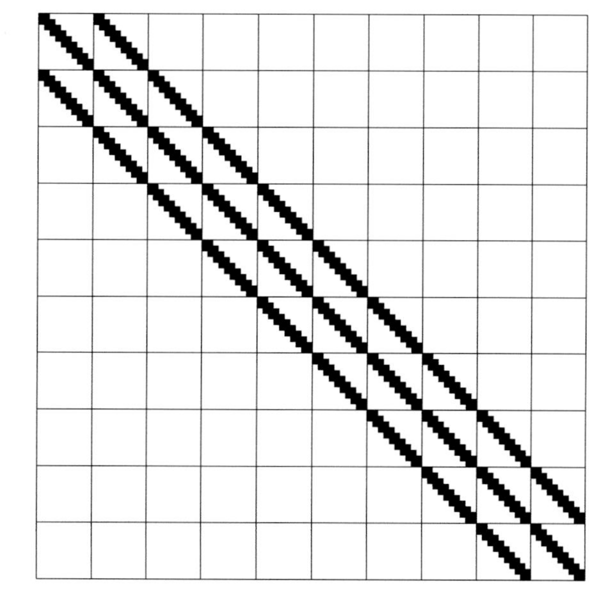

Fig. 6.16 The block tridiagonal structure of A (here $N_1 = N_2 = 10$ so A is 100×100). *Black squares* indicate nonzero matrix coefficients, *white* areas zero. The coarse grid shows the 10×10 blocks

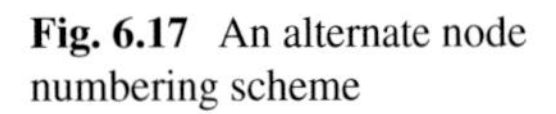

Fig. 6.17 An alternate node numbering scheme

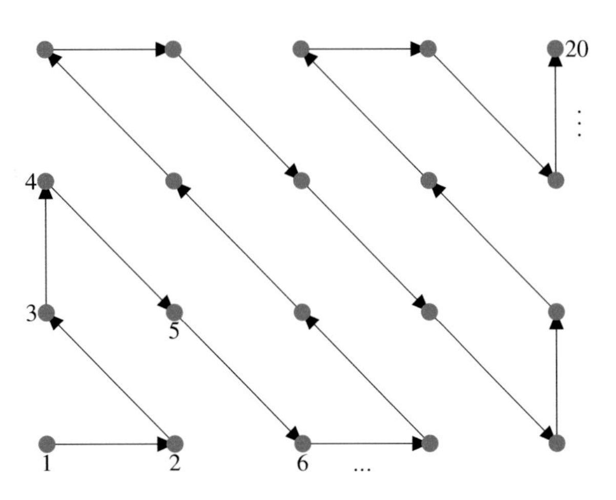

For the same 100×100 case, we obtain a matrix structure that looks like Fig. 6.18. Of course, the entries of the above matrix are the same as the previous ones after a permutation, since both matrices are similar to each other via a permutation matrix.

In the case when Ω is not a rectangle, the structure of the matrix is not as regular. For instance, the matrix associated with the mesh depicted in Fig. 6.1, with the sweep columns then rows numbering, looks like

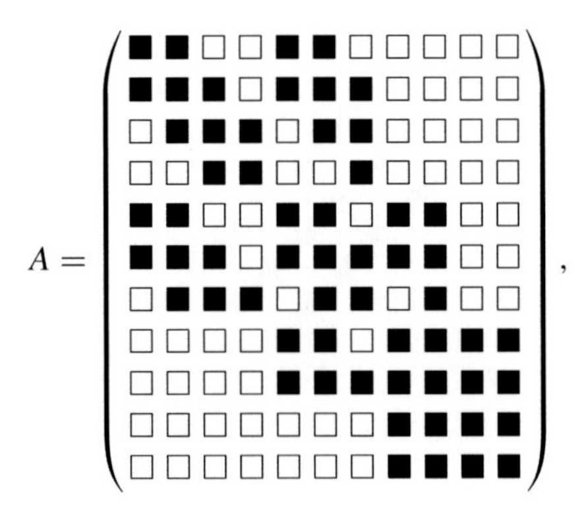

where black squares denote nonzero entries and white squares zero entries.

Let us notice that the structure of the matrix depends solely on the numbering of nodes, and not on the numbering of elements.

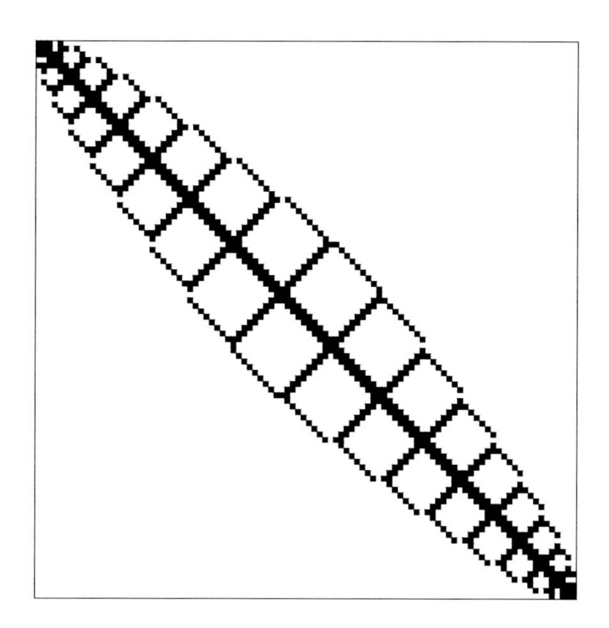

Fig. 6.18 Structure of the alternate matrix

6.5 General Definition of a Finite Element

It is now time to look back and see what are the general characteristics of the finite elements we have seen so far, so as to finally define what a finite element is! We let $\mathbb{P} = \mathbb{R}[x_1, x_2]$ denote the space of polynomials in two indeterminates.

Definition 6.4 A two-dimensional *finite element* is a triple $(T, P(T), \{\varphi_1, \dots, \varphi_d\})$ where

(1) T is a compact polygon.
(2) $P(T)$ is a finite dimensional subspace of $\mathbb{P}$, considered as a function space on T.
(3) φ_i, $i = 1, \dots, d$, are linear forms on $\mathbb{P}$, which are called the *degrees of freedom* of the finite element.

Remark 6.6 In practice, T is either a triangle or a rectangle. The same definition applies in dimensions one and three (finite elements are rarely used in dimensions higher than four, although it happens). In dimension one, T is an interval. There is more variety in dimension three, starting with tetrahedra.

The degrees of freedom are attached to T one way or another.

In the literature, in particular the engineering literature, finite elements are always presented this way, and not starting with the discrete space V_h and so on, as we have done up to now. We thus start from the top, i.e. from the discrete space, down to the finite element, instead of starting from the bottom. □

Definition 6.5 We say that a finite element is *unisolvent* if for all d-uples of scalars $(\alpha_1, \dots, \alpha_d)$, there exists one and only one polynomial $p \in P(T)$ such that $\varphi_i(p) = \alpha_i$, $i = 1, \dots, d$.

Unisolvence is a generalization of the interpolation property for all kinds of degrees of freedom.

Proposition 6.5 *If a finite element is unisolvent, then,* $d = \dim P(T)$.

Proof This is fairly obvious. Assume we want to solve the d equations $\varphi_i(p) = \alpha_i$. Since φ_i are linear forms, these equations are linear equations in $\dim P(T)$ unknowns, once we choose a basis of $P(T)$. Hence if the number of equations and the number of unknowns are different, the system of equations certainly cannot be solved uniquely for all right-hand sides. □

Remark 6.7 If we do not have the same number of degrees of freedom as the dimension of the finite element space, then the element in question is not unisolvent.

Be careful that unisolvence is not just a question of dimensions, as the following example shows.

Take $T = \{|x_1| + |x_2| \leq 1\}$, $P(T) = Q_1$ and φ_i the values at the four vertices of T. We have $\dim Q_1 = 4$ but this element is not unisolvent, since $p(x) = x_1 x_2$ is in Q_1 and $\varphi_i(p) = 0$ for all i even though $p \neq 0$.

If on the other hand, $T = [0, 1]^2$, $P(T) = Q_1$ and φ_i the values at the four vertices of T, then the finite element is unisolvent. This is the element we have been using so far in 2d.

Therefore, unisolvence somehow reflects the adequacy of the duality between the polynomial space and the degrees of freedom. □

In practice, unisolvence is checked using the following result.

Proposition 6.6 *A finite element is unisolvent if and only if* $d = \dim P(T)$ *and there exists a basis* $(p_j)_{j=1,\dots,d}$ *of* $P(T)$ *such that* $\varphi_i(p_j) = \delta_{ij}$ *for all* i, j.

Proof If the element is unisolvent, we already know that $d = \dim P(T)$. Moreover, choosing $\alpha_i = \delta_{ij}$ for $j = 1, \dots, d$ yields the existence of p_j by the very definition. The family (p_j) is linearly independent, for if

$$\sum_{j=1}^{d} \lambda_j p_j = 0,$$

applying the linear form φ_i, we obtain

$$0 = \sum_{j=1}^{d} \lambda_j \varphi_i(p_j) = \sum_{j=1}^{d} \lambda_j \delta_{ij} = \lambda_i$$

for all i. Thus it is a basis of $P(T)$.

Conversely, assume that $d = \dim P(T)$ and that we have a basis p_j with the above property. Let us be given scalars α_i. Then the polynomial $p = \sum_{j=1}^{d} \alpha_j p_j$ is the only element of $P(T)$ such that $\varphi_i(p) = \alpha_i$ by the same argument. □

Remark 6.8 The polynomials p_j are called the *basis polynomials* or *shape functions* of the finite element. They are dual to the degrees of freedom. They are also used to construct the basis functions of the discrete approximation spaces, as we have seen already in 1d with the P_1 Lagrange and P_3 Hermite approximations, and in 2d with the Q_1 Lagrange approximation. In the latter case, the shape functions were already given in Eq. (6.3) for $T = [0, 1]^2$.

This also indicates that unisolvence is far from being the end of the story in terms of finite elements. The basis polynomials must also be such that they can be combined into globally continuous functions so as to give rise to a conforming approximation. □

Speaking of duality, we can also introduce $\Sigma(T) = \text{vect}\,\{\varphi_1, \ldots, \varphi_d\}$, the vector subspace of $\mathbb{P}^*$ spanned by the degrees of freedom. In a similar vein as Proposition 6.5, we also have

Proposition 6.7 *If a finite element is unisolvent, then,* $d = \dim \Sigma(T)$.

Proof Clear. □

The basis polynomials and the degrees of freedom are obviously dual bases of their respective spanned spaces. In the counterexample shown above, the four linear forms are linearly independent as elements of $\mathbb{P}^*$, but not as elements of Q_1^*.

6.6 Q_2 and Q_3 Finite Elements

Let us briefly discuss what happens if we want to use higher degree polynomials. We start with Q_2 Lagrange elements for second order problems. The discrete approximation space is then

$$V_h = \{v_h \in C^0(\bar{\Omega}); \forall R_k \in \mathscr{T}, v_{h|R_k} \in Q_2, v_h = 0 \text{ on } \partial\Omega\},$$

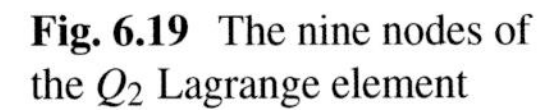

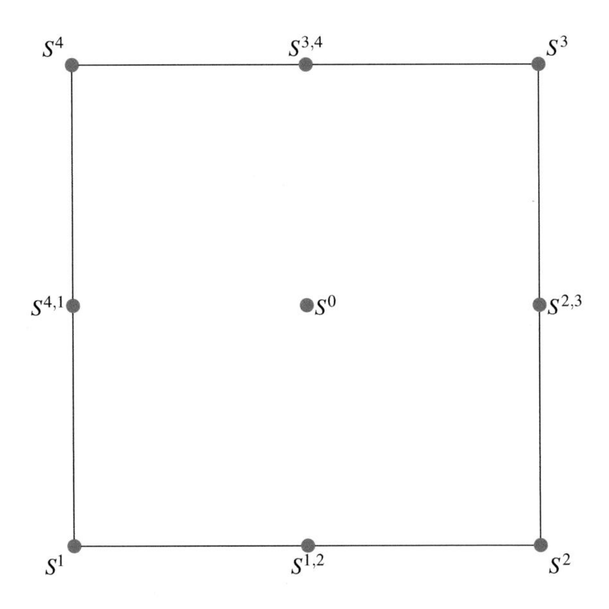

Fig. 6.19 The nine nodes of the Q_2 Lagrange element

on the same rectangular mesh as before and the general approximation theory applies (note that this space is larger than the previous one). We concentrate on the description of the finite element first. We set $R = [0, 1]^2$, $P(R) = Q_2$ and we need to describe the degrees of freedom. Since we are going to use Lagrange interpolation, these degrees of freedom are going to be values at some points of the element. The dimension of Q_2 is nine, therefore nine degrees of freedom are required to define a unisolvent finite element, i.e., nine points or nodes in R. The choice of points must also be guided by the necessity of defining a global C^0 interpolation based on the nodal values on the mesh. The set of points depicted in Fig. 6.19 turns out to satisfy both requirements.

We thus take the four vertices S^k as before, plus the four middles of the edges S^{k,k_+}, where $k+ = k + 1$ for $k = 1, 2, 3$ and $k+ = 1$ for $k = 4$, plus the center of gravity S^0. A slight misuse of notation: we let $p(S)$ denote the linear form $p \mapsto p(S)$.

Proposition 6.8 *The finite element* $(R, Q_2, \{p(S^k), p(S^{k,k_+}), p(S^0), k = 1, \dots, 4\})$ *is unisolvent.*

Proof The number of degrees of freedom matches the dimension of the space. It is thus sufficient to construct the basis polynomials. We will number them the same way as the node they correspond to. There are three polynomials to be constructed: p^1 from which the other p^k are deduced by symmetry, $p^{1,2}$ from which the other p^{k,k_+} are deduced by symmetry, and p^0.

Let us show how to compute p^0. The interpolation conditions to be satisfied are $p^0(S^0) = 1$ and $p^0 = 0$ on all other eight nodes. Now p^0 is zero at points $(0, 0)$, $(0, \frac{1}{2})$ and $(0, 1)$. The restriction of a Q_2 polynomial to the line $x_2 = 0$ is a second degree polynomial in the variable x_1, and we have just seen that this polynomial has three roots. Therefore it vanishes and $p^0 = 0$ on the straight line $x_2 = 0$. It follows that

p^0 is divisible by x_2. The same argument shows that it is divisible by x_1, $(1-x_1)$ and $(1-x_2)$. These polynomials are relatively prime, thus p^0 is divisible by their product,

$$p^0(x) = q(x)x_1x_2(1-x_1)(1-x_2),$$

for some polynomial q. Now $x_1x_2(1-x_1)(1-x_2) \in Q_2$, thus the partial degree of q is less than 0, i.e., q is a constant C. Evaluating now p^0 at point $S^0 = (\frac{1}{2}, \frac{1}{2})$, we obtain

$$1 = C \times \frac{1}{2} \times \frac{1}{2} \times \frac{1}{2} \times \frac{1}{2} = \frac{C}{16}.$$

Finally, we find that

$$p^0(x) = 16x_1x_2(1-x_1)(1-x_2).$$

Conversely, it is clear that this particular polynomial is in Q_2 and satisfies the required interpolation conditions.

The same arguments, that we leave as an exercise, show that

$$\begin{aligned}
p^1(x) &= (1-x_1)(1-2x_1)(1-x_2)(1-2x_2),\\
p^2(x) &= -x_1(1-2x_1)(1-x_2)(1-2x_2),\\
p^3(x) &= x_1(1-2x_1)x_2(1-2x_2),\\
p^4(x) &= -(1-x_1)(1-2x_1)x_2(1-2x_2),
\end{aligned}$$

and

$$\begin{aligned}
p^{1,2}(x) &= 4x_1(1-x_1)(1-x_2)(1-2x_2),\\
p^{2,3}(x) &= -4x_1(1-2x_1)x_2(1-x_2),\\
p^{3,4}(x) &= -4x_1(1-x_1)x_2(1-2x_2),\\
p^{4,1}(x) &= 4(1-2x_1)(1-x_1)x_2(1-x_2).
\end{aligned}$$

The latter three of each group are obtained by considerations of symmetry from the first one of the group. □

We draw the graphs of the different basis polynomials in Figs. 6.20, 6.21 and 6.22.

Let us now consider the whole mesh. The nodes no longer are just the element vertices, but also the middles of the edges and the centers of gravity of the elements, see Fig. 6.23.

We then have the exact analog of Propositions 6.3 and 6.4.

Proposition 6.9 *A function of V_h is uniquely determined by its values at the interior nodes of the mesh and all sets of values are interpolated by one and only one element of V_h.*

Proof By unisolvence, nine values for the nine nodes of an element determine one and only one Q_2 polynomial that interpolates these nodal values (we take the value

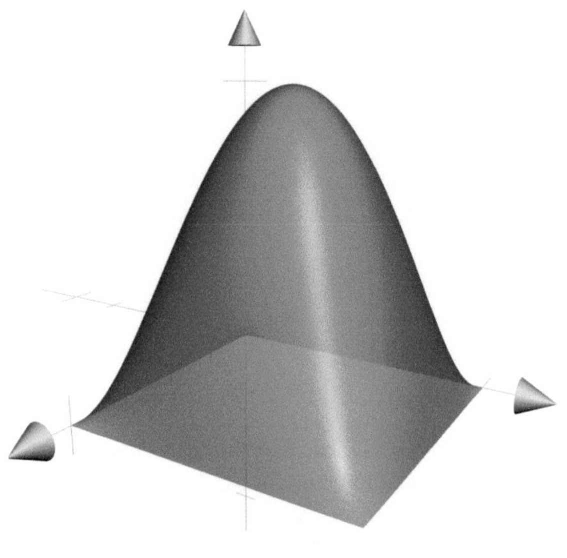

Fig. 6.20 The graph of p^0

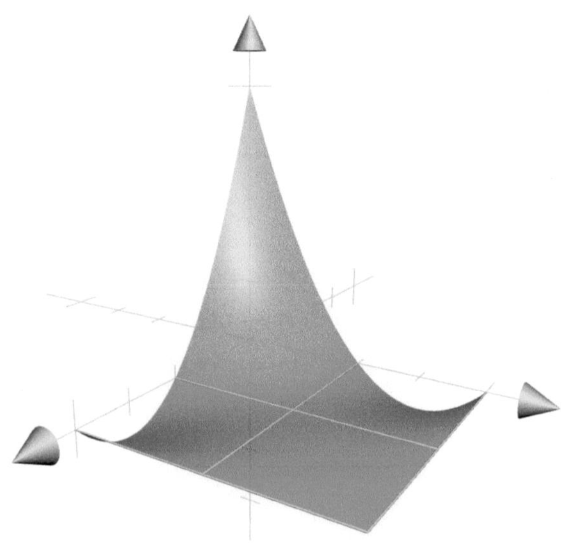

Fig. 6.21 The graph of p^1, with the segments where p^1 vanishes

0 for the nodes located on the boundary). Therefore, if we are given a set of values for each node in the mesh, this set determines one Q_2 polynomial per element. Let us check that these polynomials combine into a globally C^0 function.

Let us thus consider the situation depicted in Fig. 6.24, without loss of generality.

We thus have two Q_2 polynomials q and q' such that $q(S^1) = q'(S^1)$, $q(S^{1,2}) = q'(S^{1,2})$ and $q(S^2) = q'(S^2)$. On the segment $[S^1, S^2]$, which is parallel to the x_2 axis, $q - q'$ is a second degree polynomial in the variable x_2 that has three roots.

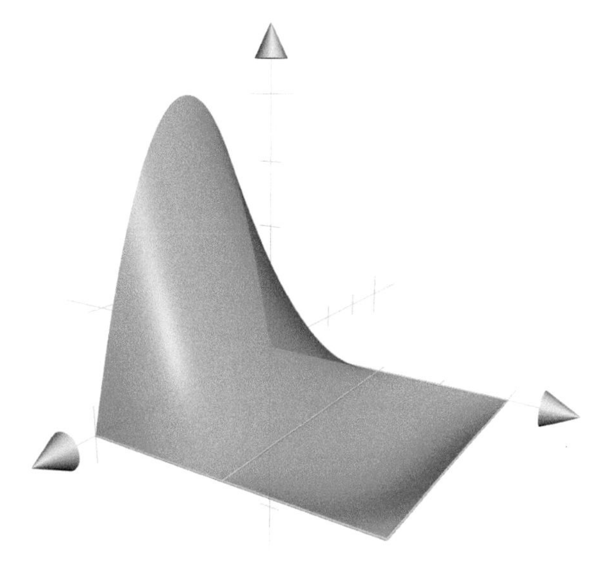

Fig. 6.22 The graph of $p^{1,2}$, with the segments where $p^{1,2}$ vanishes

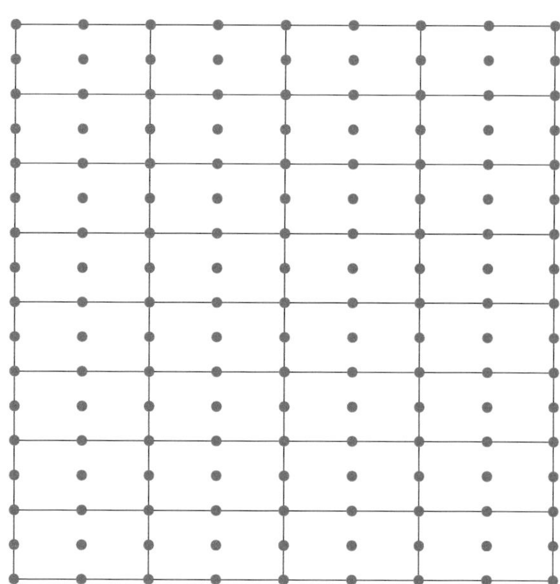

Fig. 6.23 Same mesh as Fig. 6.7, 32 elements, 153 nodes

Therefore, $q - q' = 0$ on this segment, and the function defined by $q(x)$ if $x \in R_k$, $q'(x)$ if $x \in R_{k'} \setminus R_k$ is continuous on $R_k \cup R_{k'}$. □

Corollary 6.3 *Let us be given a numbering of the nodes S^i, $i = 1, \dots, N = (2N_1 + 1)(2N_2 + 1)$. There is a basis of V_h composed of the functions w_h^i defined by $w_h^i(S^j) = \delta_{ij}$ and for all $v_h \in V_h$, we have*

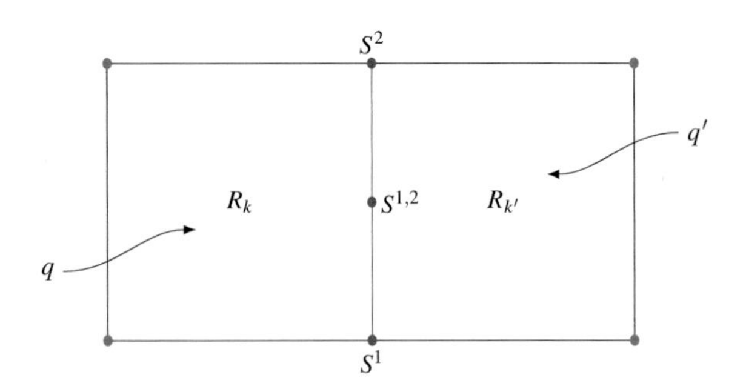

Fig. 6.24 Continuity across an internal edge, Q_2 case

$$v_h = \sum_{i=1}^{N} v_h(S^i) w_h^i.$$

Proof Same as before. □

Figures 6.25, 6.26 and 6.27 show pictures of the different types of basis functions, depending on which kind of node, i.e., center of gravity of an element, element vertex or edge middle, they are attached to.

Note that the last two basis functions change sign in Ω. This was not the case for Q_1 basis functions.

We do not pursue here matrix assembly and node numbering issues. It is to be expected that the structure of the matrix is more complicated than in the Q_1 case.

The question arises as to why introduce Q_2 elements and deal with the added complexity compared with the Q_1 case. One reason is that we thus obtain a higher order approximation method. Indeed, if $u \in H^3(\Omega)$, then we have (exercise) a better error estimate

$$\|u - u_h\|_{H^1(\Omega)} \leq Ch^2 |u|_{H^3(\Omega)},$$

than with Q_1 elements. The estimate is better in the sense that $h^2 \ll h$ when h is small, even though we do not have any idea of the order of magnitude of the constants and

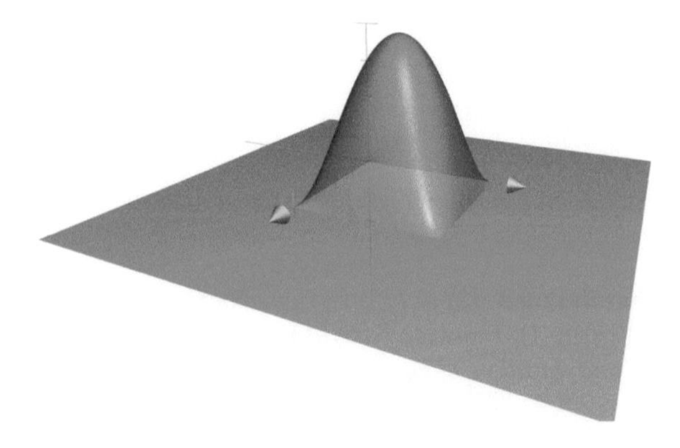

Fig. 6.25 Basis function corresponding to an element center of gravity

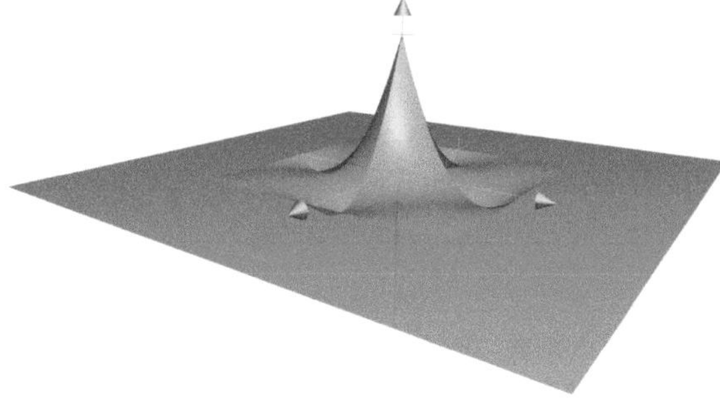

Fig. 6.26 Basis function corresponding to an element vertex

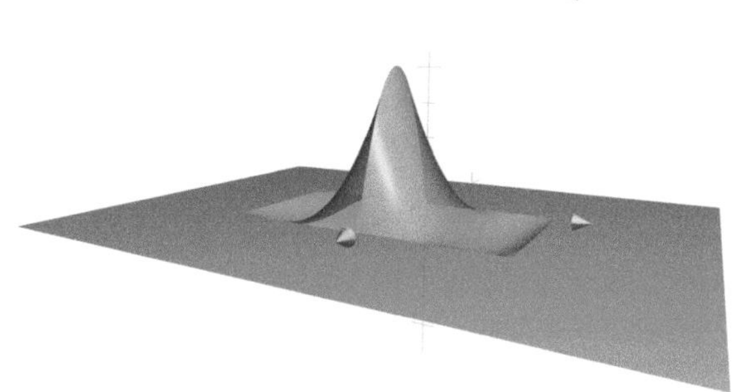

Fig. 6.27 Basis function corresponding to an edge middle

the norms in the right-hand side. So the extra implementation and computational costs induced by the extra degree must be balanced against the increased accuracy that is expected from the higher degree finite element approximation. For instance, a cheaper computation may be achieved with Q_2 elements with the same accuracy by taking less elements than with a Q_1 computation.

Let us say a few words about Q_3 finite elements. We could define Q_3-Lagrange elements by taking 16 nodes per element, since $\dim Q_3 = 16$. We would need four nodes per edge to ensure global continuity, hence the four vertices plus two points on the thirds of each edge. Four more points must be chosen inside, with obvious simple possibilities.

We can also use Q_3 finite elements for Hermite interpolation that result in C^1 functions suitable for conforming approximation of fourth order problems, such as the plate equation (1.10) for example. In this case, the degrees of freedom must also include partial derivative values. We would thus take as degrees of freedom the 4 vertex values and the 8 first partial derivatives values at the vertices. This would seem to be enough, as we recognize 1d P_3 Hermite interpolation on each edge, and there is a tensor product structure $Q_3[X, Y] = P_3[X] \otimes P_3[Y]$.

Surprisingly, this is not enough. Indeed, this choice would only provide 12 degrees of freedom for a 16-dimensional space, and there would be infinitely many different possible basis polynomials, in the sense that the interpolation relations would be satisfied, since there is infinitely many different ways of adding four more degrees of freedom. Moreover, it is not clear which choice would guarantee global C^1 regularity. Surprisingly again, if we complete the set of degrees of freedom with the four vertex values of the *second* derivatives $\frac{\partial^2 p}{\partial x_1 \partial x_2}$, we obtain a unisolvent element that generates a C^1 approximation. The Q_3-Hermite element is well adapted to the approximation of such fourth order problems.

We have seen an interesting example of the same polynomial space used with two completely different sets of degrees of freedom and yielding two completely different approximation spaces, Q_3-Lagrange and Q_3-Hermite, which are used in also different contexts.

Let us now switch to triangular finite elements, which are better adapted for problems that are posed in open sets that are not just rectangles. First we need a quick review of affine geometry.

6.7 Barycentric Coordinates

Triangular finite elements are much easier to work with using a system of coordinates in the plane that is quite different from the usual Cartesian system, namely barycentric coordinates. Actually, barycentric coordinates are natural systems of coordinates for affine geometry.

We will be given three points A^1, A^2 and A^3 in the plane. We first define weighted barycenters of these points.

Definition 6.6 Let λ_1, λ_2 and λ_3 be three scalars such that $\lambda_1 + \lambda_2 + \lambda_3 = 1$. The barycenter of the points A^j with weights λ_j is the unique point M in the plane such that $\overrightarrow{OM} = \sum_{j=1}^3 \lambda_j \overrightarrow{OA^j}$, where O is a given point. This point does not depend on the choice of O and we thus write

$$M = \sum_{j=1}^{3} \lambda_j A^j.$$

One statement in this definition needs to be checked, namely that M does not depend on O. Indeed, let O' be another choice of point, and M' be such that $\overrightarrow{O'M'} = \sum_{j=1}^3 \lambda_j \overrightarrow{O'A^j}$. We have

$$\overrightarrow{O'M'} = \sum_{j=1}^{3} \lambda_j (\overrightarrow{O'O} + \overrightarrow{OA^j}) = \Big(\sum_{j=1}^{3} \lambda_j\Big)\overrightarrow{O'O} + \sum_{j=1}^{3} \lambda_j \overrightarrow{OA^j} = \overrightarrow{O'O} + \overrightarrow{OM} = \overrightarrow{O'M}$$

hence $M' = M$.

Now of course, barycenters are likewise defined for any finite family of points and weights of sum equal to 1, and in any affine space, but we will only use three points in the plane.

From now on, we assume that the three points A^j are *not aligned*, in which case they constitute what is called an *affine basis* of the plane. In this case, we have the following basic result.

Proposition 6.10 *For all points M in the plane, there exists a unique triple* $(\lambda_1, \lambda_2, \lambda_3)$ *of real numbers with* $\lambda_1 + \lambda_2 + \lambda_3 = 1$ *such that*

$$M = \sum_{j=1}^{3} \lambda_j A^j.$$

The scalars $\lambda_i = \lambda_i(M)$ *are called the* barycentric coordinates *of M, with respect to points* A^1, A^2, A^3.

Proof We use Cartesian coordinates. Let (x_1^j, x_2^j) be the Cartesian coordinates of A^j in some Cartesian coordinate system, and (x_1, x_2) be the Cartesian coordinates of point M. We have $M = \sum_{j=1}^{3} \lambda_j A^j$ if and only if $x_k = \sum_{j=1}^{3} \lambda_j x_k^j$ for $k = 1, 2$. Moreover, we have the condition $1 = \sum_{j=1}^{3} \lambda_j$. We thus find a system of three linear equations in the three unknowns λ_j

$$\begin{cases} \lambda_1 + \quad \lambda_2 + \quad \lambda_3 = 1, \\ x_1^1\lambda_1 + x_1^2\lambda_2 + x_1^3\lambda_3 = x_1, \\ x_2^1\lambda_1 + x_2^2\lambda_2 + x_2^3\lambda_3 = x_2. \end{cases}$$

The determinant of this system is

$$\Delta = \begin{vmatrix} 1 & 1 & 1 \\ x_1^1 & x_1^2 & x_1^3 \\ x_2^1 & x_2^2 & x_2^3 \end{vmatrix} = \begin{vmatrix} 1 & 0 & 0 \\ x_1^1 & x_1^2 - x_1^1 & x_1^3 - x_1^1 \\ x_2^1 & x_2^2 - x_2^1 & x_2^3 - x_2^1 \end{vmatrix} = (x_1^2 - x_1^1)(x_2^3 - x_2^1) - (x_1^3 - x_1^1)(x_2^2 - x_2^1) \neq 0,$$

since it is equal to $\det(\overrightarrow{A^2A^1}, \overrightarrow{A^3A^1}) = 2\,\text{area}(T)$, where T is the triangle with vertices A^1 A^2 and A^3, and $\text{area}(T)$ is its algebraic area which is nonzero since the points are not aligned.

Therefore, for any right-hand side, i.e., for any point M, the system has one and only one solution. □

Remark 6.9 Going from barycentric coordinates to Cartesian coordinates is just done by applying the definition. Conversely, to compute barycentric coordinates from Cartesian coordinates, we just need to solve the above linear system.

If the three points are aligned, then we get a system which has a solution only if M is on the line spanned by the points, and there are infinitely many solutions, and if the three points are equal, the system only has a solution if M is equal to the other points, again with an infinity of solutions. □

Let us give a few miscellaneous properties of barycentric coordinates.

Proposition 6.11 *We have*

(i) $\lambda_i(A^j) = \delta_{ij}$ *for all i and j.*
(ii) *The functions* λ_i *are affine in* (x_1, x_2) *and conversely,* (x_1, x_2) *are affine functions of* $(\lambda_1, \lambda_2, \lambda_3)$.
(iii) *Let* (A^i, A^j) *denote the straight line passing through* A^i *and* A^j *for* $i \neq j$. *Then* $(A^i, A^j) = \{M; \lambda_k(M) = 0, k \neq i, k \neq j\}$.
(iv) *Let* T *be the closed triangle determined by the three points* A^j. *Then* $T = \{M, 0 \leq \lambda_i(M) \leq 1, i = 1, 2, 3\}$.

Proof (i) We have $A^1 = 1 \times A^1 + 0 \times A^2 + 0 \times A^3$ with $1 + 0 + 0 = 1$, hence by uniqueness of the barycentric coordinates, $\lambda_i(A^1) = \delta_{i1}$.
(ii) Use Cramer's rule for solving the above linear system.
(iii) The function λ_k is a nonzero affine function by (i) and (ii), thus it vanishes on a straight line. By (i), this straight line contains A^i and A^j, so it is equal to (A^i, A^j).
(iv) We have just seen by (iii) that $\lambda_k(M) = 0$ is the equation of the straight line opposite to vertex A^k. Moreover, by (i) the half-plane containing A_k is the half-plane $\{M; \lambda_k(M) \geq 0\}$. The triangle T is the intersection of these three half-planes, so it is the set of points whose barycentric coordinates are all nonnegative. Since their sum is equal to 1, they are also less than or equal to 1. □

Figure 6.28 shows the signs of the barycentric coordinates in the plane. Note that there is no $---$ region, it would be hard to have $\sum \lambda_i = 1$ in such a region …

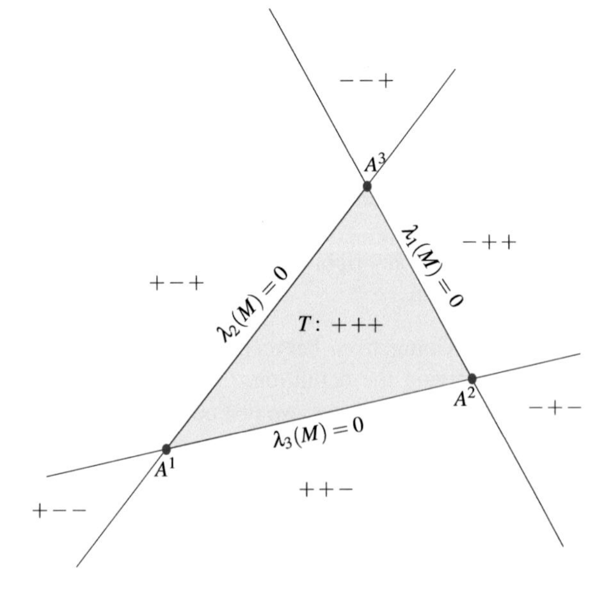

Fig. 6.28. Signs of the barycentric coordinates in order $\lambda_1, \lambda_2, \lambda_3$. For instance, $++-$ means that $\lambda_1(M) \geq 0$, $\lambda_2(M) \geq 0$ and $\lambda_3(M) \leq 0$, and so on

Let us give the barycentric coordinates of a few points of interest in a triangle:

- Middle of $[A^1A^2]$: $\left(\frac{1}{2}, \frac{1}{2}, 0\right)$,
- Middle of $[A^2A^3]$: $\left(0, \frac{1}{2}, \frac{1}{2}\right)$,
- Middle of $[A^1A^3]$: $\left(\frac{1}{2}, 0, \frac{1}{2}\right)$,
- Center of gravity of the triangle: $\left(\frac{1}{3}, \frac{1}{3}, \frac{1}{3}\right)$.

Proposition 6.12 *The equation of any straight line in barycentric coordinates is of the form*

$$\sum_{i=1}^{3} \gamma_i \lambda_i(M) = 0,$$

where the constants γ_i are not all equal.

Proof Let A and B be two distinct points with barycentric coordinates $(\alpha_1, \alpha_2, \alpha_3)$ and $(\beta_1, \beta_2, \beta_3)$. This implies not only that $(\alpha_1, \alpha_2, \alpha_3) \neq (\beta_1, \beta_2, \beta_3)$, but also that the matrix $\begin{pmatrix} \alpha_1 & \alpha_2 & \alpha_3 \\ \beta_1 & \beta_2 & \beta_3 \end{pmatrix}$ is of rank 2. Indeed if this matrix was of rank one, the two row vectors would be proportional, and since the sum of their coefficients is equal to one, the proportionality coefficient would also be equal to one, i.e., $A = B$.

A point M is on the straight line (A, B) passing through A and B if and only if it is a barycenter of A and B, i.e., if and only if there exists a scalar μ such that

$$M = \mu A + (1 - \mu)B.$$

It follows immediately that

$$\lambda_i(M) = \mu\alpha_i + (1 - \mu)\beta_i, \; i = 1, 2, 3,$$

which is a parametric representation of the straight line in barycentric coordinates with $\mu \in \mathbb{R}$.

Due to the rank remark above, the existence of μ is then clearly equivalent to the equation

$$\begin{vmatrix} \lambda_1(M) & \lambda_2(M) & \lambda_3(M) \\ \alpha_1 & \alpha_2 & \alpha_3 \\ \beta_1 & \beta_2 & \beta_3 \end{vmatrix} = 0,$$

which reads

$$\gamma_1\lambda_1(M) + \gamma_2\lambda_2(M) + \gamma_3\lambda_3(M) = 0,$$

where $\gamma_1 = \alpha_2\beta_3 - \alpha_3\beta_2$ and so on. Indeed, the vanishing of the above determinant implies that the first line is a linear combination of the other two, or that $\lambda_i(M) = \mu_1\alpha_i + \mu_2\beta_i$, $i = 1, 2, 3$. If we sum over i, we obtain $1 = \mu_1 + \mu_2$.

It remains to show that the γ_i are not all equal. This is clear since $A \in (A, B)$ so that $\sum_{i=1}^{3} \gamma_i\alpha_i = 0$. If we had $\gamma_i = \gamma$ for all i, this would imply that $0 = \sum_{i=1}^{3} \gamma_i\alpha_i = \gamma \sum_{i=1}^{3} \alpha_i = \gamma$. This would in turn imply that $(\alpha_1, \alpha_2, \alpha_3) = (\beta_1, \beta_2, \beta_3)$ or $A = B$.

Conversely, let us be given three scalars γ_i not all equal. The affine function $f: M \mapsto \sum_{i=1}^{3} \gamma_i \lambda_i(M)$ is non constant. Indeed, $f(A^i) = \gamma_i$. It thus vanishes on a straight line. □

The above equation is homogeneous, multiplying it by a nonzero constant yields another equation that obviously describes the same straight line. Conversely, two such homogeneous equations describe the same straight line if and only if their coefficients are proportional. Indeed, assume that γ_i and γ_i' describe the same straight line. This implies that the linear system

$$\begin{cases} \lambda_1 + \quad \lambda_2 + \quad \lambda_3 = 1, \\ \gamma_1\lambda_1 + \gamma_2\lambda_2 + \gamma_3\lambda_3 = 0, \\ \gamma_1'\lambda_1 + \gamma_2'\lambda_2 + \gamma_3'\lambda_3 = 0, \end{cases}$$

has infinitely many solutions. Hence its determinant is zero.

Let us give an example. Consider the line passing through the middle of $[A^1A^2]$ and the middle of $[A^1A^3]$. One equation for this line is thus

$$\begin{vmatrix} \lambda_1(M) & \lambda_2(M) & \lambda_3(M) \\ \frac{1}{2} & \frac{1}{2} & 0 \\ \frac{1}{2} & 0 & \frac{1}{2} \end{vmatrix} = 0,$$

which reads after multiplication by 4

$$\lambda_1(M) - \lambda_2(M) - \lambda_3(M) = 0.$$

We can rewrite the equation in nonhomogeneous form by using the fact that $-\lambda_2(M) - \lambda_3(M) = \lambda_1(M) - 1$, which yields

$$2\lambda_1(M) - 1 = 0,$$

in other words, this line is the locus of points such that $\lambda_1(M) = \frac{1}{2}$, which is quite visible on a figure.

Remark 6.10 Of course, if we are given the equation of a straight line in Cartesian coordinates, it is immediate to derive an equation for that same line in barycentric coordinates. Indeed, we have seen that the Cartesian coordinates are affine functions of the barycentric coordinates, $x_1(\lambda_1, \lambda_2, \lambda_3)$, $x_2(\lambda_1, \lambda_2, \lambda_3)$. Substituting these expressions in a Cartesian equation $ax_1 + bx_2 + c = 0$, we obtain an expression $\alpha\lambda_1 + \beta\lambda_2 + \gamma\lambda_3 + \delta = 0$, which we can rewrite in homogeneous form $(\alpha + \delta)\lambda_1 + (\beta + \delta)\lambda_2 + (\gamma + \delta)\lambda_3 = 0$. It is as easy to pass from an equation in barycentric coordinates to an equation in Cartesian coordinates. □

An important feature of barycentric coordinates is their invariance under affine transformations. For this we modify the notation a bit by indicating the dependence on the points A^j by writing $\lambda_i^{A^1,A^2,A^3}(M)$, which is admittedly cumbersome, and will thus not be used after this.

Proposition 6.13 *Let F be an bijective affine transformation of the plane. Then we have*

$$\lambda_i^{F(A^1),F(A^2),F(A^3)}(F(M)) = \lambda_i^{A^1,A^2,A^3}(M)$$

for $i = 1, 2, 3$ *and all* M.

Proof This is clear since affine transformations conserve barycenters. □

The barycentric coordinates also have a nice geometrical interpretation. We choose an orientation of the plane such that the loop $A^1 \to A^2 \to A^3 \to A^1$ runs counter-clockwise. Then, the algebraic area of T, which is equal to $\frac{1}{2}\det(\overrightarrow{A^1A^2}, \overrightarrow{A^1A^3})$, is strictly positive. For $i = 1$, we let $i_+ = 2$, for $i = 2$, we let $i_+ = 3$, and for $i = 3$, we let $i_+ = 1$. We also let $i_{++} = (i_+)_+$. For any point M in the plane, we denote by $T_i(M)$ the possibly degenerate, oriented triangle $MA^{i_+}A^{i_{++}}$, see Fig. 6.29. Its algebraic area is area $T_i(M) = \frac{1}{2}\det(\overrightarrow{MA^{i_+}}, \overrightarrow{MA^{i_{++}}})$.

Proposition 6.14 *We have*

$$\lambda_i(M) = \frac{\text{area } T_i(M)}{\text{area } T}$$

for $i = 1, 2, 3$ *and all* M.

Proof Taking $O = M$ in the definition of barycentric coordinates, we see that

$$0 = \sum_{j=1}^{3} \lambda_j(M)\overrightarrow{MA^j}.$$

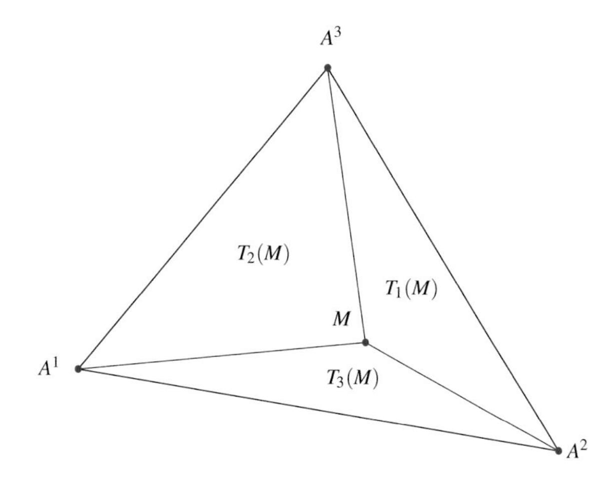

Fig. 6.29 Algebraic areas and barycentric coordinates

In particular

$$\lambda_1(M)\overrightarrow{MA^1} = -\lambda_2(M)\overrightarrow{MA^2} - \lambda_3(M)\overrightarrow{MA^3}$$

for instance, so that

$$\begin{aligned}\lambda_1(M)\det(\overrightarrow{MA^1}, \overrightarrow{MA^2}) &= \det(-\lambda_2(M)\overrightarrow{MA^2} - \lambda_3(M)\overrightarrow{MA^3}, \overrightarrow{MA^2})\\ &= -\lambda_3(M)\det(\overrightarrow{MA^3}, \overrightarrow{MA^2}).\end{aligned}$$

Therefore, we have $\lambda_1(M)$area $T_3(M) = \lambda_3(M)$area $T_1(M)$ and likewise for the other two possible choices. Hence, there exists a scalar μ such that

$$\begin{pmatrix}\lambda_1(M)\\ \lambda_2(M)\\ \lambda_3(M)\end{pmatrix} = \mu\begin{pmatrix}\text{area } T_1(M)\\ \text{area } T_2(M)\\ \text{area } T_3(M)\end{pmatrix}.$$

Summing over the three lines on both sides, we obtain $1 = \mu$ area T and the proposition is proved. □

Remark 6.11 In the context of the finite element method, in each triangle of a mesh, we will use the barycentric coordinates associated with the vertices of this particular triangle to compute all the quantities that concern the triangle in question, such as basis functions and so on. □

6.8 Triangular P_1 Lagrange Elements

Let us return to the model problem (6.1), on a polygonal domain Ω. Let us be given a triangular mesh $\mathscr{T}$ on Ω. We remind the reader that P_1 denotes the space of polynomials of total degree less or equal to 1, i.e., affine functions. We define the corresponding approximation spaces

$$W_h = \{v_h \in C^0(\bar{\Omega}), v_{h|T_k} \in P_1 \text{ for all } T_k \in \mathscr{T}\},$$

without boundary conditions and

$$V_h = \{v_h \in W_h, v_h = 0 \text{ on } \partial\Omega\},$$

with boundary conditions. The general approximation theory applies and we thus just need to describe the approximation spaces in terms of finite elements and basis functions.

Let T be a triangle with non aligned vertices A^1, A^2 and A^3. We allow the same misuse of notation as before for the degrees of freedom.

Proposition 6.15 *The finite element* $(T, P_1, \{p(A^1), p(A^2), p(A^3)\})$ *is unisolvent.*

Proof We have $\dim P_1 = 3$ so the numbers match. The basis polynomials are obvious: $\lambda_1, \lambda_2, \lambda_3$, by Proposition 6.11, (i) and (ii). □

Proposition 6.16 *A function of V_h is uniquely determined by its values at the internal nodes of the mesh and conversely, any set of values for the internal nodes is interpolated by one and only one element of V_h.*

Proof By unisolvence, three values for the three nodes of an element determine one and only one P_1 polynomial that interpolates these nodal values (we take the value 0 for the nodes located on the boundary). Therefore, if we are given a set of values for each node in the mesh, this set determines one P_1 polynomial per element. Let us check that they combine into a globally C^0 function.

Since the mesh is admissible, an edge common to two triangles T_k and $T_{k'}$ is delimited by two vertices A^1 and A^2 which are also common to both triangles, see Fig. 6.30. We thus have two P_1 polynomials p and p' such that $p(A^1) = p'(A^1)$ and $p(A^2) = p'(A^2)$. We parametrize the segment $[A^1, A^2]$ as $M = \mu A^1 + (1 - \mu)A^2$ with $\mu \in [0, 1]$. Then the restriction of $p - p'$ to this segment is a first degree polynomial in the variable μ that has two roots, $\mu = 0$ and $\mu = 1$. Therefore, $p - p' = 0$ on this segment, and the function defined by $p(x)$ if $x \in T_k$, $p'(x)$ if $x \in T_{k'}$ is continuous on $T_k \cup T_{k'}$. □

Corollary 6.4 *Let us be given a numbering of the internal nodes S^i, $i = 1, \ldots, N_{\text{int}}$. There is a basis of V_h composed of the functions w_h^i defined by $w_h^i(S^j) = \delta_{ij}$ and for all $v_h \in V_h$, we have*

$$v_h = \sum_{i=1}^{N_{\text{int}}} v_h(S^i) w_h^i.$$

Proof Same as before, see Fig. 6.31. □

Let us now talk a little bit about matrix assembly. We will not touch on the node numbering issue, which is clearly more complicated in a triangular mesh than in a rectangular mesh, especially in an *unstructured* triangular mesh, such as that shown in Fig. 6.2, in which there is no apparent natural numbering.

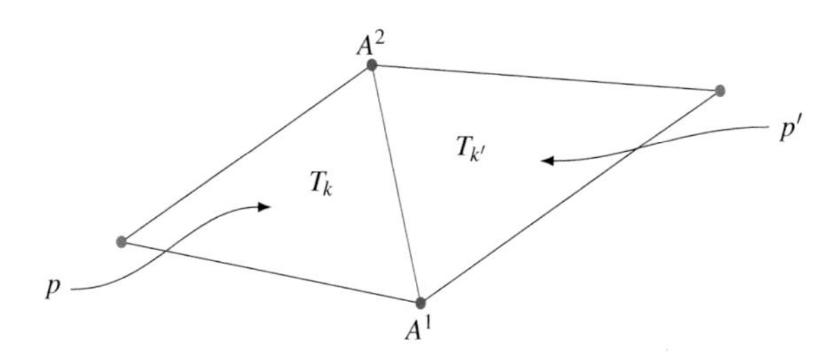

Fig. 6.30 Continuity across an internal edge, P_1 case

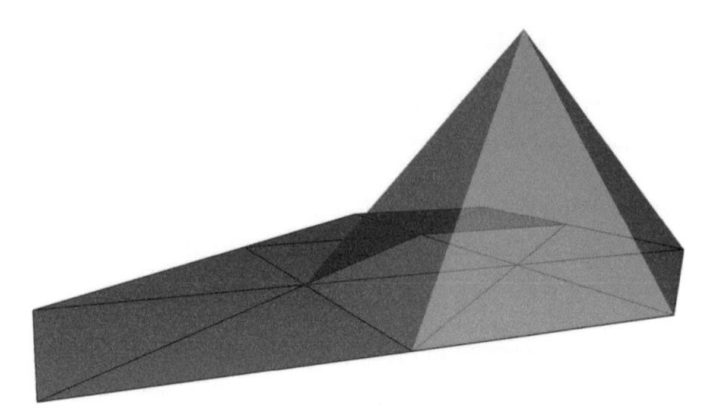

Fig. 6.31 A P_1 basis function on a triangular mesh

We will however see how the use of a reference triangle and of barycentric coordinates simplifies the computation of matrix coefficients. We have the same element-wise decomposition as in the rectangular case

$$A_{ij} = \sum_{k=1}^{N_{\mathcal{T}}} A_{ij}(T_k),$$

with

$$A_{ij}(T_k) = \int_{T_k} (\nabla w_h^j \cdot \nabla w_h^i + c w_h^j w_h^i)(x)\, dx.$$

On each triangle T_k, the basis functions either vanish or are equal to one barycentric coordinate. So we need to compute the integral of the product of two barycentric coordinates (for c constant) and the integral of the scalar product of their gradient.

We thus introduce a reference triangle

$$\widehat{T} = \{(\hat{x}_1, \hat{x}_2) \in \mathbb{R}^2, \hat{x}_1 \geq 0, \hat{x}_2 \geq 0, \hat{x}_1 + \hat{x}_2 \leq 1\}.$$

Let $\hat{A}^1 = (0,0)$, $\hat{A}^2 = (1,0)$ and $\hat{A}^3 = (0,1)$ be its vertices and $\hat{\lambda}_i$ the corresponding barycentric coordinates. Let T_k be a generic triangle in the mesh, with vertices A_k^1, A_k^2, A_k^3. Now, there exists one and only one affine mapping F_k such that $F_k(\hat{A}^j) = A_k^j$, $j = 1, 2, 3$. Indeed, since affine mappings conserve barycenters, we simply have

$$F_k(\widehat{M}) = \hat{\lambda}_1(\widehat{M})A_k^1 + \hat{\lambda}_2(\widehat{M})A_k^2 + \hat{\lambda}_3(\widehat{M})A_k^3,$$

or in other words, $\lambda_i(F_k(\widehat{M})) = \hat{\lambda}_i(\widehat{M})$, where the first barycentric coordinates are taken relative to the vertices of T_k in increasing superscript order.

Now the expression of barycentric coordinates in the reference triangle in terms of Cartesian coordinates is particularly simple:

$$\hat{\lambda}_1 = 1 - \hat{x}_1 - \hat{x}_2, \quad \hat{\lambda}_2 = \hat{x}_1, \quad \hat{\lambda}_3 = \hat{x}_2,$$

whereas they are fairly disagreeable in the generic triangle, see Fig. 6.32.

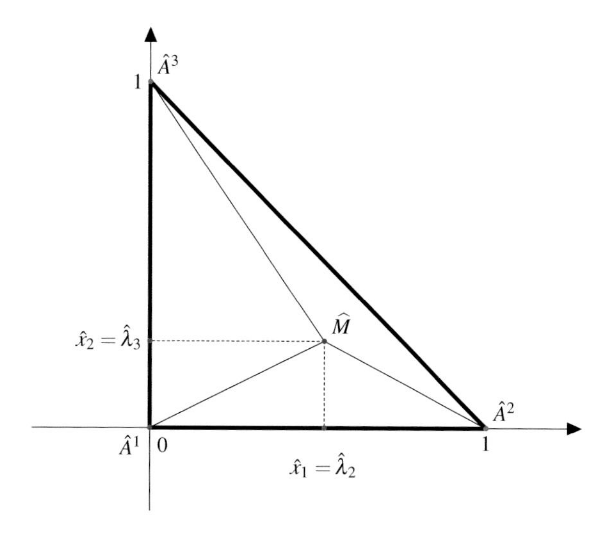

Fig. 6.32 Barycentric coordinates in the reference triangle $\widehat{T}$

Let us give an example of computation with the integral $\int_{T_k} \lambda_2^2\,dx$. We are going to use the change of variables $x = F_k(\hat{x})$. Since this change of variable is affine, its Jacobian J is constant, and we have

$$\text{area}\, T_k = \int_{T_k} dx = \int_{\widehat{T}} J\,d\hat{x} = \frac{J}{2}$$

therefore $J = 2\,\text{area}\, T_k$. Now we can compute

$$\begin{aligned}
\int_{T_k} \lambda_2^2(x)\,dx &= \int_{\widehat{T}} \hat{\lambda}_2^2(\hat{x}) J\,d\hat{x} \\
&= 2\,\text{area}\, T_k \int_{\widehat{T}} \hat{x}_1^2\,d\hat{x} \\
&= 2\,\text{area}\, T_k \int_0^1 \hat{x}_1^2 \Big(\int_0^{1-\hat{x}_1} d\hat{x}_2 \Big) d\hat{x}_1 \\
&= 2\,\text{area}\, T_k \int_0^1 \hat{x}_1^2 (1 - \hat{x}_1)\,d\hat{x}_1 \\
&= 2\,\text{area}\, T_k \Big(\frac{1}{3} - \frac{1}{4} \Big) \\
&= \frac{\text{area}\, T_k}{6}.
\end{aligned}$$

Exchanging the vertices, we find $\int_{T_k} \lambda_1^2(x)\,dx = \int_{T_k} \lambda_3^2(x)\,dx = \frac{\text{area}\, T_k}{6}$. A similar computation shows that $\int_{T_k} \lambda_i(x)\lambda_j(x)\,dx = \frac{\text{area}\, T_k}{12}$ for all $i \neq j$. Such terms are thus of the order of h^2.

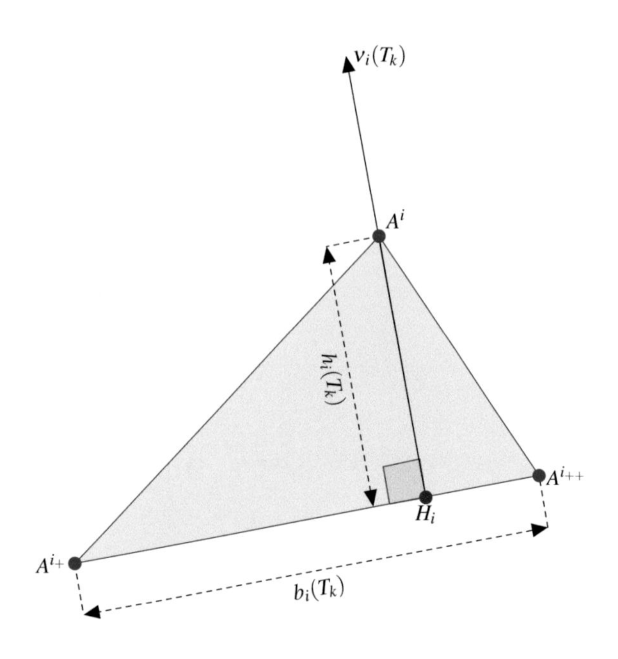

Fig. 6.33 Geometric elements of a generic triangle

Let us now turn to the gradient terms. We first need to compute $\nabla\lambda_i$, which is a constant vector.

We introduce $h_i(T_k)$ and $b_i(T_k)$ respectively the height and base of T_k relative to A^i, H_i the foot of the altitude of A^i and $\nu_i(T_k)$ the unit vector perpendicular to the base and pointing from the base toward A^i, see Fig. 6.33. Since λ_i is affine, we have for all points M

$$\lambda_i(M) = \lambda_i(H_i) + \nabla\lambda_i \cdot \overrightarrow{H_iM}.$$

Now H_i lies on the straight line (A^{i+}, A^{i++}), thus $\lambda_i(H_i) = 0$. Since λ_i vanishes on this straight line, it follows that $\nabla\lambda_i = \mu\nu_i(T_k)$ for some scalar μ. Taking $M = A^i$, we obtain

$$1 = \mu\nu_i(T_k) \cdot \overrightarrow{H_iM} = \mu h_i(T_k).$$

Therefore, we have

$$\nabla\lambda_i = \frac{1}{h_i(T_k)}\nu_i(T_k) = \frac{b_i(T_k)}{2\,\text{area}\,T_k}\nu_i(T_k).$$

It follows from instance that

$$\|\nabla\lambda_i\|^2 = \frac{b_i(T_k)^2}{4(\text{area}\,T_k)^2},$$

so that

$$\int_{T_k} \|\nabla\lambda_i\|^2\,dx = \frac{b_i(T_k)^2}{4\,\text{area}\,T_k}.$$

These terms are of the order of 1. We could likewise compute $\int_{T_k} \nabla\lambda_i \cdot \nabla\lambda_j\,dx$ without difficulty, with expressions that involve the angles of T_k.

6.9 Triangular P_2 Lagrange Elements

Let us go one step up in degree and consider P_2 elements. We have $\dim P_2 = 6$ as is shown by its canonical basis $(1, x_1, x_2, x_1^2, x_1x_2, x_2^2)$. This canonical basis is useless for our purposes and it is again much better to work in barycentric coordinates. The following result is meant to convince the reader of this fact.

Proposition 6.17 *Let T be a triangle with vertices A^1, A^2, A^3 and λ_1, λ_2, λ_3 be the corresponding barycentric coordinates. The family $(\lambda_1^2, \lambda_2^2, \lambda_3^2, \lambda_1\lambda_2, \lambda_2\lambda_3, \lambda_1\lambda_3)$ is a basis of P_2.*

Proof The functions λ_i are affine, thus products $\lambda_i\lambda_j$ belong to P_2. We have a family of 6 vectors in a 6-dimensional space, it thus suffices to show that it is linearly independent. Let us be given a family of 6 scalars α_{ij} such that

$$\sum_{i\le j=1}^{3} \alpha_{ij}\lambda_i\lambda_j = 0.$$

Evaluating first this relation at point A^k, we obtain

$$0 = \sum_{i\le j=1}^{3} \alpha_{ij}\delta_{ik}\delta_{jk} = \alpha_{kk}$$

for all k. We are thus left with

$$\alpha_{12}\lambda_1\lambda_2 + \alpha_{13}\lambda_1\lambda_3 + \alpha_{23}\lambda_2\lambda_3 = 0.$$

We evaluate this relation at point $\frac{A^1+A^2}{2}$, the middle of A^1 and A^2, for which $\lambda_1 = \lambda_2 = \frac{1}{2}$ and $\lambda_3 = 0$. Hence

$$\frac{\alpha_{12}}{4} = 0,$$

and similarly $\alpha_{13} = \alpha_{23} = 0$. □

We need 6 degrees of freedom of Lagrange interpolation. We take the three vertices A^i and the three edge middles $A^{i,i+}$, see Fig. 6.34. Then we have the following proposition, using the same misuse of notation as before.

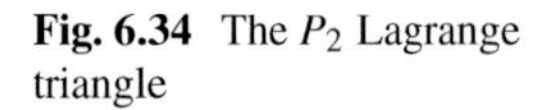

Fig. 6.34 The P_2 Lagrange triangle

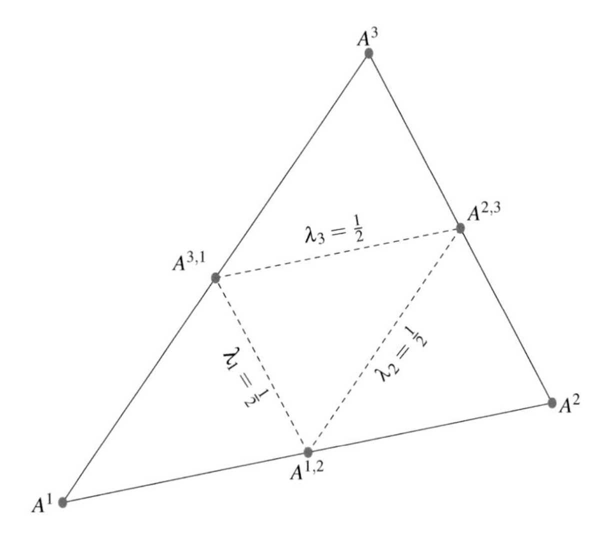

Proposition 6.18 *The finite element* $\big(T, P_2, \{p(A^i), p(A^{i,i+})\}_{i=1,2,3}\big)$ *is unisolvent.*

Proof We have the right number of degrees of freedom with respect to the dimension of the polynomial space. It is thus sufficient to construct the basis polynomials. Everything being invariant by permutation of the vertices, it is clearly sufficient to construct the basis polynomial corresponding to A^1 and that corresponding to $A^{1,2}$, for example.

Let us start with A^1. We thus need a polynomial $p_1 \in P_2$ such that $p_1(A^1) = 1$ and p_1 vanishes at all the other nodes. We will freely use the obvious fact that the restriction of a polynomial of total degree at most n in two variables to a straight line is a polynomial of degree at most n in any affine parametrization of the straight line. Here, p_1 is of degree at most 2 on (A^2, A^3), with three roots corresponding to points A^2, $A^{2,3}$ and A^3, thus it vanishes on (A^2, A^3). The equation of the straight line is $\lambda_1 = 0$, hence p_1 is divisible by λ_1, i.e., there exists a polynomial q such that $p_1 = q\lambda_1$.

Now λ_1 is of degree 1, therefore q is of degree at most one. Moreover, since $\lambda_1(A^{1,2}) = \lambda_1(A^{3,1}) = \frac{1}{2} \neq 0$, we have $q(A^{1,2}) = q(A^{3,1}) = 0$. Therefore, by the same token, q vanishes on the straight line $(A^{1,2}, A^{3,1})$, of equation $\lambda_1 - \frac{1}{2} = 0$. Thus q is divisible by $\lambda_1 - \frac{1}{2}$, so that $q = c(\lambda_1 - \frac{1}{2})$ with c of degree at most 0, i.e., a constant. Finally, the relation $p_1(A^1) = 1$ yields $1 = \frac{c}{2}$, hence $p_1 = \lambda_1(2\lambda_1 - 1)$. Conversely, it is easy—but necessary—to check that this polynomial is in P_2 and satisfies the required interpolation relations.

To sum up, we have

$$p_1 = \lambda_1(2\lambda_1 - 1), \quad p_2 = \lambda_2(2\lambda_2 - 1), \quad p_3 = \lambda_3(2\lambda_3 - 1),$$

for the basis polynomials associated with the vertices. The basis polynomials are equivalently rewritten in homogeneous form as

$$p_i = \lambda_i(\lambda_i - \lambda_{i_+} - \lambda_{i_{++}}),$$

for $i = 1, 2, 3$.

Next we deal with $A^{1,2}$. The polynomial $p_{1,2}$ has three roots on the line (A^1, A^3), where it must thus vanish. Hence it is divisible by λ_2 so that there exists q such that $p_{1,2} = q\lambda_2$. Likewise, the polynomial $p_{1,2}$ must also vanish on the line (A^2, A^3), hence be divisible by λ_1. Now the polynomials λ_1 and λ_2 are relatively prime, therefore $p_{1,2} = c\lambda_1\lambda_2$ where c is a constant. Using $p_{1,2}(A^{1,2}) = 1$, we obtain $c = 4$. Conversely, this polynomial is in P_2 and satisfies the required interpolation relations.

To sum up, we have

$$p_{1,2} = 4\lambda_1\lambda_2, \quad p_{2,3} = 4\lambda_2\lambda_3, \quad p_{3,1} = 4\lambda_1\lambda_3,$$

for the basis polynomials associated with the middles of the edges.

We have found six basis polynomials, therefore the P_2 Lagrange triangular element is unisolvent. □

Figure 6.35 shows the graphs of the different P_2 basis polynomials.

The approximation space

$$V_h = \{v_h \in C^0(\bar{\Omega});\, v_{h|T_k} \in P_2, \forall T_k \in \mathscr{T}, v_h = 0 \text{ on } \partial\Omega\}$$

is of course endowed with a set of basis functions that interpolate values at all nodes (vertices and middles). Let us quickly check the continuity across an edge. We thus have two polynomials of degree at most two, one on each side of the edge, that coincide at the vertices and the middle, see Fig. 6.36. Their restriction to the edge is

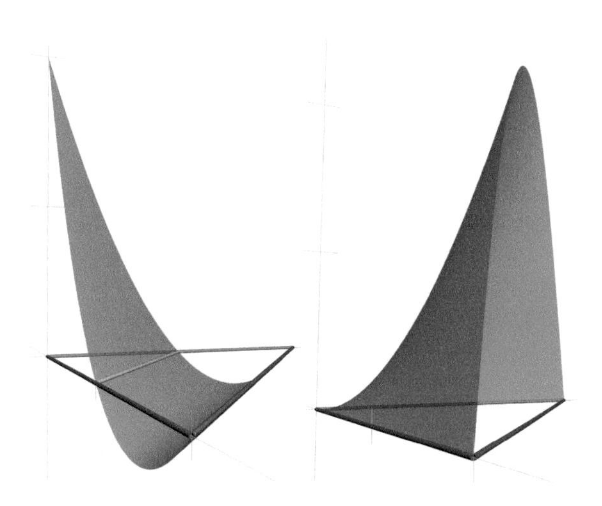

Fig. 6.35 The two different kinds of P_2 basis polynomials

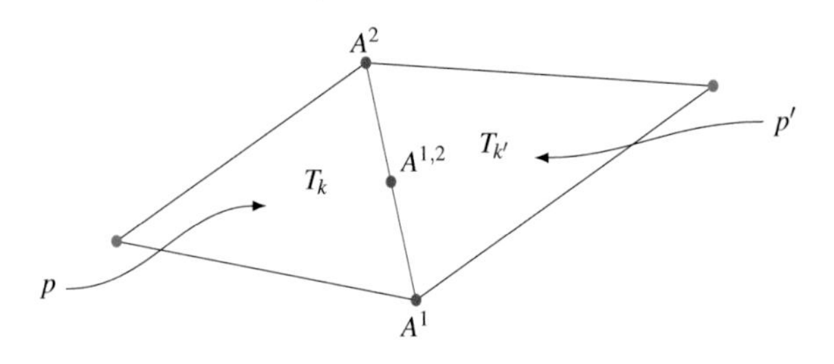

Fig. 6.36 Continuity across an internal edge, P_2 case

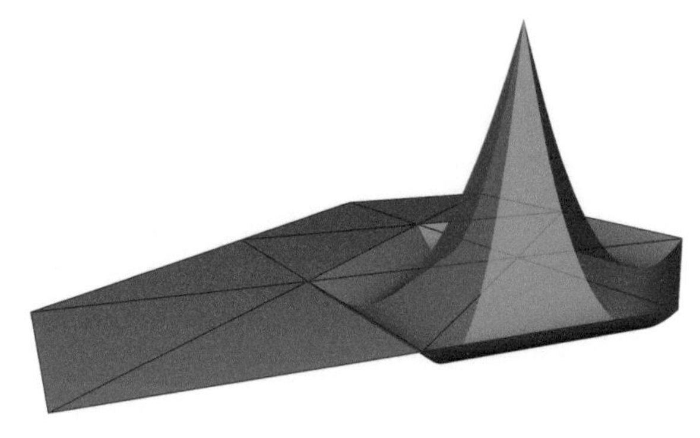

Fig. 6.37 A basis function associated with a vertex

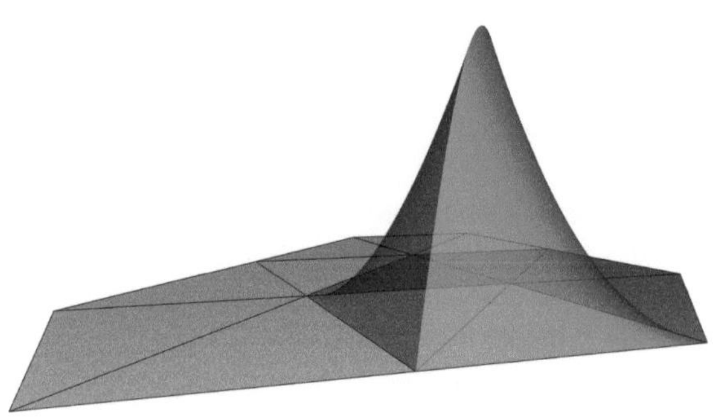

Fig. 6.38 A basis function associated with a middle

of degree two in one variable, their difference has three roots, hence they are equal on the edge. The rest follows as before. Figures 6.37 and 6.38 show the graphs of typical P_2 basis functions.

Let us say a few words about P_3 Lagrange triangles. We have $\dim P_3 = 10$, thus 10 interpolation points are needed. We take the 3 vertices plus 2 points per edge, located at the thirds (this will obviously imply global continuity). That makes 9 points. A simple choice for the tenth point is then the center of gravity, see Fig. 6.39.

Naturally, this finite element is unisolvent. We list the basis polynomials:

$$p_0 = 27\lambda_1\lambda_2\lambda_3,$$

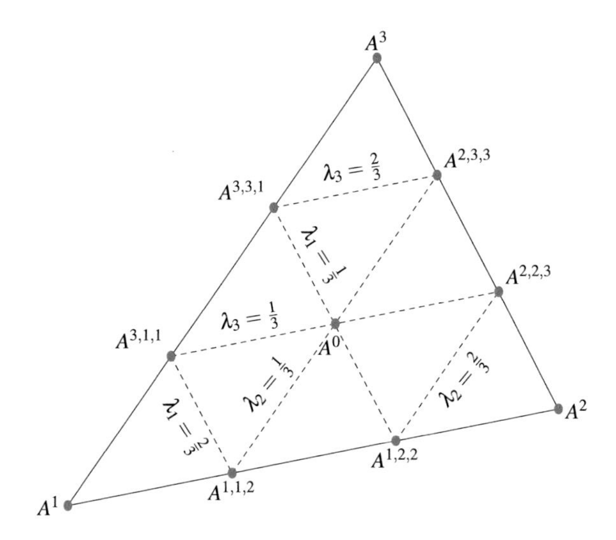

Fig. 6.39 The 10 nodes of a P_3 Lagrange triangle. We use the notation $A^{i,i,i_+} = \frac{2}{3}A^i + \frac{1}{3}A^{i_+}$ and $A^{i,i_+,i_+} = \frac{1}{3}A^i + \frac{2}{3}A^{i_+}$. The center of gravity is of course $A^0 = \frac{1}{3}(A^1 + A^2 + A^3)$

corresponding to the center of gravity, also called a *bubble* due to the shape of its graph,

$$p_i = \frac{1}{2}\lambda_i(3\lambda_i - 1)(3\lambda_i - 2), i = 1, 2, 3,$$

associated with the three vertices, and

$$p_{i,i,i_+} = \frac{9}{2}\lambda_i\lambda_{i+}(3\lambda_i - 1),\ p_{i,i_+,i_+} = \frac{9}{2}\lambda_i\lambda_{i+}(3\lambda_{i_+} - 1), i = 1, 2, 3,$$

associated with the six edge nodes. All these formulas can be rewritten in homogeneous form. Figure 6.40 shows the graphs of the different P_3 basis polynomials.

Also of course, the approximation space

$$V_h = \{v_h \in C^0(\bar{\Omega}); v_{h|T_k} \in P_3, \forall T_k \in \mathscr{T}, v_h = 0 \text{ on } \partial\Omega\}$$

has the usual basis made of basis functions which we picture in Fig. 6.41.

As in the rectangular case, the reason for facing the added complexity of using higher degree polynomials is to achieve faster convergence. Indeed, we have the following general result [19, 66], for P_k-Lagrange triangular elements corresponding to the approximation spaces

$$V_h = \{v_h \in C^0(\bar{\Omega}); v_{h|T_l} \in P_k, \forall T_l \in \mathscr{T}, v_h = 0 \text{ on } \partial\Omega\}$$

with $k \geq 1$.

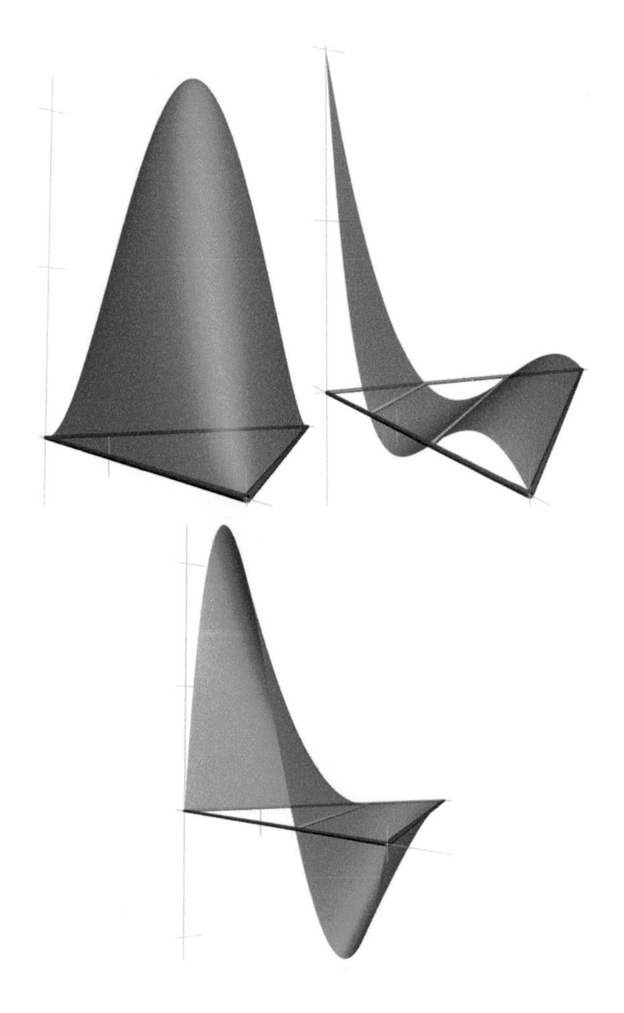

Fig. 6.40 The three different kinds of P_3 basis polynomials

Theorem 6.3 *Let us be given a regular family of triangulations indexed by h. We consider P_k Lagrange elements on the triangulations. If $u \in H^{k+1}(\Omega)$, then we have*

$$\|u - u_h\|_{H^1(\Omega)} \leq Ch^k |u|_{H^{k+1}(\Omega)}.$$

The proof is along the same lines as the proof in the Q_1 case, but with a lot more technicality due to the affine changes of variables between the reference triangle and the generic triangle.

All the above Lagrange triangular elements are adequate for H^1 approximation and are adapted to C^0 approximation spaces. It is also possible to define C^1 Hermite triangular elements for fourth order problems. One possible construction uses P_5 polynomials, hence 21 degrees of freedom. This is the Argyris finite element, [14, 19, 21, 27, 82]. More generally, there is a very large diversity of triangular

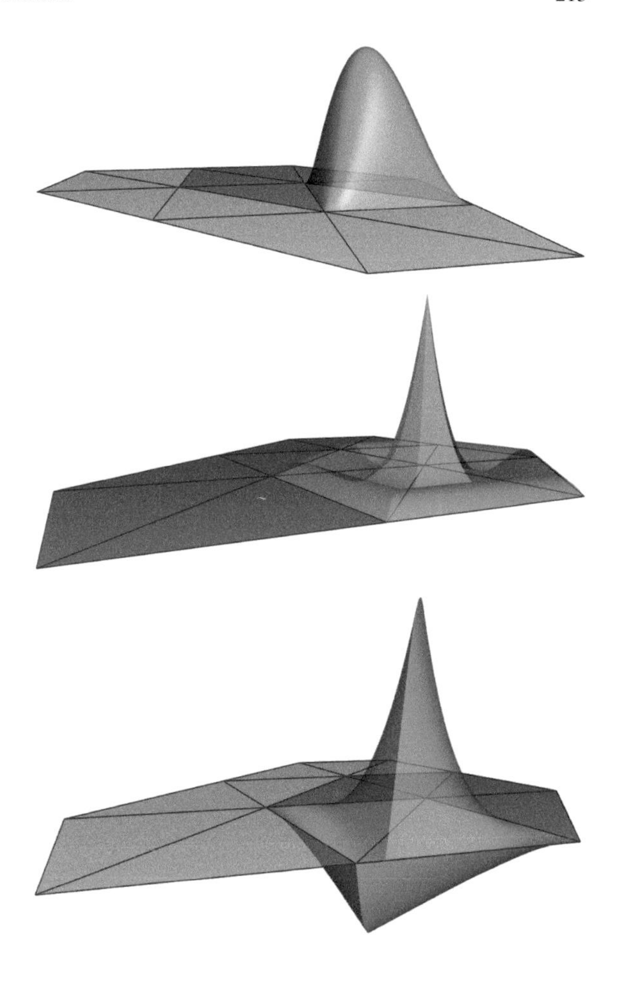

Fig. 6.41 A few P_3 basis functions

elements in the literature, sometimes especially crafted for one (class of) boundary value problem(s). Other generalizations include curvilinear triangles that are used to approximate curved boundaries. These are called isoparametric elements, see [19, 21].

6.10 An Example of 2d-Computation

To conclude this chapter, we show an example of computation made with the FreeFem++ software (a user-friendly free finite element software package available at http://www.freefem.org/). We solve the Laplace equation $-\Delta u = 1$ with a homogeneous Dirichlet boundary condition in the polygonal domain shown in Fig. 6.42, using P_1 and P_2 Lagrange elements on the same mesh.

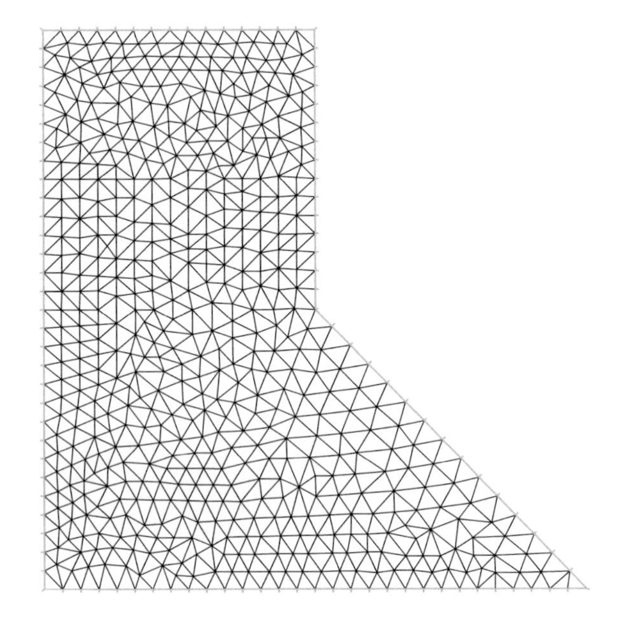

Fig. 6.42 Domain and triangular mesh

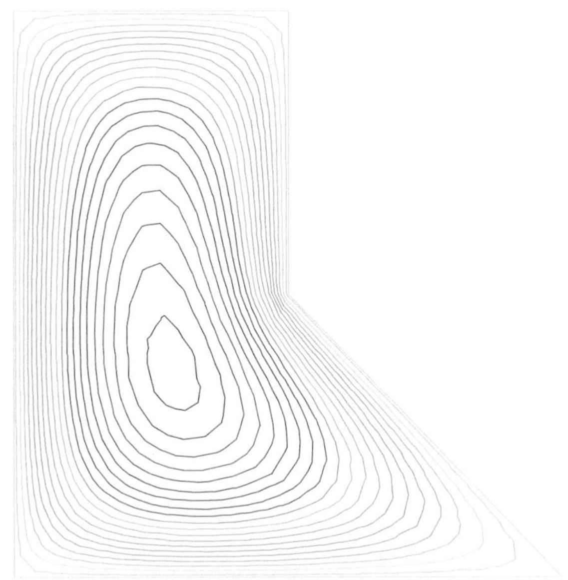

Fig. 6.43 Isovalue lines of the approximated solution u_h in the P_1 case

In this example, see Figs. 6.42, 6.43, 6.44, 6.45 and 6.46, FreeFem++ is given the boundary nodes (105 such nodes) and constructs a "good" mesh in the domain based on these boundary nodes using an automatic mesh generator. The mesh has 1,107 triangles and 607 vertices. FreeFem++ then assembles the matrix. It then proceeds to solve the linear system, then displays the solution and exports various files for

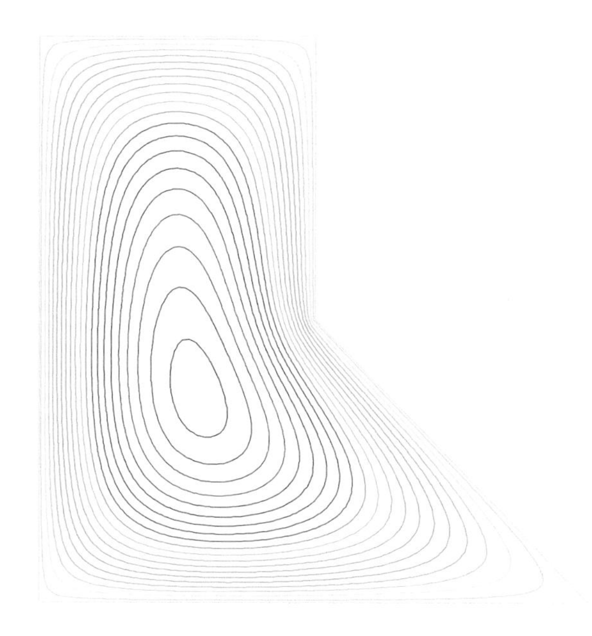

Fig. 6.44 Isovalue lines of the approximated solution u_h in the P_2 case

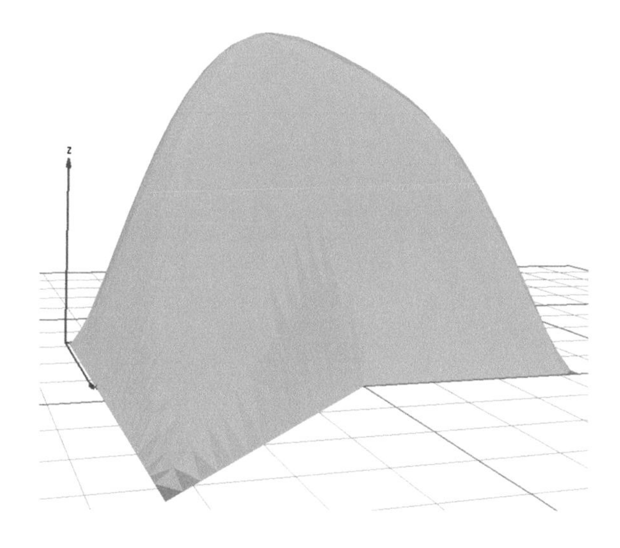

Fig. 6.45 3d visualization of the graph of u_h in the P_1 case

further use. The second computation uses P_2 elements and has 2,320 degrees of freedom, hence a 2,320×2,320 matrix. Both computations only take a small fraction of a second on a laptop computer.

The 3d visualizations of the graphs are done with medit (free software available at http://www.ljll.math.upmc.fr/~frey/software.html).

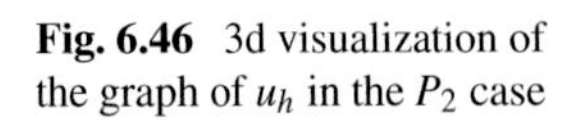

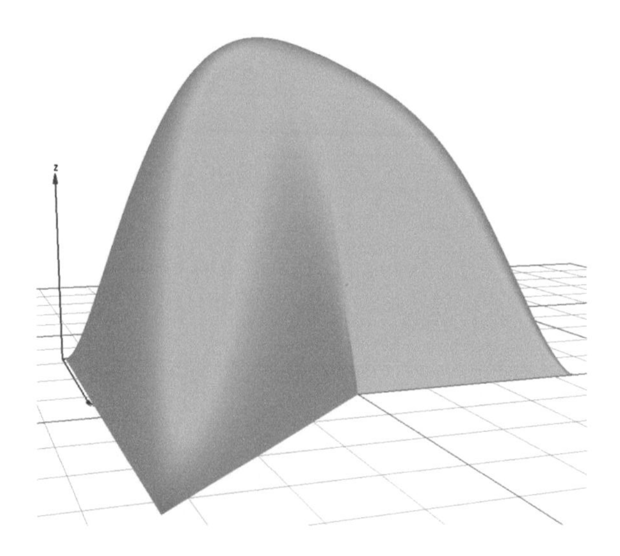

Fig. 6.46 3d visualization of the graph of u_h in the P_2 case

This example confirms that the finite element method is much more versatile than the finite difference method in two dimensions seen in Chap. 2. In fact, its range of applications is much wider.

Chapter 7
The Heat Equation

We have so far studied elliptic problems, i.e., stationary problems. We now turn to evolution problems, starting with the archetypal parabolic equation, namely the heat equation.

In this chapter, we will present a brief and far from exhaustive theoretical study of the heat equation. We will mostly work in one dimension of space, some of the results having an immediate counterpart in higher dimensions, others not. The study of numerical approximations of the heat equation will be the subject of the next chapter.

7.1 Overview

In Chap. 1, we presented the historical derivation of the heat equation by Fourier. In the general d-dimensional case, the heat equation is as follows. Let us be given Ω be an open subset of $\mathbb{R}^d$ and $T \in \mathbb{R}_+^*$. We note $Q = \Omega \times]0, T[$. When all the physical constants are set to 1, the heat equation with source term f reads

$$\frac{\partial u}{\partial t}(x,t) - \Delta u(x,t) = f(x,t) \text{ in } Q,$$

together with an initial condition

$$u(x,0) = u_0(x) \text{ in } \Omega,$$

and boundary values, for instance Dirichlet boundary values

$$u(x,t) = g(x,t) \text{ on } \partial\Omega \times]0, T[,$$

where f, u_0 and g are given functions. The unknown u is a function from $\bar{Q}$ to $\mathbb{R}$. This is called an initial-boundary value problem.

H. Le Dret and B. Lucquin, *Partial Differential Equations: Modeling, Analysis and Numerical Approximation*, International Series of Numerical Mathematics 168, DOI 10.1007/978-3-319-27067-8_7

When $d = 1$ and Ω is bounded, we can take $\Omega =]0, 1[$ without loss of generality. The problem reads

$$\begin{cases} \dfrac{\partial u}{\partial t}(x,t) - \dfrac{\partial^2 u}{\partial x^2}(x,t) = f(x,t) \text{ in } Q, \\ u(x,0) = u_0(x) \text{ in } \Omega, \\ u(0,t) = g(0,t), u(1,t) = g(1,t) \text{ in }]0,T[. \end{cases} \quad (7.1)$$

Other possible boundary conditions are the Neumann condition and the Fourier condition. For brevity, we limit ourselves to the Dirichlet case.

7.2 The Maximum Principle for the Heat Equation

We have seen a version of the maximum principle for a second order elliptic equation. Parabolic equations such as the heat equation also satisfy their own version of the maximum principle, see for example [5, 35].

Proposition 7.1 *We assume that u is a solution of problem (7.1) that belongs to $C^0(\bar{Q}) \cap C^2(Q \cup (\Omega \times \{T\}))$. If $f \geq 0$ in Q, then u attains its minimum on $(\Omega \times \{0\}) \cup (\partial\Omega \times [0,T])$.*

Proof We write the proof for $d = 1$. Let us first assume that $f > 0$ on $Q \cup (\Omega \times \{T\})$. The set $\bar{Q}$ is compact and the function u is continuous on $\bar{Q}$, thus it attains its minimum somewhere in $\bar{Q}$, say at point (x_0, t_0).

If $(x_0, t_0) \in Q$ which is an open set, then $\frac{\partial u}{\partial t}(x_0, t_0) = 0$.[1] Moreover, $\frac{\partial^2 u}{\partial x^2}(x_0, t_0) \geq 0$ since u is C^2 in a neighborhood of (x_0, t_0). Therefore $\big(\frac{\partial u}{\partial t} - \frac{\partial^2 u}{\partial x^2}\big)(x_0, t_0) \leq 0$, which contradicts $f(x_0, t_0) > 0$.

Therefore $(x_0, t_0) \in \partial Q = \big((\Omega \times \{0\}) \cup (\partial\Omega \times [0,T])\big) \cup (\Omega \times \{T\})$. Assume that $(x_0, t_0) \in \Omega \times \{T\}$, i.e., that $x_0 \in \Omega$ and $t_0 = T$. Since as a function in the variable x for $t = T$, u is also C^2, it follows again that $\frac{\partial^2 u}{\partial x^2}(x_0, T) \geq 0$ so that $\frac{\partial u}{\partial t}(x_0, T) = \frac{\partial^2 u}{\partial x^2}(x_0, T) + f(x_0, T) > 0$. Thus there exists $t < T$ such that $u(x_0, t) < u(x_0, T)$, which is consequently not a minimum value for u.

The only possibility left is that $(x_0, t_0) \in K = (\Omega \times \{0\}) \cup (\partial\Omega \times [0,T])$.

Consider now the case $f \geq 0$. Let $\varepsilon > 0$ and $u_\varepsilon(x,t) = u(x,t) + \varepsilon x(1-x)$. In particular $u(x,t) \leq u_\varepsilon(x,t)$ in $\bar{Q}$. We have

$$\frac{\partial u_\varepsilon}{\partial t} - \frac{\partial^2 u_\varepsilon}{\partial x^2} = \frac{\partial u}{\partial t} - \frac{\partial^2 u}{\partial x^2} + 2\varepsilon = f + 2\varepsilon > 0.$$

[1] The fact that $\frac{\partial u}{\partial x}(x_0, t_0) = 0$ is not useful here.

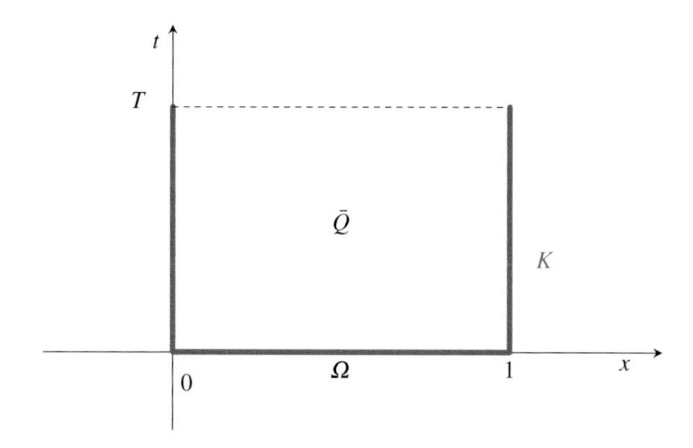

Fig. 7.1 The set K where u attains its minimum (*thicker line*)

By the previous argument, u_ε attains its minimum at a point $(x_\varepsilon, t_\varepsilon)$ of K. We have

$$u(x_0, t_0) \le u(x_\varepsilon, t_\varepsilon) \le u_\varepsilon(x_\varepsilon, t_\varepsilon) \le u_\varepsilon(x_0, t_0) = u(x_0, t_0) + \varepsilon x_0(1 - x_0).$$

Therefore,

$$u(x_0, t_0) \le u(x_\varepsilon, t_\varepsilon) \le u(x_0, t_0) + \varepsilon x_0(1 - x_0).$$

We now let $\varepsilon \to 0$. Since K is compact, we may extract a subsequence still denoted by ϵ such that $(x_\varepsilon, t_\varepsilon) \to (\bar{x}, \bar{t}) \in K$. Passing to the limit in the above inequalities and using the continuity of u, we obtain

$$u(x_0, t_0) = u(\bar{x}, \bar{t}),$$

with $(\bar{x}, \bar{t}) \in K$ and the minimum is therefore attained on K (see Fig. 7.1). □

Remark 7.1 The meaning of the maximum principle is that the minimum temperature is either attained at $t = 0$ or on the boundary of Ω at some other time $t \in]0, T]$, but in general not in $\Omega \times]0, T]$. It is also valid in any dimension d, using the same proof and the maximum principle for the Laplacian, that we have not proved here. The result cannot be refined any further since $u = 0$ is a solution for $f = 0$, $u_0 = 0$ that attains its minimum at any point in K. □

The maximum principle has many consequences, some of which we now list.

Corollary 7.1 *Under the hypotheses of Proposition 7.1, if $f \ge 0$, $g \ge 0$ and $u_0 \ge 0$, then $u \ge 0$ in $\bar{Q}$.*

Proof This is clear since the minimum of u is either of the form $u_0(x_0)$, $g(0, t_0)$ or $g(1, t_0)$. □

Remark 7.2 This form of the maximum principle is again a monotonicity result. The interesting physical interpretation is that if you heat up a room, the walls are kept at a nonnegative temperature and the initial temperature is nonnegative, then the temperature in the room stays nonnegative everywhere and at any time. □

We also have a stability result in the C^0 norm.

Corollary 7.2 *Under the hypotheses of Proposition 7.1, if $f = 0$ and $g = 0$, then $\|u\|_{C^0(\bar{Q})} = \|u_0\|_{C^0(\bar{\Omega})}$.*

Proof Let $v_+ = u + \|u_0\|_{C^0(\bar{\Omega})}$. We have

$$\frac{\partial v_+}{\partial t} - \frac{\partial^2 v_+}{\partial x^2} = 0,$$

since $f = 0$,

$$v_+(0, t) = v_+(1, t) = \|u_0\|_{C^0(\bar{\Omega})} \geq 0,$$

since $g = 0$ and

$$v_+(x, 0) = u_0 + \|u_0\|_{C^0(\bar{\Omega})} \geq 0.$$

By Corollary 7.1, $v_+ \geq 0$ in $\bar{Q}$, or in other words $u(x, t) \geq -\|u_0\|_{C^0(\bar{\Omega})}$ in $\bar{Q}$. Changing u in $-u$, we also have $u(x, t) \leq \|u_0\|_{C^0(\bar{\Omega})}$ in $\bar{Q}$, hence the result. □

Such a stability result immediately entails a uniqueness result.

Proposition 7.2 *Problem (7.1) has at most one solution in $C^0(\bar{Q}) \cap C^2(Q)$.*

Proof Indeed, if u_1 and u_2 are two solutions, then $v = u_1 - u_2$ satisfies the hypotheses of Corollary 7.2 on $\Omega \times [0, T - \eta])$ for all $\eta > 0$ with an initial value $u_0 = 0$. □

7.3 Construction of a Regular Solution

We will see several different ways of constructing solutions to the heat equation. Let us start with an elementary construction using Fourier series. It should be recalled that Joseph Fourier invented what became Fourier series in the 1800s, exactly for the purpose of solving the heat equation, see Chap. 1, Sect. 1.7.

We consider the case when $f = 0$, no heat source, and $g = 0$, homogeneous Dirichlet boundary condition, the only nonzero data being the initial condition u_0.

Proposition 7.3 *Let $u_0 \in C^0([0,1])$ be piecewise C^1 and such that $u_0(0) = u_0(1) = 0$. There exists a sequence $(b_k)_{k\in\mathbb{N}^*}$ of real numbers such that we have*

$$u_0(x) = \sum_{k=1}^{+\infty} b_k \sin(k\pi x)$$

for all $x \in [0,1]$. Moreover, $\sum_{k=1}^{+\infty} |b_k| < +\infty$.

Proof We first extend u_0 by imparity by setting $\widetilde{u}_0(x) = u_0(x)$ for $x \in [0,1]$ and $\widetilde{u}_0(x) = -u_0(-x)$ for $x \in [-1,0[$. The resulting function is odd and continuous on $[-1,1]$ by construction since $u_0(0) = 0$ and still piecewise C^1.

Secondly, we extend $\widetilde{u}_0$ to $\mathbb{R}$ by 2-periodicity by setting $\overset{\approx}{u}_0(x) = \widetilde{u}_0(x - 2\lfloor\frac{x+1}{2}\rfloor)$, where $\lfloor\cdot\rfloor$ denotes the floor function. This function is continuous since $\widetilde{u}_0(-1) = \widetilde{u}_0(1) = 0$, piecewise C^1 and 2-periodic by construction. Therefore, by Dirichlet's theorem, it can be expanded in Fourier series

$$\overset{\approx}{u}_0(x) = \frac{a_0}{2} + \sum_{k=1}^{+\infty} a_k \cos(k\pi x) + \sum_{k=1}^{+\infty} b_k \sin(k\pi x),$$

with $\sum_{k=1}^{+\infty}(|a_k| + |b_k|) < +\infty$, hence the series is normally convergent. Now $\overset{\approx}{u}_0$ is also odd by construction, so that all a_k Fourier coefficients vanish. Restricting the above expansion to $x \in [0,1]$, we obtain the result. □

Theorem 7.1 *Let u_0 be as above. Then the function defined by*

$$u(x,t) = \sum_{k=1}^{+\infty} b_k \sin(k\pi x) e^{-k^2\pi^2 t}$$

belongs to $C^0(\mathbb{R} \times [0,+\infty[) \cap C^\infty(\mathbb{R} \times]0,+\infty[)$. Its restriction to $\bar{Q}$ solves the initial-boundary value problem (7.1) with data $f = 0$, $g = 0$.

Proof We first need to show that the series above is convergent in some sense and that its sum belongs to the function spaces indicated in the theorem. Normal convergence on $\mathbb{R} \times [0,+\infty[$ is obvious since $|b_k \sin(k\pi x)e^{-k^2\pi^2 t}| \le |b_k|$, thus u exists and is continuous on $\mathbb{R} \times [0,+\infty[$.

Let us now consider differentiability. Now if u is supposed to coincide with u_0 at $t = 0$, and u_0 is only piecewise C^1, we cannot expect u to be C^∞ up to $t = 0$, hence the exclusion of $t = 0$ in the theorem. In order to use theorems on the differentiation of series, we actually need to stay away from $t = 0$ as will become clear in the proof. Let us thus chose $\varepsilon > 0$ and work for $t \ge \varepsilon$. It is convenient to notice that

$$\sin(k\pi x)e^{-k^2\pi^2 t} = \operatorname{Im}(e^{ik\pi x - k^2\pi^2 t}),$$

where $\operatorname{Im} z$ denotes the imaginary part of a complex number z. Therefore, for any nonnegative integers p and q, we have

$$\frac{\partial^{p+q}}{\partial x^p \partial t^q}\big(\sin(k\pi x)e^{-k^2\pi^2 t}\big) = (k\pi)^p(-k^2\pi^2)^q \operatorname{Im}\,(i^p e^{ik\pi x-k^2\pi^2 t}).$$

Thus

$$\begin{aligned}\left|b_k \frac{\partial^{p+q}}{\partial x^p \partial t^q}\big(\sin(k\pi x)e^{-k^2\pi^2 t}\big)\right| &\le |b_k|\pi^{p+2q}k^{p+2q}e^{-k^2\pi^2 t} \\ &\le |b_k|\pi^{p+2q}k^{p+2q}e^{-k^2\pi^2 \varepsilon}\end{aligned}$$

for $t \ge \varepsilon$. Since $b_k = \frac{1}{2}\int_{-1}^{1} \sin(k\pi x)\tilde{u}_0(x)\,dx$, we have $|b_k| \le \|u_0\|_{C^0([0,1])}$, thus

$$\left|b_k \frac{\partial^{p+q}}{\partial x^p \partial t^q}\big(\sin(k\pi x)e^{-k^2\pi^2 t}\big)\right| \le C_{p,q}k^{p+2q}e^{-k^2\pi^2\varepsilon},$$

for some constant $C_{p,q}$, because $t \ge \varepsilon$. The right-hand side is the general term of a convergent series due to the $e^{-k^2\pi^2\varepsilon}$ term with $\varepsilon > 0$, thus the left-hand side is the general term of a normally, thus uniformly convergent series, for any p and q. Therefore, u is of class C^∞ on $\mathbb{R} \times]\varepsilon, +\infty[$, for all $\varepsilon > 0$, thus belongs to $C^\infty(\mathbb{R} \times]0, +\infty[)$. Moreover, we have

$$\frac{\partial^{p+q} u}{\partial x^p \partial t^q}(x,t) = \sum_{k=1}^{+\infty} b_k \frac{\partial^{p+q}}{\partial x^p \partial t^q}\big(\sin(k\pi x)e^{-k^2\pi^2 t}\big)$$

for all $(x,t) \in \mathbb{R} \times]0, +\infty[$ and all p, q. In particular, we have

$$\frac{\partial u}{\partial t}(x,t) = \sum_{k=1}^{+\infty} b_k \frac{\partial}{\partial t}\big(\sin(k\pi x)e^{-k^2\pi^2 t}\big) = -\sum_{k=1}^{+\infty} b_k k^2\pi^2 \sin(k\pi x)e^{-k^2\pi^2 t}$$

and

$$\frac{\partial^2 u}{\partial x^2}(x,t) = \sum_{k=1}^{+\infty} b_k \frac{\partial^2}{\partial x^2}\big(\sin(k\pi x)e^{-k^2\pi^2 t}\big) = -\sum_{k=1}^{+\infty} b_k k^2\pi^2 \sin(k\pi x)e^{-k^2\pi^2 t}$$

so that

$$\frac{\partial u}{\partial t} - \frac{\partial^2 u}{\partial x^2} = 0 \text{ on } \mathbb{R} \times]0, +\infty[,$$

and u solves the heat equation.

Concerning the boundary conditions, we note that for all integers $k \ge 1$, we have $\sin(k\pi \times 0) = \sin(k\pi \times 1) = 0$, so that

$$u(0,t) = u(1,t) = 0$$

for all $t \in \mathbb{R}_+$. Finally,

$$u(x,0) = \sum_{k=1}^{+\infty} b_k \sin(k\pi x) e^{-k^2\pi^2 \times 0} = \sum_{k=1}^{+\infty} b_k \sin(k\pi x) = u_0(x),$$

and the initial condition is satisfied. □

Remark 7.3 It is worth noticing that both boundary conditions and initial condition make sense because u is continuous on $\bar{Q}$. Moreover, the regularity of u is such that the previous uniqueness result applies, thus we have found the one and only one solution in that class.

An important feature of the heat equation, and more generally of parabolic equations, is that whatever regularity u_0 may have, if $f = 0$, then the solution u becomes C^∞ instantly for $t > 0$. This is a *smoothing* effect.

For $t \geq 0$ fixed, the series that gives the function $x \mapsto u(x,t)$ is also the Fourier series of the odd and 2-periodic $\mathbb{R}$-extension of this function. The exponential term $e^{-k^2\pi^2 t}$ makes the corresponding Fourier coefficients decrease rapidly, which indicates that the sum is smooth (with respect to x), but we knew that already, both in x and t.

The smoothing effect also tells us why the backward heat equation is ill-posed. Indeed, there can be no solution to the backward heat equation with an initial condition that is not C^∞, since an initial condition for the backward heat equation is a final condition for the forward heat equation. It is not even clear that all C^∞ functions can be reached by the evolution of the heat equation. Therefore, time is irreversible in the heat equation.

We can see the same effect in the series, since for $t < 0$, $-k^2\pi^2 t > 0$ and the exponential terms become explosive instead of ensuring extremely fast convergence of the series. The only way the series can converge for $t < 0$ is for the Fourier coefficients b_k of the initial condition to be rapidly decreasing, so as to compensate for the exponential term. Again, a function with rapidly decreasing Fourier coefficients is very smooth. □

The above solution of the heat equation exhibits rapid uniform decay in time.

Proposition 7.4 *There exists a constant C such that*

$$|u(x,t)| \leq C e^{-\pi^2 t}.$$

In particular, $u(x,t) \to 0$ when $t \to +\infty$, uniformly with respect to x.

Proof Indeed, $e^{-k^2\pi^2 t} \leq e^{-\pi^2 t}$ for all k and all $t \geq 0$, so that

$$|u(x,t)| \leq \sum_{k=1}^{+\infty} |b_k| e^{-k^2\pi^2 t} \leq e^{-\pi^2 t} \sum_{k=1}^{+\infty} |b_k|,$$

hence the result since $\sum_{k=1}^{+\infty} |b_k| < +\infty$. □

Remark 7.4 If we remember the physical interpretation of the heat equation, keeping the walls of a room at 0 degree is tantamount to having paper-thin walls and a huge ice cube surrounding the room. If there is no heat source inside the room, it is not contrary to physical intuition that the temperature inside should drop to 0 degree pretty quickly, if it was positive at $t = 0$. All the heat inside the room eventually flows outside into the ice since the heat flux is proportional to the opposite of the temperature gradient. □

Apart from proving the existence of a solution in a particular case, the Fourier series expansion can also be used as a very precise numerical method, provided the Fourier coefficients of the initial condition are known with good accuracy.

In effect, we have a coarse error estimate

$$\Big|u(x,t) - \sum_{k=1}^{N} b_k \sin(k\pi x)e^{-k^2\pi^2 t}\Big| \leq \Big(\sum_{k=N+1}^{+\infty} |b_k|\Big)e^{-(N+1)^2\pi^2 t},$$

so that truncating the series and retaining only a few terms, we can expect to achieve excellent precision as soon as $t > 0$ is noticeably nonzero. Of course, the sine and exponential functions are already implemented in all computer languages.

The use of Fourier series in a numerical context is the simplest example of spectral method. Let us give an example of the numerical application of Fourier series. We consider a simple continuous, piecewise affine initial condition such as depicted in Fig. 7.2. Six terms in the Fourier series already provide a very good approximation of the solution. Figure 7.3 shows several views of the graph of u plotted in (x, t) space. The grey stripes show the graphs of $x \mapsto u(x, t)$ for a discrete sample of values of t.

Fig. 7.2 An admissible initial value u_0

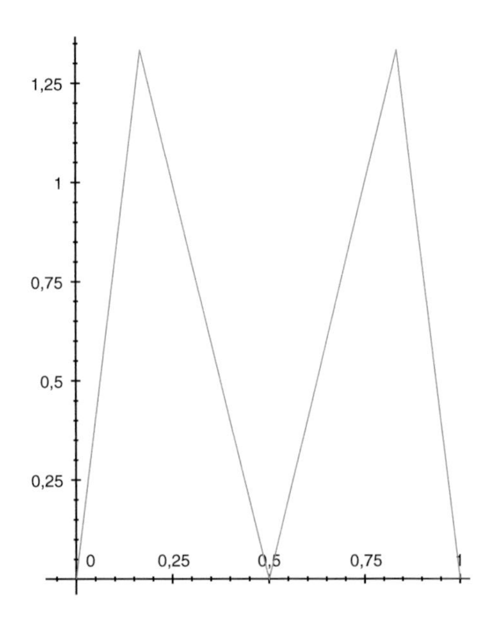

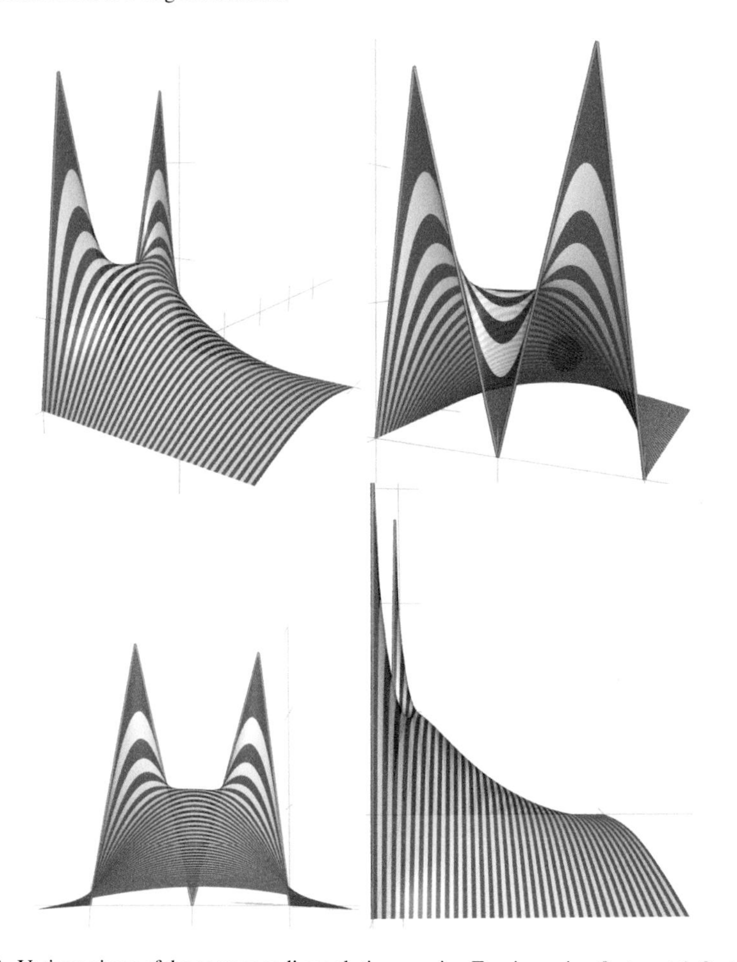

Fig. 7.3 Various views of the corresponding solution u, using Fourier series, $0 \leq x \leq 1, 0 \leq t \leq T$

We see the exponential decay in time, the smoothing effect, and also the fact that the first nonzero term in the series rapidly becomes dominant as t increases, as can be expected from the exponential terms. Note also the continuity as $t \to 0^+$, and the fact that the time derivative goes to $\pm\infty$ when (x, t) tends to a point $(x_0, 0)$ where the second space derivative of the initial condition is in a sense infinite,[2] i.e., the first space derivative is discontinuous. We also see that the minimum is attained where the maximum principle says it must be attained.[3]

The Fourier expansion even gives quite good results for cases that are not covered by the preceding analysis, for instance for an initial condition that does not satisfy

[2]Or more accurately a Dirac mass.

[3]Which is reassuring.

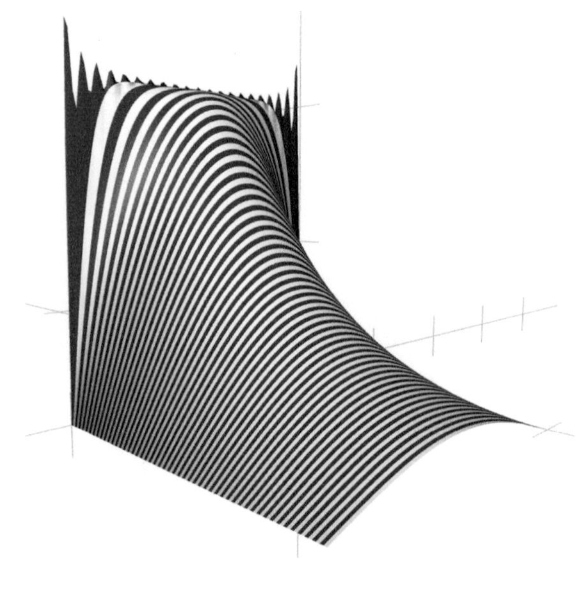

Fig. 7.4 Fourier series and discontinuous solutions, 20 terms. The Gibbs phenomenon is visible in the neighborhood of $(x, t) = (0, 0)$ and $(x, t) = (1, 0)$

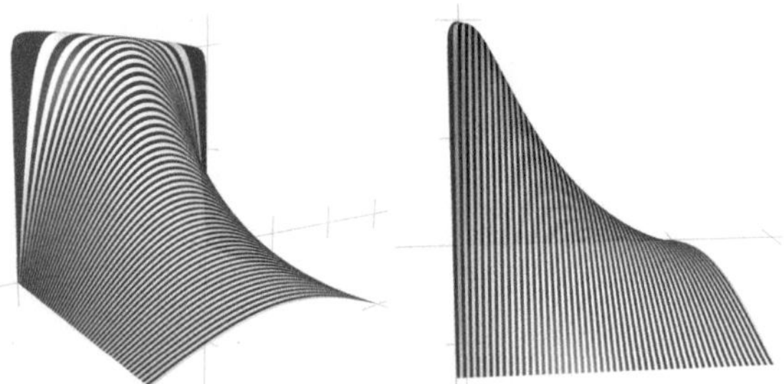

Fig. 7.5 Fourier series and discontinuous solutions, 100 terms

the Dirichlet boundary condition, such as $u_0(x) = 1$! Figure 7.4 shows 20 terms in the series, and of course a Gibbs phenomenon, i.e., localized oscillations around the discontinuities.

We also show the same computation with 100 terms in the Fourier series. The Gibbs phenomenon is still there, see Figs. 7.6 and 7.7, but does not show on Fig. 7.5 for sampling reasons: It occurs on a length scale that is too small to be captured by the graphics program. Recall that high frequency oscillations in space are damped extremely rapidly in time by the exponential term.

Figure 7.7 features a few close-ups of the 100 term Fourier series expansion of u near $(x, t) = (0, 0)$.

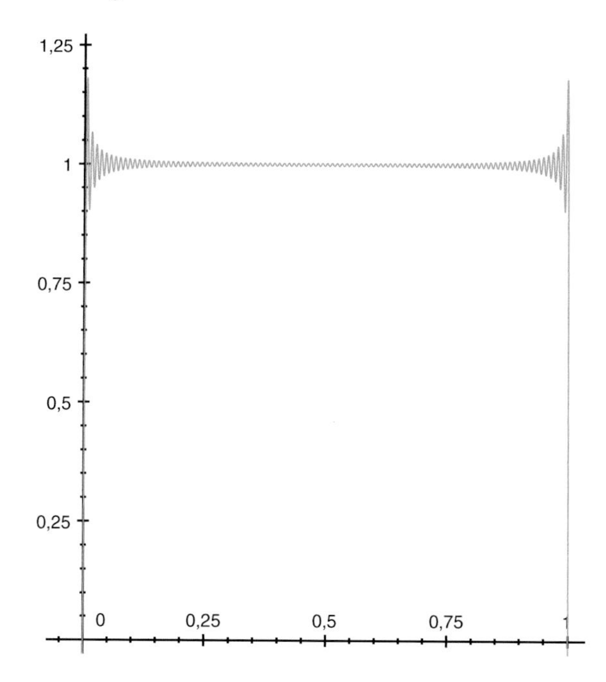

Fig. 7.6 One hundred terms in the Fourier series of $\widetilde{u}_0$, with Gibbs phenomenon around 0 and 1

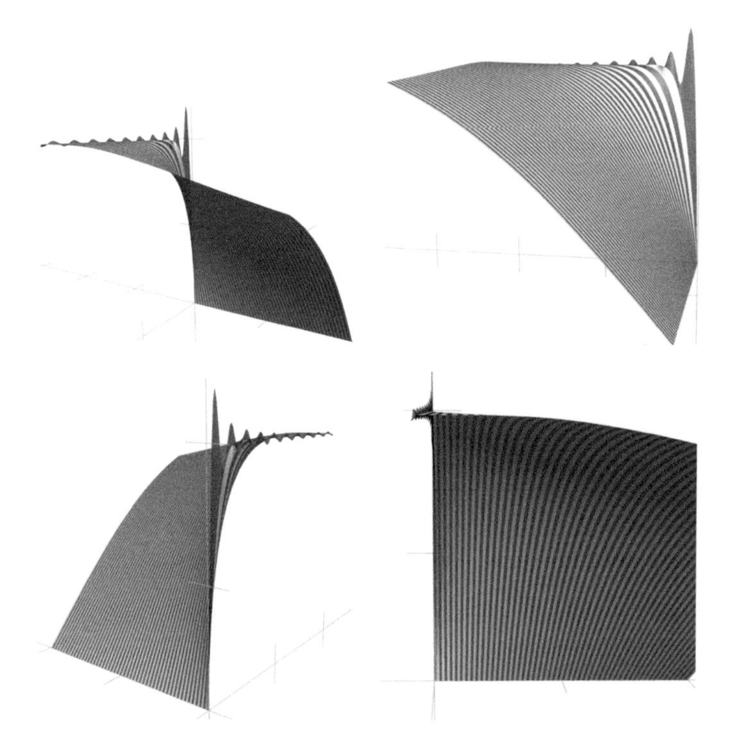

Fig. 7.7 Gibbs phenomenon and bad approximation of discontinuity, up close

7.4 Spaces of Hilbert Space-Valued Functions

In order to work with more general solutions and less smooth data, we need to introduce a few new function spaces. We will actually consider real-valued functions in the two variables x and t, as functions in the variable t with values in a space of functions in the variable x, $u(t) = u(\cdot, t)$.[4] This is because the time and space variables do not play the same role for the heat equation.

Let V denote a separable Hilbert space with norm $\|\cdot\|_V$. Let $T > 0$ be given. The space $C^0([0, T]; V)$ of continuous functions from $[0, T]$ with values in V is a Banach space for its natural norm

$$\|f\|_{C^0([0,T];V)} = \max_{t\in[0,T]} \|f(t)\|_V.$$

A V-valued function on $]0, T[$ is differentiable at point $t \in]0, T[$ if there exists a vector $f'(t) \in V$ such that

$$\left\|\frac{f(t+h) - f(t)}{h} - f'(t)\right\|_V \to 0 \text{ when } h \to 0.$$

Of course, $f'(t)$ is called the derivative of f at point t. A function is clearly continuous at all of its points of differentiability. If f is differentiable at all points t, then its derivative becomes a V-valued function. We can define

$$C^1([0, T]; V) = \{f \in C^0([0, T]; V); f' \in C^0([0, T]; V)\}$$

in the sense that f' has a continuous extension at 0 and T. When equipped with its natural norm

$$\|f\|_{C^1([0,T];V)} = \max\big(\|f\|_{C^0([0,T];V)}, \|f'\|_{C^0([0,T];V)}\big),$$

$C^1([0, T]; V)$ is a Banach space. More generally, we can define $C^k([0, T]; V)$ for all positive integers k. All of these notions are perfectly classical and work the same as in the real-valued case.

Measurability (and integrability) issues are a little trickier in the infinite dimensional valued case than in the finite dimensional valued case. There are different types of measurability and integrals when V is a Banach space or a more general topological vector space. We stick to the simplest notions. Besides we will not use V-valued integrals here. We equip $[0, T]$ with the Lebesgue σ-algebra.

Definition 7.1 A function $f : [0, T] \to V$ is called a *simple function* if there exists a finite measurable partition of $[0, T]$, $(E_i)_{i=1,\dots,k}$, and a finite set of vectors $v_i \in V$ such that

---

[4]Beware of the slightly ambiguous notation.

$$f(t) = \sum_{i=1}^{k} \mathbf{1}_{E_i}(t) v_i,$$

for all $t \in [0, T]$.

In other words, f only takes a finite number of values in V and is equal to v_i exactly on the Lebesgue measurable set E_i. It should be noted that for each t, there is one and only one nonzero term $\mathbf{1}_{E_i}(t)$ in the sum, due to the fact that the sets E_i form a partition of $[0, T]$.

Definition 7.2 A function $f : [0, T] \to V$ is said to be *measurable* if there exists a negligible set $N \subset [0, T]$ and a sequence of simple functions f_n such that $\|f_n(t) - f(t)\|_V \to 0$ when $n \to +\infty$ for all $t \notin N$.

We also say that f is an almost everywhere limit of simple functions. When $V = \mathbb{R}$, this notion of measurability coincides with the usual one. It is easy to see that a continuous function is measurable.

Proposition 7.5 *Let $f : [0, T] \to V$ be a measurable function. Then the function $N_V f : [0, T] \to \mathbb{R}_+$, $N_V f(t) = \|f(t)\|_V$, is a measurable function in the usual sense.*

Proof Let $f_n(t) = \sum_{i=1}^{k_n} \mathbf{1}_{E_{n,i}}(t) v_{n,i}$ be a sequence of simple functions that converges a.e. to f. Since the norm of V is continuous from V into $\mathbb{R}$, we have $N_V f_n \to N_V f$ a.e. Now due to the fact that for all t, there is at most one nonzero term in the sum, we also have $\|f_n(t)\|_V = \sum_{i=1}^{k_n} \mathbf{1}_{E_{n,i}}(t) \|v_{n,i}\|_V$, hence $N_V f_n$ is a real-valued simple function. Therefore $N_V f$ is measurable. □

Definition 7.3 We say that two measurable functions $f_1, f_2 : [0, T] \to V$ are *equal almost everywhere* if there exists a negligible set $N \subset [0, T]$ such that $f_1(t) = f_2(t)$ for all $t \notin N$.

Almost everywhere equality is an equivalence relation and from now on, we will not distinguish between a function and its equivalence class. The V-valued Lebesgue spaces are defined as would be expected [58]

$$L^p(0, T; V) = \{f : [0, T] \to V, \text{ measurable and such that } N_V f \in L^p(0, T)\},$$

for all $p \in [1, +\infty]$. When equipped with the norms

$$\|f\|_{L^p(0,T;V)} = \|N_V f\|_{L^p(0,T)},$$

these spaces are Banach spaces. For $p = 2$, $L^2(0, T; V)$ is a Hilbert space for the scalar product

$$(f|g)_{L^2(0,T;V)} = \int_0^T \big(f(t)|g(t)\big)_V \, dt.$$

The Hilbert norm reads explicitly

$$\|f\|_{L^2(0,T;V)} = \Big(\int_0^T \|f(t)\|_V^2\,dt\Big)^{1/2}.$$

Obviously,

$$C^0([0,T];V) \hookrightarrow L^p(0,T;V),$$

for all $p \in [1,+\infty]$.

It is also possible to define V-valued Sobolev spaces and V-valued distributions, but we will not use these spaces here.

In the case when V is itself a function space on an open set Ω of $\mathbb{R}^d$, there is a natural connection between V-valued functions on $[0,T]$ and real valued functions on $\bar{Q} = \bar{\Omega} \times [0,T]$ in $d+1$ variables. Let us give an example of this.

Proposition 7.6 *The spaces $L^2(0,T;L^2(\Omega))$ and $L^2(Q)$ are canonically isometric.*

Proof We leave aside the measurability questions, which are delicate. First of all, let us take $f \in L^2(Q)$. We thus have $\int_Q f(x,t)^2\,dxdt < +\infty$. By Fubini's theorem applied to f^2, we thus have that

$$\int_\Omega f(x,t)^2\,dx < +\infty \text{ for almost all } t \in [0,T]$$

and

$$\int_Q f(x,t)^2\,dxdt = \int_0^T \Big(\int_\Omega f(x,t)^2\,dx\Big)\,dt.$$

Therefore, if we set $\widetilde{f}(t) = f(\cdot,t)$, then we see that $\widetilde{f}(t) \in L^2(\Omega)$ for almost all t. Thus we can let $\widetilde{f}(t) = 0$ for those t for which the initial $\widetilde{f}$ is not in $L^2(\Omega)$ and $\widetilde{f}$ is then an $L^2(\Omega)$-valued function. Moreover, the second relation then reads

$$\|f\|^2_{L^2(\Omega)} = \|\widetilde{f}\|^2_{L^2(0,T;L^2(\Omega))},$$

hence the isometry.

Conversely, taking $\widetilde{f} \in L^2(0,T;L^2(\Omega))$, then for almost all t, $\widetilde{f}(t)$ is a function in the variable $x \in \Omega$ that belongs to $L^2(\Omega)$. If we thus set $f(x,t) = \widetilde{f}(t)(x)$, we define a function on Q which is such that $\int_0^T \big(\int_\Omega f(x,t)^2\,dx\big)\,dt < +\infty$. By Fubini's theorem again, it follows that $f \in L^2(Q)$ and we have the isometry. □

It is thus possible to switch between the two points of view: function in one variable with values in a function space on a d dimensional domain and real valued function in $d+1$ variables. If $\widetilde{f} \in C^1([0,T];L^2(\Omega))$, then the associated f is in $L^2(Q)$ and it can be shown that its distributional derivative $\frac{\partial f}{\partial t}$ is in $C^0([0,T];L^2(\Omega))$ and $\widetilde{\frac{\partial f}{\partial t}} = (\widetilde{f})'$. From now on, we will drop the tilde notation for simplicity.

We will also encounter such situations as $f \in C^0([0,T]; H) \cap L^2(0,T; V)$ with two (or more) different spaces $V \subset H$, meaning that $f(t)$ is unambiguously defined as an element of H for all t, and continuous with values in H, and the same $f(t)$ is in V for almost all t and square integrable with values in V. It is allowed to exit V on a negligible subset of $[0,T]$.

Let us now be given two Hilbert spaces V_1 and V_2 and A a continuous linear operator from V_1 to V_2. Let $\|A\|$ denote its operator norm. Given any V_1-valued function f, we can define a V_2-valued function Af by $(Af)(t) = A(f(t))$. This definition commutes with all previous notions.

Proposition 7.7 *If $f \in C^k([0,T]; V_1)$ then $Af \in C^k([0,T]; V_2)$ with $(Af)^{(j)} = A(f^{(j)})$ for $j \leq k$, and if $f \in L^2(0,T; V_1)$ then $Af \in L^2(0,T; V_2)$.*

Proof We start with the continuity. We have

$$\|Af(t+h)-Af(t)\|_{V_2} = \|A(f(t+h)-f(t))\|_{V_2} \leq \|A\|\|f(t+h)-f(t)\|_{V_1} \underset{h\to 0}{\longrightarrow} 0.$$

Therefore $Af \in C^0([0,T]; V_2)$. Moreover, we have $\|Af(t)\|_{V_2} \leq \|A\|\|f(t)\|_{V_1}$ so that taking the maximum for $t \in [0,T]$ on both sides, we obtain

$$\|Af\|_{C^0([0,T];V_2)} \leq \|A\|\|f\|_{C^0([0,T];V_1)}.$$

Similarly

$$\left\|\frac{Af(t+h)-Af(t)}{h} - Af'(t)\right\|_{V_2} \leq \|A\|\left\|\frac{f(t+h)-f(t)}{h} - f'(t)\right\|_{V_1} \underset{h\to 0}{\longrightarrow} 0$$

and so on for the successive derivatives and their norms. Finally,

$$\int_0^T \|Af(t)\|_{V_2}^2\,dt \leq \|A\|^2 \int_0^T \|f(t)\|_{V_1}^2\,dt < +\infty,$$

leaving aside measurability issues, which are not difficult here. Of course, the above inequality is nothing but

$$\|Af\|_{L^2(0,T;V_2)} \leq \|A\|\|f\|_{L^2(0,T;V_1)},$$

as with the C^k spaces. □

To get an idea of how this can be used, just take $V_1 = H^2(\Omega)$, $V_2 = L^2(\Omega)$ and $A = -\Delta$.

Later on, we will also use Hilbert bases of V, i.e., total orthonormal families in V (recall that a total family is a family that spans a dense vector subspace). Such bases are countable since we only consider separable Hilbert spaces.

Proposition 7.8 *Let $(e_n)_{n\in\mathbb{N}}$ be a Hilbert basis of V. Let f be a V-valued function and for all $n \in \mathbb{N}$, $f_n(t) = (f(t)|e_n)_V$.*

(*i*) *If* $f \in C^k([0,T];V)$ *then* $f_n \in C^k([0,T])$ *for all* n *and* $f_n^l(t) = (f^{(l)}(t)|e_n)_V$ *for all* $l \leq k$.
(*ii*) *If* $f \in L^2(0,T;V)$ *then* $f_n \in L^2(0,T)$ *for all* n.

Proof Let $f \in C^0([0,T];V)$ and take a sequence $t_p \to t$ in $[0,T]$. Then $f(t_p) \to f(t)$ in V and taking the scalar product with e_n, which is continuous, we see that $(f(t_p)|e_n)_V \to (f(t)|e_n)_V$ in $\mathbb{R}$. Hence, f_n is continuous.

Assume now that f is C^1. It similarly follows from the fact that $\frac{f(t_p)-f(t)}{t_p-t} \to f'(t)$ in V, that $\frac{(f(t_p)|e_n)_V-(f(t)|e_n)_V}{t_p-t} \to (f'(t)|e_n)_V$ by linearity and continuity of the scalar product. Therefore, f_n has a derivative at t for all n and $(f_n)' = (f')_n$, so that $(f_n)'$ is continuous by the previous case and thus $f_n \in C^1([0,T])$. The general case of k derivatives follows by induction on k.

Let now $f \in L^2(0,T;V)$. Since $\|f(t)\|_V^2 = \sum_n |f_n(t)|^2$ by Parseval's identity, see [68], we see that

$$\|f\|_{L^2(0,T;V)}^2 = \int_0^T \|f(t)\|_V^2\,dt = \sum_n \int_0^T |f_n(t)|^2\,dt.$$

In particular, $\int_0^T |f_n(t)|^2\,dt < +\infty$ for all n, or in other words, $f_n \in L^2(0,T)$. □

7.5 Energy Estimates, Stability, Uniqueness

In this section, we consider solutions of problem (7.1) with data that is considerably less smooth than in the previous sections. We assume that the solutions considered are regular enough so that all computations are justified. As the proof in arbitrary dimension of space d works the same as in one dimension, we will let $\Omega \subset \mathbb{R}^d$ bounded and $Q = \Omega \times]0,T[$.

We start with a lemma.

Lemma 7.1 *Let* $u \in C^1([0,T];L^2(\Omega))$. *Then the function* $t \mapsto \frac{1}{2}\int_\Omega (u(t)(x))^2\,dx$ *is of class* $C^1([0,T])$ *and its derivative is given by* $t \mapsto \int_\Omega [u(t)u'(t)](x)\,dx$.

Proof Let $E(t) = \frac{1}{2}\int_\Omega u(x,t)^2\,dx$. We write

$$\frac{E(t+h)-E(t)}{h} = \frac{1}{2}\int_\Omega \big(u(x,t+h)+u(x,t)\big)\Big(\frac{u(x,t+h)-u(x,t)}{h}\Big)\,dx.$$

Now, by L^2-valued continuity, $u(t+h) \to u(t)$ in $L^2(\Omega)$ when $h \to 0$. By L^2-valued differentiability, $\frac{u(t+h)-u(t)}{h} \to u'(t)$ in $L^2(\Omega)$ when $h \to 0$. Therefore,

$$\frac{E(t+h)-E(t)}{h} \to \int_\Omega [u(t)u'(t)](x)\,dx$$

when $h \to 0$ by the Cauchy–Schwarz inequality. By the same inequality, the right-hand side is a continuous function of t. □

Remark 7.5 This result can be construed as a kind of differentiation under the integral sign, since

$$\frac{d}{dt}\Big(\int_\Omega u(x,t)^2\,dx\Big) = 2\int_\Omega \frac{\partial u}{\partial t}(x,t)u(x,t)\,dx = \int_\Omega \frac{\partial(u^2)}{\partial t}(x,t)\,dx,$$

with the identification $\frac{\partial u}{\partial t} = u'$. It can be generalized with weaker assumptions on u, namely $u \in L^2(0,T; H_0^1(\Omega))$, with $u' \in L^2(0,T; H^{-1}(\Omega))$, see [35]. □

Proposition 7.9 *Assume that $g = 0$ (homogeneous Dirichlet condition), $u_0 \in L^2(\Omega)$ and $f \in L^2(Q)$. Then, if $u \in C^1([0,T]; L^2(\Omega)) \cap L^2(0,T; H_0^1(\Omega))$ is a solution of the problem, then*

$$\|u\|_{C^0([0,T];L^2(\Omega))} \le \|u_0\|_{L^2(\Omega)} + C\|f\|_{L^2(Q)}, \tag{7.2}$$

where C is the Poincaré inequality constant.

Proof Since $u \in C^1([0,T]; L^2(\Omega)) \cap L^2(0,T; H_0^1(\Omega))$, we have that $u(\cdot,t)$ and $\frac{\partial u}{\partial t}(\cdot,t)$ belong to $L^2(\Omega)$ for all t and that $u(\cdot,t)$ belongs to $H_0^1(\Omega)$ for almost all t. The meaning of the partial differential equation in this context is thus that $u' - \Delta u = f$ where $u' \in C^0([0,T]; L^2(\Omega))$, $f \in L^2(0,T; L^2(\Omega))$ and $\Delta u \in L^2(0,T; H^{-1}(\Omega))$, see Corollary 3.4 of Chap. 3, so that the equation is well defined in this sense. Of course, it also coincides with the distributional equation on Q.

For almost all $s \in [0,T]$, both sides of the equation are in $H^{-1}(\Omega)$. We thus take the duality bracket with u and obtain

$$\begin{aligned}\frac{1}{2}\frac{d}{ds}\Big(\int_\Omega u(x,s)^2\,dx\Big) + \int_\Omega \|\nabla u(x,s)\|^2\,dx &= \int_\Omega f(x,s)u(x,s)\,dx \\ &\le \Big(\int_\Omega f(x,s)^2\,dx\Big)^{\frac{1}{2}}\Big(\int_\Omega u(x,s)^2\,dx\Big)^{\frac{1}{2}}\end{aligned}$$

by Lemma 7.1 and by the Cauchy–Schwarz inequality. Because of the homogeneous Dirichlet condition, we have Poincaré's inequality

$$\Big(\int_\Omega u(x,s)^2\,dx\Big)^{\frac{1}{2}} \le C\Big(\int_\Omega \|\nabla u(x,s)\|^2\,dx\Big)^{\frac{1}{2}},$$

and using Young's inequality $ab \le \frac{1}{2}a^2 + \frac{1}{2}b^2$, we obtain

$$\begin{aligned}&\frac{1}{2}\frac{d}{ds}\Big(\int_\Omega u(x,s)^2\,dx\Big) + \int_\Omega \|\nabla u(x,s)\|^2\,dx \\ &\qquad\le \frac{C^2}{2}\int_\Omega f(x,s)^2\,dx + \frac{1}{2}\int_\Omega \|\nabla u(x,s)\|^2\,dx,\end{aligned}$$

so that

$$\frac{1}{2}\frac{d}{ds}\Big(\int_\Omega u(x,s)^2\,dx\Big) \le \frac{1}{2}\frac{d}{ds}\Big(\int_\Omega u(x,s)^2\,dx\Big) + \frac{1}{2}\int_\Omega \|\nabla u(x,s)\|^2\,dx$$
$$\le \frac{C^2}{2}\int_\Omega f(x,s)^2\,dx.$$

We integrate the above inequality between 0 and t with respect to s and obtain

$$\int_\Omega u(x,t)^2\,dx - \int_\Omega u(x,0)^2\,dx \le C^2\int_0^t\int_\Omega f(x,s)^2\,dx\,ds \le C^2\int_0^T\int_\Omega f(x,s)^2\,dx\,ds,$$

for all $t \in [0,T]$ due to Lemma 7.1, and since $u(x,0) = u_0(x)$, it follows that

$$\|u(\cdot,t)\|_{L^2(\Omega)} \le \Big(\|u_0\|^2_{L^2(\Omega)} + C^2\|f\|^2_{L^2(Q)}\Big)^{1/2}$$

hence the result, since $\sqrt{a^2+b^2} \le a+b$ for all a, b positive. □

Remark 7.6 The quantity $E(t) = \frac{1}{2}\int_\Omega u(x,t)^2\,dx$ is called the *energy* (up to physical constants), hence the term "energy estimate". It follows from the proof that the energy is decreasing when $f = 0$. In addition, it is quite clear also from the proof that if $f \in L^2(\Omega \times \mathbb{R}_+)$, then the energy estimate remains valid for all times, i.e.,

$$\sup_{t\in\mathbb{R}_+} \|u(\cdot,t)\|_{L^2(\Omega)} \le \|u_0\|_{L^2(\Omega)} + C\|f\|_{L^2(\Omega\times\mathbb{R}_+)},$$

provided such a solution exists.

Let us note that the energy estimate can be proved under lower regularity hypotheses, namely that $u \in C^0([0,T]; L^2(\Omega)) \cap L^2(0,T; H^1_0(\Omega))$. The first space in the intersection gives a precise meaning to the initial condition in L^2. □

As in the case of the maximum principle, the energy estimate has consequences in terms of uniqueness and stability.

Corollary 7.3 *There is at most one solution u belonging to $C^1([0,T]; L^2(\Omega)) \cap L^2(0,T; H^1(\Omega))$ to the heat equation with initial data $u_0 \in L^2(\Omega)$, right-hand side $f \in L^2(Q)$ and Dirichlet boundary condition $g \in L^2(0,T; H^{1/2}(\partial\Omega))$.*

Proof Let u_1 and u_2 be two such solutions. Then, their difference $u_1 - u_2$ belongs to $C^1([0,T]; L^2(\Omega)) \cap L^2(0,T; H^1_0(\Omega))$ and is a solution of the heat equation with zero right-hand side and initial condition. By estimate (7.2), it follows that we have $u_1 - u_2 = 0$. □

Again this also holds in $C^0([0,T]; L^2(\Omega)) \cap L^2(0,T; H^1(\Omega))$. Stability or continuous dependence on the data is straightforward. We just consider here the homogeneous Dirichlet condition $g = 0$.

Corollary 7.4 *Let* $u_i \in C^1([0,T]; L^2(\Omega)) \cap L^2(0,T; H_0^1(\Omega))$, $i = 1,2$, *be solutions corresponding to initial conditions* $u_{0,i} \in L^2(\Omega)$ *and right-hand sides* $f_i \in L^2(Q)$. *Then*

$$\|u_1 - u_2\|_{C^0([0,T];L^2(\Omega))} \leq \|u_{0,1} - u_{0,2}\|_{L^2(\Omega)} + C\|f_1 - f_2\|_{L^2(Q)}.$$

Proof Clear. □

When in addition there is no heat source, i.e., $f = 0$, we can expect some kind of exponential decay as in the regular case. Here, the energy is the relevant quantity.

Proposition 7.10 *If* $f = 0$, *then we have*

$$E(t) \leq e^{-\frac{2t}{C^2}} E(0) = \frac{e^{-\frac{2t}{C^2}}}{2}\|u_0\|^2_{L^2(\Omega)},$$

where C *is the Poincaré inequality constant.*

Proof As before, we have

$$\frac{d}{dt}\Big(\frac{1}{2}\int_\Omega u(x,t)^2\,dx\Big) + \int_\Omega \|\nabla u(x,t)\|^2\,dx = 0.$$

Thus

$$\frac{dE}{dt}(t) = -\int_\Omega \|\nabla u(x,t)\|^2\,dx \leq -\frac{1}{C^2}\int_\Omega u(x,t)^2\,dx = -\frac{2}{C^2}E(t),$$

by Poincaré's inequality. Solving this differential inequality, we obtain the announced result. □

Remark 7.7 A function in L^2 is not bounded in general, thus we cannot expect uniform decay of the temperature as in the regular case. However, it can be shown that u is of class C^∞ as soon as $t > 0$, which is the same smoothing effect as before. Thus, there is also a uniform exponential decay but starting away from $t = 0$. In fact, it can be shown that u is C^∞ on any open subset where f is C^∞, in particular when it is equal to 0. This property of the heat operator is called *hypoellipticity*, see [20]. □

7.6 Variational Formulation and Existence of Weak Solutions

So far, we still have no existence result for the initial-boundary value problem when $f \neq 0$ or $f = 0$ and $u_0 \in L^2(\Omega)$. For this, we need to recast the problem in variational form. We only consider the homogeneous Dirichlet boundary condition, since a non homogeneous Dirichlet condition can be transformed into a homogeneous one via an appropriate lift of the boundary data. We start with regularity hypotheses that are a little too strong, but not by much. As in the elliptic case, we define the bilinear form a by $a(u, v) = \int_\Omega \nabla u \cdot \nabla v \, dx$.

Proposition 7.11 *Let $u_0 \in L^2(\Omega)$, $f \in L^2(Q)$. Consider u in $C^1([0, T]; L^2(\Omega)) \cap L^2(0, T; H_0^1(\Omega))$ such that $u' - \Delta u = f$ for almost all t and $u(0) = u_0$. Then we have, for all $v \in H_0^1(\Omega)$,*

$$\frac{d}{dt}\big((u(t)|v)_{L^2(\Omega)}\big) + a(u(t), v) = (f(t)|v)_{L^2(\Omega)}$$

almost everywhere in $[0, T]$, and

$$(u(0)|v)_{L^2(\Omega)} = (u_0|v)_{L^2(\Omega)}.$$

Conversely, a solution in $C^1([0, T]; L^2(\Omega)) \cap L^2(0, T; H_0^1(\Omega))$ of the above two variational equations is a solution of the initial-boundary value problem for the heat equation with homogeneous Dirichlet boundary condition, initial data u_0 and right-hand side f.

Proof We have already seen that each term in the equation $u' - \Delta u = f$ is at worst[5] in $L^2(0, T; H^{-1}(\Omega))$. It is therefore meaningful to take the duality bracket of each one of them with an arbitrary $v \in H_0^1(\Omega)$, so that we have

$$\langle u'(t), v\rangle_{H^{-1}(\Omega),H_0^1(\Omega)} - \langle \Delta u(t), v\rangle_{H^{-1}(\Omega),H_0^1(\Omega)} = \langle f(t), v\rangle_{H^{-1}(\Omega),H_0^1(\Omega)},$$

for almost all t.

Arguing as in the proof of Lemma 7.1, we see that the real-valued function $t \mapsto (u(t)|v)_{L^2(\Omega)}$ is of class C^1 and that

$$\frac{d}{dt}\big((u(t)|v)_{L^2(\Omega)}\big) = \int_\Omega u'(t)(x)v(x)\, dx = \langle u'(t), v\rangle_{H^{-1}(\Omega),H_0^1(\Omega)},$$

since $u'(t) \in L^2(\Omega)$. Similarly,

$$\langle f(t), v\rangle_{H^{-1}(\Omega),H_0^1(\Omega)} = \int_\Omega f(x, t)v(x)\, dx.$$

Finally, by Corollary 3.4 of Chap. 3, we have

$$-\langle \Delta u(t), v\rangle_{H^{-1}(\Omega),H_0^1(\Omega)} = a(u(t), v),$$

so that the first equation is established. The second equation is trivial.

[5]In the sense of space regularity.

Conversely, let us be given a solution u of the variational problem. Since $H_0^1(\Omega)$ is dense in $L^2(\Omega)$, the second equation implies that $u(0) = u_0$. Moreover, the above calculations can be carried out backwards, so that

$$\langle u'(t) - \Delta u(t) - f(t), v\rangle_{H^{-1}(\Omega), H_0^1(\Omega)} = 0,$$

for almost all t and all $v \in H_0^1(\Omega)$. Consequently, for almost all t, $u'(t) - \Delta u(t) - f(t) = 0$ as an element of $H^{-1}(\Omega)$, hence the heat equation with right-hand side f is satisfied in this sense. □

As we said above, the regularity in time assumed above is a bit too high. Indeed, the variational formulation makes sense in a slightly less regular context. This leads to the following definition.

Definition 7.4 The variational formulation of the heat equation with homogeneous Dirichlet boundary condition, initial data $u_0 \in L^2(\Omega)$ and right-hand side $f \in L^2(Q)$ consists in looking for $u \in C^0([0, T]; L^2(\Omega)) \cap L^2(0, T; H_0^1(\Omega))$ such that, for all $v \in H_0^1(\Omega)$,

$$\begin{cases} \big((u|v)_{L^2(\Omega)}\big)' + a(u, v) = (f|v)_{L^2(\Omega)} \text{ in the sense of } \mathcal{D}'(]0, T[), \\ \qquad (u(0)|v)_{L^2(\Omega)} = (u_0|v)_{L^2(\Omega)}. \end{cases} \tag{7.3}$$

Remark 7.8 Let us check that this definition makes sense. First of all, since $u \in C^0([0, T]; L^2(\Omega))$ and v does not depend on t, we see that the function $t \mapsto (u|v)_{L^2(\Omega)}$ is continuous on $[0, T]$, hence its derivative is a distribution on $]0, T[$. Likewise, since $u \in L^2(0, T; H_0^1(\Omega))$, the function $t \mapsto a(u, v)$ is in $L^1(0, T)$ by the Cauchy–Schwarz inequality, hence a distribution on $]0, T[$ and the same holds for $t \mapsto (f|v)_{L^2(\Omega)}$. Therefore, the first equation in (7.3) is well defined in the distributional sense.

We have already seen that the second equation is equivalent to $u(0) = u_0$, and the continuity of u with respect to t with values in $L^2(\Omega)$ makes this initial condition relevant. □

We use the variational formulation to prove existence and uniqueness of solutions. We will write the proof in the 1d case, $\Omega =]0, 1[$. The general case is entirely similar. For all $k \in \mathbb{N}^*$, we let $\phi_k(x) = \sqrt{2}\sin(k\pi x)$ and $\lambda_k = k^2\pi^2$. It is well-known that the family $(\phi_k)_{k\in\mathbb{N}^*}$ is a Hilbert basis of $L^2(0, 1)$ as well as a total orthogonal family in $H_0^1(]0, 1[)$ [15, 51]. Moreover, for all $w \in H_0^1(\Omega)$, we have

$$a(w, \phi_k) = \int_0^1 \frac{dw}{dx}\frac{d\phi_k}{dx}\,dx = -\int_0^1 w\frac{d^2\phi_k}{dx^2}\,dx = \lambda_k (w|\phi_k)_{L^2(\Omega)}. \tag{7.4}$$

Theorem 7.2 *Let $u_0 \in L^2(\Omega)$, $f \in L^2(Q)$. There exists a unique solution $u \in C^0([0,T]; L^2(\Omega)) \cap L^2(0,T; H_0^1(\Omega))$ of problem (7.3), which is given by*

$$u(t) = \sum_{k=1}^{+\infty} u_k(t)\phi_k, \tag{7.5}$$

where

$$u_k(t) = (u_0|\phi_k)_{L^2(\Omega)} e^{-\lambda_k t} + \int_0^t (f(s)|\phi_k)_{L^2(\Omega)} e^{-\lambda_k(t-s)}\,ds \tag{7.6}$$

and the series converges in $C^0([0,T]; L^2(\Omega)) \cap L^2(0,T; H_0^1(\Omega))$.

Proof We start with the uniqueness. Let $u \in C^0([0,T]; L^2(\Omega)) \cap L^2(0,T; H_0^1(\Omega))$ be a solution of (7.3). For all $t \in [0,T]$, $u(t)$ is thus an element of $L^2(\Omega)$ and can therefore be expanded on the Hilbert basis $(\phi_k)_{k\in\mathbb{N}^*}$. Consequently, we have for all t

$$u(t) = \sum_{k=1}^{+\infty} u_k(t)\phi_k$$

with

$$u_k(t) = (u(t)|\phi_k)_{L^2(\Omega)}$$

for all $k \in \mathbb{N}^*$ and the series converges in $L^2(\Omega)$. Now $\phi_k \in H_0^1(\Omega)$ is a legitimate test-function in problem (7.3). In particular, since $u(t) \in H_0^1(\Omega)$ almost everywhere, we have

$$a(u(t), \phi_k) = \lambda_k u_k(t)$$

almost everywhere by (7.4), hence everywhere since the right-hand side is continuous (the left-hand side is L^1). We thus have by Proposition 7.8,

$$\begin{cases} u_k'(t) + \lambda_k u_k(t) = (f(t)|\phi_k)_{L^2(\Omega)} \text{ in the sense of } \mathcal{D}'(]0,T[), \\ u_k(0) = (u_0|\phi_k)_{L^2(\Omega)}, \end{cases}$$

for each $k \in \mathbb{N}^*$. Now this is a Cauchy problem for a linear ordinary differential equation, and there are no other distributional solutions than the usual solution obtained by variation of the constant, or Duhamel's formula:

$$u_k(t) = (u_0|\phi_k)_{L^2(\Omega)} e^{-\lambda_k t} + \int_0^t (f(s)|\phi_k)_{L^2(\Omega)} e^{-\lambda_k(t-s)}\,ds,$$

which is exactly formula (7.6).[6] Hence the uniqueness.

[6]Observe that the function u_k is continuous in t.

We now use the above series to prove existence. Since $u_0 \in L^2(\Omega)$, we have

$$\|u_0\|^2_{L^2(\Omega)} = \sum_{k=1}^{+\infty} (u_0|\phi_k)^2_{L^2(\Omega)}$$

by Parseval's identity. Similarly, $f \in L^2(Q)$ and

$$\|f\|^2_{L^2(Q)} = \int_0^T \sum_{k=1}^{+\infty} (f(t)|\phi_k)^2_{L^2(\Omega)}\,dt.$$

Let us set $u_{0,k} = (u_0|\phi_k)_{L^2(\Omega)}$ and $f_k(t) = (f(t)|\phi_k)_{L^2(\Omega)}$. We are going to show that the series in formula (7.5) converges in both spaces $C^0(0,T;L^2(\Omega))$ and $L^2(0,T;H^1_0(\Omega))$ and that its sum u is a solution of the variational problem. To do this, we will show that the partial sums $U_n(t) = \sum_{k=1}^n u_k(t)\phi_k$ are Cauchy sequences for both norms. Let $p < q$ be two given integers and let us estimate $U_p - U_q$.

First of all, due to the continuity of $t \mapsto u_k(t)$, the partial sums U_n are continuous with values in $L^2(\Omega)$. Moreover, for all $t \in [0,T]$, we have

$$\begin{aligned}
\|U_p(t) - U_q(t)\|^2_{L^2(\Omega)} &= \Big\| \sum_{k=p+1}^{q} u_k(t)\phi_k \Big\|^2_{L^2(\Omega)} \\
&= \sum_{k=p+1}^{q} u_k(t)^2 \\
&\leq 2 \sum_{k=p+1}^{q} \Big[u_{0,k}^2 + \Big(\int_0^t |f_k(s)|\,ds \Big)^2 \Big] \\
&\leq 2 \sum_{k=p+1}^{q} \Big[u_{0,k}^2 + t \int_0^t f_k(s)^2\,ds \Big] \\
&\leq 2 \sum_{k=p+1}^{q} u_{0,k}^2 + 2T \sum_{k=p+1}^{q} \int_0^T f_k(s)^2\,ds
\end{aligned}$$

since all the exponential terms are less than 1 and by the Cauchy–Schwarz inequality. Therefore

$$\begin{aligned}
\|U_p - U_q\|^2_{C^0([0,T];L^2(\Omega))} &= \max_{t\in[0,T]} \|U_p(t) - U_q(t)\|^2_{L^2(\Omega)} \\
&\leq 2 \sum_{k=p+1}^{q} u_{0,k}^2 + 2T \sum_{k=p+1}^{q} \int_0^T f_k(s)^2\,ds
\end{aligned}$$

can be made as small as we wish by taking p large enough, due to the hypotheses on u_0 and f, and the sequence is consequently Cauchy in $C^0(0,T;L^2(\Omega))$.

Similarly, the partial sums are obviously in $L^2(0, T; H_0^1(\Omega))$, in fact they even are continuous with values in $H_0^1(\Omega)$, although this continuity will not persist in the limit. We use the H_0^1 seminorm, so that $|v|^2_{H_0^1(\Omega)} = a(v, v)$ (for a more general parabolic equation, H_0^1-ellipticity of the bilinear form would here come into play[7]).

The family $(\phi_k)_{k\in\mathbb{N}^*}$ is also orthogonal in $H_0^1(\Omega)$ and we have $a(\phi_k, \phi_k) = \lambda_k$ by (7.4). Therefore, for all $v \in H_0^1(\Omega)$, it follows that

$$|v|^2_{H_0^1(\Omega)} = \sum_{k=1}^{+\infty} \lambda_k v_k^2, \tag{7.7}$$

where $v_k = (v|\phi_k)_{L^2(\Omega)}$. In particular, we have

$$|U_p(t) - U_q(t)|^2_{H_0^1(\Omega)} = \sum_{k=p+1}^{q} \lambda_k u_k(t)^2,$$

so that integrating between 0 and T, we obtain

$$\|U_p - U_q\|^2_{L^2(0,T;H_0^1(\Omega))} = \sum_{k=p+1}^{q} \int_0^T \lambda_k u_k(t)^2\,dt.$$

Let us estimate each term in the sum on the right. We have

$$\lambda_k u_k(t)^2 \le 2\lambda_k\Big(u_{0,k}^2 e^{-2\lambda_k t} + T\int_0^t f_k(s)^2 e^{-2\lambda_k(t-s)}\,ds\Big),$$

so that

$$\begin{aligned}
\int_0^T \lambda_k u_k(t)^2\,dt &\le 2\lambda_k\Big(u_{0,k}^2\int_0^T e^{-2\lambda_k t}\,dt + T\int_0^T\Big(\int_0^t f_k(s)^2 e^{-2\lambda_k(t-s)}\,ds\Big)dt\Big)\\
&= (1-e^{-2\lambda_k T})u_{0,k}^2 + 2\lambda_k T\int_0^T f_k(s)^2\Big(\int_s^T e^{-2\lambda_k(t-s)}\,dt\Big)ds\\
&= (1-e^{-2\lambda_k T})u_{0,k}^2 + T\int_0^T (1-e^{-2\lambda_k(T-s)})f_k(s)^2\,ds\\
&\le u_{0,k}^2 + T\int_0^T f_k(s)^2\,ds.
\end{aligned}$$

---

[7] Or even more generally, Gårding's inequality, which reads: for all $v \in H_0^1(\Omega)$, $a(v,v) \ge \alpha|v|^2_{H_0^1(\Omega)} - \beta\|v\|^2_{L^2(\Omega)}$ with $\alpha > 0$.

Therefore

$$\|U_p - U_q\|^2_{L^2(0,T;H^1_0(\Omega))} \le \sum_{k=p+1}^{q} u^2_{0,k} + T\sum_{k=p+1}^{q}\int_0^T f_k(s)^2\,ds,$$

which can again be made as small as we wish by taking p large enough, and the sequence is Cauchy in $L^2(0,T;H^1_0(\Omega))$.

Finally, it remains to be seen that the function u defined by the series and which belongs to $C^0(0,T;L^2(\Omega))\cap L^2(0,T;H^1_0(\Omega))$ is a solution of the variational problem (7.3). The initial condition is obvious even in non variational form since

$$u(0) = \sum_{k=1}^{+\infty}\Big(u_{0,k}e^0 + \int_0^0 f_k(s)e^{\lambda_k s}\,ds\Big)\phi_k = \sum_{k=1}^{+\infty} u_{0,k}\phi_k = u_0.$$

Regarding the evolution equation, we obtain from the ordinary differential equations for u_k that for all $v\in \mathrm{span}((\phi_k)_{k\in\mathbb{N}^*})$

$$\big((u|v)_{L^2(\Omega)}\big)' + a(u,v) = (f(t)|v)_{L^2(\Omega)} \text{ in the sense of } \mathcal{D}'(]0,T[),$$

since any such v is a linear combination of the ϕ_k.

Let now $v\in H^1_0(\Omega)$ be arbitrary and $v_n\in \mathrm{span}((\phi_k)_{k\in\mathbb{N}^*})$ be such that $v_n\to v$ in $H^1_0(\Omega)$. For any $\varphi\in\mathcal{D}(]0,T[)$, we thus have

$$-\int_0^T (u(t)|v_n)_{L^2(\Omega)}\varphi'(t)\,dt + \int_0^T a(u(t),v_n)\varphi(t)\,dt = \int_0^T (f(t)|v_n)_{L^2(\Omega)}\varphi(t)\,dt.$$

It is then fairly obvious that each term in the above relation passes to the limit as $n\to+\infty$, thus establishing the evolution equation. □

Remark 7.9 Note that we do not need any compatibility condition between the initial condition u_0 and the Dirichlet boundary condition, as opposed to the regular case. Indeed, the expansion $u_0 = \sum_{k=1}^{+\infty} u_{0,k}\phi_k$ only holds in the L^2 sense. □

Remark 7.10 Formula (7.5)–(7.6) clearly generalizes the expansion obtained in Theorem 7.1. □

Remark 7.11 The recovery of a bona-fide solution of the heat equation from the above variational solution would require the use of Hilbert space valued distributions and integrals. Let us just say that it can be done. There are other approaches to the heat equation, for instance using semigroups (see for example [28, 58]). □

Remark 7.12 The d-dimensional heat equation can be solved along the exact same lines, replacing the functions ϕ_k and scalars λ_k by the eigenfunctions and eigenvalues of the minus Laplacian in $H^1_0(\Omega)$, i.e., the solutions of $-\Delta\phi_k = \lambda_k\phi_k$, $\phi_k\in H^1_0(\Omega)$, $\phi_k\neq 0$, see [5, 26, 28]. This eigenvalue problem for Ω bounded only has solutions for λ_k in a sequence $0 < \lambda_1 < \lambda_2 \leq \lambda_3 \leq \cdots$ such that $\lambda_k \to +\infty$ when $k \to +\infty$. Of course, the eigenvalues and eigenfunctions depend on the shape of Ω, see Chap. 1, Sect. 1.6. □

We also have an energy decay and stability estimate in the present context.

Proposition 7.12 *The solution u of problem (7.3) satisfies*

$$\|u(t)\|_{L^2(\Omega)} \leq \|u_0\|_{L^2(\Omega)} e^{-\lambda_1 t} + \int_0^t \|f(s)\|_{L^2(\Omega)} e^{-\lambda_1(t-s)}\, ds, \tag{7.8}$$

for all $t \in [0, T]$.

Proof This is a consequence of the series expansion. We first observe the following fact. Let g be a L^1-function from $[0, T]$ to a Euclidean space E (i.e., a finite dimensional Hilbert space). Then the integral $\int_0^t g(s)\, ds$ is well defined as a vector of E by choosing a basis of E and integrating g componentwise. Moreover, since E is Euclidean, there exists a unit vector e such that

$$\Big\| \int_0^t g(s)\, ds \Big\|_E = \Big(\int_0^t g(s)\, ds \Big) \cdot e = \int_0^t g(s) \cdot e\, ds \leq \int_0^t \|g(s)\|_E\, ds,$$

by the Cauchy–Schwarz inequality in E.

We now turn to estimate (7.8). We have

$$u(t) = \sum_{k=1}^{+\infty} \Big(u_{0,k} e^{-\lambda_k t} + \int_0^t f_k(s) e^{-\lambda_k(t-s)}\, ds \Big) \phi_k$$

so that by the triangle inequality

$$\|u(t)\|_{L^2(\Omega)} \leq \Big\| \sum_{k=1}^{+\infty} u_{0,k} e^{-\lambda_k t} \phi_k \Big\|_{L^2(\Omega)} + \Big\| \sum_{k=1}^{+\infty} \Big(\int_0^t f_k(s) e^{-\lambda_k(t-s)}\, ds \Big) \phi_k \Big\|_{L^2(\Omega)}.$$

For the first term, we note that

$$\Big\| \sum_{k=1}^{+\infty} u_{0,k} e^{-\lambda_k t} \phi_k \Big\|_{L^2(\Omega)} = \Big(\sum_{k=1}^{+\infty} u_{0,k}^2 e^{-2\lambda_k t} \Big)^{\frac{1}{2}} \leq \Big(\sum_{k=1}^{+\infty} u_{0,k}^2 e^{-2\lambda_1 t} \Big)^{\frac{1}{2}} = \|u_0\|_{L^2(\Omega)} e^{-\lambda_1 t},$$

since the sequence of eigenvalues λ_k is increasing. For the second term, we resort to the observation above with $E = \text{span}(\phi_1, \ldots, \phi_n)$ equipped with the L^2-norm, and deduce that

$$\begin{aligned}\Big\|\sum_{k=1}^{n}\Big(\int_0^t f_k(s)e^{-\lambda_k(t-s)}\,ds\Big)\phi_k\Big\|_{L^2(\Omega)} &= \Big\|\int_0^t \Big(\sum_{k=1}^{n} f_k(s)e^{-\lambda_k(t-s)}\phi_k\Big)ds\Big\|_{L^2(\Omega)}\\ &\le \int_0^t \Big\|\sum_{k=1}^{n} f_k(s)e^{-\lambda_k(t-s)}\phi_k\Big\|_{L^2(\Omega)}\,ds\\ &= \int_0^t \Big(\sum_{k=1}^{n} f_k(s)^2 e^{-2\lambda_k(t-s)}\Big)^{\frac{1}{2}}\,ds\\ &\le \int_0^t \Big(\sum_{k=1}^{n} f_k(s)^2\Big)^{\frac{1}{2}} e^{-\lambda_1(t-s)}\,ds.\end{aligned}$$

We now let $n \to +\infty$ and conclude by the convergence of the left-hand side series in $L^2(\Omega)$ and by the Lebesgue monotone convergence theorem for the right-hand side term. □

Remark 7.13 We recover the exponential decay of the energy when $f = 0$. □

Remark 7.14 The energy estimates and existence of weak solutions can be generalized to parabolic problems that are more general than the heat equation, see [58, 66]. □

7.7 The Heat Equation on $\mathbb{R}$

Even though it is unphysical, the heat equation on $\mathbb{R}^d$ is nonetheless interesting from the point of view of mathematics. For simplicity, we will only consider the case $d = 1$. Let us thus consider the initial value problem

$$\begin{cases} \dfrac{\partial u}{\partial t}(x,t) - \dfrac{\partial^2 u}{\partial x^2}(x,t) = f(x,t) \text{ in } \mathbb{R} \times]0,T[,\\ u(x,0) = u_0(x) \text{ on } \mathbb{R}. \end{cases} \tag{7.9}$$

Note that there is no boundary data since $\mathbb{R}$ has no boundary. They may be replaced by some kind of asymptotic behavior at infinity.

Let us now introduce an extremely important function [51, 74].

Definition 7.5 The function defined on $\mathbb{R}^2$ by

$$E(x,t) = \begin{cases} \dfrac{1}{\sqrt{4\pi t}} e^{-\frac{x^2}{4t}} \text{ for } t > 0,\\ 0 \text{ for } t \le 0, \end{cases}$$

is called the *(one-dimensional) heat kernel*.

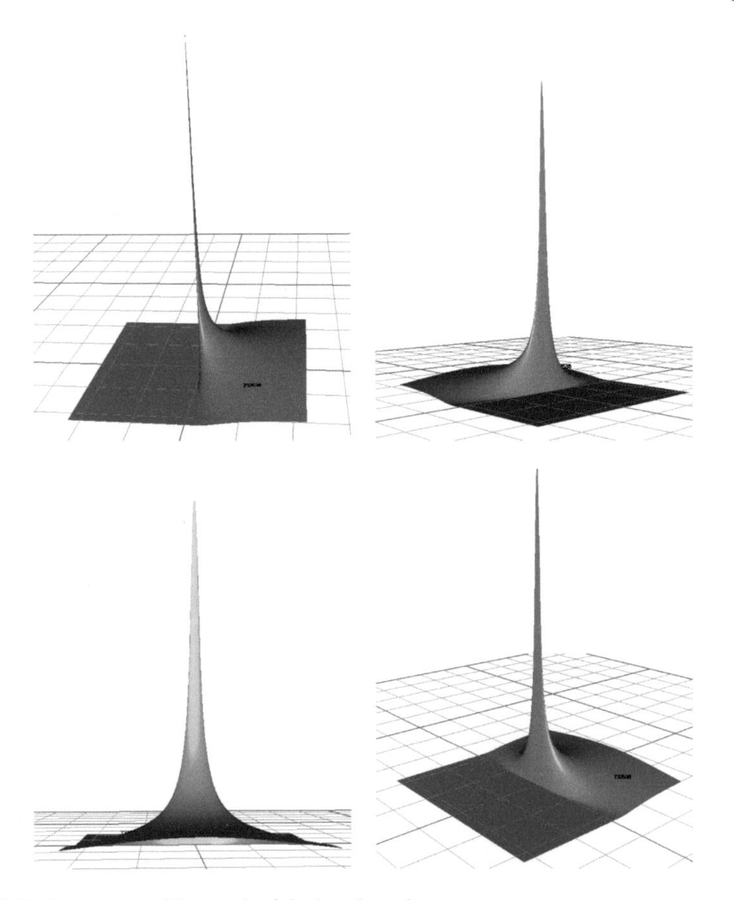

Fig. 7.8 Various views of the graph of the heat kernel

We note that for $t > 0$ fixed, the function $x \mapsto E(x,t)$ is a Gaussian. When $t \to 0^+$, the Gaussian becomes increasingly spiked. Indeed, we see that $E(x,t) = \frac{1}{\sqrt{4t}} E\big(\frac{x}{\sqrt{4t}}, \frac{1}{4}\big)$. In particular, $E(0,t) \to +\infty$, whereas $E(x,t) \to 0$ for all $x \neq 0$ when $t \to 0^+$, see Figs. 7.8 and 7.9.

Proposition 7.13 *We have $E \in L^1_{\text{loc}}(\mathbb{R}^2)$, hence $E \in \mathcal{D}'(\mathbb{R}^2)$.*

Proof Clearly $E \in C^\infty(\mathbb{R}^2 \setminus \{(0,0)\})$, therefore the only potential local integrability problem is in a compact neighborhood of $(0,0)$. It suffices to integrate $|E|$ on the square $[-a,a]^2$ for some $a > 0$. Since E vanishes for $t \leq 0$, only the upper half square is left. We have

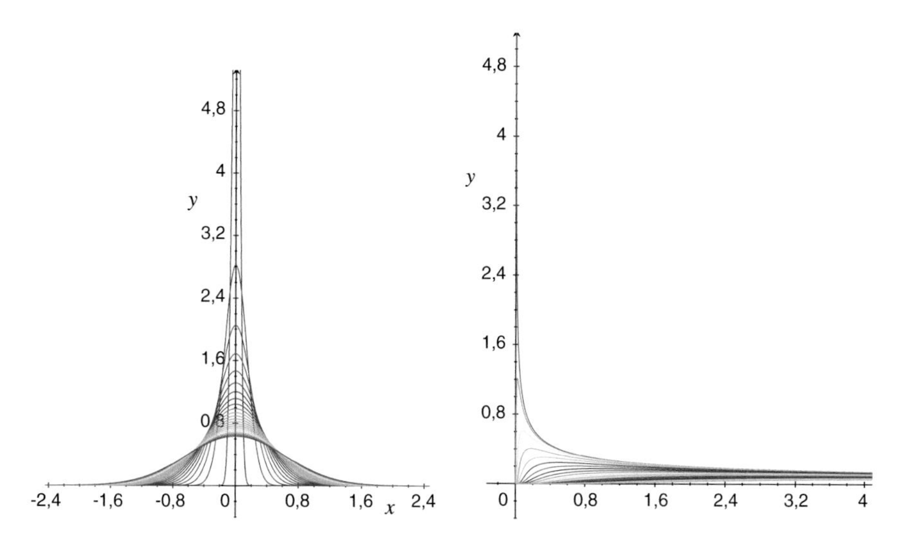

Fig. 7.9 The heat kernel at t fixed for different values of t (*left*) and x fixed for different values of x (*right*)

$$\begin{aligned}\int_{-a}^{a}\int_{0}^{a}|E(x,t)|\,dxdt &\le \int_{-\infty}^{+\infty}\int_{0}^{a}|E(x,t)|\,dxdt\\ &= \frac{1}{2\sqrt{\pi}}\int_{-\infty}^{+\infty}\Big(\int_{0}^{a}\frac{1}{\sqrt{t}}e^{-\frac{x^2}{4t}}\,dx\Big)dt\\ &= \frac{1}{\sqrt{\pi}}\int_{-\infty}^{+\infty}\Big(\int_{0}^{a}e^{-y^2}\,dy\Big)dt = a < +\infty,\end{aligned}$$

where we have performed the change of variables $x = 2\sqrt{t}\,y$ and because of the well-known value of the Gaussian integral, $\int_{-\infty}^{+\infty} e^{-y^2}\,dy = \sqrt{\pi}$. □

The heat kernel is the *fundamental solution* or *elementary solution* of the heat equation in the following sense.

Proposition 7.14 *We have*

$$\frac{\partial E}{\partial t} - \frac{\partial^2 E}{\partial x^2} = \delta_0,$$

where δ_0 is the Dirac distribution at $(x,t) = (0,0)$.

Proof Given $\varphi \in \mathcal{D}(\mathbb{R}^2)$, our goal is to show that

$$\Big\langle \frac{\partial E}{\partial t} - \frac{\partial^2 E}{\partial x^2}, \varphi \Big\rangle = \varphi(0,0).$$

We have already noticed that E is of class C^∞ everywhere except at $(x,t) = (0,0)$. Its distributional derivatives thus coincide with its classical derivatives on

$\mathbb{R}^2\setminus\{(0,0)\}$. Let us first compute these derivatives using brute force for $t > 0$ (only mild force is needed for $t < 0$). We thus have

$$\begin{aligned}\frac{\partial E}{\partial t} &= \frac{1}{2\sqrt{\pi}}\Big(-\frac{1}{2t^{3/2}} + \frac{x^2}{4t^{5/2}}\Big)e^{-\frac{x^2}{4t}},\\ \frac{\partial E}{\partial x} &= -\frac{1}{4\sqrt{\pi}}\frac{x}{t^{3/2}}e^{-\frac{x^2}{4t}},\\ \frac{\partial^2 E}{\partial x^2} &= -\frac{1}{4\sqrt{\pi}}\Big(\frac{1}{t^{3/2}} - \frac{x^2}{2t^{5/2}}\Big)e^{-\frac{x^2}{4t}},\end{aligned}$$

so that $\frac{\partial E}{\partial t} - \frac{\partial^2 E}{\partial x^2} = 0$ on $\mathbb{R}^2\setminus\{(0,0)\}$. Therefore the support of the distribution $\frac{\partial E}{\partial t} - \frac{\partial^2 E}{\partial x^2}$ is included in $\{(0,0)\}$.

Let us now work in the distributional sense. We take a test-function $\varphi \in \mathcal{D}(\mathbb{R}^2)$. We have

$$\begin{aligned}\Big\langle \frac{\partial E}{\partial t} - \frac{\partial^2 E}{\partial x^2}, \varphi\Big\rangle &= -\Big\langle E, \frac{\partial \varphi}{\partial t} + \frac{\partial^2 \varphi}{\partial x^2}\Big\rangle\\ &= -\int_{-\infty}^{+\infty}\Big(\int_0^{+\infty} E(x,t)\Big(\frac{\partial \varphi}{\partial t} + \frac{\partial^2 \varphi}{\partial x^2}\Big)(x,t)\,dt\Big)dx,\end{aligned}$$

since E is L^1_{loc} and vanishes for $t \le 0$. The derivatives $\frac{\partial \varphi}{\partial t}$ and $\frac{\partial^2 \varphi}{\partial x^2}$ have compact support, hence $E\big(\frac{\partial \varphi}{\partial t} + \frac{\partial^2 \varphi}{\partial x^2}\big)$ is in $L^1(\mathbb{R}^2)$ and the Lebesgue dominated convergence theorem implies that

$$\Big\langle \frac{\partial E}{\partial t} - \frac{\partial^2 E}{\partial x^2}, \varphi\Big\rangle = -\lim_{n\to+\infty}\int_{-\infty}^{+\infty}\Big(\int_{\frac{1}{n}}^{+\infty} E(x,t)\Big(\frac{\partial \varphi}{\partial t} + \frac{\partial^2 \varphi}{\partial x^2}\Big)(x,t)\,dt\Big)dx.$$

Now on the set $\mathbb{R}\times[\frac{1}{n}, +\infty[$, all the functions are C^∞ and we can integrate by parts, so that

$$\begin{aligned}\Big\langle \frac{\partial E}{\partial t} - \frac{\partial^2 E}{\partial x^2}, \varphi\Big\rangle = \lim_{n\to+\infty}\Big\{&\int_{-\infty}^{+\infty}\Big(\int_{\frac{1}{n}}^{+\infty}\Big(\frac{\partial E}{\partial t} - \frac{\partial^2 E}{\partial x^2}\Big)(x,t)\varphi(x,t)\,dt\Big)dx\\ &+\int_{-\infty}^{+\infty} E(x,n^{-1})\varphi(x,n^{-1})\,dx\Big\}.\end{aligned}$$

We have already seen that $\frac{\partial E}{\partial t} - \frac{\partial^2 E}{\partial x^2} = 0$ on $\mathbb{R}\times[\frac{1}{n}, +\infty[$ so that the first integral vanishes. Let us study the second integral. We perform the change of variables $y = \frac{\sqrt{n}x}{2}$. This yields

$$\int_{-\infty}^{+\infty} E(x,n^{-1})\varphi(x,n^{-1})\,dx = \frac{1}{\sqrt{\pi}}\int_{-\infty}^{+\infty} e^{-y^2}\varphi(2n^{-1/2}y, n^{-1})\,dy \to \varphi(0,0)$$

by the dominated convergence theorem. Therefore, we have shown that

$$\Big\langle \frac{\partial E}{\partial t} - \frac{\partial^2 E}{\partial x^2}, \varphi \Big\rangle = \varphi(0,0) = \langle \delta_0, \varphi \rangle,$$

and the proposition is proved. □

The heat kernel can be used to express the solution in various function spaces. Let us give an example.

Proposition 7.15 *Let $u_0 \in L^1(\mathbb{R})$ and $f \in L^1(\mathbb{R} \times \mathbb{R}_+)$. Then*

$$u(x,t) = \int_{-\infty}^{+\infty} E(x-y,t)u_0(y)\,dy + \int_{-\infty}^{+\infty} \Big(\int_0^t E(x-y,t-s) f(y,s)\,ds \Big) dy$$

is a solution of problem (7.9).

Proof We write the proof for u_0 and f continuous and bounded, for simplicity. First of all, all the integrals make sense and define a function on $\mathbb{R} \times \mathbb{R}_+^*$. Moreover, it is easy to check that all partial derivatives of the heat kernel are integrable on $\mathbb{R} \times [a, +\infty[$ for all $a > 0$. Therefore, we can differentiate under the integral signs without any problems as soon as the second argument of E stays bounded away from 0. Let us set, for all $(x,t) \in \mathbb{R} \times \mathbb{R}_+^*$,

$$v(x,t) = \int_{-\infty}^{+\infty} E(x-y,t)u_0(y)\,dy$$

and

$$w(x,t) = \int_{-\infty}^{+\infty} \Big(\int_0^t E(x-y,t-s) f(y,s)\,ds \Big) dy.$$

By the observation above, we have that

$$\Big(\frac{\partial v}{\partial t} - \frac{\partial^2 v}{\partial x^2} \Big)(x,t) = \int_{-\infty}^{+\infty} \Big(\frac{\partial}{\partial t} - \frac{\partial^2}{\partial x^2} \Big) E(x-y,t)u_0(y)\,dy = 0$$

for all $t > 0$. We need to exert a little more care to deal with w. Setting

$$w_n(x,t) = \int_{-\infty}^{+\infty} \Big(\int_0^{t-1/n} E(x-y,t-s) f(y,s)\,ds \Big) dy,$$

we see that $w_n \to w$ uniformly. Indeed,

$$|w(x,t) - w_n(x,t)| \le \|f\|_{L^\infty(\mathbb{R}\times\mathbb{R}_+)} \int_{t-\frac{1}{n}}^t \Big(\int_{-\infty}^{+\infty} E(x-y,t-s)\,dy \Big) ds = \frac{\|f\|_{L^\infty(\mathbb{R}\times\mathbb{R}_+)}}{n}.$$

Consequently, $w_n \to w$ in the sense of $\mathcal{D}'(\mathbb{R} \times \mathbb{R}_+^*)$ and therefore

$$\frac{\partial w_n}{\partial t} - \frac{\partial^2 w_n}{\partial x^2} \to \frac{\partial w}{\partial t} - \frac{\partial^2 w}{\partial x^2} \text{ in } \mathcal{D}'(\mathbb{R} \times \mathbb{R}_+^*).$$

We have no problem computing the effect of the heat operator on w_n:

$$\begin{aligned}\Big(\frac{\partial w_n}{\partial t} - \frac{\partial^2 w_n}{\partial x^2}\Big)(x,t) &= \int_{-\infty}^{+\infty} \Big(\int_0^{t-1/n} \Big(\frac{\partial}{\partial t} - \frac{\partial^2}{\partial x^2}\Big) E(x-y,t-s) f(y,s)\,ds\Big) dy \\ &\quad + \int_{-\infty}^{+\infty} E(x-y,n^{-1}) f(y,t-n^{-1})\,dy \\ &= \int_{-\infty}^{+\infty} E(x-y,n^{-1}) f(y,t-n^{-1})\,dy \underset{n\to\infty}{\to} f(x,t)\end{aligned}$$

simply, and even uniformly on compact sets, as we have essentially already seen before. Hence

$$\frac{\partial w}{\partial t} - \frac{\partial^2 w}{\partial x^2} = f.$$

Concerning the initial condition, we obviously have $w(x,0) = 0$ and $w(x,t) \to 0$ when $t \to 0$. Indeed, $\big|\int_{-\infty}^{+\infty} E(x-y,t-s) f(y,s)\,dy\big| \le \|f\|_{L^\infty(\mathbb{R}\times\mathbb{R}_+)}$. On the other hand, we have $v(x,t) \to u_0(x)$ when $t \to 0$ by the same change of variable as above. □

Remark 7.15 The analysis is a little more difficult when u_0 or f are not continuous. We have not made precise in which space the above solution is unique. □

Remark 7.16 The solution u is C^∞ on any open set of $\mathbb{R} \times \mathbb{R}_+^*$ where f is C^∞. This is again hypoellipticity. □

Remark 7.17 When u_0 is in L^∞, the result remains valid. In particular, when u_0 is periodic and $f = 0$, then u is also periodic in x for all t. Moreover if u_0 is the odd periodic extension of an initial condition for the heat equation on a bounded interval with Dirichlet boundary conditions, we obtain the same solution as the one obtained via Fourier series by restricting u to a space-time strip based on the interval in question. □

Remark 7.18 Notice an interesting phenomenon: When u_0 is positive with compact support and $f = 0$, we have $u(x,t) > 0$ for all $x \in \mathbb{R}$ and all $t > 0$, however small. In other words, a compactly supported initial distribution of temperature instantly spreads to the whole of $\mathbb{R}$. Thus, the heat equation propagates energy at infinite speed, which is strongly non physical. However, the validity of the heat equation as a model of temperature evolution is still extremely good for all classical physics and engineering applications. Indeed, assuming that the support of u_0 is included in $[-A, A]$ with $A > 0$, we have the following coarse estimate for $|x| > A$:

$$|u(x,t)| \le \frac{A}{\sqrt{\pi t}} e^{-\frac{(|x|-A)^2}{4t}} \|u_0\|_{L^\infty(\mathbb{R})},$$

so that $u(x,t)$ may be nonzero, but it is nonetheless extremely rapidly decreasing at infinity in x for $t > 0$. □

Again, the previous study is far from being exhaustive. There is a lot more to say about the heat equation, and more generally about parabolic equations, linear or nonlinear. For the ensuing numerical approximation theory, we are however now confident that solutions exist, are unique and sufficiently regular under reasonable hypotheses. Moreover, the energy estimate (7.2) will have important discrete counterparts.

Chapter 8
The Finite Difference Method for the Heat Equation

We now turn to numerical methods that can be used to approximate the solution of the heat equation. We develop the finite difference method in great detail, with particular emphasis on stability issues, which are delicate. We concentrate on the heat equation in one dimension of space, with homogeneous Dirichlet boundary conditions.

We also give some indications about finite difference (in time)-finite element (in space) approximation.

8.1 The Explicit Euler Three Point Finite Difference Scheme for the Heat Equation

We thus consider the problem

$$\begin{cases} \dfrac{\partial u}{\partial t}(x,t) - \dfrac{\partial^2 u}{\partial x^2}(x,t) = f(x,t) \text{ in } Q =]0,1[\times]0,T[, \\ u(x,0) = u_0(x) \text{ in } \Omega, \\ u(0,t) = u(1,t) = 0 \text{ in }]0,T[. \end{cases} \tag{8.1}$$

We assume that the solution u is as regular as we need it to be.

The general idea of finite difference methods for evolution equations is the same as for stationary equations that we have already seen in Chap. 2. It is simply to replace derivatives by difference quotients involving approximate discrete unknowns, and to solve for these unknowns. In the case of an evolution equation, we need a space-time grid. Let us thus be given two positive integers N and M. We set $h = \delta x = \frac{1}{N+1}$ and $x_n = nh$ for $n = 0, 1, \dots, N+1$. Similarly, we set $k = \delta t = \frac{T}{M+1}$ and $t_j = jk$ for $j = 0, 1, \dots, M+1$. The parameter h is called the space grid step and the parameter k the time grid step, or time step. The grid points are the points (x_n, t_j). Eventually, we will let N and M go to infinity, or equivalently, h and k go to 0. Note that there

H. Le Dret and B. Lucquin, *Partial Differential Equations: Modeling, Analysis and Numerical Approximation*, International Series of Numerical Mathematics 168, DOI 10.1007/978-3-319-27067-8_8

are two independent discretization parameters. This makes the analysis significantly more complicated than in the elliptic case of Chap. 2.

The discrete unknowns are scalars u_n^j for the above values of n and j, and it is hoped that u_n^j will be an approximation of $u(x_n, t_j)$, that should become better and better as N and M are increased. The boundary condition can be enforced exactly by requiring that

$$u_0^j = u_{N+1}^j = 0$$

for all $j = 0, \dots, M+1$. The initial condition is naturally discretized by requiring that

$$u_n^0 = u_0(x_n)$$

for all $n = 1, \dots, N$. If the initial data is consistent with the boundary condition, the same is true for the discrete unknowns in the sense that $u_0^0 = u_0(0) = u_0(1) = u_{N+1}^0 = 0$. The right-hand side of the equation is discretized by setting $f_n^j = f(x_n, t_j)$.

The only values that are left unknown at this stage are thus u_n^j for $n = 1, \dots, N$ and $j = 1, \dots, M+1$. We also use the notation[1]

$$U^j = \begin{pmatrix} u_1^j \\ u_2^j \\ \vdots \\ u_N^j \end{pmatrix} \in \mathbb{R}^N$$

to denote the vector of approximate values on the space grid at time t_j. Note again a fundamental difference with variational approximation methods such as the finite element method, which is that the computed approximation is not a function, but a finite set of values.

In the explicit Euler three point scheme, the partial derivatives are approximated in the same way as the simple derivatives were in the stationary case studied in Chap. 2. For the time derivative of the exact solution, we can use for example the forward differential quotient approximation

$$\frac{\partial u}{\partial t}(x_n, t_j) \approx \frac{u(x_n, t_{j+1}) - u(x_n, t_j)}{k},$$

and for the second order space derivative, by combining a forward and a backward differential quotient , we obtain the central approximation, see Remark 2.2 of Chap. 2,

[1] For consistency with the notation used in the stationary case, we should denote these vectors $U_{h,k}^j$. We drop the h, k index for brevity, but of course, these vectors crucially depend on h and k.

$$\frac{\partial^2 u}{\partial x^2}(x_n, t_j) \approx \frac{\frac{u(x_{n+1},t_j)-u(x_n,t_j)}{h} - \frac{u(x_n,t_j)-u(x_{n-1},t_j)}{h}}{h} = \frac{u(x_{n+1}, t_j) - 2u(x_n, t_j) + u(x_{n-1}, t_j)}{h^2},$$

where the $\approx$ sign can be given a precise meaning by using Taylor expansions as in Chap. 2 and we will come back to that later. The finite difference method mimics these approximations by replacing the exact values of the solution at the grid points by the discrete unknowns. In this particular case, we end up with the following explicit Euler three point finite difference scheme:

$$\begin{cases} \frac{u_n^{j+1} - u_n^j}{k} - \frac{u_{n+1}^j - 2u_n^j + u_{n-1}^j}{h^2} = f_n^j \text{ for } n = 1, \dots, N, j = 0, \dots, M, \\ u_n^0 = u_0(x_n) \text{ for } n = 1, \dots, N, \\ u_0^j = u_{N+1}^j = 0 \text{ for } j = 0, \dots, M+1. \end{cases} \tag{8.2}$$

Of course, at this point, there is no indication that (8.2) has anything to do with (8.1). The name explicit or forward Euler comes from the fact that the time derivative is approximated in the same way as it is approximated in the case of the forward Euler method for ordinary differential equations, whereas the three point name comes from the three point centered approximation of the second order space derivative already used for second order elliptic problems in Chap. 2.

We may rewrite the first N equations of the scheme in vector form as

$$\frac{U^{j+1} - U^j}{k} + A_h U^j = F^j \text{ for } j = 0, \dots, M,$$

where A_h is the same $N \times N$ tridiagonal matrix as in Chap. 2 with $c = 0$,

$$A_h = \frac{1}{h^2} \begin{pmatrix} 2 & -1 & 0 & \cdots & 0 \\ -1 & 2 & -1 & \cdots & 0 \\ \vdots & \ddots & \ddots & \ddots & \vdots \\ 0 & \cdots & -1 & 2 & -1 \\ 0 & \cdots & 0 & -1 & 2 \end{pmatrix},$$

and F^j is the vector

$$F^j = \begin{pmatrix} f_1^j \\ f_2^j \\ \vdots \\ f_N^j \end{pmatrix} \in \mathbb{R}^N.$$

The discrete initial condition is also a vector

$$U_0 = \begin{pmatrix} u_0(x_1) \\ u_0(x_2) \\ \vdots \\ u_0(x_N) \end{pmatrix} \in \mathbb{R}^N.$$

With this notation, the numerical scheme is equivalent to

$$\begin{cases} U^{j+1} = (I - kA_h)U^j + kF^j \text{ for } j = 0, \dots, M, \\ U^0 = U_0, \end{cases}$$

where I is the $N \times N$ identity matrix.[2] This simple recurrence relation shows that the scheme is well-defined. As opposed to the stationary case of Chap. 2, there is no linear system to be solved. We can also note the appearance of the factor $\frac{k}{h^2} = \frac{\delta t}{\delta x^2}$ which will play an important role in the sequel.

Let us now introduce, or reintroduce from Chap. 2 as the case may be, a few notions of interest. We will use the notation u_t for the function $x \mapsto u(x, t)$, for fixed $t \in \mathbb{R}_+$, and u_x for the function $t \mapsto u(x, t)$, for fixed $x \in [0, 1]$.

Definition 8.1 Let v be a function defined on $[0, 1]$. We define the space grid sampling operator S_h by

$$S_h(v) = \begin{pmatrix} v(x_1) \\ v(x_2) \\ \vdots \\ v(x_N) \end{pmatrix} \in \mathbb{R}^N.$$

Let now u be the solution of problem (8.1). We define the *truncation error* of the present finite difference method to be the sequence of vectors

$$\varepsilon_{h,k}(u)^j = \frac{S_h(u_{t_{j+1}}) - S_h(u_{t_j})}{k} + A_h S_h(u_{t_j}) - F^j.$$

To obtain the truncation error, we just take the finite difference scheme and replace the discrete unknowns with the corresponding grid samplings of the solution of the heat equation. Its name stems from the fact that, if we were to fictitiously apply the numerical scheme with one time step starting from the exact sampling values at t_j, i.e., let

$$\widetilde{U}^{j+1} = S_h(u_{t_j}) - kA_h S_h(u_{t_j}) + kF^j,$$

then we would make an error $S_h(u_{t_{j+1}}) - \widetilde{U}^{j+1} = k\varepsilon_{h,k}(u)^j$. The truncation error is not however directly related to the actual error between the sampling of the solution and the discrete unknown, as we will see later. This is because errors accumulate

[2] Here also, the vectors F^j and U_0 depend respectively on h and k, and on h.

from the start at each iteration of the scheme. The truncation error is however an important intermediate ingredient in the convergence analysis.

In order to analyze the convergence of finite difference methods, we need to introduce the function spaces

$$C^{m,n}(\bar{Q}) = \{u; \forall t \in [0,T], u_t \in C^m([0,1]) \text{ and } \forall x \in [0,1], u_x \in C^n([0,T]) \\ \text{with all derivatives uniformly bounded on } \bar{Q}\}.$$

For the time being, we equip the space $\mathbb{R}^N$ with the infinity norm as before, except that the dependence on h is here made explicit,

$$\|U\|_{\infty,h} = \max_{1\leq n\leq N} |U_n|,$$

where $U_1, \dots, U_N$ are the components of $U \in \mathbb{R}^N$.

We have the following easy estimate concerning the truncation error for the explicit Euler three point numerical scheme.

Proposition 8.1 *Assume that $u \in C^{4,2}(\bar{Q})$. Then we have*

$$\max_{0\leq j\leq M} \|\varepsilon_{h,k}(u)^j\|_{\infty,h} \leq C(h^2 + k),$$

where the constant C depends only on u.

We say that the forward Euler three point scheme is consistent for the ∞, h norms, of order 1 in time and order 2 in space, all these terms to be made precise later on.

Proof We use Taylor–Lagrange expansions exactly as in the one-dimensional elliptic case. First we use the fact that u_x is of class C^2. Therefore, for all n and j, there exists $\theta_n^j \in]t_j, t_{j+1}[$ such that

$$u(x_n, t_{j+1}) = u(x_n, t_j) + k\frac{\partial u}{\partial t}(x_n, t_j) + \frac{k^2}{2}\frac{\partial^2 u}{\partial t^2}(x_n, \theta_n^j),$$

which we rewrite as

$$\frac{u(x_n, t_{j+1}) - u(x_n, t_j)}{k} = \frac{\partial u}{\partial t}(x_n, t_j) + \frac{k}{2}\frac{\partial^2 u}{\partial t^2}(x_n, \theta_n^j).$$

Similarly, u_t is of class C^4. Therefore, for all n and j, there exists $\xi_n^{j,+} \in]x_n, x_{n+1}[$, $\xi_n^{j,-} \in]x_{n-1}, x_n[$ such that

$$u(x_{n+1}, t_j) = u(x_n, t_j) + h\frac{\partial u}{\partial x}(x_n, t_j) + \frac{h^2}{2}\frac{\partial^2 u}{\partial x^2}(x_n, t_j) + \frac{h^3}{6}\frac{\partial^3 u}{\partial x^3}(x_n, t_j) + \frac{h^4}{24}\frac{\partial^4 u}{\partial x^4}(\xi_n^{j,+}, t_j),$$

and

$$u(x_{n-1},t_j)=u(x_n,t_j)-h\frac{\partial u}{\partial x}(x_n,t_j)+\frac{h^2}{2}\frac{\partial^2 u}{\partial x^2}(x_n,t_j)-\frac{h^3}{6}\frac{\partial^3 u}{\partial x^3}(x_n,t_j)+\frac{h^4}{24}\frac{\partial^4 u}{\partial x^4}(\xi_n^{j,-},t_j),$$

which we rewrite as

$$\frac{u(x_{n+1},t_j)-2u(x_n,t_j)+u(x_{n-1},t_j)}{h^2}=\frac{\partial^2 u}{\partial x^2}(x_n,t_j)+\frac{h^2}{12}\frac{\partial^4 u}{\partial x^4}(\xi_n^j,t_j),$$

where $\xi_n^j\in\,]x_{n-1},x_{n+1}[$, thanks to the intermediate value theorem.[3] Taking into account the boundary conditions $u(x_0,t_j)=u(x_{N+1},t_j)=0$, we see that

$$\varepsilon_{h,k}(u)^j=S_h\Big(\Big(\frac{\partial u}{\partial t}-\frac{\partial^2 u}{\partial x^2}\Big)_{t_j}\Big)-F^j+R^j$$

with

$$R_n^j=\frac{k}{2}\frac{\partial^2 u}{\partial t^2}(x_n,\theta_n^j)-\frac{h^2}{12}\frac{\partial^4 u}{\partial x^4}(\xi_n^j,t_j).$$

Now since u is a regular solution of the heat equation, we have $S_h\big(\big(\frac{\partial u}{\partial t}-\frac{\partial^2 u}{\partial x^2}\big)_{t_j}\big)-F^j=0$ due to the definition of the sampling operator, even for $j=0$ by continuity. Moreover

$$|R_n^j|\le\max\Big(\frac{1}{2}\max_{\bar Q}\Big|\frac{\partial^2 u}{\partial t^2}\Big|,\frac{1}{12}\max_{\bar Q}\Big|\frac{\partial^4 u}{\partial x^4}\Big|\Big)(k+h^2),$$

for all n and j, which concludes the proof of the proposition. □

Remark 8.1 A natural question is to wonder whether the above order is optimal and to make sure that we are not missing a better approximation rate. To settle this question, it is sufficient to pursue the Taylor–Lagrange expansions up to third order in time and sixth order in space (assuming enough regularity for u) and see what comes up. The outcome is

$$\begin{aligned}\varepsilon_{h,k}(u)_n^j&=\frac{k}{2}\frac{\partial^2 u}{\partial t^2}(x_n,t_j)-\frac{h^2}{12}\frac{\partial^4 u}{\partial x^4}(x_n,t_j)\\&\quad+\frac{k^2}{6}\frac{\partial^3 u}{\partial t^3}(x_n,\tilde\theta_n^j)-\frac{h^4}{720}\Big(\frac{\partial^6 u}{\partial x^6}(\tilde\xi_n^{j,-},t_j)+\frac{\partial^6 u}{\partial x^6}(\tilde\xi_n^{j,+},t_j)\Big).\end{aligned}$$

Now, the function $(h,k)\mapsto Ak+Bh^2$ is identically 0 if and only if $A=B=0$. Of course, in general both $\frac{\partial^2 u}{\partial t^2}(x_n,t_j)$ and $\frac{\partial^4 u}{\partial x^4}(x_n,t_j)$ are nonzero, thus the truncation error is not of a higher order. □

[3] This is the same computation as in Chap. 2.

8.2 The Implicit Euler and Leapfrog Schemes

Before describing and analyzing general finite difference schemes, we give two more simple examples. The first example is the *implicit* or *backward Euler three point scheme*, which is associated with the backward differential quotient approximation of the time derivative

$$\frac{\partial u}{\partial t}(x_n, t_j) \approx \frac{u(x_n, t_j) - u(x_n, t_{j-1})}{k},$$

also used under the same name in the context of the numerical approximation of ordinary differential equations. In vector form, this scheme reads

$$\frac{U^j - U^{j-1}}{k} + A_h U^j = F^j \text{ for } j = 1, \dots, M+1,$$

or equivalently

$$\frac{U^{j+1} - U^j}{k} + A_h U^{j+1} = F^{j+1} \text{ for } j = 0, \dots, M.$$

This scheme is called implicit, because the above formula is not a simple recurrence relation. Indeed, U^{j+1} appears as the solution of an equation once U^j is known. It is not a priori clear that this equation is solvable. In this particular case, we have

$$\begin{cases} U^{j+1} = (I + kA_h)^{-1}(U^j + kF^{j+1}) \text{ for } j = 0, \dots, M, \\ U^0 = U_0, \end{cases}$$

since it is not hard to see that the matrix $I + kA_h$ is symmetric, positive definite, hence invertible. In practical terms, the implementation of the backward Euler method entails the solution of a linear system at each time step, whereas the explicit method is simply a matrix-vector product and vector addition at each time step. The implicit method is thus more computationally intensive than the explicit method, but it has other benefits as we will see later.

The analysis of the truncation error of the implicit Euler scheme is basically the same as in the explicit case. The method is likewise consistent, of order 1 in time and order 2 in space.

The second example is the *leapfrog* or *Richardson method*, which is associated with the central differential quotient approximation of the time derivative

$$\frac{\partial u}{\partial t}(x_n, t_j) \approx \frac{u(x_n, t_{j+1}) - u(x_n, t_{j-1})}{2k},$$

which leaps over time t_j. In vector form, this scheme reads

$$\frac{U^{j+1} - U^{j-1}}{2k} + A_h U^j = F^j \text{ for } j = 1, \dots, M.$$

This scheme is an explicit two-step method since U^{j+1} is explicitly given in terms of U^j and U^{j-1}.

$$\begin{cases} U^{j+1} = U^{j-1} - 2kA_h U^j + 2kF^j \text{ for } j = 1, \dots, M, \\ U^0 = U_0, \, U^1 = U_1. \end{cases}$$

Of course, since this is a two-step method, a given value U_1 supposed to approximate $S_h(u_{t_1})$ must somehow be ascribed to U^1 in order to initialize the recurrence, in addition to U_0.

The idea behind the leapfrog scheme is that the truncation error is of order 2 in time and order 2 in space, i.e., the truncation error is bounded from above by a quantity of the form $C(h^2 + k^2)$, which would seem to be advantageous as compared to both Euler schemes. Unfortunately, we will see that the improved truncation error is accompanied by instability, which prevents the method from being convergent. It is not usable in practice for the heat equation, and this example shows that a naive approach to finite difference schemes may very well badly fail.

8.3 General Finite Difference Schemes, Consistency, Stability, Convergence

In this section, we introduce a general framework for dealing with finite difference schemes. A finite difference scheme for the heat equation, or for any other linear evolution partial differential equation, is constructed by forming linear combinations of partial differential quotients and replicating these linear combinations on the purely discrete level. It can be cast in the following form: Let us be given two positive integers l and m with $l + m \geq 1$, and a set of $l + m + 1$ matrices B_i, $-m \leq i \leq l$, each of size $N \times N$, the entries of which are functions of h and k. We assume that B_l is invertible.

A general $l + m$ step finite difference scheme is then a recurrence relation for a sequence of vectors $U^j \in \mathbb{R}^N$, of the form

$$B_l U^{j+l} + B_{l-1} U^{j+l-1} + \cdots + B_0 U^j + \cdots + B_{-m} U^{j-m} = \widetilde{F}^j, \, m \leq j \leq (T/k) - l \tag{8.3}$$

with given initial data

$$U^0 = U_0, \, U^1 = U_1, \dots, U^{l+m-1} = U_{l+m-1}.$$

The right-hand side vector $\widetilde{F}^j$ is to be constructed from f, but is not necessarily equal to F^j. As before, the intended meaning of U^j with components u_n^j is that u_n^j is expected to provide an approximation of $u(x_n, t_j)$.

Definition 8.2 We say that the scheme (8.3) is *explicit* if the leading matrix B_l is diagonal. Otherwise, the scheme is called *implicit*.

In an explicit method, the next vector U^{j+l} is thus directly obtained from previously computed vectors by matrix-vector multiplications and vector additions, whereas an implicit method entails the actual resolution of a linear system at each time step.

Let us see how the previously introduced schemes fit into this general picture. For forward Euler, we have

$$\begin{cases} \frac{1}{k}U^{j+1} + \Big(-\frac{1}{k}I + A_h\Big)U^j = F^j, \\ U^0 = U_0, \end{cases} \tag{8.4}$$

so that $l = 1, m = 0, B_1 = \frac{1}{k}I, B_0 = -\frac{1}{k}I + A_h$ and $\widetilde{F^j} = F^j$. It is obviously one step and explicit. Of course, we can also write it with for example $B_1 = I, B_0 = -I + kA_h$ and $\widetilde{F^j} = kF^j$, there is no uniqueness of the general form for a given scheme. The backward Euler method is

$$\begin{cases} \Big(\frac{1}{k}I + A_h\Big)U^j - \frac{1}{k}U^{j-1} = F^j, \\ U^0 = U_0, \end{cases} \tag{8.5}$$

so that $l = 0$, $m = 1$, $B_0 = \frac{1}{k}I + A_h$, $B_{-1} = -\frac{1}{k}I$ and $\widetilde{F^j} = F^j$. It is obviously one step and implicit (recall that $\frac{1}{k}I + A_h$ is invertible). Finally, the leapfrog scheme is

$$\begin{cases} \frac{1}{2k}U^{j+1} + A_hU^j - \frac{1}{2k}U^{j-1} = F^j, \\ U^0 = U_0, U^1 = U_1. \end{cases} \tag{8.6}$$

so that $l = 1$, $m = 1$, $B_1 = \frac{1}{2k}I$, $B_0 = A_h$, $B_{-1} = -\frac{1}{2k}I$ and $\widetilde{F^j} = F^j$. It is obviously two step and explicit.

Remark 8.2 As mentioned before, there is no uniqueness of a general form for a given scheme. Indeed, given a general form, we can obtain another one by multiplying everything by an arbitrary function of h and k, or even by an arbitrary $N \times N$ invertible matrix function of h and k. So the definition of explicit or implicit scheme as stated before is attached to a general form and not to the scheme under consideration. However, it should be quite clear that writing the backward Euler scheme as

$$U^j - (I + kA_h)^{-1}U^{j-1} = k(I + kA_h)^{-1}F^j$$

and thus declaring it explicit, is somehow cheating. Indeed, the matrix $(I + kA_h)^{-1}$ is not know explicitly.[4] Thus the real issue is an implementation issue: do we need to numerically solve a nontrivial linear system to compute the scheme, or not? In the former case, the scheme is implicit and the latter case, it is explicit.

Different general forms give rise to different truncation errors, see Definition 8.3 below. As a general rule, it is better to choose the form that naturally comes from the discretization of the partial derivatives by differential quotients. □

There is nothing in the definition of a general finite difference scheme given above that even alludes to a particular partial differential equation that we might be interested in approximating. We therefore need a way of comparing the vectors $U^j \in \mathbb{R}^N$ and the function u solution of problem (8.1). As in the stationary case of Chap. 2, an obvious idea is to use the sampling operator already introduced in Definition 8.1.

Even then, quantitatively comparing two vectors of $\mathbb{R}^N$ involves the choice of a norm on $\mathbb{R}^N$. We are ultimately interested in letting $N \to +\infty$, thus we need a norm for each value of N. There is no reason at this point to do anything else that to choose an arbitrary norm $\|\cdot\|_N$ on $\mathbb{R}^N$ for each N. Two popular choices are

$$\|U\|_{\infty,h} = \max_{1\le n\le N} |U_n| \text{ and } \|U\|_{2,h} = \sqrt{h}\Big(\sum_{n=1}^{N} U_n^2\Big)^{1/2},$$

(recall that $h = \frac{1}{N+1}$). The reason for the $\sqrt{h}$ factor in the second norm is for comparison with the L^2 norm in the limit $h \to 0$. Of course, it is well-known that any two norms on $\mathbb{R}^N$ are equivalent, but the constants in the norm equivalence depend on N. For instance,

$$\|U\|_{2,h} \le \|U\|_{\infty,h} \le \frac{1}{\sqrt{h}}\|U\|_{2,h}$$

with basically optimal constants. This shows that consistency and convergence (see Definitions 8.4 and 8.6 below) in the ∞, h norms imply consistency and convergence in the $2, h$ norms, but not the converse.

We can now give a few definitions.

Definition 8.3 Let u be a sufficiently regular solution of problem (8.1). The *truncation error* of the general finite difference method (8.3) is the sequence of vectors

$$\varepsilon_{h,k}(u)^j = B_l S_h(u_{t_{j+l}}) + \cdots + B_0 S_h(u_{t_j}) + \cdots + B_{-m} S_h(u_{t_{j-m}}) - \widetilde{F}^j,$$

for $m \le j \le (T/k) - l$.

Again, we just replace the discrete unknown with the grid sampling of the solution in the finite difference scheme formula. Note that, since for any given scheme,

[4] Well, actually it may well be known somewhere in the literature, but let us assume it is not known for the sake of the argument.

there are infinitely many different general formulas describing the same scheme, the truncation error of a given scheme depends on how it is written in general form. Fortunately, this is totally irrelevant for the ensuing analysis. We just need to be careful in the application of the general results in each particular case.

Definition 8.4 We say that the scheme in general form (8.3) is *consistent for the family of norms* $\|\cdot\|_N$ if

$$\max_{m\le j\le (T/k)-l}\|\varepsilon_{h,k}(u)^j\|_N \to 0 \text{ when } (h,k)\to(0,0).$$

We say that it is of order p in space and q in time for the family of norms $\|\cdot\|_N$ if

$$\max_{m\le j\le (T/k)-l}\|\varepsilon_{h,k}(u)^j\|_N \le C(h^p+k^q),$$

where the constant C only depends on u.

Consistency means that the scheme is trying its best to locally approximate the right partial differential equation problem in the norm $\|\cdot\|_N$. Of course, the above definitions depend on the choice of norm and on the choice of general form. A given scheme may well be consistent for one family of norms and not for another, or be of some order in one general form and of another order in another general form. It is up to us to choose the best norm/general form combination. As in the particular cases that we have already seen, checking consistency and computing time and space orders is just a matter of having enough patience to write down the relevant Taylor–Lagrange expansions.

A significantly subtler notion is that of *stability*.

Definition 8.5 Let $\mathscr{S}\subset\mathbb{R}_+^*\times\mathbb{R}_+^*$ be such that $(0,0)\in\bar{\mathscr{S}}$. We say that the scheme in general form (8.3) is *stable for the family of norms* $\|\cdot\|_N$ *under condition* $\mathscr{S}$, if there exists two constants $C_1(T)$ and $C_2(T)$ which only depend on T such that, for all $(h,k)\in\mathscr{S}$, all initial data $U_{j'}$, $0\le j'\le l+m-1$, and all right-hand sides $\widetilde{F}^{j''}$, $m\le j''\le (T/k)-l$, we have

$$\max_{j\le T/k}\|U^j\|_N \le C_1(T)\max_{0\le j'\le l+m-1}\|U_{j'}\|_N + C_2(T)\max_{m\le j''\le (T/k)-l}\|\widetilde{F}^{j''}\|_N. \tag{8.7}$$

If $\mathscr{S}\supset\,]0,a[\,\times\,]0,a[$ for some $a>0$, we say that the scheme is *unconditionally stable*.

Stability makes no reference to the partial differential equation. It is just a property of the recurrence relation which controls the growth of its solutions in terms of the initial data and right-hand side. Stability in the present sense is very different from stability in the stationary case.

Note that in spite of the previous fact, the introduction of stability must be put in perspective with such continuous energy estimates as estimate (7.2). It is in fact a discrete version of such estimates in the case of the $2,h$ norms. We also refer to

Sect. 8.9 for a particularly striking parallel between the stability in the continuous and discrete cases.

Definition 8.6 We say that the scheme is *convergent for the family of norms* $\|\cdot\|_N$ if

$$\max_{j\le T/k} \|U^j - S_h(u_{t_j})\|_N \to 0 \text{ when } (h,k)\to(0,0),\ (h,k)\in\mathscr{S}.$$

Note that this definition is independent of the general form under which the scheme is written, as opposed to the two previous definitions. Remembering that $S_h(u_{t_j})$ is just a notation for $u(x_n,t_j)$, $n=1,\dots,N$, we see that convergence of the scheme means that $|u_n^j - u(x_n,t_j)|$ tends to 0 (at least if the choice of norms $\|\cdot\|_N$ is reasonable enough), or that the computed discrete unknowns u_n^j are in effect approximations of the value of the solution at gridpoints.

The relevance of the above definitions is clarified by means of the Lax theorem:

Theorem 8.1 (Lax Theorem) *Assume that*

$$\max_{0\le j'\le l+m-1} \|U_{j'} - S_h(u_{t_{j'}})\|_N \to 0 \textit{ when } (h,k)\to(0,0),\ (h,k)\in\mathscr{S}.$$

If the scheme in general form (8.3) *is consistent and stable under condition* $\mathscr{S}$ *for the family of norms* $\|\cdot\|_N$*, then it is convergent for that same family of norms.*

Proof Let us compare the formulas for the truncation error and for the scheme.

$$B_l S_h(u_{t_{j+l}}) + \cdots + B_0 S_h(u_{t_j}) + \cdots + B_{-m} S_h(u_{t_{j-m}}) = \varepsilon_{h,k}(u)^j + \widetilde{F}^j,$$
$$B_l U^{j+l} + \cdots + B_0 U^j + \cdots + B_{-m} U^{j-m} = \widetilde{F}^j.$$

Setting $V^j = U^j - S_h(u_{t_j})$ and subtracting the above two formulas, we see that

$$B_l V^{j+l} + \cdots + B_0 V^j + \cdots + B_{-m} V^{j-m} = -\varepsilon_{h,k}(u)^j,$$

with the initial data

$$V^{j'} = U_{j'} - S_h(u_{t_{j'}}) \text{ for } 0\le j'\le l+m-1.$$

By the stability hypothesis, it follows that

$$\max_{j\le T/k}\|V^j\|_N \le C_1 \max_{0\le j'\le l+m-1} \|U_{j'} - S_h(u_{t_{j'}})\|_N + C_2 \max_{m\le j''\le (T/k)-l} \|\varepsilon_{h,k}(u)^{j''}\|_N,$$

and the right-hand side goes to 0 by consistency and the hypothesis on the initial data for the scheme. □

Remark 8.3 We have written here the most useful part of the Lax theorem, i.e., consistency plus stability imply convergence, see [7, 29, 67, 81]. There is a less useful converse part, see Proposition 8.4 later on. Therefore, we have not missed anything by focusing on consistency and stability. □

Corollary 8.1 *Assume that the scheme is stable under condition $\mathscr{S}$ and of order p in space and q in time for the family of norms $\|\cdot\|_N$, and that*

$$\max_{0\leq j'\leq l+m-1} \|U_{j'} - S_h(u_{t_{j'}})\|_N \leq C(h^p + k^q),$$

for some constant C. Then

$$\max_{j\leq T/k} \|U^j - S_h(u_{t_j})\|_N \leq C'(h^p + k^q),$$

where C' only depends on T and u.

Remark 8.4 High order stable schemes thus result in (in principle) more accurate approximations than low order schemes. This is conditional on the initial data for the scheme not destroying this accuracy. If the scheme uses several time steps, the corresponding initial data must therefore be computed by using some other, equally accurate method. If the scheme is one time step, then we are at liberty to have exact initial data (discounting round-off errors). □

Remark 8.5 Let us emphasize again that all this is highly dependent on the choice of norms. Assume that, for one outrageous reason or another, we had chosen $\|U\|_N = 2^{-N}\|U\|_{\infty,h}$. Then, it is likely that even the most wildly non consistent scheme for the ∞, h norms would become consistent for the new norms! Since stability is not affected by multiplication of the norm by a constant, if the scheme was stable for the ∞, h norms, then it would also be stable for the $\|\cdot\|_N$ norms.[5] Hence, by the Lax theorem, it would be convergent for that family of norms. There is however no contradiction. Indeed, saying that $2^{-N}|u_n^j - u(x_n, t_j)| \to 0$ tells us next to nothing about what we really are interested in, namely, is u_n^j a good approximation of $u(x_n, t_j)$ or not. In this respect, the two choices $\|\cdot\|_{\infty,h}$ and $\|\cdot\|_{2,h}$ are much more natural. □

Remark 8.6 It should also be noted that the fact that the underlying partial differential equation is the heat equation plays no role in the Lax theorem. The theorem holds true for any finite difference scheme devised to approximate the solution of any evolution partial differential equation problem in one or several dimensions of space. We will thus also use it for hyperbolic equations in Chaps. 9 and 10. □

Let us apply the previous results to the explicit Euler scheme. Let $\|\cdot\|_N$ be a norm on $\mathbb{R}^N$ and $|||\cdot|||_N$ the associated induced matrix norm, see Chap. 2. For any $N \times N$ real matrix A and vector $U \in \mathbb{R}^N$, we have

$$\|AU\|_N \leq |||A|||_N \|U\|_N.$$

[5] Note that stability is also affected by the general form used to write the scheme, via the term $\widetilde{F}^j$, see Proposition 8.3 and Remark 8.12 for a more precise statement.

In the particular case of the ∞, h norm, we have seen in Proposition 2.3 of Chap. 2 that

$$|||A|||_{\infty,h} = \max_{1 \le n \le N} \Big(\sum_{j=1}^{N} |A_{nj}| \Big).$$

Let us choose the general form

$$\frac{1}{k} U^{j+1} - \frac{1}{k} U^j + A_h U^j = F^j,$$

for which we have consistency in the $\|\cdot\|_{\infty,h}$ norms and $\widetilde{F}^j = F^j$.

Proposition 8.2 *Let $\mathscr{S} = \big\{(h,k) \in \mathbb{R}_+^* \times \mathbb{R}_+^*; \frac{k}{h^2} \le \frac{1}{2}\big\}$. The explicit three-point Euler scheme in this general form is stable under condition $\mathscr{S}$ for the norms $\|\cdot\|_{\infty,h}$, hence convergent for these norms.*

Proof The above general form can be rewritten as

$$U^{j+1} = \mathscr{A}_{h,k} U^j + kF^j,$$

where $\mathscr{A}_{h,k} = I - kA_h$. Therefore

$$\|U^{j+1}\|_{\infty,h} \le |||\mathscr{A}_{h,k}|||_{\infty,h} \|U^j\|_{\infty,h} + k\|F^j\|_{\infty,h}.$$

Let us set $r = \frac{k}{h^2}$. By direct inspection, we see that

$$\mathscr{A}_{h,k} = \begin{pmatrix} 1-2r & r & 0 & \cdots & 0 \\ r & 1-2r & r & \cdots & 0 \\ \vdots & \ddots & \ddots & \ddots & \vdots \\ \vdots & \ddots & \ddots & \ddots & r \\ 0 & \cdots & 0 & r & 1-2r \end{pmatrix}.$$

It follows that

$$|||\mathscr{A}_{h,k}|||_{\infty,h} = |1-2r| + 2r = \begin{cases} 1 & \text{if } r \le \frac{1}{2}, \\ 4r-1 & \text{if } r > \frac{1}{2}. \end{cases}$$

Therefore, if $r \le \frac{1}{2}$, we have that

$$\begin{aligned} \|U^{j+1}\|_{\infty,h} &\le \|U^j\|_{\infty,h} + k\|F^j\|_{\infty,h} \\ &\le \|U^j\|_{\infty,h} + k \max_{n \le T/k} \|F^n\|_{\infty,h}. \end{aligned}$$

Iterating backwards, we obtain that for all j such that $j \le \frac{T}{k}$,

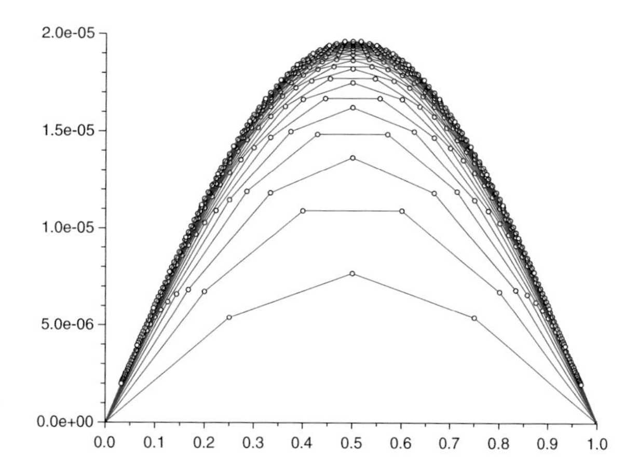

Fig. 8.1 Convergence of the explicit Euler method when stability is satisfied

$$\|U^j\|_{\infty,h} \le \|U^0\|_{\infty,h} + jk \max_{n\le T/k} \|F^n\|_{\infty,h}$$
$$\le \|U^0\|_{\infty,h} + T \max_{n\le T/k} \|F^n\|_{\infty,h},$$

hence the stability of the scheme for the norm ∞, h under condition $\mathscr{S}$. □

We plot[6] in Fig. 8.1 the discrete values U^{M+1} with M corresponding to $T = 1.1$, for the initial value $u_0(x) = 4x(1-x)$. Each curve corresponds to different values of N going from 3 to 30, and choosing M in such a way that $\frac{k}{h^2} \approx 0.49$.

Remark 8.7 The above estimates are not sufficient to conclude that the scheme is not stable when $r > \frac{1}{2}$. However, numerical experiments with $r > \frac{1}{2}$ quickly show that the explicit Euler scheme is non convergent for the ∞, h norm. Since it is consistent, this means it must be unstable. In particular, round-off errors are amplified exponentially fast.

To illustrate this, we plot in Fig. 8.2 the same sequence of computations as above, with $\frac{k}{h^2} \approx 0.51$.

Note the extreme and erratic variations in the vertical scale of the above plots depending on N (and its parity). The explicit Euler scheme appears to be wildly non convergent for such values of the discretization parameters, due to the failure of stability even by a very small amount. □

Remark 8.8 When $(h, k) \in \mathscr{S}$, we thus have $k \le \frac{h^2}{2} \ll h$ for h small. For instance, if we want a modest amount of 1000 points in the space grid, then the time step must be smaller than 5×10^{-7}, i.e., to compute up to a final time of $T = 1$ s, we need at least 2×10^6 iterations in time. Such stability requirements can rapidly make the scheme too computationally expensive, in spite of its otherwise simplicity. □

[6]For an easier visualization, we also plot a linear interpolation of the finite difference discrete values.

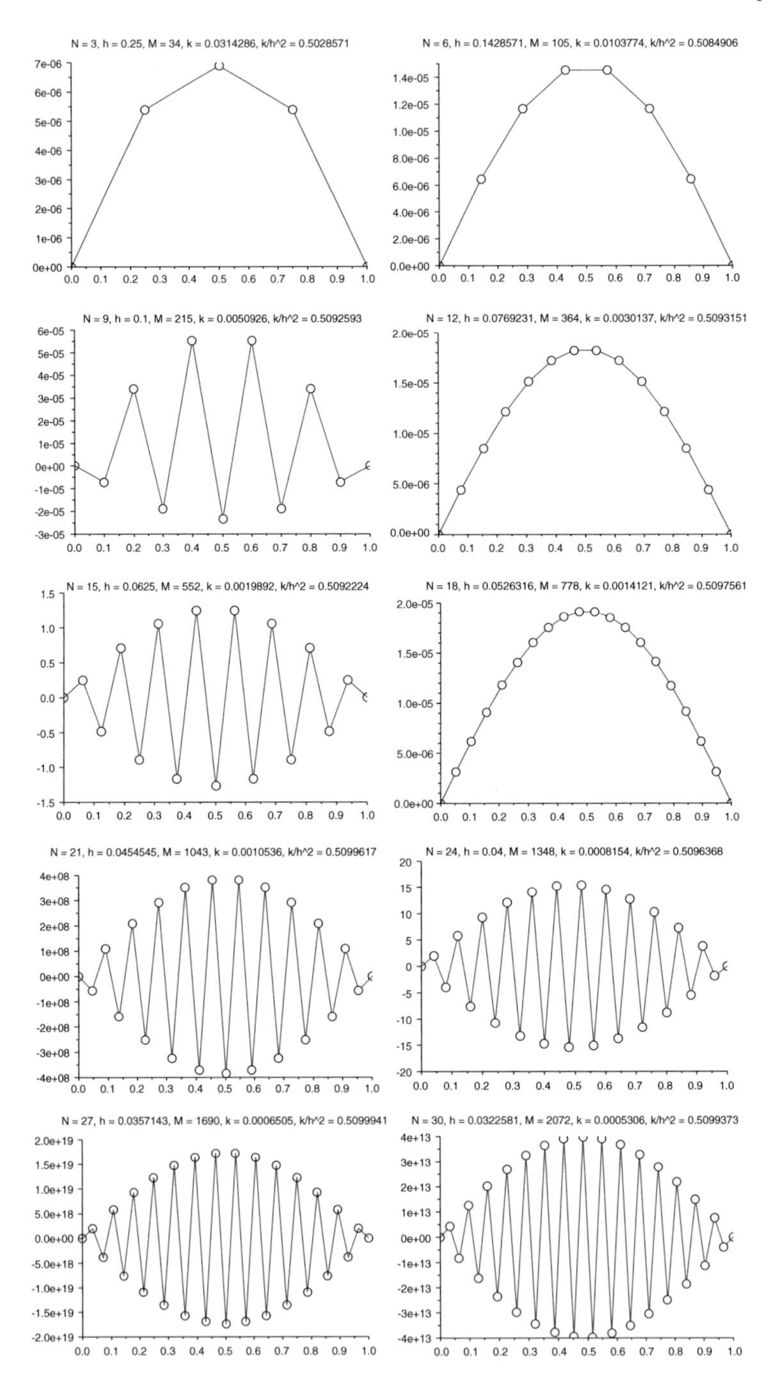

Fig. 8.2 Divergence of the explicit Euler method when stability is violated

Remark 8.9 The above example shows that it is fairly easy to give sufficient conditions of stability in the ∞, h norms for explicit schemes, which is thus a reasonable choice of norms for such schemes. In the case of an implicit scheme, determining stability in the ∞, h norm family can in some cases be obtained by using the discrete maximum principle, in the absence of the explicit knowledge of the inverse matrix involved. Another case when the ∞, h norm of the inverse matrix can be estimated is if there exists $\delta > 0$ such that $|A_{nn}| \geq \delta + \sum_{j \neq n} |A_{nj}|$ for all n. In this case, it is easy to show that $\|A^{-1}\|_{\infty,h} \leq \frac{1}{\delta}$. If $\delta \geq 1$, then the implicit scheme is stable in the ∞, h norms. □

Remark 8.10 We will see later that the $2, h$ norms are more adapted to implicit schemes since it is possible in certain cases to compute the $2, h$ norm of an inverse matrix without computing the inverse in question explicitly. □

Corollary 8.2 *When $\frac{k}{h^2} \leq \frac{1}{2}$, the explicit three-point Euler scheme satisfies the error estimate*

$$\max_{n,j} |u_n^j - u(x_n, t_j)| \leq Ch^2.$$

Proof This is a consequence of Corollary 8.1, Proposition 8.1, and the fact that we have $k \leq \frac{h^2}{2}$. □

The second order accuracy is however obtained at the expense of a lot of iterations in time, see Remark 8.8.

8.4 General Criteria for Stability

We have seen that proving consistency is always a matter of combining several Taylor–Lagrange expansions together, which can be tedious but does not pose much difficulty in principle. Stability is another matter.

Let us consider a general scheme (8.3). We rewrite it as a one time step scheme of the form

$$V^{j+1} = \mathscr{A}_{h,k} V^j + G^j, \tag{8.8}$$

where $V^j \in \mathbb{R}^{(l+m)N}$ is defined as

$$V^j = \begin{pmatrix} U^{j+l-1} \\ \vdots \\ U^j \\ \vdots \\ U^{j-m} \end{pmatrix} \text{ for } m \leq j \leq M + 2 - l,$$

the matrix $\mathscr{A}_{h,k}$ is the $(l+m)N \times (l+m)N$ matrix

$$\mathscr{A}_{h,k} = \begin{pmatrix} -B_l^{-1}B_{l-1} & -B_l^{-1}B_{l-2} & \cdots & \cdots & -B_l^{-1}B_{-m} \\ I & 0 & \cdots & \cdots & 0 \\ 0 & I & 0 & \cdots & 0 \\ \vdots & \ddots & \ddots & \ddots & \vdots \\ 0 & \cdots & 0 & I & 0 \end{pmatrix}$$

and

$$G^j = \begin{pmatrix} B_l^{-1}\widetilde{F}^j \\ 0 \\ \vdots \\ 0 \end{pmatrix}.$$

The matrix $\mathscr{A}_{h,k}$ is called the *amplification matrix* of the scheme. It depends on k and h through its coefficients and it must not be forgotten that its size also depends on $h = 1/(N+1)$. Note that the amplification matrix $\mathscr{A}_{h,k}$ and the term G^j do not depend on the general form (8.3), but only on the scheme itself, in the sense that two general forms for the same scheme will lead to the same iteration (8.8).

In view of the Lax theorem, it is important to prove stability for a general form (8.3) for which we also have consistency. We give a first stability criterion for the family of norms $\|\cdot\|_N$. We associate to this family of norms the family

$$\|V\|_{l+m,N} = \max_{-m\leq k\leq l-1} \|U^k\|_N,$$

with an obvious definition of $V \in \mathbb{R}^{(l+m)N}$ in terms of $U^k \in \mathbb{R}^N$. For each N, this is a norm on $\mathbb{R}^{(l+m)N}$. We denote the induced matrix norm by $|\!|\!|\cdot|\!|\!|_{l+m,N}$.

Proposition 8.3 *Let us consider a general scheme in the form* (8.3) *for which there exists a constant C_0 independent of h and k such that $|\!|\!|B_l^{-1}|\!|\!|_N \leq C_0 k$. Such a scheme is stable under condition $\mathscr{S}$ if and only if there exists a constant $C(T)$ depending only on T such that for all $(h,k) \in \mathscr{S}$,*[7]

$$\max_{j\leq (T/k)+1-(l+m)} |\!|\!|\mathscr{A}_{h,k}^j|\!|\!|_{l+m,N} \leq C(T). \tag{8.9}$$

Proof We remark that, by definition of the $\|\cdot\|_{l+m,N}$ norm, we have

$$\max_{j\leq T/k} \|U^j\|_N = \max_{m\leq j\leq (T/k)+1-l} \|V^j\|_{l+m,N}. \tag{8.10}$$

Let us first assume that the general form of the scheme is stable. This means that there exist two constants $C_1(T)$ and $C_2(T)$ such that for any U_0 and $\widetilde{F}^j$ and all $(h,k) \in \mathscr{S}$

[7] Beware of the notation: up to now V^j meant the jth vector in the sequence, but here $\mathscr{A}^j$ means the jth power of the matrix $\mathscr{A}$.

$$\max_{j\le T/k}\|U^j\|_N \le C_1(T)\max_{0\le j'\le l+m-1}\|U_{j'}\|_N + C_2(T)\max_{m\le j''\le (T/k)-l}\|\widetilde{F}^{j''}\|_N.$$

We take $\widetilde{F}^{j''}=0$ for all j''. In this case, $V^j=\mathscr{A}_{h,k}^{j-m}V^m$ so that we have

$$\max_{m\le j\le (T/k)+1-l}\|\mathscr{A}_{h,k}^{j-m}V^m\|_{l+m,N}\le C_1(T)\|V^m\|_{l+m,N}.$$

Since this is true for all $V^m\in\mathbb{R}^{(l+m)N}$, it follows that

$$\max_{m\le j\le (T/k)+1-l}|\!|\!|\mathscr{A}_{h,k}^{j-m}|\!|\!|_{l+m,N}\le C_1(T).$$

Changing $j-m$ into j, we obtain estimate (8.9).

Conversely, assume that estimate (8.9) holds true for all $(h,k)\in\mathscr{S}$. We can write, for all $m+1\le j\le (T/k)+1-l$

$$\begin{aligned}
V^j &= \mathscr{A}_{h,k}V^{j-1}+G^{j-1}\\
\mathscr{A}_{h,k}V^{j-1} &= \mathscr{A}_{h,k}^2V^{j-2}+\mathscr{A}_{h,k}G^{j-2}\\
\vdots\quad &\quad \vdots\\
\mathscr{A}_{h,k}^{j-m-1}V^{m+1} &= \mathscr{A}_{h,k}^{j-m}V^m+\mathscr{A}_{h,k}^{j-m-1}G^m,
\end{aligned}$$

so that summing these equations, we obtain

$$V^j=\mathscr{A}_{h,k}^{j-m}V^m+\sum_{n=m}^{j-1}\mathscr{A}_{h,k}^{j-n-1}G^n.$$

Therefore

$$\begin{aligned}
\|V^j\|_{l+m,N} &\le \|\mathscr{A}_{h,k}^{j-m}V^m\|_{l+m,N}+\sum_{n=m}^{j-1}\|\mathscr{A}_{h,k}^{j-n-1}G^n\|_{l+m,N}\\
&\le C(T)\|V^m\|_{l+m,N}+C(T)\sum_{n=m}^{j-1}\|G^n\|_{l+m,N}\\
&\le C(T)\|V^m\|_{l+m,N}+(j-m)C(T)\max_{m\le n\le j-1}\|G^n\|_{l+m,N}\\
&\le C(T)\|V^m\|_{l+m,N}+\frac{T}{k}C(T)\max_{n\le (T/k)-l}\|G^n\|_{l+m,N}
\end{aligned}$$

whenever $m+1\le j\le T/k$. In addition, the final estimate holds trivially for $j=m$. Recall now that $\|G^n\|_{l+m,N}=\|B_l^{-1}\widetilde{F}^n\|_N$. Therefore

$$\|G^n\|_{l+m,N}\le |\!|\!|B_l^{-1}|\!|\!|_N\|\widetilde{F}^n\|_N\le C_0k\|\widetilde{F}^n\|_N$$

in view of the hypothesis made on B_l^{-1}. We thus obtain the stability under condition $\mathscr{S}$, with $C_1(T) = C(T)$ and $C_2(T) = TC_0C(T)$, because of Eq. (8.10) and the fact that $\|V^m\|_{l+m,N} = \max_{0\le j'\le l+m-1} \|U_{j'}\|_N$. □

Remark 8.11 The hypothesis $|||B_l^{-1}|||_N \le C_0 k$ is essential for the above estimate even though it seems to be often overlooked in the literature, in which the definition of stability is also often different and weaker than the one we present here. Of course, as was said before, the usefulness of stability is only for general forms for which consistency holds. In particular, this hypothesis is clearly satisfied for the forward Euler scheme in the form (8.4) for any family of norms since $|||I|||_N = 1$. □

Remark 8.12 Under the above hypothesis on B_l^{-1}, it follows that stability does not depend on the right-hand side of the equation. In other words, it is enough to test it for a zero right-hand side. More precisely, stability in the sense of Definition 8.5 is implied by the less demanding estimate

$$\max_{j\le T/k} \|U^j\|_N \le C_1(T) \max_{0\le j'\le l+m-1} \|U_{j'}\|_N,$$

for all $(h, k) \in \mathscr{S}$, where U^j is any solution of the scheme with zero right-hand side. This is called the Dahlquist zero-stability condition, see [49] for example. □

The criterion given in Proposition 8.3 is not too practical in general, since the quantity $\max_j |||\mathscr{A}_{h,k}^j|||_{l+m,N}$ is not necessarily easy to estimate. Nonetheless, we have a sufficient condition as an immediate corollary.

Corollary 8.3 *If* $|||\mathscr{A}_{h,k}|||_{l+m,N} \le 1$ *for all* $(h, k) \in \mathscr{S}$, *then the scheme is stable under condition* $\mathscr{S}$.

Proof An operator norm is submultiplicative, i.e., $|||\mathscr{A}\mathscr{B}|||_{l+m,N} \le |||\mathscr{A}|||_{l+m,N} |||\mathscr{B}|||_{l+m,N}$ for any $\mathscr{A}$ and $\mathscr{B}$. Therefore, $|||\mathscr{A}_{h,k}^j|||_{l+m,N} \le |||\mathscr{A}_{h,k}|||_{l+m,N}^j \le 1$ for all j, thus in particular for all j smaller than $(T/k) + 1 - (l + m)$. □

This is what we actually did for the forward Euler scheme and the ∞, h norms in the proof of Proposition 8.2. In the case of the $2, h$ norms, we will see in the next section that it is possible to be a little more precise and give a *necessary and sufficient* condition of stability for a certain class of amplification matrices. This is one of the main reasons for using these norms, see Sect. 8.5.

Let us now show the converse of the Lax theorem.

Proposition 8.4 *Let us be given a scheme that is convergent under condition* $\mathscr{S}$ *for the family of norms* $\|\cdot\|_N$. *Then any general form* (8.3) *such that* $|||B_l^{-1}|||_N \le C_0 k$ *is stable under condition* $\mathscr{S}$ *in the sense of Definition* 8.5.

Proof We first rewrite the scheme in the form (8.8). We argue by contradiction. We assume that there is no constant $C(T)$ such that inequality (8.9) holds in a neighborhood of $(0, 0)$ in $\mathscr{S}$.[8] We are thus given a sequence $(h_n, k_n) \in \mathscr{S}$ such that

[8]This is the only relevant case.

$(h_n, k_n) \to (0,0)$ (so that $N_n \to +\infty$). Let $\mathscr{A}_{h_n,k_n} = \mathscr{A}_n$ be the corresponding sequence of amplification matrices. By hypothesis, there exists a subsequence n' and a sequence $j_{n'}$ such that $j_{n'} \leq (T/k_{n'}) + 1 - (l+m)$ with

$$|||\mathscr{A}_{n'}^{j_{n'}}|||_{l+m,N_{n'}} = \lambda_{n'} \to +\infty \text{ when } n' \to +\infty.$$

By compactness in finite dimensional spaces, there exists $V_{n'} \in \mathbb{R}^{(l+m)N_{n'}}$ such that

$$\|V_{n'}\|_{l+m,N_{n'}} = 1 \text{ and } \|\mathscr{A}_{n'}^{j_{n'}} V_{n'}\|_{l+m,N_{n'}} = \lambda_{n'}.$$

Let us set $W_{n'} = V_{n'}/\lambda_{n'}$. It follows that

$$\|W_{n'}\|_{l+m,N_{n'}} \to 0 \text{ and } \|\mathscr{A}_{n'}^{j_{n'}} W_{n'}\|_{l+m,N_{n'}} = 1.$$

The first convergence shows that the vectors $W_{n'}$ are appropriate discrete initial conditions for the scheme applied to the case $u_0 = 0$ and $f = 0$. The scheme being convergent in this family of norms, it follows that $\|\mathscr{A}_{n'}^{j_{n'}} W_{n'}\|_{l+m,N_{n'}} \to 0$, which contradicts the second relation above.

Finally, the hypothesis $|||B_l^{-1}|||_N \leq C_0 k$ combined with estimate (8.9) implies the stability of the general form by Proposition 8.3. □

From now on, we will concentrate on one time step schemes in the case of the $2,h$ norms.

8.5 Stability for One Time Step Schemes in the 2, *h* Norms

Let us consider a general scheme (8.3) with one time step, i.e., $l = 1, m = 0$ or $l = 0$, $m = 1$. As before, we rewrite the scheme as

$$U^{j+1} = \mathscr{A} U^j + G^j,$$

where

$$\mathscr{A} = \begin{cases} -B_1^{-1} B_0 & \text{if } l = 1, m = 0, \\ -B_0^{-1} B_{-1} & \text{if } l = 0, m = 1, \end{cases}$$

and

$$G^j = \begin{cases} B_1^{-1} \widetilde{F}^j & \text{if } l = 1, m = 0, \\ B_0^{-1} \widetilde{F}^{j+1} & \text{if } l = 0, m = 1. \end{cases}$$

The amplification matrix $\mathscr{A}$ is now a $N \times N$ matrix. We henceforth omit the h, k subscript in amplification matrices for notational brevity.

We first need to determine the $2, h$ induced matrix norms. We let $\rho(B)$ denote the *spectral radius* of a matrix B, i.e., $\rho(B) = \max\{|\lambda_p|, p = 1, \ldots, N\}$, where $\lambda_p \in \mathbb{C}$ are the eigenvalues of B.

Proposition 8.5 *Let A be a real $N \times N$ matrix. We have*

$$|||A|||_{2,h} = \sqrt{\rho(A^T A)}.$$

Proof For all $X \in \mathbb{R}^N$, we have $\|AX\|_{2,h}^2 = h(AX)^T AX = hX^T(A^T A)X$. The matrix $A^T A$ is symmetric, hence it is orthogonally diagonalizable: there exists an orthogonal matrix Q, i.e., a real matrix such that $Q^T Q = QQ^T = I$, such that $A^T A = Q^T DQ$ where D is a diagonal matrix, the diagonal entries of which are the eigenvalues d_p of $A^T A$. This matrix is also nonnegative, so that these eigenvalues are all nonnegative. Therefore, $\rho(A^T A)$ is simply the largest eigenvalue of $A^T A$.

We can thus write

$$\|AX\|_{2,h}^2 = hX^T(Q^T DQ)X = h(QX)^T D(QX).$$

If we set $Y = QX$, then $\|Y\|_{2,h} = \|X\|_{2,h}$ since Q is orthogonal and

$$\|AX\|_{2,h}^2 = hY^T DY = h\sum_{p=1}^N d_p Y_p^2$$
$$\leq h\rho(A^T A)\sum_{p=1}^N Y_p^2 = \rho(A^T A)\|Y\|_{2,h}^2 = \rho(A^T A)\|X\|_{2,h}^2.$$

Taking the square root and dividing by $\|X\|_{2,h}$ for $X \neq 0$, we thus obtain

$$|||A|||_{2,h} \leq \sqrt{\rho(A^T A)}.$$

Let now X be a unit eigenvector of $A^T A$ associated with the eigenvalue $\rho(A^T A)$, $A^T AX = \rho(A^T A)X$ and $\|X\|_{2,h}^2 = hX^T X = 1$. For this vector, we have

$$\|AX\|_{2,h}^2 = hX^T(A^T AX) = h\rho(A^T A)X^T X = \rho(A^T A),$$

from which we infer that

$$|||A|||_{2,h} \geq \sqrt{\rho(A^T A)},$$

and the Proposition is proved. □

A real matrix A is said to be *normal* if $A^T A = AA^T$. In particular, a real symmetric matrix is normal.

Proposition 8.6 *If A is a normal matrix, then $\rho(A) = \rho(A^T A)^{1/2} = |||A|||_{2,h}$.*

Proof A normal matrix A is unitarily similar to a diagonal matrix Λ. Thus there exists a unitary matrix U, i.e., a complex matrix satisfying $U^*U = UU^* = I$,[9] such that $A = U^*\Lambda U$ and therefore the diagonal entries of Λ are the eigenvalues $\lambda_p \in \mathbb{C}$ of A. It follows that

$$A^T A = U^*\Lambda^*(UU^*)\Lambda U = U^*\Lambda^*\Lambda U,$$

so that the eigenvalues of $A^T A$ are the diagonal entries of $\Lambda^*\Lambda$, namely $|\lambda_p|^2$, $p = 1, \dots, N$. Thus $\rho(A)^2 = \big(\max|\lambda_p|\big)^2 = \rho(A^T A)$. □

Of course, for a general non normal matrix A, we only have $\rho(A) \le |||A|||_{2,h}$, with often a strict inequality (take for instance a nonzero nilpotent matrix A for which $\rho(A) = 0$). Let us apply the above considerations to finite difference schemes. In the sequel, when we say that a scheme is stable, we refer to a scheme written in a general form satisfying the hypotheses of Proposition 8.3.

Proposition 8.7 *If the amplification matrix $\mathscr{A}$ is normal, then the scheme is stable for the norms $\|\cdot\|_{2,h}$ if and only of there exists a constant $C'(T) \ge 0$ depending only on T such that*

$$\rho(\mathscr{A}) \le 1 + C'(T)k. \tag{8.11}$$

Proof Let us first assume that there exists $C'(T) \ge 0$ such that $\rho(\mathscr{A}) \le 1 + C'(T)k$. By hypothesis, $\mathscr{A}$ is normal, therefore $\mathscr{A}^j$ is also normal and $|||\mathscr{A}^j|||_{2,h} = \rho(\mathscr{A}^j) = \rho(\mathscr{A})^j$. Consequently, for all $j \le T/k$,

$$|||\mathscr{A}^j|||_{2,h} \le (1 + C'(T)k)^j \le e^{C'(T)kj} \le e^{C'(T)T},$$

and the constant $e^{C'(T)T}$ depends only on T. Therefore, the scheme is stable according to Proposition 8.3.

Conversely, assume that the scheme is stable. By Proposition 8.3 again, this implies that $\rho(\mathscr{A})^j \le C(T)$ or $\rho(\mathscr{A}) \le C(T)^{1/j}$ for all $j \le T/k$. There are two cases. Either $C(T) \le 1$ and thus $\rho(\mathscr{A}) \le 1$ and we are done with $C'(T) = 0$, or $C(T) > 1$. In this case, we take $j = T/k$ so that

$$\rho(\mathscr{A}) \le C(T)^{\frac{k}{T}} = e^{\frac{k}{T}\ln(C(T))},$$

with $\ln(C(T)) > 0$. This implies that the function $s \mapsto e^{s\ln(C(T))}$ is convex on $[0, 1]$, which in turn implies that for all $s \in [0, 1]$, $e^{s\ln(C(T))} \le (1 - s) + se^{\ln(C(T))} = 1 + s(C(T) - 1)$. In particular, for $s = k/T$, we obtain

[9] For any matrix A, A^* denotes the adjoint matrix of A, i.e., its conjugate transpose.

$$\rho(\mathscr{A}) \leq 1 + \frac{C(T) - 1}{T} k,$$

hence estimate (8.11) with $C'(T) = \frac{C(T)-1}{T}$. □

Remark 8.13 Inspection of the above proof shows that the fact that the matrix is normal is not used in the converse part of Proposition 8.7. Therefore, condition (8.11) is a necessary condition of stability for any matrix $\mathscr{A}$. □

Remark 8.14 It is important to stress again that the matrix $\mathscr{A}$ is a function of k and h, and so is its spectral radius. Therefore, the above estimates are by no means obvious.

Note that a sufficient condition for stability in the $2, h$ norm in the case of a normal amplification matrix, often used in practice, is thus that $\rho(\mathscr{A}) \leq 1$. This is particularly indicated if we are interested in the computation of long term behavior of the solution, i.e., T large. Indeed, the less demanding condition (8.11) allows for exponential growth with T.

We now see that the reason for Remark 8.10 is that the eigenvalues of an inverse matrix are just the inverses of the eigenvalues of that matrix. □

Let us apply all of the above to both Euler schemes and to the leapfrog scheme. We first need to determine the eigenvalues of the kind of tridiagonal matrices involved in these schemes.

Lemma 8.1 *Consider the $N \times N$ matrix*

$$A = \begin{pmatrix} a & b & 0 & \cdots & 0 \\ b & a & b & \cdots & 0 \\ \vdots & \ddots & \ddots & \ddots & \vdots \\ 0 & \cdots & b & a & b \\ 0 & \cdots & 0 & b & a \end{pmatrix},$$

with $a, b \in \mathbb{R}$. The eigenvalues of A are given by

$$\lambda_p = a + 2b \cos\Bigl(\frac{p\pi}{N+1}\Bigr), p = 1, \ldots, N.$$

Proof We have $A = aI + bB$ with $B = \begin{pmatrix} 0 & 1 & \cdots & 0 \\ 1 & \ddots & \ddots & \vdots \\ \vdots & \ddots & \ddots & 1 \\ 0 & \cdots & 1 & 0 \end{pmatrix}$. Of course, $AV = \lambda V$ is equivalent to $BV = \frac{\lambda - a}{b} V$. It is thus sufficient to find the eigenvalues of B.

Let $V \in \mathbb{R}^N \setminus \{0\}$ be an eigenvector of B associated with the eigenvalue λ. We thus have

$$
\begin{aligned}
V_2 &= \lambda V_1, \\
V_1 + V_3 &= \lambda V_2, \\
&\vdots \\
V_{j-1} + V_{j+1} &= \lambda V_j, \\
&\vdots \\
V_{N-1} &= \lambda V_N.
\end{aligned}
$$

If we set $V_0 = V_{N+1} = 0$, we see that the components of V are a solution of the linear homogeneous recurrence relation with constant coefficients

$$V_{j-1} - \lambda V_j + V_{j+1} = 0, \ \text{for}\, j = 0, \dots, N.$$

It is well known that the general solution of such a recurrence relation depends on the roots of its characteristic equation $r^2 - \lambda r + 1 = 0$. There are two roots r_1 and r_2 which are such that $r_1 + r_2 = \lambda$ and $r_1 r_2 = 1$. If the roots are simple, we have $V_j = \alpha r_1^j + \beta r_2^j$, and if there is a double root, $V_j = (\alpha + \beta j)r^j$, where α and β are constants to be determined.

Now the question is: given λ, can we find α and β such that $V_0 = V_{N+1} = 0$ with $V \neq 0$ (this is a kind of boundary value problem in a sense)? In the case of a double root, clearly these conditions imply $\alpha = \beta = 0$, hence such a λ cannot be an eigenvalue. Let us thus consider the case of simple roots. We are thus requiring that

$$0 = V_0 = \alpha + \beta \text{ and } 0 = V_{N+1} = \alpha r_1^{N+1} + \beta r_2^{N+1}.$$

Therefore, we see that λ is an eigenvalue if and only if $r_1^{N+1} = r_2^{N+1}$. Multiplying this equation by r_1^{N+1}, it follows that $r_1^{2(N+1)} = 1$. Consequently, r_1 and r_2 must be of the form

$$r_1 = e^{i\frac{p\pi}{N+1}}, \quad r_2 = e^{-i\frac{p\pi}{N+1}}$$

for some integer p. It follows that the eigenvalues are necessarily of the form

$$\lambda_p = e^{i\frac{p\pi}{N+1}} + e^{-i\frac{p\pi}{N+1}} = 2\cos\Big(\frac{p\pi}{N+1}\Big).$$

Conversely, it is easy to check that the vector V_p given by $(V_p)_j = \sin\big(\frac{jp\pi}{N+1}\big)$, $j = 1, \dots, N$, is an associated eigenvector. For $p = 1, \dots, N$, we have thus obtained N distinct eigenvalues, hence we have found them all. □

Let us remark that the vectors V_p are eigenvectors for all matrices A independently of the values of a and b.

Corollary 8.4 *The eigenvalues of A_h are*

$$\lambda_p = \frac{4}{h^2}\sin^2\Big(\frac{p\pi}{2(N+1)}\Big), p = 1, \ldots, N.$$

Proof Apply the previous Lemma with $a = \frac{2}{h^2}$ and $b = -\frac{1}{h^2}$. □

We now return to the explicit Euler scheme (8.4).

Proposition 8.8 *Let $\mathscr{S} \subset \mathbb{R}_+^* \times \mathbb{R}_+^*$. The explicit three-point Euler scheme is stable for the norms $\|\cdot\|_{2,h}$ under condition $\mathscr{S}$ if and only if*

$$\mathscr{S} \subset \Big\{(h,k); \frac{k}{h^2}\cos^2\Big(\frac{\pi h}{2}\Big) \leq \frac{1}{2} + Ck\Big\}, \tag{8.12}$$

for some $C \geq 0$. In this case, it is convergent for these norms and we have the error estimate

$$\max_{j \leq T/k} \|U^j - S_h(u_{t_j})\|_{2,h} \leq Ch^2,$$

where C depends only on u and T.

Proof First of all, since $B_1^{-1} = kI$ so that $|||B_1^{-1}|||_{2,h} = k$, we can apply Proposition 8.3. We have $\mathscr{A} = I - kA_h$. It is a real symmetric matrix, hence a normal matrix. We may thus apply Proposition 8.7. We have

$$\begin{aligned}\rho(\mathscr{A}) &= \max_{1 \leq p \leq N}\Big|1 - \frac{4k}{h^2}\sin^2\Big(\frac{p\pi}{2(N+1)}\Big)\Big| \\ &= \max\Big(1 - \frac{4k}{h^2}\sin^2\Big(\frac{\pi}{2(N+1)}\Big), \frac{4k}{h^2}\sin^2\Big(\frac{N\pi}{2(N+1)}\Big) - 1\Big).\end{aligned}$$

The first expression in the maximum is always between 0 and 1. We thus just need to consider the second expression. Condition (8.12) is then a simple rewriting of condition (8.11). □

Remark 8.15 Note that if $\frac{k}{h^2} \leq \frac{1}{2}$, then $\frac{k}{h^2}\cos^2\big(\frac{\pi h}{2}\big) \leq \frac{1}{2}$ for all h and k. The region $\big\{(h,k); \frac{k}{h^2} \leq \frac{1}{2}\big\}$ is thus a stability region, as was the case for the ∞, h norm. Besides, convergence in the ∞, h norm implies convergence in the $2, h$ norm, so that nothing new is gained in this case.

We know a little more about instability in the $2, h$ norm, which on the other hand does not directly imply instability in the ∞, h norm. However, using the converse part of the Lax theorem, if the scheme is unstable in the $2, h$ norm, it is not convergent in the $2, h$ norm. Therefore, it is not convergent in the ∞, h norm. Since it is consistent in the ∞, h norms, it is thus unstable for that same norm.

In this respect, it is in any case better to choose h and k such that $\frac{k}{h^2} \leq \frac{1}{2}$, since in this case, we are assured that $|||\mathscr{A}|||_{2,h} = \rho(\mathscr{A}) < 1$, which is numerically a good thing. In particular, a spectral radius which is strictly larger than 1 manifests itself

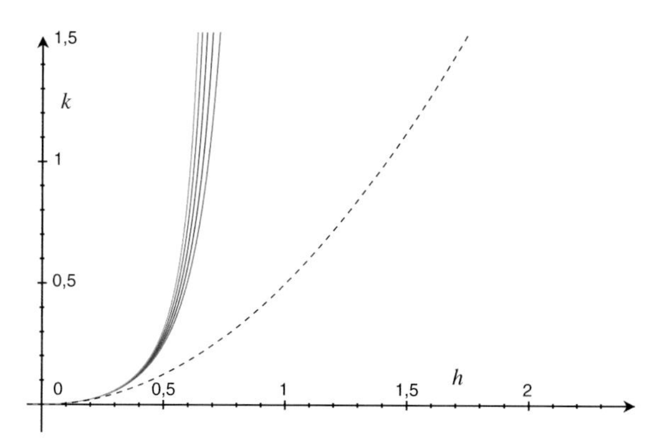

Fig. 8.3 The boundaries of a few stability regions defined in the $(h; k)$ plane by formula (8.12) for different values of C ($C = 0, 0.1, 0.2, 0.3, 0.4$) and of the "safe" region $k \leq h^2/2$ *dashed*. They are all tangent at $(0, 0)$

as a numerical instability, with oscillations that become very large exponentially fast with the number of iterations as we will see in a few examples later, even though this particular choice of values for h and k belongs to a stability region in the previous sense. □

We see from Fig. 8.3 that there is no practical difference between the different stability regions where it counts, that is to say in a neighborhood of $(0, 0)$. Let us now turn to the implicit Euler scheme (8.5).

Proposition 8.9 *The implicit Euler three-point scheme is unconditionally stable for the norms* $\|\cdot\|_{2,h}$. *It is convergent for these norms and we have the error estimate*

$$\max_{j\leq T/k} \|U^j - S_h(u_{t_j})\|_{2,h} \leq C(h^2 + k),$$

where C *depends only on* u *and* T.

Proof We have $\mathscr{A} = (I + kA_h)^{-1}$, which is real symmetric, hence normal. Its eigenvalues are

$$\lambda_p = \frac{1}{1 + \frac{4k}{h^2}\sin^2\left(\frac{p\pi}{2(N+1)}\right)}, p = 1, \ldots, N,$$

and are all between 0 and 1. Hence $\rho(\mathscr{A}) < 1$ for all h and k, and since $B_0^{-1} = k\mathscr{A}$, the scheme is unconditionally stable. □

Remark 8.16 We see here the great advantage of the implicit Euler scheme over the explicit Euler scheme. The number of time iterations needed to reach a given time T is not constrained by the space step.

In this example, we also see why the 2, h norms are easier to work with than the ∞, h norms, for such an implicit scheme.

We plot in Fig. 8.4 the solution of the implicit Euler scheme corresponding to $u_0(x) = 4x(1 - x)$ at $T = 1.1$ and $N = M$ ranging from 120 to 240 by increments of 12. Only the discrete values are drawn, no linear interpolation.

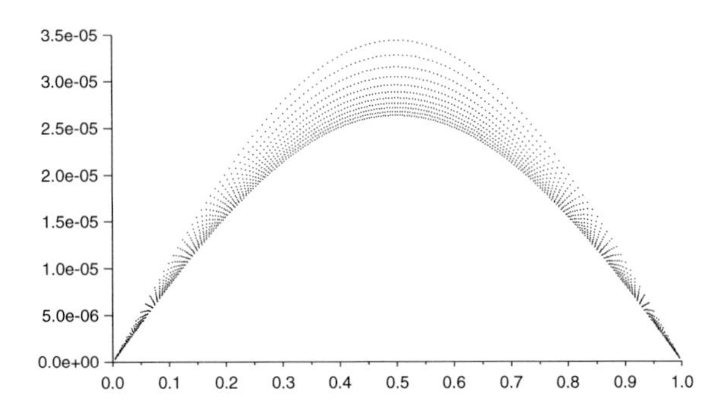

Fig. 8.4 Convergence of the implicit Euler scheme

In the above computations, $\frac{k}{h^2}$ range from approximately 100 to 200, with no ill effect of course. □

We finally deal with the leapfrog scheme (8.6). The leapfrog scheme is a two time step scheme with the general form

$$B_1 u^{j+1} + B_0 U^j + B_{-1} U^{j-1} = \widetilde{F}^j$$

with

$$B_1 = \frac{1}{2k} I, B_0 = A_h, B_{-1} = -\frac{1}{2k} I \text{ and } \widetilde{F}^j = F^j.$$

As in the general case, we rewrite it as a single time step scheme by setting

$$V^j = \begin{pmatrix} U^j \\ U^{j-1} \end{pmatrix} \in \mathbb{R}^{2N},$$

so that

$$V^{j+1} = \mathscr{A} V^j + G^j,$$

where

$$\mathscr{A} = \begin{pmatrix} -2kA_h & I \\ I & 0 \end{pmatrix} \tag{8.13}$$

is the $2N \times 2N$ symmetric amplification matrix and $G^j \in \mathbb{R}^{2N}$. Since $|||B_1^{-1}|||_{2,h} = 2k$, Proposition 8.3 applies. However, instead of using the norm used in this proposition, we use the matrix norm induced by the canonical Euclidean norm on $\mathbb{R}^{2N}$, which is equivalent. Indeed,

$$\frac{1}{\sqrt{2}}(\|U_1\|_{2,h}^2 + \|U_2\|_{2,h}^2)^{1/2} \leq \max(\|U_1\|_{2,h}, \|U_2\|_{2,h}) \leq (\|U_1\|_{2,h}^2 + \|U_2\|_{2,h}^2)^{1/2}$$

for all $U_1, U_2 \in \mathbb{R}^N$. We thus just need to find the spectral radius of the matrix $\mathscr{A}$.

Lemma 8.2 *Let C be a $N \times N$ complex matrix and B the $2N \times 2N$ complex matrix defined by blocks as*

$$B = \begin{pmatrix} C & I \\ I & 0 \end{pmatrix},$$

where I (resp. 0) is the $N \times N$ identity (resp. zero) matrix. If $\lambda \in \mathbb{C}$ is an eigenvalue of B, then $\lambda \neq 0$ and $\lambda - \frac{1}{\lambda}$ is an eigenvalue of C. Conversely, if $\mu \in \mathbb{C}$ is an eigenvalue of C, then there exists an eigenvalue λ of B such that $\mu = \lambda - \frac{1}{\lambda}$.

Proof Let $\lambda \in \mathbb{C}$ be such that there exists a vector Y in $\mathbb{C}^{2N}$, $Y = \begin{pmatrix} Y_1 \\ Y_2 \end{pmatrix}$, $Y \neq 0$, such that $BY = \lambda Y$. Using the block structure of B, we see that this is equivalent to

$$\begin{cases} CY_1 + Y_2 = \lambda Y_1, \\ Y_1 = \lambda Y_2. \end{cases}$$

If $Y_1 = 0$, the first equation implies that $Y_2 = 0$, which is impossible. Thus $Y_1 \neq 0$, which implies $\lambda \neq 0$ by the second equation. We may thus divide by λ so that $Y_2 = \frac{1}{\lambda} Y_1$ and replacing in the first equation $CY_1 = \left(\lambda - \frac{1}{\lambda}\right) Y_1$. Since we have already seen that $Y_1 \neq 0$, this implies that $\lambda - \frac{1}{\lambda}$ is an eigenvalue of C.

Conversely, let $\mu \in \mathbb{C}$ be an eigenvalue of C with eigenvector $Y_1 \in \mathbb{C}^N$, $Y_1 \neq 0$. The polynomial $X^2 - \mu X - 1$ has two roots in $\mathbb{C}$, which are nonzero since their product is -1. Let λ be one of these roots. Dividing by λ, we see that $\lambda - \mu - \frac{1}{\lambda} = 0$, hence $\mu = \lambda - \frac{1}{\lambda}$. Furthermore

$$B \begin{pmatrix} Y_1 \\ \frac{1}{\lambda} Y_1 \end{pmatrix} = \begin{pmatrix} CY_1 + \frac{1}{\lambda} Y_1 \\ Y_1 \end{pmatrix} = \begin{pmatrix} (\mu + \frac{1}{\lambda}) Y_1 \\ Y_1 \end{pmatrix} = \lambda \begin{pmatrix} Y_1 \\ \frac{1}{\lambda} Y_1 \end{pmatrix},$$

so that λ is an eigenvalue of B. □

Remark 8.17 If λ is an eigenvalue of B, then $-\frac{1}{\lambda}$ is also an eigenvalue of B. This pair corresponds to the same eigenvalue μ of C. □

Let us now apply this to the leapfrog scheme.

Proposition 8.10 *The leapfrog scheme is unstable for the norms $\|\cdot\|_{2,h}$, hence not convergent for these norms.*

Proof The matrix $\mathscr{A}$ defined by (8.13) is real symmetric, hence normal. We may thus apply Proposition 8.7 with the equivalence of matrix norms noted before.

The eigenvalues of the matrix $-2kA_h$ are

$$\mu_p = -\frac{8k}{h^2} \sin^2\left(\frac{p\pi}{2(N+1)}\right), p = 1, \ldots, N,$$

and those of the matrix $\mathscr{A}$

$$\lambda_p^{\pm} = \frac{\mu_p \pm \sqrt{\mu_p^2 + 4}}{2}$$

according to Lemma 8.2. In particular, for $p = N$, we have

$$\sin^2\Bigl(\frac{N\pi}{2(N+1)}\Bigr) = \cos^2\Bigl(\frac{\pi}{2(N+1)}\Bigr) \geq \frac{1}{2}$$

since $\frac{\pi}{2(N+1)} \leq \frac{\pi}{4}$. Therefore

$$-\mu_N \geq \frac{4k}{h^2}.$$

It follows that

$$\rho(\mathscr{A}) \geq \left|\frac{\mu_N - \sqrt{\mu_N^2 + 4}}{2}\right| = \frac{\sqrt{\mu_N^2 + 4} - \mu_N}{2} \geq \frac{2 + \frac{4k}{h^2}}{2} = 1 + \frac{2k}{h^2}.$$

Consequently, there is no constant $C \geq 0$ such that $\rho(\mathscr{A}) \leq 1 + Ck$. Indeed, assume there was such a constant, for (h, k) in some region $\mathscr{S}$. Then we would have $\frac{2}{h^2} \leq C$, which precludes $h \to 0$. This is inconsistent with $(0, 0) \in \bar{\mathscr{S}}$. □

Remark 8.18 The leapfrog scheme is thus not usable in practice for solving the heat equation. Numerical experiments show that it may diverge very rapidly. We remark in addition that

$$\rho(\mathscr{A}^{M+1}) \geq \Bigl(1 + \frac{2k}{h^2}\Bigr)^{M+1} \geq 1 + \frac{2(M+1)k}{h^2} = 1 + \frac{2T}{h^2} \to +\infty \text{ when } h \to 0.$$

The same is true for any number of iterations needed to reach a fixed time $T > 0$. □

Figure 8.5 shows a sequence of plots corresponding to the leapfrog method applied to $u_0(x) = 4x(1 - x)$ at $T = 1.1$ for $N = M$ ranging from 3 to 18. Notice the vertical scale on the plots. The maximum of the exact solution u at time T is of the order of 2×10^{-5}.

8.6 Stability via the Discrete Fourier Transform

From now on, and for the rest of this chapter, we concentrate on stability criteria without reference to consistency and convergence, but nonetheless keeping the Lax theorem in mind. Let us thus present a slightly, although not fundamentally different way of dealing with stability using the discrete Fourier transform. We consider here periodic boundary conditions instead of Dirichlet conditions. More precisely, we are interested in the restriction to $[0, 1] \times \mathbb{R}_+^*$ of functions that are 1-periodic in

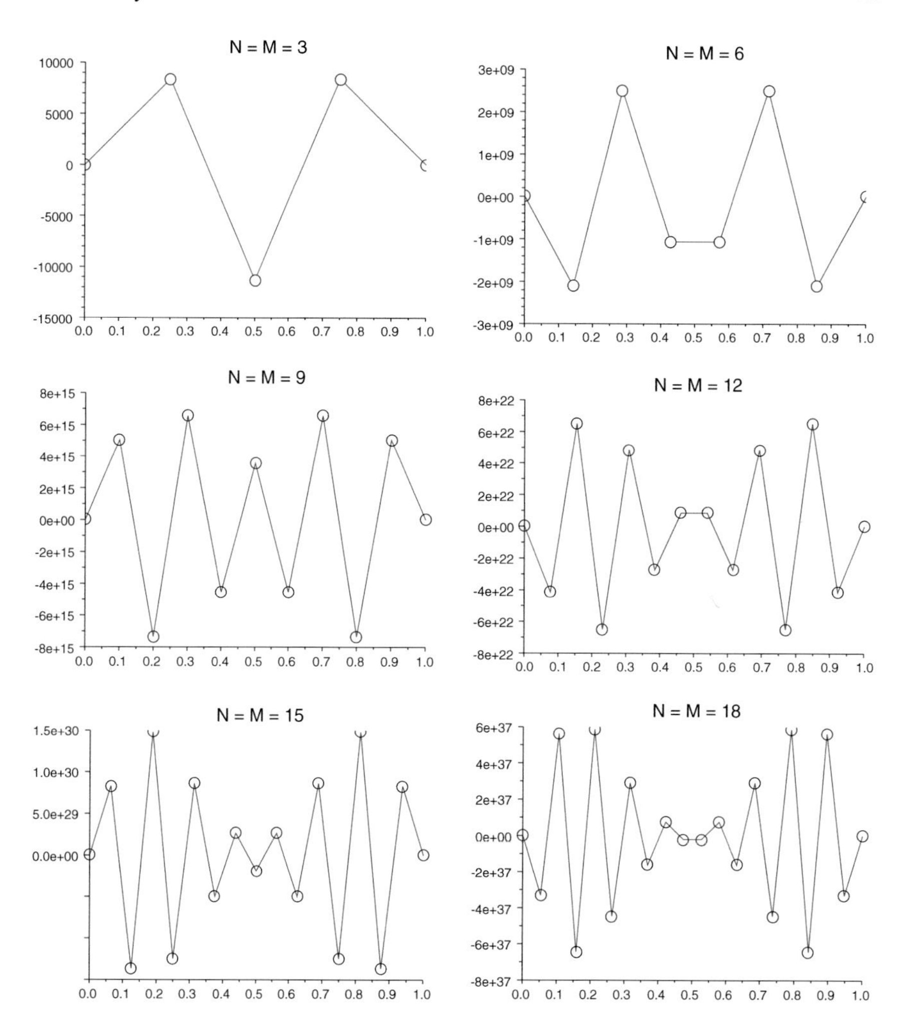

Fig. 8.5 Divergence of the leapfrog method

space and that satisfy the heat equation on $\mathbb{R}_+^* \times \mathbb{R}$ with zero right-hand side. Such functions are regular and thus obey the following system

$$\begin{cases} \dfrac{\partial u}{\partial t}(x,t) - \dfrac{\partial^2 u}{\partial x^2}(x,t) = 0 \text{ in } Q, \\ u(x,0) = u_0(x) \text{ in } \Omega, \\ u(0,t) = u(1,t) \text{ in }]0,T[, \\ \frac{\partial u}{\partial x}(0,t) = \frac{\partial u}{\partial x}(1,t) \text{ in }]0,T[. \end{cases} \tag{8.14}$$

It is easy to check that if u_0 satisfies the same boundary conditions, it is given by its Fourier series expansion $u_0(x) = \frac{a_0}{2} + \sum_{l=1}^{+\infty} a_l \cos(2l\pi x) + \sum_{l=1}^{+\infty} b_l \sin(2l\pi x)$. In this case, the unique solution to (8.14) is given by

$$u(x,t) = \frac{a_0}{2} + \sum_{l=1}^{+\infty} e^{-4l^2\pi^2 t} a_l \cos(2l\pi x) + \sum_{l=1}^{+\infty} e^{-4l^2\pi^2 t} b_l \sin(2l\pi x),$$

using similar arguments as in Sect. 7.3. It is also possible to write a weak formulation for this periodic problem.

Up to now, the vectors U^j were always computed in the canonical basis of $\mathbb{R}^N$. It turns out that there is another basis that is better adapted to the study of finite difference schemes, provided that we are willing to replace $\mathbb{R}$ by $\mathbb{C}$. In this basis, the Fourier basis, computations are straightforward.

We use the same space-time grid as before, but due to the change of boundary conditions, the forward Euler scheme involves unknowns u_n^j, $-1 \le n \le N+1$, $0 \le j \le M$, with

$$\begin{cases} \dfrac{u_n^{j+1} - u_n^j}{k} - \dfrac{u_{n+1}^j - 2u_n^j + u_{n-1}^j}{h^2} = 0 \text{ for } n = 0,\dots,N, j = 0,\dots,M, \\ u_n^0 = u_0(x_n) \text{ for } n = 0,\dots,N+1, \\ u_0^j = u_{N+1}^j \text{ for } j = 0,\dots,M, \\ u_{-1}^j = u_N^j \text{ for } j = 0,\dots,M. \end{cases} \tag{8.15}$$

The relation $u_0^j = u_{N+1}^j$ is a discretization of the periodicity condition $u(0,t) = u(1,t)$, and the relation $u_{-1}^j = u_N^j$ combined with the previous one discretizes the other periodicity condition $\frac{\partial u}{\partial x}(0,t) = \frac{\partial u}{\partial x}(1,t)$. Indeed, $\frac{\partial u}{\partial x}(0,t) \approx \frac{u_0^j - u_{-1}^j}{h}$ and $\frac{\partial u}{\partial x}(1,t) \approx \frac{u_{N+1}^j - u_N^j}{h}$.

We assume of course that u_0 is periodic. The corresponding vectors U^j now live in $\mathbb{R}^{N+1}$ with components $(u_0^j, u_1^j, \dots, u_N^j)$ in the canonical basis. We embed $\mathbb{R}^{N+1}$ into $\mathbb{C}^{N+1}$ in the canonical fashion. We let $(U|V) = \sum_{m=0}^{N} u_m \bar{v}_m$ denote the canonical scalar product on $\mathbb{C}^{N+1}$ and introduce the (discrete) Fourier basis of $\mathbb{C}^{N+1}$ as $(\omega_n)_{n=0,\dots,N}$ where

$$(\omega_n)_m = e^{2i\pi \frac{nm}{N+1}}, \text{ for } m = 0,\dots,N.$$

Proposition 8.11 *The family* $(\omega_n)_{n=0,\dots,N}$ *is an orthogonal basis of* $\mathbb{C}^{N+1}$. *In fact, we have*

$$(\omega_l|\omega_n) = (N+1)\delta_{ln}. \tag{8.16}$$

Proof It is enough to compute the scalar products (8.16). Indeed

$$(\omega_l|\omega_n) = \sum_{m=0}^{N} e^{2i\pi \frac{lm}{N+1}} e^{-2i\pi \frac{nm}{N+1}} = \sum_{m=0}^{N} e^{2i\pi \frac{(l-n)m}{N+1}} = \begin{cases} N+1 & \text{if } l = n, \\ \frac{1-e^{2i\pi(l-n)}}{1-e^{2i\pi \frac{l-n}{N+1}}} = 0 & \text{if } l \neq n. \end{cases}$$

We therefore have an orthogonal, hence linearly independent family of vectors, the cardinal of which is equal to the dimension of the $\mathbb{C}$-vector space $\mathbb{C}^{N+1}$. It is thus a basis. □

We now are at liberty to decompose U^j on this basis

$$U^j = \sum_{n=0}^{N} c_n^j \omega_n \text{ with } c_n^j = \frac{1}{N+1}(U^j|\omega_n) = \frac{1}{N+1} \sum_{m=0}^{N} u_m^j e^{2i\pi \frac{nm}{N+1}},$$

and read the scheme on the discrete Fourier coefficients c_n^j.

Proposition 8.12 *We have*

$$c_n^{j+1} = \Big(1 - \frac{4k}{h^2} \sin^2\Big(\frac{\pi n}{N+1}\Big)\Big) c_n^j,$$

and therefore

$$c_n^j = \Big(1 - \frac{4k}{h^2} \sin^2\Big(\frac{\pi n}{N+1}\Big)\Big)^j c_n^0,$$

where c_n^0 are the discrete Fourier coefficients of the initial condition.

Proof The key observation here is that the space finite difference operator acts by multiplication on the Fourier basis vectors. In other words, these vectors are eigenvectors of the finite difference operator. Indeed, for $m = 0, \dots, N$,

$$\begin{aligned} \frac{-(\omega_l)_{m-1} + 2(\omega_l)_m - (\omega_l)_{m+1}}{h^2} &= \frac{-e^{-2i\pi \frac{l}{N+1}} + 2 - e^{2i\pi \frac{l}{N+1}}}{h^2} e^{2i\pi \frac{lm}{N+1}} \\ &= \frac{2}{h^2}\Big(1 - \cos\Big(\frac{2\pi l}{N+1}\Big)\Big)(\omega_l)_m, \end{aligned}$$

with $(\omega_l)_{-1} = (\omega_l)_N$ and $(\omega_l)_{N+1} = (\omega_l)_0$ by convention, and the corresponding eigenvalue is

$$\lambda_l = \frac{4}{h^2} \sin^2\Big(\frac{\pi l}{N+1}\Big).$$

The result now follows from the first equation in system (8.15) by linearity. □

Now it is clear that we can analyze other schemes in the case of periodic boundary conditions, such as the backward Euler scheme or the leapfrog scheme, with the discrete Fourier transform.

Definition 8.7 We say that a scheme is *stable in the sense of von Neumann* if $|c_n^{j+1}| \leq |c_n^j|$ for all relevant n and j and all initial condition u^0.

We have $\|\omega_n\|_{2,h} = 1$ for all n, therefore $\|U^j\|_{2,h}^2 = \sum_{n=0}^N |c_n^j|^2$. Thus, stability in the sense of von Neumann implies stability for the 2, h norms for all T and uniformly with respect to T.

Proposition 8.13 *The forward Euler scheme is stable in the sense of von Neumann if* $\frac{k}{h^2} \leq \frac{1}{2}$.

Proof Assume that $\frac{k}{h^2} \leq \frac{1}{2}$. In view of Proposition 8.12, von Neumann stability occurs if and only if

$$\left|1 - \frac{4k}{h^2}\sin^2\Big(\frac{\pi n}{N+1}\Big)\right| \leq 1$$

for all n, that is to say if

$$\frac{2k}{h^2}\sin^2\Big(\frac{\pi n}{N+1}\Big) \leq 1$$

for all n. This is obviously the case under our hypothesis. □

Similarly, we find that the backward Euler scheme is unconditionally von Neumann stable.

8.7 Stability via Fourier Series

Stability for the 2, h norms is closely related to the spectral radius of the amplification matrix, at least when the latter is normal. Unfortunately, it is not always easy to compute the eigenvalues of a given matrix. We now present an alternate way using Fourier series, which is not directly applicable to the previously introduced schemes—in fact it applies to a slightly different context—but that still gives stability information in a much more workable fashion.

We thus now work with the heat equation on $\mathbb{R}$, therefore without boundary conditions. For definiteness, let us consider the forward Euler scheme

$$\begin{cases} \dfrac{u_n^{j+1} - u_n^j}{k} - \dfrac{u_{n+1}^j - 2u_n^j + u_{n-1}^j}{h^2} = 0, n \in \mathbb{Z}, \\ u_n^0 = u_0(nh), n \in \mathbb{Z}. \end{cases} \tag{8.17}$$

As in the previously considered cases, the discrete unknowns u_n^j are intended to be approximations of the exact values of the solution $u(x_n, t_j), n \in \mathbb{Z}, j = 1, \ldots, M$. We also write $u^0 = S_h(u_0)$ to denote the sampling operator on all grid points indexed by $\mathbb{Z}$.

If we assume that $(u_n^0)_{n\in\mathbb{Z}} \in \ell^2(\mathbb{Z})$, i.e., that $\sum_{n\in\mathbb{Z}} |u_n^0|^2 < +\infty$, then it is quite clear that $(u_n^j)_{n\in\mathbb{Z}}$ is well defined and belongs to $\ell^2(\mathbb{Z})$ for all j. We equip $\ell^2(\mathbb{Z})$ with the norm

$$\|(v_n)_{n\in\mathbb{Z}}\|_{2,h} = \sqrt{h}\Big(\sum_{n\in\mathbb{Z}} |v_n|^2\Big)^{1/2}$$

for which it is a Hilbert space, using the same notation as in the bounded interval case. Of course, the forward Euler scheme is also defined on other spaces of $\mathbb{Z}$-indexed sequences, but we concentrate here on ℓ^2. It should be noted that such schemes are not implementable in practice, since they involve an infinite number of unknowns. Their interest is purely theoretical.

We introduce the operator $\mathscr{T} : \ell^2(\mathbb{Z}) \to \ell^2(\mathbb{Z})$ defined by

$$(\mathscr{T}v)_n = \lambda v_{n+1} + (1 - 2\lambda)v_n + \lambda v_{n-1},$$

with $\lambda = \frac{k}{h^2}$. Then the scheme reads

$$u^{j+1} = \mathscr{T}u^j, \quad u^0 = S_h(u_0), \tag{8.18}$$

or

$$u^j = \mathscr{T}^j u^0,$$

for $j = 0, \dots, M$.[10] We introduce a concept of stability adapted to the present context.

Definition 8.8 We say that the scheme (8.17) is stable in $\ell^2(\mathbb{Z})$ if there exists a constant $C(T)$ such that

$$\max_{j\le T/k} \|u^j\|_{2,h} \le C(T)\|u^0\|_{2,h},$$

for all $u^0 \in \ell^2(\mathbb{Z})$.

Now for all $v \in \ell^2(\mathbb{Z}; \mathbb{C})$, we define $\mathscr{F}v \in L^2(0, 2\pi; \mathbb{C})$ by $\mathscr{F}v(s) = \sum_{n\in\mathbb{Z}} v_n e^{ins}$. It is well-known that the series converges in $L^2(0, 2\pi; \mathbb{C})$ and that operator $\mathscr{F}$ is an isometry between $\ell^2(\mathbb{Z}; \mathbb{C})$ and $L^2(0, 2\pi; \mathbb{C})$, when we equip the latter with the norm

$$\|f\|_{L^2(0,2\pi;\mathbb{C}),h} = \sqrt{\frac{h}{2\pi}}\Big(\int_0^{2\pi} |f(s)|^2\, ds\Big)^{1/2},$$

due to Parseval's formula, see [68]. The coefficients v_k are just the Fourier coefficients of the 2π-periodic L^2-function $\mathscr{F}v$. Conversely, any 2π-periodic L^2-function gives rise to an element of $\ell^2(\mathbb{Z}; \mathbb{C})$ by considering its Fourier coefficients. For brevity, from now on we omit the reference to $\mathbb{C}$ in the notation. The next proposition follows directly from Parseval's formula.

---

[10] Again, beware of the notation: u^j is the jth element in the sequence, whereas $\mathscr{T}^j$ is the jth iterate of the operator $\mathscr{T}$.

Proposition 8.14 *The scheme (8.17) is stable in $\ell^2(\mathbb{Z})$ if and only if there exists a constant $C(T)$ such that*

$$\max_{j \le T/k} \|\mathscr{F}u^j\|_{L^2(0,2\pi),h} \le C(T)\|\mathscr{F}u^0\|_{L^2(0,2\pi),h},$$

for all $u^0 \in \ell^2(\mathbb{Z})$.

We let $\mathscr{G} = \mathscr{F} \circ \mathscr{T} \circ \mathscr{F}^{-1}$. Since $u^j = \mathscr{T}^j u_0$, it follows that $\mathscr{F}u^j = \mathscr{G}^j(\mathscr{F}u^0)$. Therefore we have the following proposition.

Proposition 8.15 *The scheme* (8.17) *is stable in $\ell^2(\mathbb{Z})$ if there exists a constant $C(T)$ such that*

$$\max_{j \le T/k} \|\mathscr{G}^j\|_{\mathscr{L}(L^2(0,2\pi))} \le C(T).$$

Before continuing further, let us discuss the relationship between the initial data of the continuous problem, the function u_0, and the initial data of the discrete scheme, the sequence u^0, in the L^2/ℓ^2 framework. We need to associate a function in $L^2(\mathbb{R})$ with each sequence of numbers belonging to $\ell^2(\mathbb{Z})$.

Proposition 8.16 *For all $v \in \ell^2(\mathbb{Z})$, we define a piecewise constant interpolation $I_h v$ by*

$$\forall i \in \mathbb{Z}, \forall x \in \left]x_n - \frac{h}{2}, x_n + \frac{h}{2}\right[, \quad I_h v(x) = v_n.$$

The interpolation operator I_h is an isometry between $\ell^2(\mathbb{Z})$ equipped with the $\|\cdot\|_{2,h}$ norm and $L^2(\mathbb{R})$ equipped with its usual norm.

Proof Indeed,

$$\|I_h v\|^2_{L^2(\mathbb{R})} = \int_{\mathbb{R}} I_h v(x)^2\,dx = \sum_{n\in\mathbb{Z}} \int_{x_n-\frac{h}{2}}^{x_n+\frac{h}{2}} I_h v(x)^2\,dx = \sum_{n\in\mathbb{Z}} h v_n^2 = \|v\|^2_{2,h},$$

and the proof is complete. □

We can now see in which sense the stability definition (Definition 8.8) relates to the continuous problem, in the sense that $\mathscr{F}u^0 \in L^2(0, 2\pi)$ contains enough information about the function u_0 defined on $\mathbb{R}$, under some mild regularity assumption.

Proposition 8.17 *Assume that $u_0 \in H^1(\mathbb{R})$, then we have*

$$\|u_0 - I_h u^0\|_{L^2(\mathbb{R})} \le \frac{h}{2}\|u_0'\|_{L^2(\mathbb{R})}.$$

Proof Since $u_0 \in H^1(\mathbb{R})$, by formula (3.12), we have for all $x \in \mathbb{R}$ and all $n \in \mathbb{Z}$

$$u_0(x) - u_0(nh) = \int_{nh}^{x} u_0'(z)\,dz.$$

Therefore

$$\int_{nh-\frac{h}{2}}^{nh+\frac{h}{2}} |u_0(x) - u_0(nh)|^2 \, dx = \int_{nh-\frac{h}{2}}^{nh+\frac{h}{2}} \left| \int_{nh}^{x} u_0'(z) \, dz \right|^2 dx$$
$$\leq \int_{nh-\frac{h}{2}}^{nh+\frac{h}{2}} \left(|x - nh| \int_{nh}^{x} |u_0'(z)|^2 \, dz \right) dx$$
$$\leq \left(\int_{nh-\frac{h}{2}}^{nh+\frac{h}{2}} |x - nh| \, dx \right) \int_{nh-\frac{h}{2}}^{nh+\frac{h}{2}} |u_0'(z)|^2 \, dz$$
$$= \frac{h^2}{4} \int_{nh-\frac{h}{2}}^{nh+\frac{h}{2}} |u_0'(z)|^2 \, dz.$$

Summing over $n \in \mathbb{Z}$, we obtain the Proposition. □

Under the above hypothesis, we thus have

$$\|\mathscr{F} u^0\|_{L^2(0,2\pi),h} = \|S_h u_0\|_{2,h} = \|I_h u^0\|_{L^2(\mathbb{R})} \leq \sqrt{2} \|u_0\|_{H^1(\mathbb{R})}$$

for $h \leq 2$. Therefore, if the scheme is stable in $\ell^2(\mathbb{Z})$ in the sense of Definition 8.8, it follows that

$$\max_{j \leq T/k} \|u^j\|_{2,h} \leq \sqrt{2} C(T) \|u_0\|_{H^1(\mathbb{R})}.$$

The feature of the Fourier series transform that makes it so useful here, in addition to being an isometry, is that it transforms translations of indices into multiplications by exponentials. More precisely, if $v \in \ell^2(\mathbb{Z})$ and $m \in \mathbb{Z}$, letting $(\tau_m v)_n = v_{n+m}$, then

$$\mathscr{F}(\tau_m v)(s) = \sum_{n \in \mathbb{Z}} v_{n+m} e^{ins} = \sum_{n \in \mathbb{Z}} v_n e^{i(n-m)s} = e^{-ims} \mathscr{F} v(s).$$

Proposition 8.18 *Let* $a(s) = 1 - 4\lambda \sin^2\left(\frac{s}{2}\right)$. *Then we have*

$$\|\mathscr{G}^j\|_{\mathscr{L}(L^2(0,2\pi))} = \max_{s \in [0,2\pi]} |a(s)|^j.$$

Proof We apply the Fourier series transform to the scheme in the form (8.18). This yields

$$\mathscr{F}(u^{j+1})(s) = \left(1 + \lambda(e^{is} - 2 + e^{-is})\right) \mathscr{F}(u^j)(s) = a(s) \mathscr{F}(u^j)(s).$$

Iterating this relation, we obtain for all $g \in L^2(0, 2\pi)$[11]

$$(\mathscr{G}^j g)(s) = a(s)^j g(s).$$

[11] Again, beware of the notation: here $\mathscr{G}^j$ is the jth iterate of $\mathscr{G}$ and a^j is the function a to the power j.

Let now M be a multiplier operator, i.e., an operator on $L^2(0, 2\pi)$ of the form

$$(Mg)(s) = m(s)g(s),$$

with $m \in L^\infty(0, 2\pi)$, which is the case of $\mathscr{G}^j$ above. Let us show that

$$\|M\|_{\mathscr{L}(L^2(0,2\pi))} = \|m\|_{L^\infty(0,2\pi)}.$$

First of all, for all $g \in L^2(0, 2\pi)$, we have

$$\|Mg\|^2_{L^2(0,2\pi),h} = \frac{h}{2\pi}\int_0^{2\pi} |m(s)|^2 |g(s)|^2\, ds \le \|m\|^2_{L^\infty(0,2\pi)}\|g\|^2_{L^2(0,2\pi),h},$$

so that

$$\|M\|_{\mathscr{L}(L^2(0,2\pi))} \le \|m\|_{L^\infty(0,2\pi)}.$$

Next, let $\|m\|_{L^\infty(0,2\pi)} \ge \varepsilon > 0$ and $A \subset [0, 2\pi]$ be a set of strictly positive measure such that $|m(t)| \ge \|m\|_{L^\infty(0,2\pi)} - \varepsilon \ge 0$ on A, assuming $m \neq 0$ since the case $m = 0$ is not difficult. We take $g = (\frac{h\,\mathrm{meas}\,A}{2\pi})^{-1/2}\mathbf{1}_A \in L^2(0, 2\pi)$. Then $\|g\|_{L^2(0,2\pi),h} = 1$ and

$$\begin{aligned}\|Mg\|^2_{L^2(0,2\pi),h} &= \frac{h}{2\pi}\int_0^{2\pi} |m(s)|^2|g(s)|^2\, ds = \frac{1}{\mathrm{meas}\,A}\int_A |m(s)|^2\, ds \\ &\ge \frac{(\|m\|_{L^\infty(0,2\pi)} - \varepsilon)^2}{\mathrm{meas}\,A}\int_A ds = (\|m\|_{L^\infty(0,2\pi)} - \varepsilon)^2.\end{aligned}$$

Therefore

$$\|M\|_{\mathscr{L}(L^2(0,2\pi))} \ge \|m\|_{L^\infty(0,2\pi)} - \varepsilon$$

for all $\varepsilon > 0$, and the proposition is proved, since in our particular case, the function $m = a^j$ is continuous on $[0, 2\pi]$ and its L^∞ norm is just the maximum of its absolute value on $[0, 2\pi]$. □

Proposition 8.19 *The forward Euler scheme is stable in* $\ell^2(\mathbb{Z})$ *if* $\frac{k}{h^2} \le \frac{1}{2}$ *and unstable if* $\frac{k}{h^2} \ge \lambda_0 > \frac{1}{2}$.

Proof We have $a(s) = 1 - \frac{4k}{h^2}\sin^2\left(\frac{s}{2}\right) \le 1$ for all $s \in [0, 2\pi]$ and $a(0) = 1$. On the other hand, the minimum of $a(s)$ is attained for $\frac{s}{2} = \frac{\pi}{2}$ and its minimum value is $1 - \frac{4k}{h^2}$. Therefore

$$\max_{s\in[0,2\pi]} |a(s)| = \max\left(1, \left|1 - \frac{4k}{h^2}\right|\right).$$

Consequently, if $\frac{k}{h^2} \le \frac{1}{2}$, then $\max_{s\in[0,2\pi]} |a(s)| = 1$ so that $\|\mathscr{G}^j\|_{\mathscr{L}(L^2(0,2\pi))} = 1$ and the scheme is stable in $\ell^2(\mathbb{Z})$.

If, on the other hand, $\frac{k}{h^2} \ge \lambda_0 > \frac{1}{2}$, then

$$\max_{s\in[0,2\pi]} |a(s)|^j \geq (4\lambda_0 - 1)^j,$$

so that

$$\max_{j\leq T/k} \max_{s\in[0,2\pi]} |a(s)|^j \geq (4\lambda_0 - 1)^{T/k} \to +\infty \text{ when } k \to 0,$$

hence the scheme is unstable. □

We can apply the same philosophy to a general single time step finite difference scheme and obtain corresponding schemes on $\ell^2(\mathbb{Z})$ which are of the form $\mathscr{F}(u^{j+1})(s) = a(s)\mathscr{F}(u^j)(s)$, $\mathscr{F}(u^0)$ given, in Fourier space. The function a, which depends on h and k as parameters, is called the *amplification coefficient* of the scheme.

Using the same arguments as those used with matrices, it is easy to prove that a scheme is stable in ℓ^2 if and only if there exists a positive constant C that depends only on T such that $|a(s)| \leq 1 + Ck$ for all s. We now introduce the analogue of the previous notion of von Neumann stability (Definition 8.7) for schemes on $\ell^2(\mathbb{Z})$.

Definition 8.9 We say that a scheme on $\ell^2(\mathbb{Z})$ is *stable in the sense of von Neumann* if $\max_{s\in[0,2\pi]} |a(s)| \leq 1$.

Clearly, stability in the sense of von Neumann implies stability in $\ell^2(\mathbb{Z})$ for all T and uniformly with respect to T. It is thus a sufficient condition of stability. Obviously, computations in Fourier space are much easier than evaluations of spectral radii.

We now consider the example of a family of schemes, collectively known as *the θ-scheme*. Let us be given a number $\theta \in [0, 1]$. The θ-scheme is as follows:

$$\frac{u_n^{j+1} - u_n^j}{k} - \theta \frac{u_{n+1}^{j+1} - 2u_n^{j+1} + u_{n-1}^{j+1}}{h^2} - (1-\theta)\frac{u_{n+1}^j - 2u_n^j + u_{n-1}^j}{h^2} = \theta f_n^{j+1} + (1-\theta)f_n^j, \tag{8.19}$$

with initial conditions. The θ-scheme is thus a weighted average of the explicit Euler scheme ($\theta = 0$) and the implicit Euler scheme ($\theta = 1$). It is implicit as soon as $\theta > 0$. Before we can even talk about stability, it is not clear that such an implicit scheme is actually well-defined on $\ell^2(\mathbb{Z})$. The Fourier transform is also the key here. Indeed, in Fourier space, we have (with $f = 0$)

$$\frac{\mathscr{F}(u^{j+1})(s) - \mathscr{F}(u^j)(s)}{k} - \theta \frac{e^{is} - 2 + e^{-is}}{h^2} \mathscr{F}(u^{j+1})(s) - (1-\theta)\frac{e^{is} - 2 + e^{-is}}{h^2}\mathscr{F}(u^j)(s) = 0,$$

which boils down to

$$\Big(1 + \theta\frac{4k}{h^2}\sin^2\Big(\frac{s}{2}\Big)\Big)\mathscr{F}(u^{j+1})(s) = \Big(1 - (1-\theta)\frac{4k}{h^2}\sin^2\Big(\frac{s}{2}\Big)\Big)\mathscr{F}(u^j)(s).$$

Now we see that $1 + \theta \frac{4k}{h^2} \sin^2\left(\frac{s}{2}\right) \geq 1$, hence, its inverse is in L^∞ and the scheme in Fourier space can be rewritten as a multiplier operator with amplification coefficient

$$a(s) = \frac{1 - (1-\theta)\frac{4k}{h^2} \sin^2\left(\frac{s}{2}\right)}{1 + \theta \frac{4k}{h^2} \sin^2\left(\frac{s}{2}\right)} \in L^\infty(0, 2\pi).$$

The fact that the Fourier transform is an isomorphism implies that the discrete scheme is well-defined.

Now in terms of stability, clearly, $a(s) \leq 1$ for all $s \in [0, 2\pi]$. Stability in the sense of von Neumann thus depends on whether or not we have $a(s) \geq -1$ for all $s \in [0, 2\pi]$.

Proposition 8.20 *If $\theta \geq \frac{1}{2}$, then the θ-scheme is unconditionally stable in the sense of von Neumann. If $\theta < \frac{1}{2}$, it is stable in the sense of von Neumann under the condition $\frac{k}{h^2} \leq \frac{1}{2(1-2\theta)}$.*

Proof After a little bit of computation, it can be checked that $a(s) \geq -1$ if and only if $1 + (2\theta - 1)\frac{2k}{h^2} \sin^2\left(\frac{s}{2}\right) \geq 0$, hence the result. □

The case $\theta = \frac{1}{2}$ is special and is called the *Crank–Nicolson* scheme. We will go back to this scheme in detail in Sect. 8.9.

8.8 Stability via the Continuous Fourier Transform

We now present yet another approach to stability, using this time the continuous Fourier transform. Again, this approach is not directly applicable to discrete schemes, but it makes computations much easier. The ensuing stability analysis turns out to be very similar to the one done via Fourier series, and we actually basically use the same notation.

We thus consider again the heat equation on $\mathbb{R}$ with zero right-hand side and start with the forward Euler scheme (8.17) in the $\ell^2(\mathbb{Z})$ context.

Instead of working directly with the above discrete scheme, we introduce a *semi-discrete* version of it. In a semi-discrete scheme, only time is fully discretized. Space is only semi-discretized in the sense that it remains continuous even though we retain the space step h. We thus consider sequences of functions $u^j : \mathbb{R} \to \mathbb{R}$ which are such that u^j is supposed to be an approximation of the function $x \mapsto u(x, t_j)$.

The semi-discrete version of the forward Euler scheme is as follows:

$$\begin{cases} \dfrac{u^{j+1}(x) - u^j(x)}{k} - \dfrac{u^j(x+h) - 2u^j(x) + u^j(x-h)}{h^2} = 0, \\ u^0(x) = u_{0,h}(x), \end{cases} \tag{8.20}$$

where $u_{0,h}$ is some approximation of u_0. So the idea is to use the differential quotient on which the discrete scheme is based to approximate the space derivative, and the usual discrete difference quotient for the time derivative. This way, any discrete scheme admits a semi-discrete version.

A good functional setting for this is for example $L^2(\mathbb{R})$. Indeed, if $u_{0,h} \in L^p(\mathbb{R})$, then clearly, u^j is well defined and belongs to $L^p(\mathbb{R})$. In effect, if $u_{0,h} \in L^2(\mathbb{R})$, then we can write $u^{j+1} = G(u^j)$ where G is the continuous linear operator in $\mathscr{L}(L^2(\mathbb{R}), L^2(\mathbb{R}))$ defined by

$$Gv(x) = v(x) + \frac{k}{h^2}(v(x+h) - 2v(x) + v(x-h)), \tag{8.21}$$

or equivalently

$$G = \Big(1 - \frac{2k}{h^2}\Big)I + \frac{k}{h^2}\Big(\tau_h + \tau_{-h}\Big),$$

where τ_s denotes the operator of translation by s, $\tau_s u(x) = u(x+s)$. Therefore, $u^j = G^j(u_{0,h})$ and the properties of the scheme are the properties of the iterates of the operator G, provided $u_{0,h}$ remains bounded.

Let us discuss the relationship between the fully discrete and semi-discrete points of view. It turns out that the discrete scheme and the semi-discrete scheme are equivalent when the initial data of the semi-discrete scheme is in the range of the interpolation operator I_h of Proposition 8.16.

Proposition 8.21 *Let us be given $(u_n^0)_{n\in\mathbb{Z}} \in \ell^2(\mathbb{Z})$. If $u_{0,h} = I_h\big((u_n^0)_{n\in\mathbb{Z}}\big)$, then $u^j = I_h\big((u_n^j)_{n\in\mathbb{Z}}\big)$ for all $j \in \mathbb{N}$.*

Proof We prove this by induction on j. The statement is true for $j = 0$ by hypothesis. Let us thus assume that $u^j = I_h\big((u_n^j)_{n\in\mathbb{Z}}\big)$. This means that for all $x \in \big]x_n - \frac{h}{2}, x_n + \frac{h}{2}\big[$, we have $u^j(x) = u_n^j$. Therefore, in view of (8.21), for the same values of x, we have

$$\begin{aligned} Gu^j(x) &= u^j(x) + \frac{k}{h^2}(u^j(x+h) - 2u^j(x) + u^j(x-h)) \\ &= u_n^j + \frac{k}{h^2}(u_{n+1}^j - 2u_n^j + u_{n-1}^j) = u_n^{j+1}, \end{aligned}$$

so that $u^{j+1} = Gu^j = I_h\big((u_n^{j+1})_{n\in\mathbb{Z}}\big)$. □

So the idea is that, if we start the semi-discrete scheme with an initial data constructed by piecewise interpolation from the discrete scheme, the semi-discrete scheme will construct exactly the same values as the discrete scheme. The advantage is that the semi-discrete scheme works for much more general initial data, which in turn makes the study of stability considerably easier.

Definition 8.10 We say that the semi-discrete scheme is stable in $L^2(\mathbb{R})$ if there exists a constant $C(T)$ such that

$$\max_{j\leq T/k} \|u^j\|_{L^2(\mathbb{R})} \leq C(T)\|u_{0,h}\|_{L^2(\mathbb{R})},$$

for all $u_{0,h} \in L^2(\mathbb{R})$.

Here $u_{0,h}$ is no longer to be thought of as some approximation of u_0. Clearly, this is equivalent to $\|G^j\|_{\mathscr{L}(L^2(\mathbb{R}),L^2(\mathbb{R}))}$ being bounded independently of j, h and k.[12] In view of Propositions 8.16 and 8.21, stability of the semi-discrete scheme implies stability of the discrete scheme in the $\|\cdot\|_{2,h}$ norms, hence the interest of the approach.

The reason for singling out L^2 among all L^p spaces is that the continuous Fourier transform is an isometry on L^2. Let us briefly state a few facts about the continuous Fourier transform. When $u \in L^1(\mathbb{R})$, the Fourier transform of u is defined by

$$\widehat{u}(\xi) = \mathscr{F}u(\xi) = \frac{1}{\sqrt{2\pi}}\int_{-\infty}^{+\infty} e^{-ix\xi}u(x)\,dx.$$

The function $\widehat{u}$ is continuous and tends to 0 at infinity. When $u \in L^1(\mathbb{R}) \cap L^2(\mathbb{R})$, it can be shown that $\widehat{u}$ also belongs to $L^2(\mathbb{R})$ and that $\|\widehat{u}\|_{L^2(\mathbb{R})} = \|u\|_{L^2(\mathbb{R})}$, which is called the Plancherel formula, see [68]. Thus the Fourier transform extends as an isometry to the whole of L^2 by density of $L^1(\mathbb{R}) \cap L^2(\mathbb{R})$ in $L^2(\mathbb{R})$ (but not by the simple Lebesgue integral formula above, which makes no sense in the L^2 context).

In addition to being an isometry, the continuous Fourier transform transforms translations into multiplications by exponentials. More precisely, if $u \in L^1(\mathbb{R})$ and $s \in \mathbb{R}$, then

$$\widehat{\tau_s u}(\xi) = \frac{1}{\sqrt{2\pi}}\int_{-\infty}^{+\infty} e^{-ix\xi}u(x+s)\,dx = \frac{1}{\sqrt{2\pi}}\int_{-\infty}^{+\infty} e^{-i(y-s)\xi}u(y)\,dy = e^{is\xi}\widehat{u}(\xi),$$

and the equality $\widehat{\tau_s u}(\xi) = e^{is\xi}\widehat{u}(\xi)$ remains true for any $u \in L^2(\mathbb{R})$ by density.

Proposition 8.22 *Let* $a(\xi) = 1 - \frac{4k}{h^2}\sin^2\left(\frac{h\xi}{2}\right)$. *Then we have*

$$\|G^j\|_{\mathscr{L}(L^2(\mathbb{R}),L^2(\mathbb{R}))} = \sup_{\xi\in\mathbb{R}} |a(\xi)|^j.$$

Proof We apply the Fourier transform to the semi-discrete scheme (8.20). This yields

$$\frac{\widehat{w^{j+1}}(\xi) - \widehat{w^j}(\xi)}{k} - \frac{e^{ih\xi} - 2 + e^{-ih\xi}}{h^2}\widehat{w^j}(\xi) = 0,$$

or

$$\widehat{w^{j+1}}(\xi) = \widehat{w^j}(\xi) + \frac{2k}{h^2}(\cos(h\xi) - 1)\widehat{w^j}(\xi),$$

[12] Here again, G depends on h and k even though the notation does not make it plain.

or again

$$\widehat{u^{j+1}}(\xi) = a(\xi)\widehat{u^j}(\xi).$$

Iterating this relation, we obtain[13]

$$\mathscr{F}(G^j u_0)(\xi) = \widehat{u^j}(\xi) = a(\xi)^j \widehat{u_0}(\xi).$$

The conclusion follows as in the proof of Proposition 8.18. □

Proposition 8.23 *The forward Euler semi-discrete scheme is stable in $L^2(\mathbb{R})$ if $\frac{k}{h^2} \leq \frac{1}{2}$ and unstable if $\frac{k}{h^2} \geq \lambda_0 > \frac{1}{2}$.*

Proof We have $a(\xi) = 1 - \frac{4k}{h^2}\sin^2\left(\frac{h\xi}{2}\right)$ for $\xi \in \mathbb{R}$. As in the proof of Proposition 8.19, we obtain

$$\sup_{\xi\in\mathbb{R}} |a(\xi)| = \max\left(1, \left|1 - \frac{4k}{h^2}\right|\right),$$

and the conclusion follows along the same lines. □

Corollary 8.5 *The forward Euler discrete scheme is stable in the $\|\cdot\|_{2,h}$ norms if $\frac{k}{h^2} \leq \frac{1}{2}$.*

Any single time step finite difference scheme has a semi-discrete version, which is of the form $\widehat{u^{j+1}}(\xi) = a(\xi)\widehat{u^j}(\xi)$, $\widehat{u^0}$ given, in Fourier space. The function a, which depends on h and k as parameters, is again called the *amplification coefficient* of the scheme.

Of course, a scheme is stable in L^2 if and only if there exists a positive constant C that depends only on T such that $|a(\xi)| \leq 1 + Ck$ for all ξ. There is also a concept of von Neumann stability for semi-discrete schemes.

Definition 8.11 We say that a semi-discrete scheme is *stable in the sense of von Neumann* if $\sup_{\xi\in\mathbb{R}} |a(\xi)| \leq 1$.

Clearly, stability in the sense of von Neumann implies stability in $L^2(\mathbb{R})$ for all T and uniformly with respect to T. It is thus a sufficient condition of stability for both semi-discrete and discrete schemes.

We now consider the θ-scheme. The semi-discrete version of the θ-scheme (with 0 right-hand side) is

$$\frac{u^{j+1}(x) - u^j(x)}{k} - \theta\frac{u^{j+1}(x+h) - 2u^{j+1}(x) + u^{j+1}(x-h)}{h^2} \\ - (1-\theta)\frac{u^j(x+h) - 2u^j(x) + u^j(x-h)}{h^2} = 0.$$

[13] Again, beware of the notation: u^j is the jth function in the sequence, whereas G^j is the jth iterate of the operator G and a^j is the function a to the power j.

Again the scheme is implicit for $\theta > 0$. We must show that it is well-defined on $L^2(\mathbb{R})$. We use of course the Fourier transform again. Indeed, in Fourier space, we have

$$\frac{\widehat{u^{j+1}}(\xi) - \widehat{u^j}(\xi)}{k} - \theta \frac{e^{ih\xi} - 2 + e^{-ih\xi}}{h^2} \widehat{u^{j+1}}(\xi) \\ - (1-\theta) \frac{e^{ih\xi} - 2 + e^{-ih\xi}}{h^2} \widehat{u^j}(\xi) = 0,$$

and the exact same computation as in the Fourier series case yields an amplification coefficient

$$a(\xi) = \frac{1 - (1-\theta)\frac{4k}{h^2}\sin^2\left(\frac{h\xi}{2}\right)}{1 + \theta\frac{4k}{h^2}\sin^2\left(\frac{h\xi}{2}\right)} \in L^\infty(\mathbb{R}).$$

Proposition 8.24 *If $\theta \geq \frac{1}{2}$, then the semi-discrete θ-scheme is unconditionally stable in the sense of von Neumann. If $\theta < \frac{1}{2}$, it is stable in the sense of von Neumann under the condition $\frac{k}{h^2} \leq \frac{1}{2(1-2\theta)}$.*

Proof See the proof of Proposition 8.20. □

8.9 The Crank–Nicolson Scheme, Stability via the Energy Method

Let us first talk about consistency and order of the θ-scheme for problem (8.1), defined by (8.19), of which the Crank–Nicolson scheme is a special case.

Proposition 8.25 *The θ-scheme is of order 1 in time and 2 in space for $\theta \neq \frac{1}{2}$, and of order 2 in time and 2 in space for $\theta = \frac{1}{2}$, for the ∞, h norms.*

Proof Let u be a sufficiently regular solution of problem (8.1). Let us list the results of the application of Taylor–Lagrange expansions to the various terms, without writing the remainders explicitly since we know that they are uniformly bounded in terms of the relevant parameters. For the time derivative, we have

$$\begin{aligned}\frac{u(x_n, t_{j+1}) - u(x_n, t_j)}{k} &= \frac{\partial u}{\partial t}(x_n, t_j) + \frac{k}{2}\frac{\partial^2 u}{\partial t^2}(x_n, t_j) + O(k^2) \\ &= \frac{\partial u}{\partial t}(x_n, t_{j+1}) - \frac{k}{2}\frac{\partial^2 u}{\partial t^2}(x_n, t_{j+1}) + O(k^2)\end{aligned}$$

For the space derivatives, we obtain

$$\frac{u(x_{n+1}, t_j) - 2u(x_n, t_j) + u(x_{n-1}, t_j)}{h^2} = \frac{\partial^2 u}{\partial x^2}(x_n, t_j) + O(h^2)$$

and

$$\frac{u(x_{n+1},t_{j+1})-2u(x_n,t_{j+1})+u(x_{n-1},t_{j+1})}{h^2}=\frac{\partial^2 u}{\partial x^2}(x_n,t_{j+1})+O(h^2).$$

Therefore, combining these relations together, we see that

$$\begin{aligned}\varepsilon_{h,k}(u)_n^j &= \theta\Big(\frac{\partial u}{\partial t}(x_n,t_{j+1})-\frac{\partial^2 u}{\partial x^2}(x_n,t_{j+1})\Big)+(1-\theta)\Big(\frac{\partial u}{\partial t}(x_n,t_j)-\frac{\partial^2 u}{\partial x^2}(x_n,t_j)\Big)\\ &\quad-\theta\frac{k}{2}\frac{\partial^2 u}{\partial t^2}(x_n,t_{j+1})+(1-\theta)\frac{k}{2}\frac{\partial^2 u}{\partial t^2}(x_n,t_j)+O(k^2)+O(h^2)\\ &\quad-\theta f_n^{j+1}-(1-\theta)f_n^j.\end{aligned}$$

Now we can write
$$\frac{\partial^2 u}{\partial t^2}(x_n,t_{j+1})=\frac{\partial^2 u}{\partial t^2}(x_n,t_j)+O(k).$$

Canceling all cancelable terms, we thus obtain

$$\varepsilon_{h,k}(u)_n^j=k\Big(\frac{1}{2}-\theta\Big)\frac{\partial^2 u}{\partial t^2}(x_n,t_j)+O(k^2)+O(h^2),$$

and the result follows. □

Remark 8.19 The θ-scheme for $\theta=\frac{1}{2}$, or Crank–Nicolson scheme, thus appears to be particularly attractive: it is unconditionally (von Neumann) stable and of order 2 in time and space on $\mathbb{R}$. Of course, we still have to prove the unconditional stability of its discrete version on a bounded interval. We rewrite it here in full

$$\begin{cases}\dfrac{u_n^{j+1}-u_n^j}{k}-\dfrac{1}{2h^2}(u_{n+1}^{j+1}-2u_n^{j+1}+u_{n-1}^{j+1}+u_{n+1}^j-2u_n^j+u_{n-1}^j)=\dfrac{1}{2}(f_n^{j+1}+f_n^j)\\ u^0=u_0,\\ u_0^j=u_{N+1}^j=0.\end{cases}\tag{8.22}$$

The scheme is implicit, with the same computational cost as the backward Euler scheme, since evaluating U^{j+1} in terms of U^j entails solving a tridiagonal linear system with a very similar, invertible matrix. □

In order to prove the stability of the discrete Crank–Nicolson scheme, we use an argument that is the discrete analogue of the energy estimate for the continuous equation, hence the name *stability via the energy method*. We first need a lemma that is a discrete version of the integration by parts formula.

Let us introduce the forward and backward difference operators

$$(D_f v)_n=\frac{v_{n+1}-v_n}{h},\quad (D_b v)_n=\frac{v_n-v_{n-1}}{h},$$

which are a priori defined for real-valued, $\mathbb{Z}$-indexed sequences $(v_n)_{n\in\mathbb{Z}}$. We clearly have

$$D_f \circ D_b = D_b \circ D_f = D^2,$$

where D^2 is the second order centered difference operator

$$(D^2 v)_n = \frac{v_{n+1} - 2v_n + v_{n-1}}{h^2}$$

that was used to approximate the second order space derivative.

Of course, all these difference operators can be applied to finite sequences, by completing them with zeros to the right and to the left, which is appropriate when dealing with homogeneous Dirichlet boundary conditions. We will be doing this implicitly in all that follows.

Let us now establish the summation by parts formula, which is an interesting result by itself with many other applications.

Lemma 8.3 *Let $v = (v_n)_{n=1,\dots,N+1}$ and $w = (w_n)_{n=0,\dots,N+1}$. We have*

$$\sum_{n=1}^{N} (D_f v)_n w_n = -\sum_{n=1}^{N+1} v_n (D_b w)_n + \frac{1}{h}(v_{N+1}w_{N+1} - v_1 w_0). \tag{8.23}$$

In particulier, if $w_0 = w_{N+1} = 0$, then

$$\sum_{n=1}^{N} (D_f v)_n w_n = -\sum_{n=1}^{N+1} v_n (D_b w)_n. \tag{8.24}$$

Proof Let us expand the left-hand side of Eq. (8.23). We obtain

$$\begin{aligned}
\sum_{n=1}^{N} (D_f v)_n w_n &= \frac{1}{h}\sum_{n=1}^{N} (v_{n+1} - v_n) w_n \\
&= \frac{1}{h}\Big(\sum_{n=1}^{N} v_{n+1} w_n - \sum_{n=1}^{N} v_n w_n\Big) \\
&= \frac{1}{h}\Big(\sum_{n=2}^{N+1} v_n w_{n-1} - \sum_{n=1}^{N} v_n w_n\Big) \\
&= \frac{1}{h}\Big(\sum_{n=1}^{N+1} v_n (w_{n-1} - w_n) - v_1 w_0 + v_{N+1} w_{N+1}\Big) \\
&= -\sum_{n=1}^{N+1} v_n (D_b w)_n + \frac{1}{h}(v_{N+1} w_{N+1} - v_1 w_0),
\end{aligned}$$

and the result follows. □

Remark 8.20 There are other formulations of the summation by parts, for instance

$$\sum_{n=1}^{N}(D_f v)_n w_n = -\sum_{n=1}^{N} v_n (D_b w)_n + \frac{1}{h}(v_{N+1}w_N - v_1 w_0).$$

which is established in the same way. □

We also need a discrete version of the Poincaré inequality:

Lemma 8.4 *Let $w = (w_n)_{n=0,\dots,N}$ be such that $w_0 = 0$. Then we have*

$$\sum_{n=1}^{N} w_n^2 \leq \sum_{n=1}^{N} (D_b w)_n^2. \tag{8.25}$$

Proof For all $n = 1, \dots, N$, we can write

$$\begin{aligned} w_n &= w_n - w_{n-1} + w_{n-1} - w_{n-2} + \cdots + w_1 - w_0 \\ &= h\sum_{j=1}^{n}(D_b w)_j. \end{aligned}$$

Therefore, by the Cauchy–Schwarz inequality, it follows that

$$w_n^2 \leq h^2 n \sum_{j=1}^{n} (D_b w)_j^2 \leq h^2 N \sum_{j=1}^{N} (D_b w)_j^2 \leq h \sum_{j=1}^{N} (D_b w)_j^2$$

since $h = \frac{1}{N+1}$. Summing the above inequality from $n = 1$ to N, we obtain the Lemma. □

We now are in a position to mimic the continuous energy estimate at the discrete level.

Proposition 8.26 *The Crank–Nicolson scheme (8.22) is unconditionally stable for the 2, h norms.*

Proof We let $W^j = U^j + U^{j+1}$, $\widetilde{F}^j = \frac{F^j + F^{j+1}}{2}$ and rewrite the scheme as

$$\frac{U^{j+1} - U^j}{k} - \frac{1}{2}(D_f \circ D_b)W^j = \widetilde{F}^j.$$

Note that this is exactly the same form as the one for which we have already computed the orders in time and space in the ∞, h norms, hence in the 2, h norms. We multiply row n of the above relation by hW_j^i and sum with respect to n, or equivalently, take the 2, h scalar product in $\mathbb{R}^N$ with W^j, and obtain

$$\frac{1}{k}\left(\|U^{j+1}\|_{2,h}^2 - \|U^j\|_{2,h}^2\right) - \frac{1}{2}\left((D_f \circ D_b)W^j \middle| W^j\right)_{2,h} = \left(\widetilde{F}^j \middle| W^j\right)_{2,h}.$$

Let us estimate the various terms. By the Cauchy–Schwarz inequality and the discrete Poincaré inequality (8.25), we have

$$\begin{aligned}\left(\widetilde{F}^j \middle| W^j\right)_{2,h} \leq \|\widetilde{F}^j\|_{2,h}\|W^j\|_{2,h} \leq \|\widetilde{F}^j\|_{2,h}\|D_b W^j\|_{2,h} \\ \leq \frac{1}{2}\|\widetilde{F}^j\|_{2,h}^2 + \frac{1}{2}\|D_b W^j\|_{2,h}^2,\end{aligned}$$

indeed, $(W^j)_0 = u_0^j + u_0^{j+1} = 0$ due to the Dirichlet boundary condition.

For the same reason, we also have $(W^j)_{N+1} = 0$, so that the summation by parts formula (8.24) implies that

$$-\frac{1}{2}\left((D_f \circ D_b)W^j \middle| W^j\right)_{2,h} = \frac{1}{2}\left(D_b W^j \middle| D_b W^j\right)_{2,h} = \frac{1}{2}\|D_b W^j\|_{2,h}^2.$$

Putting these equalities and estimate together, we obtain

$$\frac{1}{k}\left(\|U^{j+1}\|_{2,h}^2 - \|U^j\|_{2,h}^2\right) \leq \frac{1}{2}\|\widetilde{F}^j\|_{2,h}^2,$$

so that

$$\|U^{j+1}\|_{2,h}^2 \leq \|U^j\|_{2,h}^2 + \frac{k}{2}\|\widetilde{F}^j\|_{2,h}^2.$$

Therefore, summing these inequalities from 0 to $j-1 \leq M$, we obtain

$$\begin{aligned}\|U^j\|_{2,h}^2 &\leq \|U_0\|_{2,h}^2 + \frac{jk}{2}\max_{m\leq T/k}\|\widetilde{F}^m\|_{2,h}^2 \\ &\leq \|U_0\|_{2,h}^2 + \frac{(M+1)k}{2}\max_{m\leq T/k}\|\widetilde{F}^m\|_{2,h}^2 = \|U_0\|_{2,h}^2 + \frac{T}{2}\max_{m\leq T/k}\|\widetilde{F}^m\|_{2,h}^2.\end{aligned}$$

This is exactly the stability of the scheme in the $2,h$ norms, once we apply the inequality $\sqrt{a^2+b^2} \leq a+b$ to the right-hand side. ☐

Remark 8.21 Note that when $f=0$, the discrete energy $\frac{1}{2}\|U^j\|_{2,h}^2$ is (in general strictly) decreasing with j, as it is in the continuous case with respect to continuous time. ☐

We can now put everything together.

Proposition 8.27 *The Crank–Nicolson scheme (8.22) is unconditionally convergent for the $2,h$ norms, of order 2 in time and space.*

For purposes of comparison, we plot in Fig. 8.6 the results of the backward Euler scheme, the Crank–Nicolson scheme with the same discretization parameters h and k fixed throughout, and the exact solution, for various values of j on the same graphs.

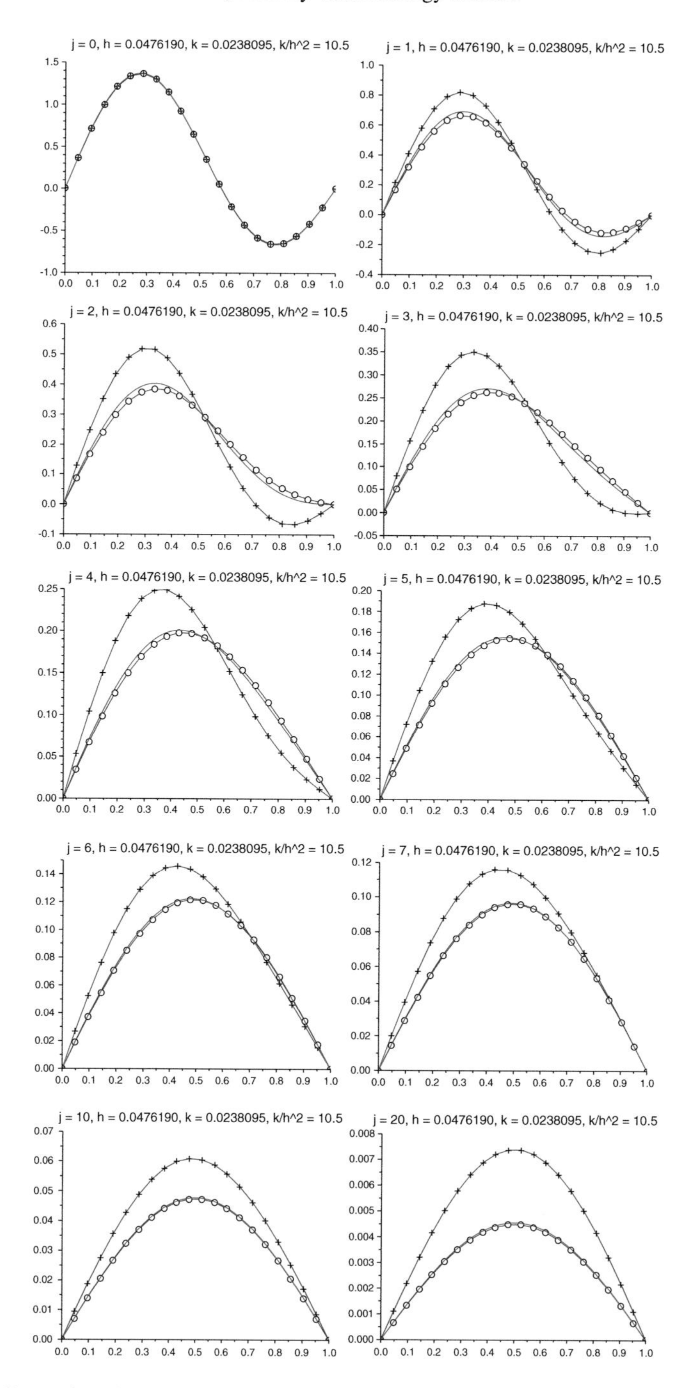

Fig. 8.6 Comparison between implicit Euler, (+) Crank–Nicolson (∘), and exact solutions (*solid line*)

The initial condition is $u_0(x) = \sin(\pi x)/2 + \sin(2\pi x)$ and the right-hand side f is zero. In this case, the exact solution is

$$u(x,t) = \frac{1}{2}\sin(\pi x)e^{-\pi^2 t} + \sin(2\pi x)e^{-4\pi^2 t},$$

which makes comparisons possible.

The backward Euler scheme solution is drawn with + marks and linearly interpolated, that of the Crank–Nicolson scheme with $\circ$ marks also linearly interpolated, and the exact solution with a solid line. The vertical scale varies from plot to plot.

Both schemes are stable and the higher order, hence better accuracy, of the Crank–Nicolson scheme is clearly visible for this particular initial data.

8.10 Other Approximations of the Heat Equation

The finite difference method seems quite satisfactory in the case of one space dimension, see also [76] for the finite difference method for general parabolic equations. Higher dimensional versions exist, see [75] for example. They however suffer from the same drawbacks as the finite difference method for elliptic problems in more than two space dimensions. Chiefly, it is difficult if not downright impossible to accommodate complex domain geometries in space $\Omega \subset \mathbb{R}^d$.

One way of going around this difficulty is to devise methods that combine a finite difference approximation in time, since there is an ordinary differential equation aspect with respect to time, with a finite element (or other) method in space, since there is a PDE boundary value aspect with respect to space. Let us quickly introduce finite difference-finite element schemes.

We start from the variational formulation (7.3). Given $u_0 \in L^2(\Omega)$ and $f \in L^2(Q)$, the solution u is such that $u \in C^0([0,T]; L^2(\Omega)) \cap L^2(0,T; H_0^1(\Omega))$ and for all $v \in H_0^1(\Omega)$,

$$\begin{cases} \big((u|v)_{L^2(\Omega)}\big)' + a(u,v) = (f|v)_{L^2(\Omega)} \text{ in the sense of } \mathscr{D}'(]0,T[), \\ (u(0)|v)_{L^2(\Omega)} = (u_0|v)_{L^2(\Omega)}. \end{cases}$$

The discretization proceeds in two consecutive steps. First comes the space discretization, exactly in the same spirit as for the abstract variational approximation methods for elliptic problems. Next, time discretization is performed. Let us start with the space discretization.

We thus assume that we are given a finite element subspace V_h of $V = H_0^1(\Omega)$. We then let $u_h \in C^0([0,T]; V_h)$ be the solution of

$$\begin{cases} \big((u_h|v_h)_{L^2(\Omega)}\big)' + a(u_h,v_h) = (f|v_h)_{L^2(\Omega)} \text{ in the sense of } \mathscr{D}'(]0,T[), \\ (u_h(0)|v_h)_{L^2(\Omega)} = (u_{0,h}|v_h)_{L^2(\Omega)}, \end{cases}$$

for all $v_h \in V_h$, where $u_{0,h} \in V_h$ is some approximation of u_0, for instance its L^2-orthogonal projection on V_h. This is a Cauchy problem for a system of linear ordinary differential equations, since V_h is finite dimensional, hence existence and uniqueness are not a real issue.

More precisely, assume that dim $V_h = N$ and let us be given a basis $(w^j)_{j=1,\dots,N}$ of V_h consisting of hat functions that are constructed from the shape functions associated with the underlying finite element. We can thus write

$$u_h(t) = \sum_{j=1}^{N} u_{h,j}(t) w^j$$

and u_h is determined by the N unknown real-valued functions $u_{h,j} \in C^0([0,T])$.

First of all, the initial condition is obviously equivalent to $u_h(0) = u_{0,h}$, that is in components $u_{h,j}(0) = u_{0,h,j}$, for all j.

Next, taking $v_h = w^i$ for $n = 1, \dots, N$, we obtain

$$\sum_{j=1}^{N} u'_{h,j}(t)(w^j|w^i)_{L^2(\Omega)} + \sum_{j=1}^{N} u_{h,j}(t) a(w^j, w^i) = (f|w^i)_{L^2(\Omega)}$$

for $n = 1, \dots, N$. Let us rewrite this in vector form by introducing the vectors

$$U_h(t) = \begin{pmatrix} u_{h,1}(t) \\ u_{h,2}(t) \\ \vdots \\ u_{h,N}(t) \end{pmatrix}, U_{0,h} = \begin{pmatrix} u_{0,h,1} \\ u_{0,h,2} \\ \vdots \\ u_{0,h,N} \end{pmatrix} \text{ and } F_h(t) = \begin{pmatrix} (f|w^1)_{L^2(\Omega)}(t) \\ (f|w^2)_{L^2(\Omega)}(t) \\ \vdots \\ (f|w^N)_{L^2(\Omega)}(t) \end{pmatrix},$$

and the $N \times N$ matrices

$$M = \begin{pmatrix} (w^1|w^1)_{L^2(\Omega)} & (w^2|w^1)_{L^2(\Omega)} & \cdots & (w^N|w^1)_{L^2(\Omega)} \\ (w^1|w^2)_{L^2(\Omega)} & (w^2|w^2)_{L^2(\Omega)} & \cdots & (w^N|w^2)_{L^2(\Omega)} \\ \vdots & \ddots & \ddots & \vdots \\ (w^1|w^{N-1})_{L^2(\Omega)} & \cdots & (w^{N-1}|w^{N-1})_{L^2(\Omega)} & (w^N|w^{N-1})_{L^2(\Omega)} \\ (w^1|w^N)_{L^2(\Omega)} & \cdots & (w^{N-1}|w^N)_{L^2(\Omega)} & (w^N|w^N)_{L^2(\Omega)} \end{pmatrix}$$

and

$$A = \begin{pmatrix} a(w^1|w^1) & a(w^2|w^1) & \cdots & a(w^N|w^1) \\ a(w^1|w^2) & a(w^2|w^2) & \cdots & a(w^N|w^2) \\ \vdots & \ddots & \ddots & \vdots \\ a(w^1|w^{N-1}) & \cdots & a(w^{N-1}|w^{N-1}) & a(w^N|w^{N-1}) \\ a(w^1|w^N) & \cdots & a(w^{N-1}|w^N) & a(w^N|w^N) \end{pmatrix},$$

or in other words, $M_{nj} = (w^j|w^i)_{L^2(\Omega)}$ and $A_{nj} = a(w^j, w^i)$. Since we are dealing with a hat-function basis, both matrices are sparse, and we can assume a good numbering of the degrees of freedom to obtain a band matrix.

The system then becomes

$$MU_h'(t) + AU(t) = F_h(t),$$

with the initial condition

$$U_h(0) = U_{0,h}.$$

The matrix M is called the *mass matrix* and the matrix A is called the *stiffness matrix*. Since M is the Gram matrix of a basis, it is nonsingular. Therefore, we can write

$$U_h'(t) + M^{-1}AU(t) = M^{-1}F_h(t),$$

from which existence and uniqueness are obvious since this is a system of linear ordinary differential equations. Conversely, it is clear that the solution of this system of ordinary differential equations gives rise to a solution of the variational problem on V_h.

So far, only space was discretized. In order to obtain a fully discrete scheme, we also need to discretize time. Any ordinary differential equation numerical scheme may a priori be used here: forward Euler, backward Euler, Crank–Nicolson, 4th order Runge–Kutta, linear multistep methods and so on, see [8, 16, 23, 60, 64]. Of course, when using ordinary differential equation schemes, stability criteria such as Dahlquist's zero-stability, L-stability or A-stability, become essential, see [12, 48, 49].

For this, we choose an integer M and let $k = \frac{T}{M+1}$ be the time step. We denote by U_h^j an approximation of $U_h(jk)$. The forward Euler scheme then reads

$$M\frac{U_h^{j+1} - U_h^j}{k} + AU_h^j = F_h(jk),$$

with the initial condition

$$U_h^0 = U_{0,h},$$

and so on for the other choices of time finite difference schemes. We have implicitly assumed some regularity of f with respect to t. Note that, even for the forward Euler scheme, we still need to solve a $N \times N$ linear system with the mass matrix in order to compute U_h^{j+1} from U_h^j.

Of course, the convergence of these methods when h and k simultaneously go to 0 must be established and we do not pursue in this direction. We refer for example to [5, 52, 62, 64, 66, 77].

After having discussed the first two main classes of problems, namely elliptic and parabolic problems, we now turn to the third class of PDE problems, hyperbolic problems. We first focus on the wave equation in Chap. 9, then on the transport equation in Chap. 10.

Chapter 9
The Wave Equation

In this chapter, we present a short and even more far from exhaustive theoretical study of the wave equation. We establish the existence and uniqueness of the solution, as well as the energy estimates. We describe the qualitative behavior of solutions, which is very different from that of the heat equation. Again, we will mostly work in one dimension of space.

In the same chapter, we introduce finite difference methods for the numerical approximation of the wave equation. Here again, stability issues are prominent, and significantly more delicate than for the heat equation.

9.1 Regular Solutions of the Wave Equation

Recall that the general wave equation reads

$$\frac{\partial^2 u}{\partial t^2} - \Delta u = f \text{ in } Q = \Omega \times]0, T[,$$

where Ω is an open subset of $\mathbb{R}^d$ and f is a given function on Q, complemented with boundary and initial conditions, see Chap. 1, Sects. 1.5 and 1.6. The propagation speed c is set to 1, which we can always assume after a change of time or length unit. There are two different settings depending on whether Ω is bounded or not. In the one-dimensional case, $d = 1$, we thus have either $\Omega =]a, b[$ or $\Omega = \mathbb{R}$ without loss of generality.[1]

[1]Admittedly, there is a third case, $\Omega = \mathbb{R}_+^*$, but we will not consider it here.

H. Le Dret and B. Lucquin, *Partial Differential Equations: Modeling, Analysis and Numerical Approximation*, International Series of Numerical Mathematics 168, DOI 10.1007/978-3-319-27067-8_9

Let us begin with the bounded case. We are thus looking for a function $u\colon [a,b]\times[0,T]\to\mathbb{R}$ which solves the initial-boundary value problem

$$\begin{cases} \dfrac{\partial^2 u}{\partial t^2} - \dfrac{\partial^2 u}{\partial x^2} = f \text{ in } Q, \\ u(a,t) = u(b,t) = 0 \text{ for } t \in [0,T], \\ u(x,0) = u_0(x),\ \dfrac{\partial u}{\partial t}(x,0) = u_1(x) \text{ for } x \in \,]a,b[, \end{cases} \tag{9.1}$$

with homogeneous Dirichlet boundary conditions for simplicity and two initial data, u_0 and u_1. If we think of the vibrating string interpretation, this means that the string is fixed at both ends, and that we are given its initial position and initial velocity. This is quite normal, since the equation is derived from Newton's law of motion and is of second order in time.

Definition 9.1 The quantity

$$E(t) = \frac{1}{2}\int_a^b \Big[\Big(\frac{\partial u}{\partial t}(x,t)\Big)^2 + \Big(\frac{\partial u}{\partial x}(x,t)\Big)^2\Big]\,dx$$

is called the *energy*.

Of course, we assume that the solution is regular enough for the above quantity to make sense. In the vibrating string interpretation, this is exactly the mechanical energy of the string at time t. The first term corresponds to the kinetic energy since it is half the square of the velocity at point x and time t, integrated along the string. The second term corresponds to the elastic energy, which can be seen by examining the work done by the exterior forces, based on the analysis in Chap. 1, Sect. 1.1. The initial energy is then

$$E(0) = \frac{1}{2}\int_a^b (u_1(x)^2 + u_0'(x)^2)\,dx.$$

The initial energy is finite for $u_0 \in H^1(]a,b[)$ and $u_1 \in L^2(a,b)$.

Proposition 9.1 *Let u be a smooth enough solution of problem (9.1), then we have*

$$\frac{dE}{dt}(t) = \int_a^b f(x,t)\frac{\partial u}{\partial t}(x,t)\,dx.$$

Proof By differentiation under the integral sign, we have

$$\frac{dE}{dt}(t) = \frac{1}{2}\int_a^b \Big[\frac{\partial}{\partial t}\Big(\Big(\frac{\partial u}{\partial t}\Big)^2\Big) + \frac{\partial}{\partial t}\Big(\Big(\frac{\partial u}{\partial x}\Big)^2\Big)\Big]\,dx$$
$$= \int_a^b \Big[\frac{\partial u}{\partial t}\frac{\partial^2 u}{\partial t^2} + \frac{\partial u}{\partial x}\frac{\partial^2 u}{\partial x \partial t}\Big]\,dx.$$

We integrate the second term by parts

$$\int_a^b \frac{\partial u}{\partial x}\frac{\partial^2 u}{\partial x \partial t}\,dx = \Big[\frac{\partial u}{\partial x}\frac{\partial u}{\partial t}\Big]_a^b - \int_a^b \frac{\partial^2 u}{\partial x^2}\frac{\partial u}{\partial t}\,dx = -\int_a^b \frac{\partial^2 u}{\partial x^2}\frac{\partial u}{\partial t}\,dx,$$

since $\frac{\partial u}{\partial t}(a,t) = \frac{\partial u}{\partial t}(b,t) = 0$ due to the Dirichlet boundary condition. Therefore,

$$\frac{dE}{dt}(t) = \int_a^b \Big[\frac{\partial u}{\partial t}\Big(\frac{\partial^2 u}{\partial t^2} - \frac{\partial^2 u}{\partial x^2}\Big)\Big]\,dx = \int_a^b f\frac{\partial u}{\partial t}\,dx,$$

and the proposition is proved. □

In the vibrating string interpretation, we thus find that the time derivative of the energy is the power of the applied forces, as is expected from physics.

Corollary 9.1 *If the right-hand side f in problem (9.1) vanishes, then the energy is constant*

$$E(t) = E(0).$$

Proof Indeed, in this case, $\frac{dE}{dt} = 0$. □

Remark 9.1 We note here a sharp contrast with the heat equation, for which the energy was exponentially decreasing for a zero right-hand side. The heat equation, which is a parabolic equation, dissipates the energy, whereas the wave equation—a hyperbolic equation—conserves the energy: a vibrating string keeps vibrating forever in the absence of dissipation. □

Corollary 9.2 *Problem (9.1) has at most one smooth solution.*

Proof Let u_1 and u_2 be solutions of problem (9.1), and $u = u_1 - u_2$. Then u is a solution of problem (9.1) with right-hand side $f = 0$, so that $E(t) = E(0)$, and zero initial data, so that $E(0) = 0$. It follows from Definition 9.1 that $u = 0$. □

In order to further exploit the energy, we need a general purpose result, known as *Gronwall's lemma* or *Gronwall's inequality*.

Theorem 9.1 (Gronwall's lemma) *Let α, β and γ be three continuous functions defined on $[0,T]$ such that α is differentiable on $]0,T[$. We assume that*

$$\alpha'(t) \le \beta(t)\alpha(t) + \gamma(t) \textit{ for all } t \in]0,T[.$$

Then, we have

$$\alpha(t) \leq e^{\int_0^t \beta(s)\,ds}\alpha(0) + \int_0^t e^{\int_s^t \beta(u)\,du}\gamma(s)\,ds.$$

Proof Let $B(t) = \int_0^t \beta(s)\,ds$ and define $\delta(t) = e^{-B(t)}\alpha(t)$. Then δ is differentiable on $]0, T[$ and

$$\begin{aligned}\delta'(t) &= e^{-B(t)}\alpha'(t) - \beta(t)e^{-B(t)}\alpha(t) = e^{-B(t)}(\alpha'(t) - \beta(t)\alpha(t)) \\ &\leq e^{-B(t)}\gamma(t).\end{aligned}$$

Therefore, by the mean value inequality,

$$\delta(t) - \delta(0) \leq \int_0^t e^{-B(s)}\gamma(s)\,ds$$

and we conclude by multiplying the above inequality by $e^{B(t)}$ and by noticing that $B(t) - B(s) = \int_s^t \beta(u)\,du$. □

Proposition 9.2 *We have the energy estimate*

$$\sup_{t\in[0,T]} E(t) \leq e^T E(0) + \frac{1}{2}\int_0^T \int_a^b e^{T-s} f(x,s)^2\,dxds.$$

Proof It follows from Proposition 9.1 that

$$E'(t) \leq \frac{1}{2}\int_a^b f(x,t)^2\,dx + \frac{1}{2}\int_a^b \Big(\frac{\partial u}{\partial t}(x,t)\Big)^2\,dx \leq \frac{1}{2}\int_a^b f(x,t)^2\,dx + E(t).$$

Thus, by Gronwall's lemma,

$$E(t) \leq e^t E(0) + \frac{1}{2}\int_0^t \int_a^b e^{t-s} f(x,s)^2\,dxds \leq e^T E(0) + \frac{1}{2}\int_0^T \int_a^b e^{T-s} f(x,s)^2\,dxds,$$

for all $t \in [0, T]$. □

Remark 9.2 The energy estimate provides a stability result in the energy norm, in the sense of establishing the continuity of the solution with respect to the initial data and right-hand side. Indeed, if u_1 and u_2 are two solutions corresponding to right-hand sides f_1 and f_2 and initial data $u_{1,0}, u_{1,1}$ and $u_{2,0}, u_{2,1}$, applying the energy estimate to $u_1 - u_2$, we obtain

$$\begin{aligned}&\sup_{t\in[0,T]}\Big(\|u_1 - u_2\|^2_{H_0^1(]a,b[)} + \Big\|\frac{\partial u_1}{\partial t} - \frac{\partial u_2}{\partial t}\Big\|^2_{L^2(a,b)}\Big) \\ &\qquad \leq e^T(\|u_{1,0} - u_{2,0}\|^2_{H_0^1(]a,b[)} + \|u_{1,1} - u_{2,1}\|^2_{L^2(a,b)} + \|f_1 - f_2\|^2_{L^2(Q)}).\end{aligned}$$

□

We now use Fourier series to construct regular solutions of problem (9.1) when $f = 0$. For simplicity, we let $a = 0$, $b = 1$ and we assume that the initial data are compatible with the Dirichlet condition, i.e., $u_0(0) = u_0(1) = u_1(0) = u_1(1) = 0$. As in the case of the heat equation, we expand both functions in Fourier series

$$u_0(x) = \sum_{k=1}^{+\infty} b_k^0 \sin(k\pi x), \quad u_1(x) = \sum_{k=1}^{+\infty} b_k^1 \sin(k\pi x).$$

Theorem 9.2 *Let $u_0 \in C^4([0,1])$ and $u_1 \in C^3([0,1])$ be such that $u_0''(0) = u_0''(1) = u_1''(0) = u_1''(1) = 0$. Then the function defined by*

$$u(x,t) = \sum_{k=1}^{+\infty} \Big(b_k^0 \cos(k\pi t) + \frac{b_k^1}{k\pi} \sin(k\pi t) \Big) \sin(k\pi x) \tag{9.2}$$

belongs to $C^2([0,1] \times [0,+\infty[)$ and solves problem (9.1) with $f = 0$.

Proof Under the hypotheses made on u_0 and u_1, it is easy to see that $|b_k^0| \le Ck^{-4}$ and $|b_k^1| \le Ck^{-3}$ for some constant C. Then the series in formula (9.2) as well as the series of all first order and second order derivatives are normally convergent. Hence, u is of class C^2. Moreover, since the functions $(x,t) \mapsto e^{ik\pi(t \pm x)}$ are solutions of the wave equation with zero right-hand side, it is clear that the normal convergence of second derivatives implies that u is also a solution of the wave equation.

For $t = 0$, we have

$$u(x,0) = \sum_{k=1}^{+\infty} b_k^0 \sin(k\pi x) = u_0(x)$$

and

$$\frac{\partial u}{\partial t}(x,0) = \sum_{k=1}^{+\infty} b_k^1 \sin(k\pi x) = u_1(x),$$

hence the initial conditions are satisfied. Finally, the Dirichlet boundary conditions are also satisfied since $\sin(k\pi) = 0$. □

Remark 9.3 We find that the solution is a superposition of harmonics, see Chap. 1, Sect. 1.5. Which harmonics are excited depend on the initial conditions. For instance, for such a musical instrument as the piano, the strings are initially at rest, $u_0 = 0$, and are hit by a hammer, $u_1 \neq 0$. In the case of a guitar or a harpsichord, the strings are typically plucked, $u_0 \neq 0$, sometimes with no initial velocity, $u_1 = 0$. Note that other combinations are possible, all resulting in different sounds, see Figs. 9.1 and 9.2. □

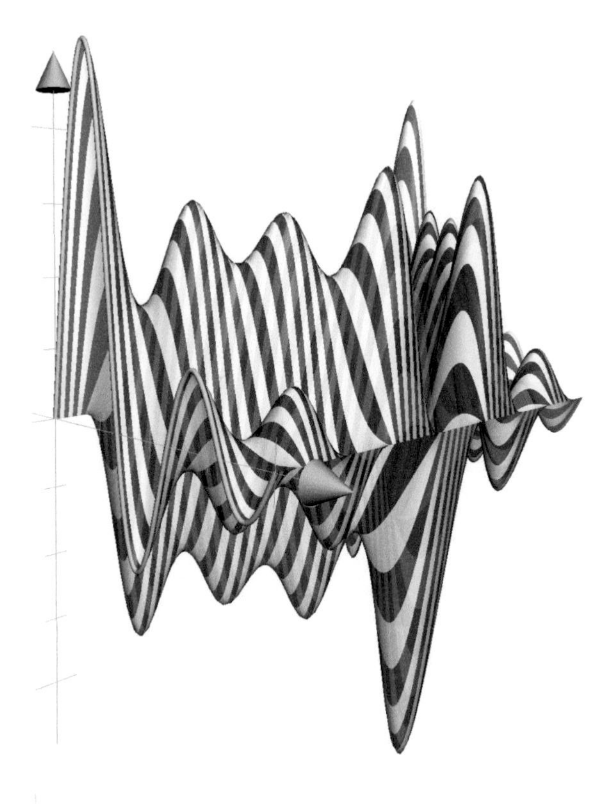

Fig. 9.1 A view of the evolution in the case of zero initial velocity u_1, u_0 has four nonzero harmonics

Fig. 9.2 A view of the evolution in the case of zero initial position, initial velocity $+1$ in $]0, \frac{1}{2}[$, -1 in $]\frac{1}{2}, 1[$, two hundred nonzero terms in the Fourier series

Remark 9.4 The regularity hypotheses made on u_0 and u_1 are just there to ensure easy convergence of the series of partial derivatives up to the second order. Indeed, if the series (9.2) converges in a much weaker sense, its sum is still going to be a solution of the wave equation in the sense of distributions at least, since differentiation is continuous in the sense of distributions. The difficulty lies in the meaning of the initial conditions, as some kind of continuity with respect to time is required for them to make sense. □

Remark 9.5 A fundamental difference with the heat equation is that the Fourier coefficients of $u(\cdot, t)$ are not rapidly damped by exponential terms for $t > 0$, which cause the solution of the heat equation to be smooth for $t > 0$, whatever the initial data. Here, the wave equation has no smoothing effect whatsoever. The regularity or lack thereof of the initial conditions is propagated in time without any gain. This is one of the main differences between parabolic and hyperbolic problems. □

9.2 Variational Formulation and Existence of Weak Solutions

We now introduce a variational formulation for the wave equation in a manner that is quite similar to the one described in Sect. 7.6 for the heat equation.

Definition 9.2 The variational formulation of the wave equation (9.1) with homogeneous Dirichlet boundary condition, initial data $u_0 \in H_0^1(\Omega)$, $u_1 \in L^2(\Omega)$ and right-hand side $f \in L^2(Q)$ is: Find $u \in C^0([0,T]; H_0^1(\Omega)) \cap C^1([0,T]; L^2(\Omega))$ such that, for all $v \in H_0^1(\Omega)$,

$$\begin{cases} \big((u|v)_{L^2(\Omega)}\big)'' + a(u,v) = (f|v)_{L^2(\Omega)} \text{ in the sense of } \mathscr{D}'(]0,T[), \\ (u(0)|v)_{L^2(\Omega)} = (u_0|v)_{L^2(\Omega)}, \\ (u'(0)|v)_{L^2(\Omega)} = (u_1|v)_{L^2(\Omega)}. \end{cases} \tag{9.3}$$

Remark 9.6 This definition clearly makes sense. The last two equations are a weak form of the initial conditions $u(0) = u_0$, $u'(0) = u_1$. □

For simplicity, we work again on $\Omega =]0,1[$ with the minus Laplacian eigenfunctions $\phi_k(x) = \sqrt{2}\sin(k\pi x)$ and eigenvalues $\lambda_k = k^2\pi^2$, and we have $a(w,\phi_k) = \lambda_k(w|\phi_k)_{L^2(\Omega)}$ for all $w \in H_0^1(\Omega)$, see Eq. (7.4).

Theorem 9.3 *Let $u_0 \in H_0^1(\Omega)$, $u_1 \in L^2(\Omega)$, $f \in L^2(Q)$. There exists a unique solution $u \in C^0([0,T]; H_0^1(\Omega)) \cap C^1([0,T]; L^2(\Omega))$ of the initial-boundary value problem (9.3), which is given by*

$$u(t) = \sum_{k=1}^{+\infty} u_k(t)\phi_k, \tag{9.4}$$

where

$$u_k(t) = (u_0|\phi_k)_{L^2(\Omega)} \cos\big(\sqrt{\lambda_k}t\big) + \frac{(u_1|\phi_k)_{L^2(\Omega)}}{\sqrt{\lambda_k}} \sin\big(\sqrt{\lambda_k}t\big) + \frac{1}{\sqrt{\lambda_k}} \int_0^t (f(s)|\phi_k)_{L^2(\Omega)} \sin\big(\sqrt{\lambda_k}(t-s)\big)\,ds. \tag{9.5}$$

Proof We start with the uniqueness. Let $u \in C^0([0,T]; H_0^1(\Omega)) \cap C^1([0,T]; L^2(\Omega))$ be a solution of (9.3). We expand $u(t)$ on the Hilbert basis $(\phi_k)_{k\in\mathbb{N}^*}$ of $L^2(\Omega)$ so that, for all t,

$$u(t) = \sum_{k=1}^{+\infty} u_k(t)\phi_k$$

with

$$u_k(t) = (u(t)|\phi_k)_{L^2(\Omega)}$$

for all $k \in \mathbb{N}^*$ and the series converges in $L^2(\Omega)$. Likewise, we set $u_0 = \sum_{k=1}^{+\infty} u_{0,k}\phi_k$, $u_1 = \sum_{k=1}^{+\infty} u_{1,k}\phi_k$ and $f(t) = \sum_{k=1}^{+\infty} f_k(t)\phi_k$. Taking $\phi_k \in H_0^1(\Omega)$ as a test-function in problem (9.3), we obtain

$$\begin{cases} u_k''(t) + \lambda_k u_k(t) = f_k(t) \text{ in the sense of } \mathscr{D}'(]0,T[), \\ u_k(0) = u_{0,k}, \\ u_k'(0) = u_{1,k}, \end{cases}$$

for all $k \in \mathbb{N}^*$. For each k, this is a Cauchy problem for an ordinary differential equation which has the unique solution

$$u_k(t) = u_{0,k}\cos\big(\sqrt{\lambda_k}t\big) + \frac{u_{1,k}}{\sqrt{\lambda_k}}\sin\big(\sqrt{\lambda_k}t\big) + \frac{1}{\sqrt{\lambda_k}}\int_0^t f_k(s)\sin\big(\sqrt{\lambda_k}(t-s)\big)\,ds,$$

hence the uniqueness.

We now use the above series to prove existence. We have that $u_0 \in H_0^1(\Omega)$ and $u_1 \in L^2(\Omega)$ by hypothesis, therefore thanks to formula (7.7),

$$\|u_0\|_{H^1(\Omega)}^2 = \sum_{k=1}^{+\infty}(1+\lambda_k)u_{0,k}^2,\ |u_0|_{H^1(\Omega)}^2 = \sum_{k=1}^{+\infty}\lambda_k u_{0,k}^2 \text{ and } \|u_1\|_{L^2(\Omega)}^2 = \sum_{k=1}^{+\infty} u_{1,k}^2. \tag{9.6}$$

Similarly, $f \in L^2(Q)$ and

$$\|f\|_{L^2(Q)}^2 = \int_0^T \sum_{k=1}^{+\infty} f_k(t)^2\,dt. \tag{9.7}$$

As before, we consider the sequence of partial sums $U_n(t) = \sum_{k=1}^n u_k(t)\phi_k$ and show that it is Cauchy for both $C^0([0,T]; H_0^1(\Omega))$ and $C^1([0,T]; L^2(\Omega))$ norms. Let $p < q$ be two given integers and let us estimate $U_p - U_q$ in these various norms.

First of all, we have for all t

$$\begin{aligned}|U_p(t) - U_q(t)|^2_{H_0^1(\Omega)} &= \sum_{k=p+1}^{q} \lambda_k u_k(t)^2 \\ &\leq 2 \sum_{k=p+1}^{q} \lambda_k \Big[u_{0,k}^2 + \frac{1}{\lambda_k} u_{1,k}^2 + \frac{1}{\lambda_k} \Big(\int_0^t |f_k(s)|\, ds \Big)^2 \Big] \\ &\leq 2 \sum_{k=p+1}^{q} \lambda_k u_{0,k}^2 + 2 \sum_{k=p+1}^{q} u_{1,k}^2 + 2T \sum_{k=p+1}^{q} \int_0^T f_k(s)^2\, ds,\end{aligned}$$

since all the trigonometric terms are less than 1 in absolute value and by the Cauchy–Schwarz inequality. Therefore

$$\|U_p - U_q\|^2_{C^0([0,T];H_0^1(\Omega))} \leq 2 \sum_{k=p+1}^{q} \lambda_k u_{0,k}^2 + 2 \sum_{k=p+1}^{q} u_{1,k}^2 + 2T \sum_{k=p+1}^{q} \int_0^T f_k(s)^2\, ds$$

can be made as small as we wish by taking p large enough, due to the hypotheses on u_0, u_1 and f and formulas (9.6)–(9.7), and the sequence is consequently Cauchy in $C^0(0, T; H_0^1(\Omega))$.

It follows from the previous estimate and the Poincaré inequality that the sequence is also Cauchy in $C^0(0, T; L^2(\Omega))$. We need to look at its time derivative. Of course, $U_n'(t) = \sum_{k=1}^{n} u_k'(t)\phi_k$ with

$$u_k'(t) = -\sqrt{\lambda_k} u_{0,k} \sin(\sqrt{\lambda_k} t) + u_{1,k} \cos(\sqrt{\lambda_k} t) + \int_0^t f_k(s) \cos(\sqrt{\lambda_k}(t-s))\, ds,$$

so that

$$\|U_p' - U_q'\|^2_{C^0([0,T];L^2(\Omega))} \leq 2 \sum_{k=p+1}^{q} \lambda_k u_{0,k}^2 + 2 \sum_{k=p+1}^{q} u_{1,k}^2 + 2T \sum_{k=p+1}^{q} \int_0^T f_k(s)^2\, ds$$

and the sequence U_n' is Cauchy in $C^0(0, T; L^2(\Omega))$, which completes the proof of the convergence of the series (9.4) in the above-mentioned spaces.

Regarding the wave equation itself, setting $F_n(t) = \sum_{k=1}^{n} f_k(t)\phi_k$, we have

$$\sum_{k=1}^{n} u_k''(t)\phi_k + \sum_{k=1}^{n} \lambda_k u_k(t)\phi_k = F_n(t).$$

For all test-functions $v \in H_0^1(\Omega)$, by taking the L^2 scalar product of the above formula with $v = \sum_{k=1}^{+\infty} v_k\phi_k$, we thus obtain

$$\sum_{k=1}^{n} u_k''(t)v_k + \sum_{k=1}^{n} \lambda_k u_k(t)v_k = (F_n(t)|v)_{L^2(\Omega)}.$$

Now $\sum_{k=1}^{n} u_k(t)v_k = (U_n(t)|v)_{L^2(\Omega)} \to (u(t)|v)_{L^2(\Omega)}$ in $C^0([0, T])$, so that

$$\sum_{k=1}^{n} u_k''(t)v_k = \big((U_n(t)|v)_{L^2(\Omega)}\big)'' \to \big((u(t)|v)_{L^2(\Omega)}\big)'' \text{ in the sense of } \mathscr{D}'(]0, T[)$$

when $n \to +\infty$. Similarly

$$\sum_{k=1}^{n} \lambda_k u_k(t)v_k = a(U_n(t), v) \to a(u(t), v) \text{ in } C^0([0, T]).$$

Finally, $F_n \to f$ in $L^2(0, T; L^2(\Omega))$ and therefore

$$(F_n(t)|v)_{L^2(\Omega)} \to (f(t)|v)_{L^2(\Omega)} \text{ in } L^2(0, T),$$

and we obtain the variational form of the wave equation in the limit $n \to +\infty$.

The initial conditions are obviously satisfied by construction. □

Remark 9.7 For this proof to work, we need the compatibility condition between the initial condition u_0 and the Dirichlet boundary condition, but no such condition is needed for the initial velocity u_1. Formulas (9.4)–(9.5) clearly generalizes the expansion obtained in Theorem 9.2. □

Remark 9.8 The d-dimensional wave equation can be solved along the exact same lines, see [5, 28]. However, here again, other approaches, such as semigroups, are possible. □

Remark 9.9 The series estimates above immediately imply stability in the energy norm for the weak solutions as well, in the sense that given two sets of data with corresponding solutions u_1 and u_2, we have

$$\|u_1 - u_2\|^2_{C^0([0,T];H^1_0(\Omega))} + \|u_1 - u_2\|^2_{C^1([0,T];L^2(\Omega))}$$
$$\leq C(|u_{1,0} - u_{2,0}|^2_{H^1_0(\Omega)} + \|u_{1,1} - u_{2,1}\|^2_{L^2(\Omega)} + \|f_1 - f_2\|^2_{L^2(Q)}).$$

This is also a continuity result of the solution with respect to the initial conditions and right-hand side. □

As a consequence, the energy equality of Proposition 9.1 is still valid here, the energy being defined by $E(t) = \frac{1}{2}\big(\|u'(t)\|^2_{L^2(\Omega)} + |u(t)|^2_{H^1_0(\Omega)}\big)$ as before. More precisely,

Proposition 9.3 *Let u be the solution given by Theorem 9.3. Then we have $E \in H^1(]0, T[)$ with*

$$\frac{dE}{dt}(t) = \big(f(t)|u'(t)\big)_{L^2(\Omega)}.$$

Proof We approximate u_0, u_1 and f by smooth functions u_0^n, u_1^n and f^n, in their respective function spaces. By the stability estimate above, the corresponding solution u^n is such that $u^n \to u$ in $C^0([0,T]; H_0^1(\Omega)) \cap C^1([0,T]; L^2(\Omega))$. We can apply Proposition 9.1 to u^n so that[2]

$$\frac{dE^n}{dt}(t) = \big(f^n(t)|(u^n)'(t)\big)_{L^2(\Omega)},$$

where $E^n(t)$ is the energy of u^n. Clearly $E^n \to E$ in $C^0([0,T])$. Moreover,

$$\begin{aligned}
&\Big|\big(f^n(t)|(u^n)'(t)\big)_{L^2(\Omega)} - \big(f(t)|u'(t)\big)_{L^2(\Omega)}\Big| \\
&\quad \le \Big|\big(f^n(t) - f(t)|(u^n)'(t)\big)_{L^2(\Omega)}\Big| + \Big|\big(f(t)|(u^n)'(t) - u'(t)\big)_{L^2(\Omega)}\Big| \\
&\quad \le \|f^n(t) - f(t)\|_{L^2(\Omega)}\|(u^n)'(t)\|_{L^2(\Omega)} \\
&\qquad + \|f(t)\|_{L^2(\Omega)}\|(u^n)'(t) - u'(t)\|_{L^2(\Omega)},
\end{aligned}$$

so that squaring and integrating in time, we obtain

$$\Big\|\frac{dE^n}{dt} - (f|u')_{L^2(\Omega)}\Big\|_{L^2(0,T)}^2 \le C\big(\|f^n - f\|_{L^2(0,T;L^2(\Omega))}^2 + \|(u^n)' - u'\|_{C^0([0,T];L^2(\Omega))}^2\big).$$

It follows from this that $\frac{dE^n}{dt} \to (f|u')_{L^2(\Omega)}$ in $L^2(0,T)$, and since $\frac{dE^n}{dt} \to \frac{dE}{dt}$ in the sense of $\mathscr{D}'(]0,T[)$, that $\frac{dE}{dt} = (f|u')_{L^2(\Omega)}$ belongs to $L^2(0,T)$. □

Remark 9.10 In the case $f = 0$, we obtain that the energy is also conserved for weak solutions. □

9.3 The Wave Equation on $\mathbb{R}$

We now consider the case of the wave equation on $\mathbb{R}$ with $f = 0$. There is no boundary condition. In this case, there is an explicit formula for the solution, similar to that obtained for the transport equation and known as the d'Alembert formula, see [35].

Theorem 9.4 *Let $u_0 \in C^1(\mathbb{R})$ and $u_1 \in C^0(\mathbb{R})$. The solution of problem (9.1) on $\mathbb{R}$ with $f = 0$ is given by*

$$u(x,t) = \frac{1}{2}\Big(u_0(x+t) + u_0(x-t) + \int_{x-t}^{x+t} u_1(s)\,ds\Big). \tag{9.8}$$

[2]We admit here that both formulations coincide in the smooth case.

Proof The function given by formula (9.8) is continuous on $\mathbb{R}\times\mathbb{R}_+$, hence is a distribution on $\mathbb{R}\times\mathbb{R}_+^*$. It is in fact of class C^1 on $\mathbb{R}\times\mathbb{R}_+$, and we can write

$$u(x,t) = F(x+t) + G(x-t) \tag{9.9}$$

with $F(y) = \frac{1}{2}\big(u_0(y) + \int_0^y u_1(s)\,ds\big)$ and $G(y) = \frac{1}{2}\big(u_0(y) + \int_y^0 u_1(s)\,ds\big)$. Let us set $U(x,t) = F(x+t)$ and $V(x,t) = G(x-t)$. Of course, we have

$$\frac{\partial U}{\partial t}(x,t) = F'(x+t), \quad \frac{\partial U}{\partial x}(x,t) = F'(x+t) \tag{9.10}$$

and

$$\frac{\partial V}{\partial t}(x,t) = -G'(x-t), \quad \frac{\partial V}{\partial x}(x,t) = G'(x-t). \tag{9.11}$$

Let us compute $\frac{\partial^2 U}{\partial t^2} - \frac{\partial^2 U}{\partial x^2}$ in the sense of distributions. We thus take $\varphi \in \mathscr{D}(\mathbb{R}\times\mathbb{R}_+^*)$ and consider the following duality bracket

$$\Big\langle \frac{\partial^2 U}{\partial t^2} - \frac{\partial^2 U}{\partial x^2}, \varphi \Big\rangle = -\Big\langle \frac{\partial U}{\partial t} - \frac{\partial U}{\partial x}, \frac{\partial \varphi}{\partial t} + \frac{\partial \varphi}{\partial x} \Big\rangle = 0,$$

by Eq. (9.10), and similarly $\frac{\partial^2 V}{\partial t^2} - \frac{\partial^2 V}{\partial x^2} = 0$ in the sense of distributions by Eq. (9.11).

Concerning the initial conditions, of course

$$u(x,0) = \frac{1}{2}\Big(u_0(x) + u_0(x) + \int_x^x u_1(s)\,ds\Big) = u_0(x)$$

and since

$$\frac{\partial u}{\partial t}(x,t) = \frac{1}{2}\big(u_0'(x+t) - u_0'(x-t) + u_1(x+t) + u_1(x-t)\big),$$

we have

$$\frac{\partial u}{\partial t}(x,0) = \frac{1}{2}\big(u_0'(x) - u_0'(x) + u_1(x) + u_1(x)\big) = u_1(x).$$

This proves the theorem. □

Remark 9.11 Formula (9.8) makes sense for much less regular data, for example u_0 and u_1 in $L^1_{\text{loc}}(\mathbb{R})$, and still gives rise to a solution of the wave equation in the sense of distributions. In fact, we may even take u_0 and u_1 in $\mathscr{D}'(\mathbb{R})$ by interpreting the integral as the sum of two primitives. The problem is thus to make sense of the initial condition in such a nonsmooth context, see Figs. 9.3 and 9.4. Some continuity with respect to time is needed, but we do not pursue in this direction.

Such an explicit formula as (9.8) is specific to the one-dimensional case. The solution is not so simple in higher dimensions of space. □

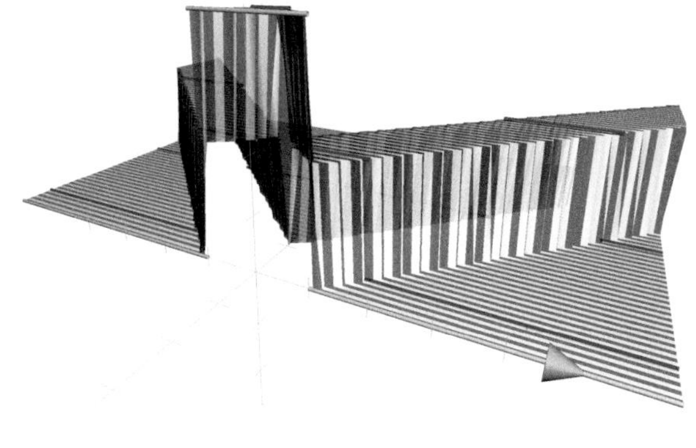

Fig. 9.3 A view of the evolution in the case of $u_0 = \mathbf{1}_{[-1/2,1/2]}$ and zero initial velocity, with the two issuing waves propagating right and left $\frac{1}{2}u_0(x-t)$ and $\frac{1}{2}u_0(x+t)$. Also pictured in *thicker red line* the solution at $t = \frac{1}{3}$ and $t = 2$

Fig. 9.4 A view of the evolution in the case of $u_1 = \mathbf{1}_{[-1,1]}$ and $u_0 = 0$. Also pictured in *thicker red line* the solution at $t = \frac{1}{2}$, $t = 1$ and $t = 2$. It is only Lipschitz in space and time

Remark 9.12 The solution pictured in Fig. 9.3 is discontinuous in space and time. It is thus meant to be understood as a solution of the wave equation in the sense of distributions. The interpretation of the initial conditions, in particular for the velocity, is admittedly a little more delicate. □

Remark 9.13 Independently of any considerations of initial data, it is easy to see that all solutions of the wave equation are of the form (9.9). Indeed, the change of variables $w = x + t$, $z = x - t$ leads to the equation $\frac{\partial^2 u}{\partial w \partial z} = 0$ whose solutions are clearly of the form $F(w) + G(z)$. The solution is thus seen as the superposition of two waves, one traveling to the left at speed -1 ($F(x+t)$) and the other traveling to the right at speed $+1$ ($G(x-t)$). In the general case, $c \neq 1$, the corresponding form is $u(x,t) = F(x+ct) + G(x-ct)$. □

Remark 9.14 We also see that the wave equation propagates waves at finite speed ($\pm c$), as opposed to the heat equation which has infinite speed of propagation. In particular, if the initial data are compactly supported in $[a,b]$, then the solution at time t is compactly supported in $[a-ct, b+ct]$. Another way of seeing this is to note that the value of the solution at point (x,t) only depends on what happens in its *backward cone of influence* $\{(y,s) \in \mathbb{R} \times \mathbb{R}_+; s \leq t, |y-x| \leq c(t-s)\}$, see Fig. 9.5. The information situated outside of the cone of influence does not have the time to propagate to point (x,t). □

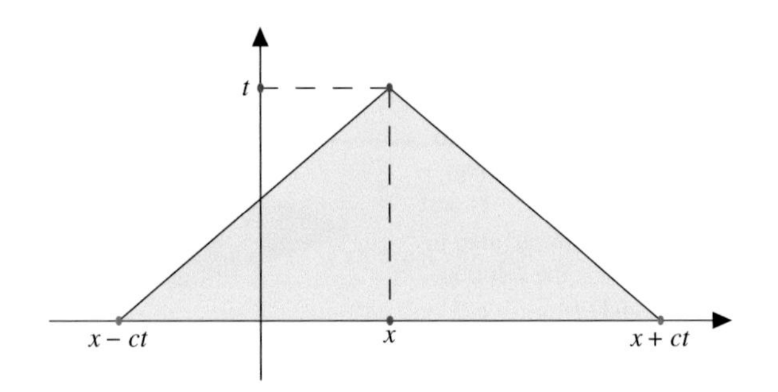

Fig. 9.5 The backward cone of influence of point (x, t)

Remark 9.15 If u_0 is compactly supported and $u_1 = 0$, an observer located at some point $x > 0$ initially outside of the support of u_0, sees a wave $\frac{1}{2}u_0(x - ct)$ reach him or her after some time, pass through, and then go back to exactly 0. This is a feature of the wave equation in odd dimensions of space. This explains why we see light and hear sounds as we do: a flash of light at some point in space-time results in a spherical wavefront expanding at the speed of light that an observer experiences as a single instantaneous flash when reached by the wavefront. The same goes for sound. This is not true in even dimensions. For example, if we throw a rock on a lake, the resulting wave on the surface of the lake expands as a circle traveling at the speed of waves on water, but never goes back to rest inside the disk, even though the solution is much smaller there. If we lived on the surface of the water, we would experience a flash followed by a never-ending afterglow... good thing we live in an odd-dimensional space. □

Remark 9.16 The wave equation is invariant under the change $t \to -t$. This means that time is reversible in the wave equation, which is another feature in sharp contrast with the heat equation. □

9.4 Finite Difference Schemes for the Wave Equation

The principle of finite difference methods for the wave equation is exactly the same as for the heat equation, and the notation is also the same. Since the wave equation is of second order in time, a natural idea is to consider two time steps finite difference schemes, even though we will see that this is not necessarily a good idea. We will assume the initial conditions U^0 and U^1 to be given in terms of u_0 and u_1. For instance, a simple choice could be

$$U^0 = S_h(u_0), \quad U^1 = U^0 + kS_h(u_1),$$

or higher order approximations for U^1.

The most obvious scheme consists in approximating the second time derivative by means of the usual central difference, which yields

$$\frac{u_n^{j+1}-2u_n^j+u_n^{j-1}}{k^2}-\frac{u_{n+1}^j-2u_n^j+u_{n-1}^j}{h^2}=f_n^j, \tag{9.12}$$

with the usual boundary conditions and initial data. This is obviously an explicit, two time steps scheme. In vector form, it reads

$$\frac{U^{j+1}-2U^j+U^{j-1}}{k^2}+A_hU^j=F^j,$$

where A_h is still given by formula (2.8) with $c=0$, p. 40 of Chap. 2. It should be quite clear that the scheme is consistent and of order 2 in space and time. Therefore, its convergence is solely a matter of stability.

We reformulate the above scheme as a single time step scheme by setting

$$V^j=\begin{pmatrix}U^j\\U^{j-1}\end{pmatrix}\in\mathbb{R}^{2N},$$

and

$$\begin{aligned}\frac{1}{k^2}V^{j+1}&=\begin{pmatrix}\frac{2}{k^2}I-A_h & -\frac{1}{k^2}I\\ \frac{1}{k^2}I & 0\end{pmatrix}\begin{pmatrix}U^j\\U^{j-1}\end{pmatrix}+\begin{pmatrix}F^j\\0\end{pmatrix}\\&=\frac{1}{k^2}\mathscr{A}V^j+G^j\end{aligned}$$

with a $2N\times 2N$ amplification matrix $\mathscr{A}=\begin{pmatrix}C & -I\\ I & 0\end{pmatrix}$ with $C=2I-k^2A_h$, and $G^j\in\mathbb{R}^{2N}$. Unfortunately, the matrix $\mathscr{A}$ is not normal. Indeed

$$\mathscr{A}^T\mathscr{A}=\begin{pmatrix}C^2+I & -C\\ -C & I\end{pmatrix}\neq\begin{pmatrix}C^2+I & C\\ C & I\end{pmatrix}=\mathscr{A}\mathscr{A}^T.$$

Therefore, we only have $\rho(\mathscr{A})\le|||\mathscr{A}|||_{2,h}$ and the condition $\rho(\mathscr{A})\le 1+C(T)k$ is just a necessary condition for stability, see Remark 8.13 in Chap. 8, whereas we also would like to have a sufficient condition for stability. In order to have a necessary stability condition that is valid for all T, it is easier to require $\rho(\mathscr{A})\le 1$. Let us see what we can say about the spectral radius of $\mathscr{A}$.

Lemma 9.1 *Let C be a $N\times N$ complex matrix and B the $2N\times 2N$ complex matrix defined by blocks as*

$$B=\begin{pmatrix}C & -I\\ I & 0\end{pmatrix}.$$

If $\lambda \in \mathbb{C}$ is an eigenvalue of B, then $\lambda \neq 0$ and $\lambda + \frac{1}{\lambda}$ is an eigenvalue of C. Conversely, if $\mu \in \mathbb{C}$ is an eigenvalue of C, then there exists an eigenvalue λ of B such that $\mu = \lambda + \frac{1}{\lambda}$.

Proof The proof is similar to that of Lemma 8.2 of Chap. 8. □

Proposition 9.4 *If $\frac{k}{h} \leq 1$, then the necessary stability condition for scheme (9.12) is satisfied.*

Proof Recall that stability is meant here in the sense of $\rho(\mathscr{A}) \leq 1$. So we need to find out when all the eigenvalues λ of $\mathscr{A}$ are such that $|\lambda| \leq 1$. According to Lemma 9.1, the eigenvalues in question are of the form $\lambda_{\pm} = \frac{\mu \pm \sqrt{\mu^2 - 4}}{2}$ where μ is an eigenvalue of $C = 2I - k^2 A_h$, hence is real. We thus see that there are two cases:

1. $|\mu| > 2$. In this case, the two eigenvalues $\lambda_{\pm}$ are real, distinct, and since their product is equal to 1, one of them is strictly larger than 1 in absolute value. Hence this is an unstable case.

2. $|\mu| \leq 2$. In this case, $\lambda_{\pm}$ are complex conjugate, and since their product is equal to 1, they are both of modulus 1. The necessary stability condition is thus satisfied.

Now we have $\mu = 2 - 4\frac{k^2}{h^2}\sin^2\left(\frac{p\pi}{2(N+1)}\right)$, $p = 1, \ldots, N$. Clearly, if $\frac{k}{h} \leq 1$, then we have $|\mu| \leq 2$. □

Remark 9.17 The condition $\frac{k}{h} \leq 1$ is called the *Courant–Friedrichs–Lewy* or *CFL* condition. In the general case, the CFL condition assumes the form $\frac{k}{h} \leq \frac{1}{c}$. In a sense, $\frac{h}{k}$ is the numerical velocity needed to reach the neighboring grid points in one time step starting from one spatial grid point, see Fig. 9.6. The CFL condition is that this numerical velocity must be larger than the propagation velocity.

In other words, the discrete backward cone of influence of a point (x_n, t_j) must contain its continuous backward cone of influence, in order for the scheme to have access to all the information needed to compute a relevant approximation at that point. Of course, this kind of requirement only applies to explicit schemes. □

Let us plot the result of the explicit scheme with + marks and the exact solution in solid line in Fig. 9.7. We take the same u_0 as for the heat equation, i.e., $u_0(x) = \sin(\pi x)/2 + \sin(2\pi x)$, and $u_1 = 0$. We have taken $U^1 = U^0$, which is actually a

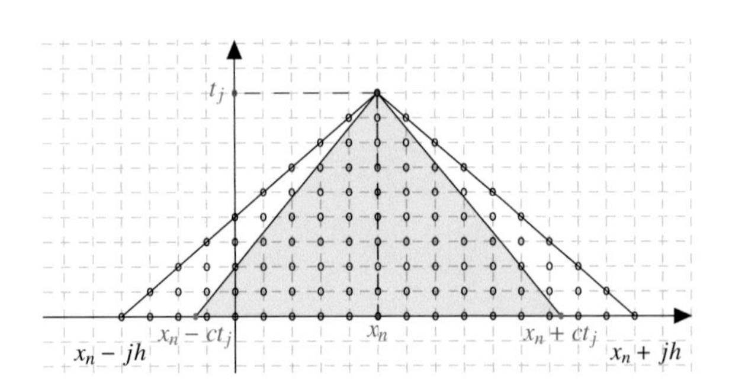

Fig. 9.6 Discrete cone of influence versus continuous cone of influence

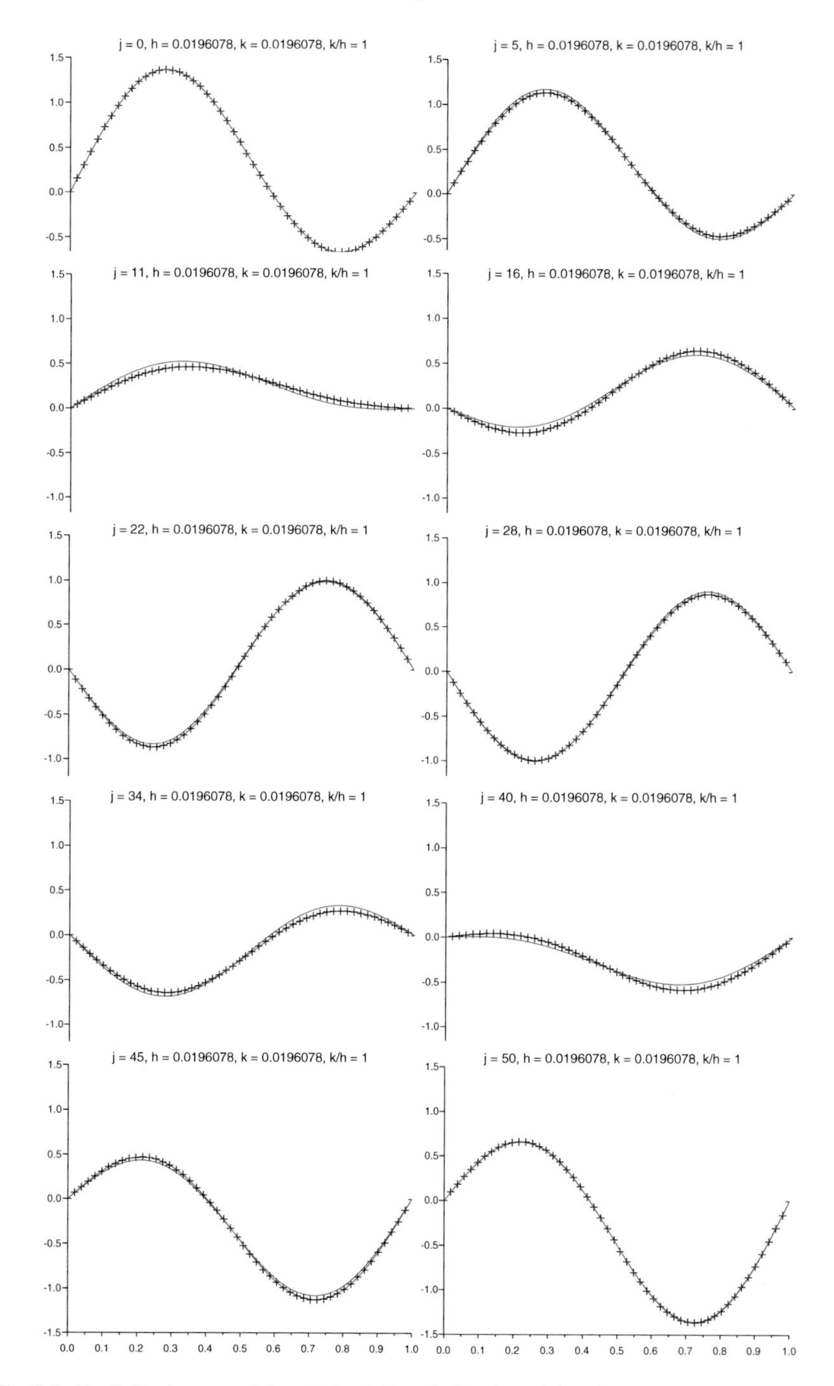

Fig. 9.7 Explicit scheme, $u_0(x) = \sin(\pi x)/2 + \sin(2\pi x)$, $u_1(x) = 0$

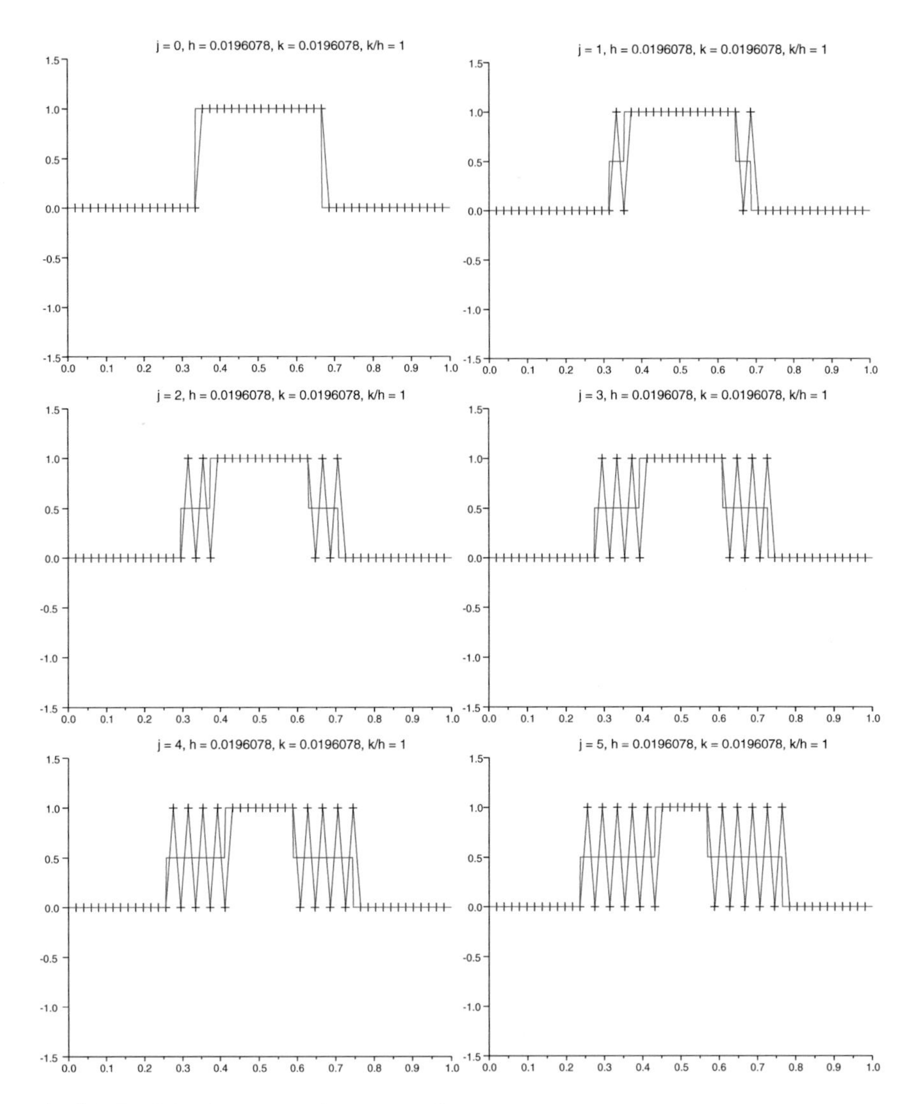

Fig. 9.8 Explicit scheme, $u_0 = \mathbf{1}_{[\frac{1}{3},\frac{2}{3}]}$, $u_1 = 0$

second order approximation of the condition $u_1 = 0$, hence the good global accuracy of the scheme in this particular case.

Of course, the initial condition is very smooth here. If we want to compute a discontinuous solution with this scheme, we run into trouble with severe unwanted oscillations, see Fig. 9.8. This kind of discontinuous solution is of physical interest in situations where *shock waves* occur. Devising numerical schemes capable of reliably capturing shocks thus requires skills that go beyond the scope of these notes.

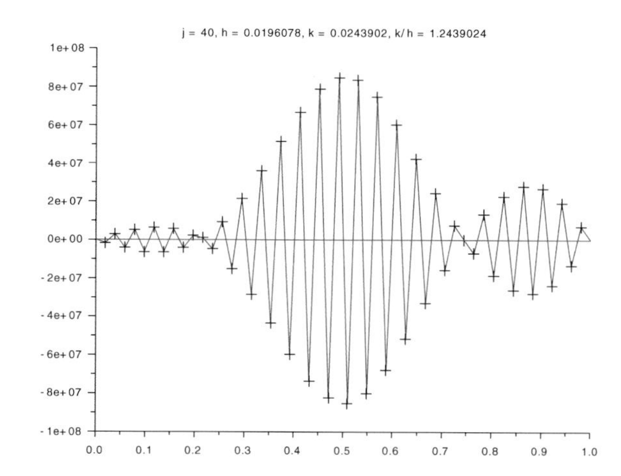

Fig. 9.9 Explicit scheme, $u_0(x) = \sin(\pi x)/2 + \sin(2\pi x)$, $u_1(x) = 0$, CFL condition not satisfied

For the record, Fig. 9.9 shows what happens after a few iterations when the CFL condition is violated, even with the very smooth initial condition used above.

The next obvious scheme is the implicit version of the former one

$$\frac{u_n^{j+1} - 2u_n^j + u_n^{j-1}}{k^2} - \frac{u_{n+1}^{j+1} - 2u_n^{j+1} + u_{n-1}^{j+1}}{h^2} = f_n^{j+1}, \tag{9.13}$$

with the usual boundary conditions and initial data. In vector form, it reads

$$\Big(\frac{1}{k^2} I + A_h\Big) U^{j+1} = \frac{2}{k^2} U^j - \frac{1}{k^2} U^{j-1} + F^{j+1}.$$

We rewrite it as a single time step scheme

$$V^{j+1} = \mathscr{A} V^j + G^j$$

with

$$\mathscr{A} = \begin{pmatrix} 2(I + k^2 A_h)^{-1} & -(I + k^2 A_h)^{-1} \\ I & 0 \end{pmatrix}.$$

Again, the matrix $\mathscr{A}$ is not normal, but we can look at its spectral radius. Setting $C = (I + k^2 A_h)^{-1}$, the same kind of arguments as before show that the eigenvalues λ of $\mathscr{A}$ are of the form $\lambda_\pm = \mu \pm \sqrt{\mu^2 - \mu}$ where μ is an eigenvalue of C. Now $\mu \in]0, 1[$, therefore $\mu^2 - \mu < 0$ and the eigenvalues λ_+ and λ_- are complex conjugate, of modulus $\sqrt{\mu}$. The necessary condition for stability is thus unconditionally satisfied.

The implicit scheme does a slightly better job of capturing shocks than the explicit scheme for the same discretization parameters, but it still has a lot of numerical diffusion that spreads out the shocks, see Fig. 9.10.

A third scheme is the θ-scheme for $\theta \in [0, \frac{1}{2}]$, written here for $f = 0$,

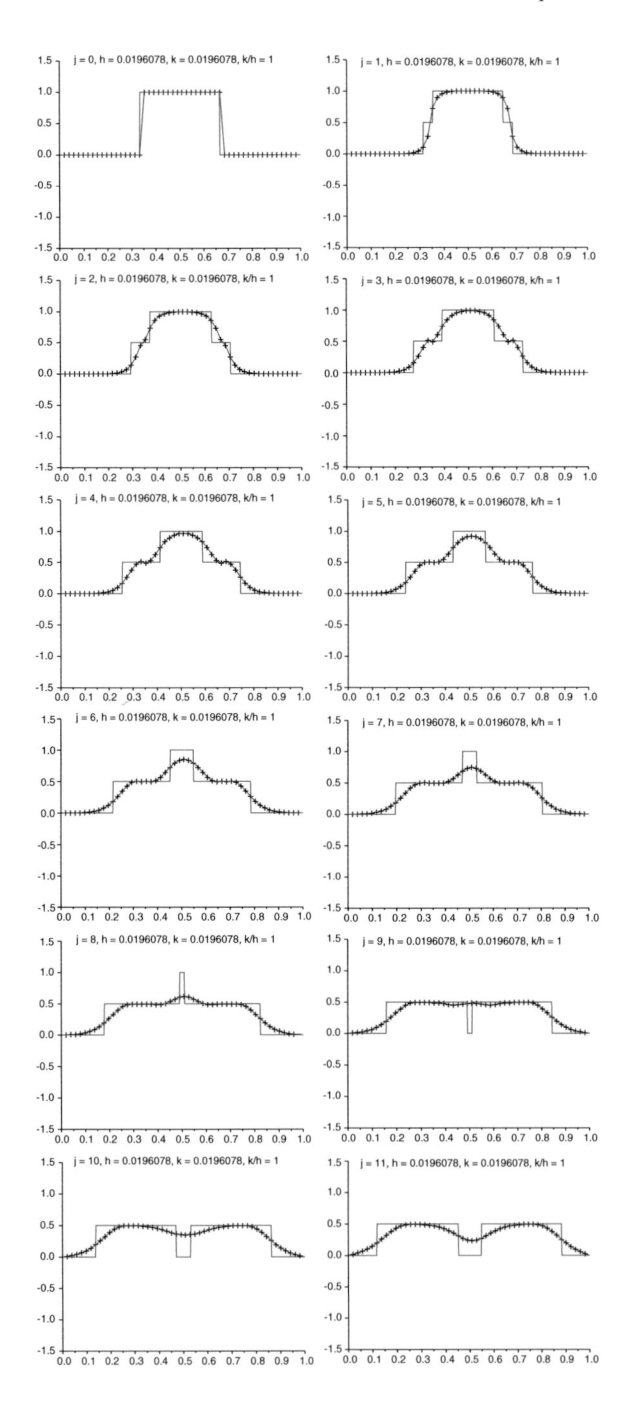

Fig. 9.10 Implicit scheme, $u_0 = \mathbf{1}_{[\frac{1}{3}, \frac{2}{3}]}, u_1 = 0$

$$\frac{u_n^{j+1} - 2u_n^j + u_n^{j-1}}{k^2} - \theta \frac{u_{n+1}^{j+1} - 2u_n^{j+1} + u_{n-1}^{j+1}}{h^2} \\ - (1-2\theta)\frac{u_{n+1}^j - 2u_n^j + u_{n-1}^j}{h^2} - \theta \frac{u_{n+1}^{j-1} - 2u_n^{j-1} + u_{n-1}^{j-1}}{h^2} = 0, \quad (9.14)$$

which reduces to the explicit scheme for $\theta = 0$ and is implicit for $\theta > 0$.

9.5 Stability via the Fourier Approach

So far we have only obtained necessary conditions for stability, because the matrices were not normal. As in the case of numerical schemes for the heat equation, we can also use the Fourier method to obtain sufficient conditions. Again, we work on the whole of $\mathbb{R}$. We take $f = 0$. As was already mentioned, there are no boundary conditions and let us forget for the moment that the solution is given by an explicit formula.

Concerning the relationship between discrete and semi-discrete schemes, every thing said for the heat equation holds true here. In addition, we note that, due to the finite speed of propagation, if the initial data is compactly supported in an interval, a finite difference scheme on the interval with boundary conditions will compute exactly the same values as the same scheme on $\mathbb{R}$ as long as the wave has not hit the ends of the interval. Therefore, so does the semi-discrete scheme, and the stability conditions obtained from the Fourier method actually apply to the scheme on an interval with boundary conditions (at least under the previous conditions).

Let us first consider the explicit scheme (9.12), for which we already have a necessary stability condition, but no sufficient condition. Both Fourier approaches are more complicated than for the heat equation, since the linear recurrence relations obtained are two-step relations, which are harder to analyze than the one-step relations in the heat equation case. Since the two Fourier approaches are very similar to each other, we first concentrate on the Fourier series point of view for brevity, see Sect. 8.7.

Since we are working on the whole space, the scheme (9.12) with zero right-hand side takes the form

$$\frac{u_n^{j+1} - 2u_n^j + u_n^{j-1}}{k^2} - \frac{u_{n+1}^j - 2u_n^j + u_{n-1}^j}{h^2} = 0, \text{ for } n \in \mathbb{Z}. \quad (9.15)$$

In Fourier space, the scheme reads

$$\frac{\mathscr{F}(u^{j+1})(s) - 2\mathscr{F}(u^j)(s) + \mathscr{F}(u^{j-1})(s)}{k^2} + \frac{4}{h^2}\sin^2\left(\frac{s}{2}\right)\mathscr{F}(u^j)(s) = 0, \quad (9.16)$$

for $s \in [0, 2\pi]$. As was mentioned above, this is a two-step linear recurrence relation, which we rewrite in vector form by introducing the $\mathbb{R}^2$-valued sequence

$Z_n^j = \begin{pmatrix} u_n^{j+1} \\ u_n^j \end{pmatrix}$, the Fourier series transform of which satisfies

$$\mathscr{F}(Z^{j+1})(s) = A(s)\mathscr{F}(Z^j)(s)$$

where

$$A(s) = \begin{pmatrix} 2 - a(s)^2 & -1 \\ 1 & 0 \end{pmatrix}$$

with $a(s) = \frac{2k}{h}\sin\left(\frac{s}{2}\right)$. The matrix A is called the *amplification matrix* of the scheme. We list without proof several properties of such schemes, since they are easy generalizations of former results.

Let $A \in C^0([0, 2\pi]; \mathbb{M}_2(\mathbb{C}))$ and consider the multiplier operator L defined on $L^2([0, 2\pi]; \mathbb{C}^2)$ with values in $L^2([0, 2\pi]; \mathbb{C}^2)$ by $(LY)(s) = A(s)Y(s)$, for almost all $s \in [0, 2\pi]$. The generalization of Eq. (8.7) in this case is

$$\|L\|_{\mathscr{L}(L^2([0,2\pi];\mathbb{C}^2))} = \max_{s\in[0,2\pi]} |||A(s)|||_2$$

(we use the standard hermitian norm on $\mathbb{C}^2$), see [67]. It follows that the scheme (9.16) is stable in L^2 if and only if there exists a constant $C(T)$ such that

$$\max_{s\in[0,2\pi]} |||A(s)^j|||_2 \leq C(T)$$

for all $j \leq T/k$. Since $\rho(A(s)) \leq |||A(s)^j|||_2^{1/j}$, a necessary condition of stability is that there exists a nonnegative constant C that does not depend on k and h such that $\rho(A(s)) \leq 1 + Ck$ for all s. We say that the scheme is *stable in the sense of von Neumann* if $\rho(A(s)) \leq 1$ for all s.

Proposition 9.5 *The scheme (9.16) is stable in the sense of von Neumann if and only if* $\frac{k}{h} \leq 1$.

Proof The characteristic polynomial of $A(s)$ is

$$P_{A(s)}(X) = X^2 - (2 - a(s)^2)X + 1.$$

The product of the two roots is 1, thus if they are both real and simple, one of them is strictly larger than 1 in absolute value. On the other hand, if they are complex conjugate, they are both of modulus 1. Consequently, von Neumann stability is equivalent to the discriminant being non positive.

We thus need to see under which condition $\Delta(s) = a(s)^2(a(s)^2 - 4) \leq 0$ for all s. Taking $s = \pi$, we see that a necessary condition is $\frac{k}{h} \leq 1$. Conversely, this condition is clearly sufficient. □

In the case of a normal amplification matrix, the former conditions are also sufficient for L^2-stability. Now the matrix $A(s)$ obtained above is not normal in general, except for $s = \pi$ if $\frac{k}{h} = 1$, thus we need a more general result. We work directly on the stability of the scheme expressed in Fourier space by the operator L above, since this is equivalent to L^2-stability for the discrete scheme, due to Parseval's formula.

Stability in L^2 is ensured if $\|\!| A(s)^j \|\!|_2 \leq C(T)$ for all s and $j \leq T/k$, where $C(T)$ does not depend on h or k. It should be noted that this condition is a priori much easier to check here than it was for (actual) finite difference schemes, since the matrix in question is always of the same size, i.e., 2×2, whereas the size of the matrix in finite difference schemes was $N \times N$ with $h = \frac{1}{N+1}$ and the norm also depended on h.

Proposition 9.6 *The discrete scheme (9.16) is not stable in L^2.*

Proof It is enough to look at what happens for $s = 0$. In this case, the amplification matrix has the double eigenvalue 1 and is clearly not diagonalizable. In effect,

$$A(0) = \begin{pmatrix} 2 & -1 \\ 1 & 0 \end{pmatrix} = P \begin{pmatrix} 1 & 1 \\ 0 & 1 \end{pmatrix} P^{-1}, \text{ where } P = \begin{pmatrix} 1 & 1 \\ 1 & 0 \end{pmatrix},$$

from which we immediately deduce that

$$A(0)^j = \begin{pmatrix} j+1 & -j \\ j & 1-j \end{pmatrix}.$$

We thus have $\|\!| A(0)^j \|\!|_2 = \sqrt{2j^2 + 1 + 2j\sqrt{j^2+1}} \geq 2j$. It follows in particular that $\|\!| A(0)^{T/k} \|\!|_2 \geq \frac{2T}{k}$, hence the instability of the discrete scheme. □

Remark 9.18 This is an unsettling result, since we could have rightfully expected the very natural scheme (9.15) to be stable under the CFL condition $\frac{k}{h} \leq 1$ or at least $\frac{k}{h} < 1$. This is not the case.

The reason for the natural scheme not to be stable is the following. If it was stable in ℓ^2, then having sequences of initial data $(u^0)_m$ and $(u^1)_m$ bounded in ℓ^2 would result in a sequence of solutions $(u^j)_m$ also bounded in ℓ^2, uniformly for $j \leq T/k$. However, this boundedness could be achieved with $(u^0)_m$ and $(u^1)_m$ having strictly nothing to do with each other. Now we need to remember that $u_n^0 = u_0(nh)$ and u_n^1 is supposed to be some approximation of $u(nh, k) \approx u_0(nh) + k\frac{\partial u}{\partial t}(nh, 0)$. Thus both initial conditions of the wave equation $u(x, 0) = u_0(x)$ and $\frac{\partial u}{\partial t}(x, 0) = u_1(x)$ must somehow be taken into account in the discrete initial data u^0 and u^1 for the scheme to have any chance to converge. This is not the case if $(u^0)_m$ and $(u^1)_m$ can be chosen independently of each other.

Note that the scheme is unstable in spite of being von Neumann stable, an unhappy effect of terminology. □

Remark 9.19 We can see the instability of the scheme on the following example. Let us consider the initial data $u^0 = 0$ and u^1 given by $u_0^1 = h^{-1/2}$ and $u_n^1 = 0$ for $n \neq 0$. We have $\|u^1\|_{2,h} = 1$. Let us show that $\|u^j\|_{2,h}$ is not bounded for $j \leq T/k$. In Fourier space, the recurrence relation reads

$$\mathscr{F}(u^{j+1})(s) = (2 - a^2(s))\mathscr{F}(u^j)(s) - \mathscr{F}(u^{j-1})(s).$$

The characteristic equation of this recurrence relation $X^2 - (2 - a^2(s))X + 1 = 0$ has two roots $r_\pm = e^{\pm i\theta(s)}$, $\theta(s) = \arccos(1 - \frac{a^2(s)}{2})$ for $s \neq 0$ and $s \neq 2\pi$, and a double root $r = 1$ for $s = 0$ or 2π.

We have $\mathscr{F}(u^0)(s) = 0$ and $\mathscr{F}(u^1)(s) = h^{-1/2}$, thus

$$\mathscr{F}(u^j)(s) = \begin{cases} h^{-1/2}\frac{\sin(j\theta(s))}{\sin(\theta(s))}, & \text{for } 0 < s < 2\pi, \\ jh^{-1/2}, & \text{for } s = 0 \text{ or } s = 2\pi. \end{cases}$$

We are interested in the L^2 norm of the above function. Clearly

$$\|\mathscr{F}(u^j)\|^2_{L^2(0,2\pi),h} \geq \int_0^{\theta^{-1}(\frac{\pi}{2j})} \frac{\sin^2(j\theta(s))}{\sin^2(\theta(s))}\,ds.$$

Now, on the interval $\left[0, \theta^{-1}\left(\frac{\pi}{2j}\right)\right]$, we have $\frac{\sin(j\theta(s))}{\sin(\theta(s))} \geq \frac{2j}{\pi}$. Consequently

$$\|\mathscr{F}(u^j)\|^2_{L^2(0,2\pi),h} \geq \frac{4j^2}{\pi^2}\theta^{-1}\left(\frac{\pi}{2j}\right).$$

After a little bit of computation, we find that $\theta^{-1}\left(\frac{\pi}{2j}\right) \sim \frac{\pi}{2\sqrt{2\lambda}j}$ when $j \to +\infty$. Therefore, for j large enough, we obtain

$$\|\mathscr{F}(u^j)\|^2_{L^2(0,2\pi),h} \geq \frac{j}{\pi\sqrt{2\lambda}} \to +\infty \text{ when } j \to +\infty.$$

Going back to the original discrete scheme, it follows that $\|u^j\|_{2,h} \to +\infty$ when $j \to +\infty$ for this particular sequence of bounded initial data. □

In order to obtain sufficient stability conditions, we actually need to change the unknowns so as to appropriately take care of the wave equation initial conditions. First, we rewrite the wave equation: Find $u\colon Q = \mathbb{R} \times [0, T] \to \mathbb{R}$ such that

$$\begin{cases} \dfrac{\partial^2 u}{\partial t^2} - \dfrac{\partial^2 u}{\partial x^2} = 0 \text{ in } Q, \\ u(x, 0) = u_0(x), \dfrac{\partial u}{\partial t}(x, 0) = u_1(x) \text{ for } x \in \mathbb{R}, \end{cases} \tag{9.17}$$

as a first order system.

Proposition 9.7 *Let $v = \frac{\partial u}{\partial t}$ and $w = \frac{\partial u}{\partial x}$. Problem (9.17) is equivalent to the system of first order PDEs*

$$\begin{cases} \dfrac{\partial v}{\partial t} - \dfrac{\partial w}{\partial x} = 0 \text{ in } Q, \\ \dfrac{\partial w}{\partial t} - \dfrac{\partial v}{\partial x} = 0 \text{ in } Q, \\ v(x,0) = u_1(x),\, w(x,0) = u_0'(x) \text{ for } x \in \mathbb{R}, \end{cases} \tag{9.18}$$

up to an additive constant.

Proof If u solves problem (9.17), then the first equation in (9.18) is just the wave equation, and the second equation is just $\frac{\partial^2 u}{\partial x \partial t} = \frac{\partial^2 u}{\partial t \partial x}$. The initial conditions are obvious.

Conversely, let (v, w) solve (9.18). Since Q is simply connected, the second equation implies that there exists $\widetilde{u}$ such that $v = \frac{\partial \widetilde{u}}{\partial t}$ and $w = \frac{\partial \widetilde{u}}{\partial x}$. The first equation is the wave equation for $\widetilde{u}$. By the second initial condition, there exists a constant c_0 such that $\widetilde{u}(x,0) = u_0(x) + c_0$. Hence, $u = \widetilde{u} - c_0$ solves (9.17). □

Remark 9.20 Let us note that, in the case of a bounded interval with Dirichlet conditions, the energy estimate of Corollary 9.1 gives an L^2 bound on the variables v and w, and not directly on u. This explains the choice of these variables for an L^2 stability analysis. □

We perform a similar operation on the finite difference scheme. We use the notation $x_\tau = \tau h$ for $\tau \in \mathbb{R}$, which agrees with the former notation x_n when $\tau = n \in \mathbb{Z}$.

Proposition 9.8 *Let*

$$v_n^j = \frac{u_n^j - u_n^{j-1}}{k} \text{ and } w_{n-1/2}^j = \frac{u_n^j - u_{n-1}^j}{h}, \tag{9.19}$$

for $n \in \mathbb{Z}$ and $j \in \mathbb{N}^*$. If u^j is a solution of the finite difference scheme (9.15), then (v^j, w^j) are solution of the finite difference scheme

$$\begin{cases} \dfrac{v_n^{j+1} - v_n^j}{k} - \dfrac{w_{n+1/2}^j - w_{n-1/2}^j}{h} = 0, \\ \dfrac{w_{n-1/2}^{j+1} - w_{n-1/2}^j}{k} - \dfrac{v_n^{j+1} - v_{n-1}^{j+1}}{h} = 0, \end{cases} \tag{9.20}$$

with v_n^1 and $w_{n-1/2}^1$ given by formula (9.19) in terms of the initial data u_n^0 and u_n^1 of (9.15).

Proof Replacing $v_n^j = \frac{u_n^j - u_n^{j-1}}{k}$ and $w_{n-1/2}^j = \frac{u_n^j - u_{n-1}^j}{h}$ into (9.20), we see that the second relation is satisfied by the very definition of v_n^j and $w_{n-\frac{1}{2}}^j$, and that the first relation reduces to the original finite difference scheme. Moreover, the initial conditions are satisfied by construction. □

Remark 9.21 We note that the scheme (9.20) is explicit. Indeed, assuming that v_n^j and $w_{n+1/2}^j$ are already known, the first relation gives $v_n^{j+1} = v_n^j + \frac{k}{h}(w_{n+1/2}^j - w_{n-1/2}^j)$ for all $n \in \mathbb{Z}$. After that, we see that $w_{n-1/2}^{j+1} = w_{n-1/2}^j + \frac{k}{h}(v_n^{j+1} - v_{n-1}^{j+1})$ for all $n \in \mathbb{Z}$ by the second relation. Hence, the solution of (9.20) with initial data v^1 and w^1 exists and is unique. □

Remark 9.22 Clearly, v_n^j is intended to be an approximation of $v(x_n, t_j) = \frac{\partial u}{\partial t}(x_n, t_j)$ and $w_{n-1/2}^j$ an approximation of $w(x_{n-1/2}, t_j) = \frac{\partial u}{\partial x}(x_{n-1/2}, t_j)$.

In view of this, other initial data are reasonable for (9.20), for example $v_n^0 = u_1(x_n)$ and $w_{n-1/2}^0 = u_0'(x_{n-\frac{1}{2}})$, yielding a different approximation from which an approximation of u must be reconstructed. These initial conditions directly take into account the initial conditions of the wave equation. □

This time, we choose to work in the continuous Fourier transform framework, see Sect. 8.8. We thus introduce the semi-discrete version of the scheme as

$$\begin{cases} \dfrac{v^{j+1}(x) - v^j(x)}{k} - \dfrac{w^j(x+h/2) - w^j(x-h/2)}{h} = 0, \\ \dfrac{w^{j+1}(x-h/2) - w^j(x-h/2)}{k} - \dfrac{v^{j+1}(x) - v^{j+1}(x-h)}{h} = 0, \end{cases} \tag{9.21}$$

for all $x \in \mathbb{R}$ and with appropriate initial data. Rewriting this in Fourier space, we obtain

$$\begin{cases} \dfrac{\widehat{v^{j+1}}(\xi) - \widehat{v^j}(\xi)}{k} - \dfrac{e^{i\frac{h\xi}{2}} - e^{-i\frac{h\xi}{2}}}{h}\widehat{w^j}(\xi) = 0, \\ e^{-i\frac{h\xi}{2}}\dfrac{\widehat{w^{j+1}}(\xi) - \widehat{w^j}(\xi)}{k} - \dfrac{1 - e^{-ih\xi}}{h}\widehat{v^{j+1}}(\xi) = 0, \end{cases}$$

for all $\xi \in \mathbb{R}$. Writing $Y^j(x) = \begin{pmatrix} v^j(x) \\ w^j(x) \end{pmatrix}$, we obtain

$$\widehat{Y^j}(\xi) = B(\xi)\widehat{Y^{j-1}}(\xi)$$

with the amplification matrix

$$B(\xi) = \begin{pmatrix} 1 & ia(\xi) \\ ia(\xi) & 1 - a(\xi)^2 \end{pmatrix}$$

where $a(\xi) = \frac{2k}{h}\sin\left(\frac{h\xi}{2}\right)$.

We see that amplification matrices are now complex matrices, we thus need to generalize the results of Chap. 8 to the complex case. First of all, when A is a complex $N \times N$ matrix, its induced matrix norm for the canonical Hermitian norm on $\mathbb{C}^N$ is defined as

$$|||A|||_{2,h} = \sup_{X \in \mathbb{C}^N, X \neq 0} \frac{\|AX\|_{2,h}}{\|X\|_{2,h}}.$$

A complex matrix A is said to be *normal* if $A^*A = AA^*$ where A^* is the adjoint matrix. The following results are proved along the same lines as the corresponding results in the real case.

Proposition 9.9 *Let A be a complex $N \times N$ matrix. We have*

$$|||A|||_{2,h} = \sqrt{\rho(A^*A)}.$$

In addition, if A is normal then $\rho(A) = \rho(A^*A)^{1/2} = |||A|||_{2,h}$.

We now return to the stability of scheme (9.21).

Proposition 9.10 *The scheme (9.21) is stable in the sense of von Neumann if and only if $\frac{k}{h} \leq 1$.*

Proof The matrix $B(\xi)$ has the same characteristic polynomial as the matrix $A(s)$ of Proposition 9.5, therefore the proof is the same. □

The matrix $B(\xi)$ is not normal in general, thus von Neumann stability is not a priori sufficient for L^2 stability. The following simple matrix result is useful in this context. Let $M \in \mathbb{M}_2(\mathbb{C})$. Every complex matrix is triangularisable, thus we can write $M = PUP^{-1}$ with $P \in GL_2(\mathbb{C})$ and U upper-triangular.

Proposition 9.11 *For all diagonalizable matrices $M \in \mathbb{M}_2(\mathbb{C})$ such that $\rho(M) \leq 1$, we have*

$$|||M^j|||_2 \leq |||P|||_2 |||P^{-1}|||_2,$$

for all $j \in \mathbb{N}$.

Proof We can write $M = PUP^{-1}$ with $P \in GL_2(\mathbb{C})$ and U diagonal,

$$U = \begin{pmatrix} \lambda_1 & 0 \\ 0 & \lambda_2 \end{pmatrix},$$

where λ_1, λ_2 are the eigenvalues of M. We have $M^j = PU^jP^{-1}$. Therefore

$$|||M^j|||_2 \leq |||P|||_2 |||U^j|||_2 |||P^{-1}|||_2.$$

Now $|||U^j|||_2 = \rho(M)^j$, which completes the proof. □

Remark 9.23 The constant $|||P|||_2 |||P^{-1}|||_2$ appearing in the estimate of M^j is nothing but the condition number (introduced in Remark 2.11 of Chap. 2) of the change of basis matrix P. □

Remark 9.24 If M is not diagonalizable, it has a double eigenvalue λ and we have $U = \Lambda + N$ with

$$\Lambda = \begin{pmatrix} \lambda & 0 \\ 0 & \lambda \end{pmatrix}, \quad N = \begin{pmatrix} 0 & 1 \\ 0 & 0 \end{pmatrix},$$

by the Jordan decomposition theorem. The matrix N is nilpotent, $N^2 = 0$, and commutes with Λ. Therefore, by the binomial identity,

$$U^j = \Lambda^j + j\lambda^{j-1}N,$$

for all $j \in \mathbb{N}$. It follows that if $\rho(M) < 1$,

$$|||U^j|||_2 \leq C(\rho(M)),$$

for all $j \in \mathbb{N}$, where C is a function of the spectral radius, and if $\rho(M) = 1$ then

$$|||M^j|||_2 \to +\infty$$

when $j \to +\infty$. □

Let us now apply Proposition 9.11 to the study of the stability of the scheme (9.21) by applying the proposition to the matrix $M = B(\xi)$, for any $\xi \in \mathbb{R}$. We let

$$B(\xi) = P(\xi)U(\xi)P^{-1}(\xi)$$

where $P(\xi) \in GL_2(\mathbb{C})$ and $U(\xi) \in \mathbb{M}_2(\mathbb{C})$ is upper-triangular for all $\xi \in \mathbb{R}$. Of course, all these matrices are also functions of h and k.

Proposition 9.12 *Let $0 < \lambda_0 < 1$ and assume that $\frac{k}{h} \leq \lambda_0$. Then, the scheme (9.21) is stable in $L^2(\mathbb{R})$.*

Proof In the case $a(\xi) = 0$, then $B(\xi) = I = U(\xi) = P(\xi)$ and there is nothing to do. Let us assume that ξ is such that $a(\xi) \neq 0$. In this case, $\Delta(\xi) < 0$, there are two simple eigenvalues and $B(\xi)$ is diagonalizable. We already know that the eigenvalues of $B(\xi)$ are of modulus 1, so that $\rho(B(\xi)) = 1$. We have the estimate

$$|||B^j(\xi)|||_2 \leq |||P(\xi)|||_2 |||P^{-1}(\xi)|||_2,$$

by Proposition 9.11. Thus, we only need to bound the condition number of the matrix $P(\xi)$. Let us note for the record that $|a(\xi)| \leq 2\lambda_0 < 2$ for all ξ.

Computing the eigenvectors of $A(\xi)$, we find that the change of basis matrix

$$P(\xi) = \begin{pmatrix} ia(\xi) & ia(\xi) \\ \lambda_+(\xi) - 1 & \lambda_-(\xi) - 1 \end{pmatrix}$$

is uniformly bounded since $|a(\xi)| < 2$. More importantly, since $\lambda_\pm(\xi) - 1 = \dfrac{-a(\xi)^2 \pm ia(\xi)\sqrt{4 - a(\xi)^2}}{2}$, we clearly have

$$|||P(\xi)|||_2 \leq C|a(\xi)|.$$

Moreover,

$$P(\xi)^{-1} = \det(P(\xi))^{-1}\begin{pmatrix}\lambda_-(\xi)-1 & -ia(\xi)\\ 1-\lambda_+(\xi) & ia(\xi)\end{pmatrix} = \det(P(\xi))^{-1}Q(\xi).$$

We likewise have $|||Q(\xi)|||_2 \le C|a(\xi)|$. Since

$$\det(P(\xi)) = ia(\xi)(\lambda_-(\xi)-\lambda_+(\xi)) = a(\xi)^2\sqrt{4-a(\xi)^2},$$

it follows that

$$|\det(P(\xi))| \ge 2\sqrt{1-\lambda_0^2}|a(\xi)|^2,$$

hence

$$|||P(\xi)^{-1}|||_2 \le C|a(\xi)|^{-1},$$

and the result follows. □

Remark 9.25 We have now established the conditional ℓ^2 stability of scheme (9.20). This means that under the CFL condition, if the initial data v^1 and w^1 remain in a bounded set of ℓ^2, so do the corresponding solutions v^j and w^j for $j \le T/k$. The unknowns v and w, which are in a sense quite natural for formulating the wave equation, are however not the initial unknown u, either continuous or discrete. The initial values v^1 and w^1 are supposed to be approximations of $\frac{\partial u}{\partial t}(x,0) = u_1(x)$ and $\frac{\partial u}{\partial x}(x,0) = u_0'(x)$ respectively. In this sense, they are independent from each other as opposed to what happened in the case of the natural scheme, cf. Remark 9.18.

A natural question now is to ask if the first order scheme provides some stability information for the original scheme. We encounter a difficulty here. Indeed, if we try to reconstruct u^j from w^j, i.e., perform a kind of discrete integration, we see that

$$u_n^j = u_0^j + h\sum_{l=0}^{n} w_{n-l-1/2}^j,$$

for $n \ge 0$. Now requiring that $u^j \in \ell^2(\mathbb{Z})$ implies that $u_n^j \to 0$ when $n \to +\infty$. So the partial sums on the right must converge and we must have

$$u_n^j = -h\sum_{l=n+1}^{\infty} w_{n-l-1/2}^j,$$

again for $n \ge 0$. This is not possible in general since $w^j \in \ell^2(\mathbb{Z}) \not\subset \ell^1(\mathbb{Z})$.

What we can say however, is that if we are given $u^0 \in \ell^2(\mathbb{Z})$ and $v^1 \in \ell^2(\mathbb{Z})$, and if we define $u^1 = u^0 + kv^1 \in \ell^2(\mathbb{Z})$ and set $w_{n-1/2}^1 = \frac{u_n^1 - u_{n-1}^1}{h}$, that is to say if w^1 is the discrete derivative in some sense of an element of $\ell^2(\mathbb{Z})$, then for all j, so is w^j. Indeed, it is simply the discrete derivative of u^j obtained by the original scheme with the above initial conditions, by uniqueness. In this sense, we can say that the original scheme is stable in $\ell^2(\mathbb{Z})$ for such initial conditions, but with a stability measured in the norms $\|\frac{u_n^j-u_n^{j-1}}{k}\|_{2,h}+\|\frac{u_n^j-u_{n-1}^j}{h}\|_{2,h}$, which are more natural in view of Remark 9.20. □

Remark 9.26 The proof requires $\lambda_0<1$ to work. Indeed, if $\frac{k}{h}=1$, then for $\xi=\frac{\pi}{h}$,

$$B(\xi)=\begin{pmatrix}1 & 2i\\ 2i & -3\end{pmatrix}$$

has the double eigenvalue -1 and is not diagonalizable. Therefore $|||B(\xi)^{T/k}|||_2\sim Ck^{-1}$ with $C>0$ for k small for this value of ξ and the semi-discrete scheme is thus unstable in this case. Note that this tells us nothing about the stability of the discrete scheme when $k=h$, and this instability occurs even though the scheme is von Neumann stable. □

Remark 9.27 If we modify the second relation of the finite difference scheme as follows

$$\frac{w_{n-1/2}^{j+1}-w_{n-1/2}^{j}}{k}-\frac{v_n^j-v_{n-1}^j}{h}=0,$$

which may seem more natural than the scheme above, we get an amplification matrix

$$B(\xi)=\begin{pmatrix}1 & ia(\xi)\\ ia(\xi) & 1\end{pmatrix}.$$

This matrix is normal and its eigenvalues are $1\pm ia(\xi)$. They are of modulus strictly larger than 1 (except when $a(\xi)=0$). Hence this scheme is not stable in the sense of von Neumann, and since the matrix is normal and given the expression of $a(\xi)$, we see that it is not stable in L^2. This shows again that finite difference schemes must be chosen with care and that seemingly natural choices may very well fail. □

Let us now study the stability of the implicit scheme (9.13). The scheme is implicit and it is not obvious in the $\ell^2(\mathbb{Z})$ context that is even well-defined.

Proposition 9.13 *The implicit scheme (9.13) is well-defined with initial data in* $\ell^2(\mathbb{Z})$.

Proof We use the Fourier series argument. For $u\in\ell^2(\mathbb{Z})$, let $\mathscr{T}$ denote the continuous linear operator on $\ell^2(\mathbb{Z})$ defined by

$$(\mathscr{T}u)_n=-\lambda u_{n+1}+(1+2\lambda)u_n-\lambda u_{n-1},\ n\in\mathbb{Z},$$

and $\lambda=\frac{k^2}{h^2}$. We rewrite the implicit scheme as follows

$$\mathscr{T}u^{j+1}=v^j,$$

with $v_n^j = 2u_n^j - u_n^{j-1}$. The question is whether or not the operator $\mathscr{T}$ is an isomorphism.

Regarding uniqueness, we note that if $\mathscr{T}u = 0$, then u is of the form

$$u_n = C_1 r_1^n + C_2 r_2^n,\ n \in \mathbb{Z}, \tag{9.22}$$

for some C_1 and C_2 in $\mathbb{C}$, where r_1 and r_2 are the roots of the characteristic equation $-\lambda r^2 + (1+2\lambda)r - \lambda = 0$. These roots are real, both positive and their product is 1, therefore the only sequence of the form (9.22) that belongs to $\ell^2(\mathbb{Z})$ is such that $C_1 = C_2 = 0$.

We now consider existence. As before, for any $v \in \ell^2(\mathbb{Z};\mathbb{C})$ we let $\mathscr{F}v \in L^2(0,2\pi;\mathbb{C})$ be defined by $\mathscr{F}v(s) = \sum_{n\in\mathbb{Z}} v_n e^{ins}$.

We have

$$\begin{aligned}
\mathscr{F}\mathscr{T}u(s) &= \sum_{n\in\mathbb{Z}} \big(-\lambda u_{n+1} + (1+2\lambda)u_n - \lambda u_{n-1}\big)e^{ins}\\
&= -\lambda \sum_{n\in\mathbb{Z}} u_{n+1}e^{ins} + (1+2\lambda)\sum_{n\in\mathbb{Z}} u_n e^{ins} - \lambda \sum_{n\in\mathbb{Z}} u_{n-1}e^{ins}\\
&= \big(-\lambda e^{-is} + (1+2\lambda) - \lambda e^{is}\big)\mathscr{F}u(s)\\
&= \Big(1 + 4\lambda \sin^2\Big(\frac{s}{2}\Big)\Big)\mathscr{F}u(s).
\end{aligned}$$

Now the function $s \mapsto \big(1 + 4\lambda\sin^2\big(\frac{s}{2}\big)\big)^{-1}$ is in $L^\infty(0,2\pi)$, therefore

$$u = \mathscr{F}^{-1}\left(\frac{\mathscr{F}v(s)}{\big(1+4\lambda\sin^2\big(\frac{s}{2}\big)\big)}\right),$$

is a solution in $\ell^2(\mathbb{Z};\mathbb{C})$ of $\mathscr{T}u = v$. Of course, by uniqueness, when v is real-valued, so is u. □

Remark 9.28 We could not use the semi-discrete version of the scheme in Fourier space, because the equivalence of this scheme with the discrete scheme for piecewise constant initial data rests on the existence of the discrete scheme. The Fourier series approach does not suffer from this drawback. □

We can now switch to the semi-discrete point of view to study the stability. If we try to work on the initial formulation of the scheme in Fourier space

$$\frac{\widehat{u^{j+1}}(\xi) - 2\widehat{u^j}(\xi) + \widehat{u^{j-1}}(\xi)}{k^2} + \frac{4}{h^2}\sin^2\Big(\frac{h\xi}{2}\Big)\widehat{u^{j+1}}(\xi) = 0,$$

we encounter the same kind of difficulties as with the explicit scheme. Namely, we obtain an amplification matrix

$$A(\xi) = \begin{pmatrix} \frac{2}{1+a(\xi)^2} & -\frac{1}{1+a(\xi)^2} \\ 1 & 0 \end{pmatrix}.$$

This matrix is never normal. For $\xi = 0$, it is the same as in Proposition 9.6, therefore the semi-discrete scheme is not stable.

Once again, we must change the unknowns and use a first order system version of the scheme in order to be able to conclude. The first order scheme is simply

$$\begin{cases} \dfrac{v_n^{j+1} - v_n^j}{k} - \dfrac{w_{n+1/2}^{j+1} - w_{n-1/2}^{j+1}}{h} = 0, \\ \dfrac{w_{n-1/2}^{j+1} - w_{n-1/2}^j}{k} - \dfrac{v_n^{j+1} - v_{n-1}^{j+1}}{h} = 0. \end{cases} \tag{9.23}$$

Writing down the semi-discrete version of this last scheme, we obtain the following amplification matrix

$$B(\xi) = \frac{1}{1 + a(\xi)^2} \begin{pmatrix} 1 & ia(\xi) \\ ia(\xi) & 1 \end{pmatrix}.$$

Now this matrix is normal. Its eigenvalues are $\lambda_\pm(\xi) = \frac{1 \pm ia(\xi)}{1+a(\xi)^2}$, so that

$$\rho(B(\xi)) = \frac{1}{\sqrt{1 + a(\xi)^2}} \leq 1$$

for all $\xi \in \mathbb{R}$. We have thus shown

Proposition 9.14 *The implicit scheme (9.23) is unconditionally von Neumann stable and L^2 stable.*

Let us close this section by saying a few words about the stability of the θ-scheme (9.14). If we write the semi-discrete version of the scheme, apply the Fourier transform and rewrite the result in vector form, we obtain an amplification matrix

$$A(\xi) = \begin{pmatrix} -b(\xi) & -1 \\ 1 & 0 \end{pmatrix},$$

with

$$b(\xi) = \frac{(1 - 2\theta)a(\xi)^2 - 2}{1 + \theta a(\xi)^2}.$$

This matrix is not normal. Its eigenvalues are the roots of the polynomial $P(X) = X^2 + b(\xi)X + 1$. The discriminant reads

$$\Delta(\xi) = \frac{a(\xi)^2\big((1 - 4\theta)a(\xi)^2 - 4\big)}{\big(1 + \theta a(\xi)^2\big)^2}.$$

If the discriminant is positive for some value of ξ, we thus have two distinct real roots, the product of which is 1, hence von Neumann instability. If on the other hand, the discriminant is nonpositive for all ξ, we have two complex conjugate roots of modulus 1, hence von Neumann stability. Recalling that $a(\xi) = \frac{2k}{h}\sin\bigl(\frac{h\xi}{2}\bigr)$, we thus obtain the following proposition:

Proposition 9.15 *The θ-scheme is unconditionally von Neumann stable for $\theta \geq \frac{1}{4}$. For $\theta < \frac{1}{4}$, it is von Neumann stable under the condition $\frac{k}{h} \leq \frac{1}{\sqrt{1-4\theta}}$.*

Of course, in terms of L^2 stability, we have the exact same problem as before for $\xi = 0$, which implies L^2 instability of the semi-discrete θ-scheme. We can try to go around this difficulty by using again a system

$$\begin{cases} \dfrac{v_n^{j+1} - v_n^j}{k} - \theta\dfrac{w_{n+1/2}^{j+1} - w_{n-1/2}^{j+1}}{h} - (1-2\theta)\dfrac{w_{n+1/2}^{j} - w_{n-1/2}^{j}}{h} \\ \qquad\qquad\qquad\qquad\qquad\qquad - \theta\dfrac{w_{n+1/2}^{j-1} - w_{n-1/2}^{j-1}}{h} = 0, \\ \dfrac{w_{n-1/2}^{j+1} - w_{n-1/2}^{j}}{k} - \dfrac{v_n^{j+1} - v_{n-1}^{j+1}}{h} = 0. \end{cases} \tag{9.24}$$

Now on the surface, this scheme appears still to be a two time step scheme, hence nothing seems to be gained. We can however rewrite it as a one time step scheme as follows. We first apply the Fourier transform to the semi-discrete version of the scheme

$$\begin{cases} \widehat{v^{j+1}}(\xi) - \widehat{v^j}(\xi) - ia(\xi)\bigl(\theta\widehat{w^{j+1}}(\xi) + (1-2\theta)\widehat{w^j}(\xi) + \theta\widehat{w^{j-1}}(\xi)\bigr) = 0, \\ \widehat{w^{j+1}}(\xi) - \widehat{w^j}(\xi) - ia(\xi)\widehat{v^{j+1}}(\xi) = 0. \end{cases}$$

In addition to a formula for $\widehat{w^{j+1}}$ in terms of $\widehat{w^j}$ and $\widehat{v^{j+1}}$, the second equation also yields

$$\widehat{w^{j-1}}(\xi) = \widehat{w^j}(\xi) - ia(\xi)\widehat{v^j}(\xi).$$

We replace these expressions in the first equation

$$\begin{aligned} \widehat{v^{j+1}}(\xi) - \widehat{v^j}(\xi) - ia(\xi)\bigl(\theta(\widehat{w^j}(\xi) + ia(\xi)\widehat{v^{j+1}}(\xi)) \\ + (1-2\theta)\widehat{w^j}(\xi) + \theta(\widehat{w^j}(\xi) - ia(\xi)\widehat{v^j}(\xi))\bigr) = 0, \end{aligned}$$

or

$$(1 + \theta a(\xi)^2)\bigl(\widehat{v^{j+1}}(\xi) - \widehat{v^j}(\xi)\bigr) - ia(\xi)\widehat{w^j}(\xi) = 0.$$

This scheme thus corresponds to the amplification matrix

$$B(\xi) = \begin{pmatrix} 1 & ia_\theta(\xi) \\ ia(\xi) & 1 - a(\xi)a_\theta(\xi) \end{pmatrix},$$

with

$$a_\theta(\xi) = \frac{a(\xi)}{1 + \theta a(\xi)^2}.$$

This matrix is not normal and has the same eigenvalues as the previous one, hence the same von Neumann stability. The case $a(\xi) = 0$ is not a problem anymore however, since the matrix is then the identity matrix.

Proposition 9.16 *The semi-discrete version of the θ-scheme (9.24) is stable in L^2 for $\theta \geq \frac{1}{4}$ under the condition $\frac{k}{h} \leq M$, for any given M. For $\theta < \frac{1}{4}$, given any $0 < \lambda_0 < \frac{1}{\sqrt{1-4\theta}}$, it is L^2 stable under the condition $\frac{k}{h} \leq \lambda_0$.*

Proof Let us consider the case $\theta \geq \frac{1}{4}$. First of all, at $\xi = \frac{2m\pi}{h}$, $m \in \mathbb{Z}$, we have $a(\xi) = 0$ so that nothing needs to be done for these values of ξ, as was already mentioned. For the other values of ξ, the matrix $B(\xi)$ is diagonalizable with two distinct, complex conjugate eigenvalues of modulus one, therefore no problem for the diagonal part either. We have the change of basis matrix

$$P(\xi) = \begin{pmatrix} -\frac{1}{2}\Big(\sqrt{\frac{4a_\theta(\xi)}{a(\xi)} - a_\theta(\xi)^2} + ia_\theta(\xi)\Big) & \frac{1}{2}\Big(\sqrt{\frac{4a_\theta(\xi)}{a(\xi)} - a_\theta(\xi)^2} - ia_\theta(\xi)\Big) \\ 1 & 1 \end{pmatrix},$$

with $\frac{4a_\theta(\xi)}{a(\xi)} - a_\theta(\xi)^2 \geq 0$ since $\theta \geq \frac{1}{4}$.

After a little bit of computer algebra aided manipulations, we obtain the following value for the condition number of $P(\xi)$:

$$\mathrm{cond}_2(P(\xi)) = \frac{\mathrm{sign}\,(a)(a+b) + \sqrt{a^2b^2 + (a-b)^2}}{\sqrt{4ab - a^2b^2}},$$

where $a = a(\xi)$ and $b = a_\theta(\xi)$ for brevity. Replacing b by its value as a function of a, we obtain

$$\begin{aligned} \mathrm{cond}_2(P(\xi)) &= \frac{2 + \theta a(\xi)^2 + |a(\xi)|\sqrt{1 + \theta^2 a(\xi)^2}}{\sqrt{(4\theta - 1)a(\xi)^2 + 4}} \\ &\leq 1 + \frac{|a(\xi)|}{2} + \frac{a(\xi)^2}{2} \\ &\leq 1 + M + 2M^2, \end{aligned}$$

hence the stability of the scheme. We leave the case $\theta < \frac{1}{4}$ as an exercise. □

Remark 9.29 Proposition 9.16 in the case $\theta \geq \frac{1}{4}$ is a bit of a disappointment. Indeed, in that case, the scheme is unconditionally von Neumann stable and we only obtain actual L^2 stability under the condition $\frac{k}{h} \leq M$ with M arbitrary. Now in practice, neither k nor h actually go to 0, and such a condition as $\frac{k}{h} \leq M$ with M arbitrary is not discernible from unconditional stability. □

Remark 9.30 Instead of using the Jordan decomposition of $B(\xi)$, we could think of using the Schur decomposition of $B(\xi)$, $B(\xi) = U(\xi)T(\xi)U(\xi)^*$, where $U(\xi)$ is unitary and $T(\xi)$ is upper triangular. The advantage of the Schur decomposition over the Jordan decomposition in this context, is that $|||B(\xi)^j|||_2 = |||T(\xi)^j|||_2$ for all j and we lose no information by passing from $B(\xi)$ to $T(\xi)$. Moreover, $T(\xi)^j$ is fairly easy to express explicitly. The disadvantage is that the expression of the upper right entry of $T(\xi)^j$ is even less user-friendly than $\text{cond}_2(P(\xi))$ when it comes to estimating it. We do not pursue this direction here. □

9.6 For a Few Schemes More

In the previous section, we rewrote the wave equation as the first order system (9.18). This system is of the form

$$\frac{\partial U}{\partial t} + \frac{\partial (f(U))}{\partial x} = 0$$

with

$$U(x,t) = \begin{pmatrix} v(x,t) \\ w(x,t) \end{pmatrix} \text{ and } f(U) = \begin{pmatrix} 0 & -1 \\ -1 & 0 \end{pmatrix} U.$$

When U is $\mathbb{R}^p$-valued and f is a general nonlinear function from $\mathbb{R}^p$ to $\mathbb{R}^p$, satisfying certain conditions, this is a *(nonlinear) hyperbolic system*. Such systems are of paramount importance in many applications, for example in gas dynamics, and there is a very large body of numerical schemes that are adapted to the approximation of the solutions of such systems, see [42] for example.

We present a few of these schemes in the case of our simple $\mathbb{R}^2$-valued, linear hyperbolic system (9.18). We will also return to some of these schemes in the next chapter. Again, we work on the whole line[3] and on the usual (nh, jk) space-time finite difference grid for the approximations. We start with the *Lax–Friedrichs scheme*, which reads in general

$$\frac{U_n^{j+1} - \frac{1}{2}(U_{n+1}^j + U_{n-1}^j)}{k} + \frac{f(U_{n+1}^j) - f(U_{n-1}^j)}{2h} = 0,$$

[3] Boundary conditions are a delicate question for such systems.

and in our particular case

$$\begin{cases} v_n^{j+1} = \frac{1}{2}(v_{n+1}^j + v_{n-1}^j) + \frac{k}{2h}(w_{n+1}^j - w_{n-1}^j), \\ w_n^{j+1} = \frac{1}{2}(w_{n+1}^j + w_{n-1}^j) + \frac{k}{2h}(v_{n+1}^j - v_{n-1}^j). \end{cases}$$

The scheme is one-step, of order one and explicit. We write the usual semi-discrete version of the scheme, then apply the Fourier transform, and we obtain

$$\begin{cases} \widehat{v^{j+1}}(\xi) = \cos(h\xi)\widehat{v^j}(\xi) + i\frac{k}{h}\sin(h\xi)\widehat{w^j}(\xi), \\ \widehat{w^{j+1}}(\xi) = \cos(h\xi)\widehat{w^j}(\xi) + i\frac{k}{h}\sin(h\xi)\widehat{v^j}(\xi). \end{cases}$$

Therefore, the amplification matrix of the Lax–Friedrichs scheme is

$$B(\xi) = \begin{pmatrix} \cos(h\xi) & i\frac{k}{h}\sin(h\xi) \\ i\frac{k}{h}\sin(h\xi) & \cos(h\xi) \end{pmatrix}.$$

This matrix is normal and its spectral radius is $\rho(B(\xi)) = \big(\cos^2(h\xi) + \frac{k^2}{h^2}\sin^2(h\xi)\big)^{1/2}$. It clearly follows that

Proposition 9.17 *The Lax–Friedrichs scheme is von Neumann stable and L^2 stable under the condition $\frac{k}{h} \leq 1$.*

We consider next the *Lax–Wendroff scheme*. In our particular case, the scheme reads

$$\begin{cases} v_n^{j+1} = v_n^j + \frac{k}{2h}(w_{n+1}^j - w_{n-1}^j) + \frac{k^2}{2h^2}(v_{n+1}^j - 2v_n^j + v_{n-1}^j), \\ w_n^{j+1} = w_n^j + \frac{k}{2h}(v_{n+1}^j - v_{n-1}^j) + \frac{k^2}{2h^2}(w_{n+1}^j - 2w_n^j + w_{n-1}^j). \end{cases}$$

The scheme is one-step, of order two and explicit. After Fourier transform, it becomes

$$\begin{cases} \widehat{v^{j+1}}(\xi) = \Big(1 - \frac{2k^2}{h^2}\sin^2\Big(\frac{h\xi}{2}\Big)\Big)\widehat{v^j}(\xi) + i\frac{k}{h}\sin(h\xi)\widehat{w^j}(\xi), \\ \widehat{w^{j+1}}(\xi) = \Big(1 - \frac{2k^2}{h^2}\sin^2\Big(\frac{h\xi}{2}\Big)\Big)\widehat{w^j}(\xi) + i\frac{k}{h}\sin(h\xi)\widehat{v^j}(\xi), \end{cases}$$

hence the amplification matrix

$$B(\xi) = \begin{pmatrix} \big(1 - \frac{2k^2}{h^2}\sin^2\big(\frac{h\xi}{2}\big)\big) & i\frac{k}{h}\sin(h\xi) \\ i\frac{k}{h}\sin(h\xi) & \big(1 - \frac{2k^2}{h^2}\sin^2\big(\frac{h\xi}{2}\big)\big) \end{pmatrix}.$$

This matrix is again normal and it follows from elementary computations that it has a spectral radius $\rho(B(\xi)) = \left(1 - 4\frac{k^2}{h^2}(1 - \frac{k^2}{h^2})\sin^4\left(\frac{h\xi}{2}\right)\right)^{1/2}$. Therefore, we see that

Proposition 9.18 *The Lax–Wendroff scheme is von Neumann stable and L^2 stable under the condition $\frac{k}{h} \leq 1$.*

We can also revisit the *leapfrog scheme*, which reads here

$$\begin{cases} v_n^{j+1} = v_n^{j-1} + \dfrac{k}{h}(w_{n+1}^j - w_{n-1}^j), \\ w_n^{j+1} = w_n^{j-1} + \dfrac{k}{h}(v_{n+1}^j - v_{n-1}^j). \end{cases}$$

Note that it leapfrogs in time as well as in space. The scheme is two-step, of order two and explicit. To write an amplification matrix for it, we need to double the dimension and consider for example the vectors $\left(\widehat{v^{j+1}}(\xi), \widehat{w^{j+1}}(\xi), \widehat{v^j}(\xi), \widehat{w^j}(\xi)\right)^T$, a choice which yields the amplification matrix

$$B(\xi) = \begin{pmatrix} 0 & 2i\frac{k}{h}\sin(h\xi) & 1 & 0 \\ 2i\frac{k}{h}\sin(h\xi) & 0 & 0 & 1 \\ 1 & 0 & 0 & 0 \\ 0 & 1 & 0 & 0 \end{pmatrix}.$$

This matrix is not normal.

Proposition 9.19 *The leapfrog scheme is von Neumann stable under the condition $\frac{k}{h} \leq 1$. It is L^2 stable under the condition $\frac{k}{h} \leq \lambda_0 < 1$.*

Proof Let $b(\xi) = \frac{k}{h}\sin(h\xi)$. If we write $B(\xi)$ as a 2×2 block matrix of four 2×2 blocks, we see by Lemma 8.2 of Chap. 8 that its eigenvalues are given by $\lambda = \pm\sqrt{1 - b^2} \pm ib$ when $\frac{k}{h} \leq 1$, so that $\rho(B(\xi)) = 1$ for all ξ.

Concerning the L^2-stability, if $\frac{k}{h} \leq \lambda_0 < 1$, we have four distinct eigenvalues of modulus 1, so we only need to estimate the condition number of the change of matrix basis

$$P(\xi) = \begin{pmatrix} 1 & 1 & 1 & 1 \\ 1 & 1 & -1 & -1 \\ \sqrt{1-b^2} - ib & -\sqrt{1-b^2} - ib & \sqrt{1-b^2} + ib & -\sqrt{1-b^2} + ib \\ \sqrt{1-b^2} - ib & -\sqrt{1-b^2} - ib & -\sqrt{1-b^2} - ib & \sqrt{1-b^2} - ib \end{pmatrix}.$$

Using computer algebra again, we find that

$$\operatorname{cond}_2(P(\xi)) = \sqrt{\frac{1 + |b(\xi)|}{1 - |b(\xi)|}} \leq \sqrt{\frac{2}{1 - \lambda_0}},$$

hence the stability of the scheme. □

Remark 9.31 The conditional stability of the leapfrog scheme is to be contrasted with the situation for the heat equation, where the leapfrog scheme is always unstable, see Proposition 8.10 in Chap. 8. □

Remark 9.32 We note that we cannot allow $\lambda_0 = 1$. Indeed, in this case, there are values of ξ for which $b(\xi) = \pm 1$. For these values of ξ, the matrix $B(\xi)$ has two double eigenvalues $\pm i$ and is not diagonalizable. Therefore $|||B(\xi)^j|||_2 \to +\infty$ as $j \to +\infty$, and the scheme is unstable in L^2. □

9.7 Concluding Remarks

To conclude this chapter, we note that there are other issues than just consistency and stability in the study of finite difference schemes for hyperbolic systems. Even though we have not mentioned them at all here, they are important in assessing the performance of a given scheme. Among these issues are *dissipativity*, i.e., the possible damping of wave amplitudes with time, and *dispersivity*, i.e., the possibility that numerically approximated waves of different frequencies could travel at different numerical speeds.

Naturally, there are other numerical methods applicable to the wave equation, for instance finite difference-finite element methods, see for example [65].

We have so far described and analyzed two major classes of numerical methods, finite difference methods and finite element methods, in the contexts of the three main classes of problems, elliptic, parabolic and hyperbolic. In the last chapter, we introduce a more recent method, the finite volume method, on a few elliptic and hyperbolic examples.

Chapter 10
The Finite Volume Method

The finite volume method is a more recent method than both finite difference and finite element methods. It is widely used in practice for example in fluid dynamics computations. We present the method in the simplest possible settings, first for one-dimensional elliptic problems, then for the transport equation in one and two dimensions. For the latter, we return to the method of characteristics already introduced in Chap. 1, to solve one-dimensional nonlinear problems as well as two-dimensional linear problems.

10.1 The Elliptic Case in One Dimension

Let us consider the elliptic problem $-u'' = f$ in $]0, 1[$, with homogeneous Dirichlet boundary conditions $u(0) = u(1) = 0$. The idea of the finite volume discretization is to subdivide the domain into a finite number of subsets called cells, or control volumes especially in higher dimension, which are disjoint up to a set of zero measure, and integrate the equation on each cell. The resulting integrals are then approximated by computable quantities. The main difference with all previous methods is that the discrete unknowns to be introduced will be meant to be approximations of the mean value of the solution on each cell. In a sense, the finite volume approximation is an approximation by piecewise constant functions.

It may seem strange to approximate an a priori smooth function, say of class C^2 if f is continuous, that solves a problem involving derivatives, by piecewise constant functions. In fact, the method does not approximate the differential equation itself in one form or another, but starts from the integration of the differential equation over each cell, followed by approximation. So this is a quite different philosophy compared to both finite difference and finite element methods which work on the PDE itself expressed in different forms.

H. Le Dret and B. Lucquin, *Partial Differential Equations: Modeling, Analysis and Numerical Approximation*, International Series of Numerical Mathematics 168, DOI 10.1007/978-3-319-27067-8_10

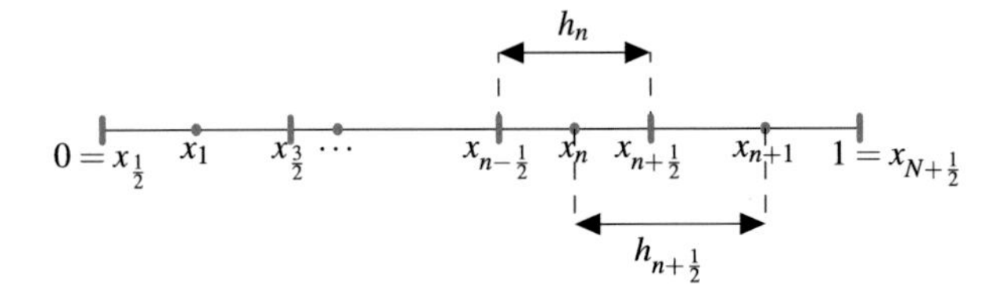

Fig. 10.1 Finite volume cells in 1D

More precisely, let $N \in \mathbb{N}^*$ be given. In the one-dimensional case, cells are just going to be subintervals C_n, $n = 1, \dots, N$. We let x_n denote the middle of C_n and $x_{n-1/2} < x_{n+1/2}$ denote its extremities so that $C_n = [x_{n-1/2}, x_{n+1/2}]$. The local cell size is $h_n = x_{n+1/2} - x_{n-1/2}$. It is not a priori a constant, see Fig. 10.1.

We also set $h_0 = h_1$, $h_{N+1} = h_N$, $x_0 = -x_1$ and $x_{N+1} = 2 - x_N$. For all $n = 0, \dots, N$, we let $h_{n+1/2} = \frac{h_{n+1}+h_n}{2} = x_{n+1} - x_n$ be the distance between the centers of two consecutive cells. If we integrate the equation $-u'' = f$ on C_n, we obtain

$$-u'\big(x_{n+\frac{1}{2}}\big) + u'\big(x_{n-\frac{1}{2}}\big) = \int_{C_n} f(x)\,dx, \tag{10.1}$$

for $n = 1, \dots, N$. We use the following approximations for the derivatives

$$u'\big(x_{n+\frac{1}{2}}\big) \approx \frac{\frac{1}{h_{n+1}}\int_{C_{n+1}} u(x)\,dx - \frac{1}{h_n}\int_{C_n} u(x)\,dx}{h_{n+\frac{1}{2}}}, \tag{10.2}$$

for $n = 1, \dots, N-1$,

$$u'\big(x_{\frac{1}{2}}\big) \approx \frac{\frac{1}{h_1}\int_{C_1} u(x)\,dx}{h_{\frac{1}{2}}}, \tag{10.3}$$

and

$$u'\big(x_{N+\frac{1}{2}}\big) \approx -\frac{\frac{1}{h_N}\int_{C_N} u(x)\,dx}{h_{N+\frac{1}{2}}}. \tag{10.4}$$

Indeed, assuming that u is of class C^2, a Taylor–Lagrange expansion shows that

$$u(x) = u\big(x_{n+\frac{1}{2}}\big) + \big(x - x_{n+\frac{1}{2}}\big)u'\big(x_{n+\frac{1}{2}}\big) + O(h^2),$$

where $h = \max_n h_n$. Therefore, if $n = 1, \dots, N-1$

$$\frac{1}{h_{n+1}}\int_{C_{n+1}} u(x)\,dx = u\big(x_{n+\frac{1}{2}}\big) + \frac{h_{n+1}}{2}u'\big(x_{n+\frac{1}{2}}\big) + O(h^2),$$
$$\frac{1}{h_n}\int_{C_n} u(x)\,dx = u\big(x_{n+\frac{1}{2}}\big) - \frac{h_n}{2}u'\big(x_{n+\frac{1}{2}}\big) + O(h^2), \tag{10.5}$$

and thus

$$\left|\frac{\frac{1}{h_{n+1}}\int_{C_{n+1}} u(x)\,dx - \frac{1}{h_n}\int_{C_n} u(x)\,dx}{h_{n+\frac{1}{2}}} - u'\big(x_{n+\frac{1}{2}}\big)\right| \le Ch, \tag{10.6}$$

where the constant C only depends on u. For $n = 0$, we have by the same token

$$\frac{1}{h_1}\int_{C_1} u(x)\,dx = u\big(x_{\frac{1}{2}}\big) + \frac{h_1}{2}u'\big(x_{\frac{1}{2}}\big) + O(h^2) = \frac{h_1}{2}u'\big(x_{\frac{1}{2}}\big) + O(h^2),$$

due to the Dirichlet boundary condition, so that

$$\left|\frac{\frac{1}{h_1}\int_{C_1} u(x)\,dx}{h_{\frac{1}{2}}} - u'\big(x_{\frac{1}{2}}\big)\right| \le Ch,$$

and likewise for $n = N$. We have thus shown that

Proposition 10.1 *The finite volume approximation (10.2)–(10.4) of first derivatives is consistent of order 1.*

Following a process that is now familiar, we introduce discrete unknowns u_n, $n = 1, \dots, N$, to take the place of the average value of the exact solution on cell C_n, $\frac{1}{h_n}\int_{C_n} u(x)\,dx$, and replace it in an analogue of Eq. (10.1). We first let

$$F_{n+\frac{1}{2}} = -\frac{u_{n+1} - u_n}{h_{n+\frac{1}{2}}} \text{ for } 2 \le n \le N-1,$$

and

$$F_{\frac{1}{2}} = -\frac{u_1}{h_{\frac{1}{2}}},\ F_{N+\frac{1}{2}} = \frac{u_N}{h_{N+\frac{1}{2}}},$$

where the latter two relations take the Dirichlet condition into account at the discrete level. The quantities $F_{n+1/2}$ are called the *numerical fluxes*, since they are meant to approximate the actual fluxes $-u'(x_{n+1/2})$. This way, we obtain the finite volume scheme

$$F_{n+\frac{1}{2}} - F_{n-\frac{1}{2}} = f_n,\ n = 1, \dots, N, \tag{10.7}$$

with

$$f_n = \int_{C_n} f(x)\,dx. \tag{10.8}$$

Rewriting the finite volume scheme line by line, we obtain for $n = 1$

$$\Big(\frac{1}{h_{\frac{1}{2}}} + \frac{1}{h_{\frac{3}{2}}}\Big)u_1 - \frac{1}{h_{\frac{3}{2}}}u_2 = f_1,$$

for $2 \le n \le N-1$,

$$-\frac{1}{h_{n-\frac{1}{2}}}u_{n-1} + \Big(\frac{1}{h_{n-\frac{1}{2}}} + \frac{1}{h_{n+\frac{1}{2}}}\Big)u_n - \frac{1}{h_{n+\frac{1}{2}}}u_{n+1} = f_n,$$

and for $n = N$,

$$-\frac{1}{h_{N-\frac{1}{2}}}u_{N-1} + \Big(\frac{1}{h_{N-\frac{1}{2}}} + \frac{1}{h_{N+\frac{1}{2}}}\Big)u_N = f_N.$$

Thus we see that the scheme assumes the form of a linear system

$$\begin{pmatrix} \frac{1}{h_{\frac{1}{2}}} + \frac{1}{h_{\frac{3}{2}}} & -\frac{1}{h_{\frac{3}{2}}} & \cdots & & \cdots & 0 \\ -\frac{1}{h_{\frac{3}{2}}} & \frac{1}{h_{\frac{3}{2}}} + \frac{1}{h_{\frac{5}{2}}} & -\frac{1}{h_{\frac{5}{2}}} & & \cdots & 0 \\ \vdots & \ddots & \ddots & & \ddots & \vdots \\ 0 & \cdots & -\frac{1}{h_{N-\frac{3}{2}}} & \frac{1}{h_{N-\frac{3}{2}}} + \frac{1}{h_{N-\frac{1}{2}}} & & -\frac{1}{h_{N-\frac{1}{2}}} \\ 0 & \cdots & \cdots & -\frac{1}{h_{N-\frac{1}{2}}} & & \frac{1}{h_{N-\frac{1}{2}}} + \frac{1}{h_{N+\frac{1}{2}}} \end{pmatrix} \begin{pmatrix} u_1 \\ u_2 \\ \vdots \\ u_N \end{pmatrix} = \begin{pmatrix} f_1 \\ f_2 \\ \vdots \\ f_N \end{pmatrix}, \tag{10.9}$$

or in short $A_h U_h = F_h$. The matrix A_h of the above linear system bears a strong resemblance to the matrix by the same name involved in the finite difference method in Chap. 2. Before we discuss the well-posedness and eventual convergence of the finite volume method, let us see that the method is nonetheless fundamentally different from the finite difference method, in spite of appearances.

Let us thus consider for an instant the linear system (10.9) as stemming from a finite difference scheme, that is to say assume that $u_n \approx u(x_n)$. Then the approximation in the left-hand side of Eq. (10.7) divided by h_n, so that the right-hand side becomes a consistent approximation of $f(x_n)$, is not a consistent approximation of $-u''(x_n)$. Indeed, if we compute the truncation error in this context, we find

$$h_n\varepsilon_h(u)_n = -\frac{1}{h_{n-\frac{1}{2}}}u(x_{n-1}) + \Big(\frac{1}{h_{n-\frac{1}{2}}} + \frac{1}{h_{n+\frac{1}{2}}}\Big)u(x_n) - \frac{1}{h_{n+\frac{1}{2}}}u(x_{n+1}) - \int_{C_n} f(x)\,dx,$$

so that using Taylor–Lagrange expansions, we obtain

$$\varepsilon_h(u)_n = \left(1 - \frac{h_{n-\frac{1}{2}} + h_{n+\frac{1}{2}}}{2h_n}\right)u''(x_n) + O(h).$$

Now, if we are given a sequence of finite volume meshes indexed by h and such that $h \to 0$, there is no reason in general for $\frac{h_{n-1/2}+h_{n+1/2}}{2h_n}$ to tend to 1, except in very special cases. Actually, it is easy to construct a sequence that does not exhibit such a convergence. Therefore, the above truncation error does not tend to 0 in general. This shows that the finite volume scheme is quite different from a finite difference scheme.

We now return to the study of the finite volume scheme (10.7)–(10.8). First of all, the scheme is well-posed.

Proposition 10.2 *The matrix A_h is invertible.*

Proof We show that the matrix A_h is actually positive definite, which implies that it is invertible. Given $X \in \mathbb{R}^N$, we have

$$X^T A_h X = \frac{X_1^2}{h_{\frac{1}{2}}} + \sum_{n=1}^{N-1} \frac{(X_{n+1} - X_n)^2}{h_{n+\frac{1}{2}}} + \frac{X_N^2}{h_{N+\frac{1}{2}}},$$

from which the result easily follows. □

Let us now show the convergence of the scheme when $h \to 0$.

Proposition 10.3 *Assume that $f \in C^0([0, 1])$. There exists a constant C that does not depend on h such that*

$$\max_{1\le i\le N} \left|u_n - \frac{1}{h_n}\int_{C_n} u(x)\,dx\right| \le Ch.$$

Proof We let $e_n = u_n - \frac{1}{h_n}\int_{C_n} u(x)\,dx$ for $n = 1, \dots, N$ and $e_0 = e_{N+1} = 0$. Let $\bar{F}_{n+1/2} = -u'(x_{n+1/2})$ denote the actual fluxes. We have by Eq. (10.1)

$$\bar{F}_{n+\frac{1}{2}} - \bar{F}_{n-\frac{1}{2}} = f_n,\ n = 1, \dots, N,$$

so that taking Eq. (10.7) into account, it follows that

$$D_{n+\frac{1}{2}} - D_{n-\frac{1}{2}} = 0,\ n = 1, \dots, N, \tag{10.10}$$

where $D_{n+1/2} = F_{n+1/2} - \bar{F}_{n+1/2}$.

Since f is continuous, it follows that u is of class C^2, hence estimate (10.6) applies and we can write

$$-\frac{\frac{1}{h_{n+1}}\int_{C_{n+1}} u(x)\,dx - \frac{1}{h_n}\int_{C_n} u(x)\,dx}{h_{n+\frac{1}{2}}} = \bar{F}_{n+\frac{1}{2}} + d_{n+\frac{1}{2}},$$

with $|d_{n+1/2}| \leq Ch$, with the obvious modifications for $n = 0$ and $n = N$. Therefore

$$D_{n+\frac{1}{2}} = -\frac{e_{n+1} - e_n}{h_{n+\frac{1}{2}}} + d_{n+\frac{1}{2}},$$

and the constancy (10.10) of $D_{n+1/2}$ implies that

$$\frac{e_n - e_{n-1}}{h_{n-\frac{1}{2}}} - d_{n-\frac{1}{2}} - \frac{e_{n+1} - e_n}{h_{n+\frac{1}{2}}} + d_{n+\frac{1}{2}} = 0.$$

We multiply the previous relation by e_n and sum for $n = 1$ to N. This yields

$$\sum_{n=1}^{N} e_n \frac{e_n - e_{n-1}}{h_{n-\frac{1}{2}}} - \sum_{n=1}^{N} e_n \frac{e_{n+1} - e_n}{h_{n+\frac{1}{2}}} = -\sum_{n=1}^{N} d_{n+\frac{1}{2}} e_n + \sum_{n=1}^{N} d_{n-\frac{1}{2}} e_n.$$

Taking into account the fact that $e_0 = e_{N+1} = 0$ and reordering the terms, we obtain

$$\sum_{n=0}^{N} \frac{(e_{n+1} - e_n)^2}{h_{n+\frac{1}{2}}} = -\sum_{n=0}^{N} d_{n+\frac{1}{2}}(e_n - e_{n+1}) \leq Ch \sum_{n=0}^{N} |e_n - e_{n+1}|.$$

By the Cauchy–Schwarz inequality, the right-hand side is estimated by

$$\sum_{n=0}^{N} |e_n - e_{n+1}| \leq \Big(\sum_{n=0}^{N} \frac{(e_{n+1} - e_n)^2}{h_{n+\frac{1}{2}}}\Big)^{1/2} \Big(\sum_{n=0}^{N} h_{n+\frac{1}{2}}\Big)^{1/2}.$$

Now of course $\sum_{n=0}^{N} h_{n+1/2} = 1$ and combining the above estimates, we obtain

$$\sum_{n=0}^{N} \frac{(e_{n+1} - e_n)^2}{h_{n+\frac{1}{2}}} \leq C^2 h^2.$$

It follows immediately from the last two estimates that

$$\sum_{n=0}^{N} |e_{n+1} - e_n| \leq Ch.$$

To conclude, we notice that $e_n = \sum_{j=0}^{n-1}(e_{j+1} - e_j)$, for $n = 1, \dots, N$, so that

$$|e_n| \le \sum_{j=0}^{n-1} |e_{j+1} - e_j| \le \sum_{j=0}^{N} |e_{j+1} - e_j| \le Ch,$$

by the previous estimate. □

The above proof shows that u_n is close to $\frac{1}{h_n}\int_{C_n} u(x)\,dx$ when h is small, uniformly with respect to n. We can easily infer a finite difference-like convergence result as follows.

Corollary 10.1 *Under the previous hypotheses, we have*

$$\max_{1\le i\le N} |u_n - u(x_n)| \le Ch,$$

where C does not depend on h.

Proof Indeed, for $x \in C_n$, we can write $u(x) = u(x_n) + (x - x_n)u'(\xi)$ for some $\xi \in C_n$. Therefore $|u(x) - u(x_n)| \le h \max_{x\in[0,1]} |u'|$. Integrating this inequality on C_n, we obtain

$$\left|\frac{1}{h_n}\int_{C_n} u(x)\,dx - u(x_n)\right| \le \frac{1}{h_n}\int_{C_n} |u(x) - u(x_n)|\,dx \le h \max_{x\in[0,1]} |u'|,$$

and the Corollary follows. □

Remark 10.1 If u is C^4, it is possible to obtain a better right-hand side of the error estimate of Corollary 10.1, namely Ch^2. □

Remark 10.2 Proposition 10.3 still holds true if f is only supposed to be in L^1, see [36]. □

10.2 The Transport Equation in One Dimension

Let us first expand a little on the presentation of the theory of the transport equation of Chap. 1, Sect. 1.4, before describing the finite volume method in this context.

The Cauchy problem for the transport, or advection, equation reads

$$\begin{cases} \dfrac{\partial u}{\partial t}(x,t) + a\dfrac{\partial u}{\partial x}(x,t) = 0, x \in \mathbb{R}, t > 0, \\ u(x,0) = u_0(x), x \in \mathbb{R}, \end{cases} \tag{10.11}$$

where $u(x,t)$ is the unknown density of some quantity at point x and time t, a quantity per unit length, and a is a given constant that represents the propagation or advection speed of u. In this interpretation, $q(x, t) = au(x, t)$ represents the flux of the quantity under consideration passing through x at time t, i.e., a quantity per unit of time. The initial data u_0 is given. We have seen in Chap. 1 that when u_0 is regular, there is a unique regular solution given by

$$u(x, t) = u_0(x - at). \tag{10.12}$$

As a matter of fact, a crucial property of the transport equation is that the solution u is constant along the characteristics, i.e., the integral curves of the vector field a (admittedly, a one-dimensional vector field in our present case). By definition, a characteristic is a curve $t \to X(t)$ which satisfies the ordinary differential equation

$$\frac{dX}{dt}(t) = a. \tag{10.13}$$

We see that when plotted in space–time, all characteristics, i.e., the curves $t \mapsto (X(t), t)$,[1] are straight lines that are parallel to each other. Moreover, given any $(y, s) \in \mathbb{R}^2$, there is one and only one characteristic passing through point y at time s, $X(t; y, s) = y + a(t - s), t \in \mathbb{R}$. Note that if $x = X(t; y, s)$, then $y = X(s; x, t)$, a relation which holds true for more general right-hand sides in Eq. (10.13), of the form $a(X(t), t)$. With this notation, the solution of the transport equation (10.12) becomes

$$u(x, t) = u_0(X(0; x, t)).$$

The initial data is simply propagated on the characteristics, at constant speed a, to the right if $a > 0$, to the left if $a < 0$, and it is stationary if $a = 0$. To find the value of the solution u at (x, t), it is enough to look at the characteristic passing through x at time t, and pick the value of u_0 at the point $x_0 = X(0; x, t)$ where the characteristic was at time 0, see Fig. 1.10 in Chap. 1 where the characteristics are drawn in a space–time diagram.

The definition of characteristics can be extended to the case of transport equations with non constant speed, i.e., a is a given function of x and t,

$$\frac{\partial u}{\partial t}(x, t) + a(x, t)\frac{\partial u}{\partial x}u(x, t) = 0.$$

The characteristics are again defined by

$$\begin{cases} \dfrac{d}{dt}X(t; y, s) = a(X(t; y, s), t), \\ X(s; y, s) = y. \end{cases} \tag{10.14}$$

[1]We call both curves characteristics, even though they do not live in the same space, and switch between the two meanings without notice.

The mapping $t \mapsto X(t; y, s)$ is the characteristic going through y at time s. The characteristics no longer are straight lines in space–time. Local existence of such curves is ensured by the Picard–Lindelöf or Cauchy–Lipschitz theorem, as soon as a is sufficiently regular (continuous with respect to (x, t), Lipschitz with respect the space variable x uniformly with respect to t). It is easily seen that the solution u is still constant on each characteristic. Moreover, due to Cauchy–Lipschitz uniqueness, different characteristics do not intersect and therefore $u(x, t) = u_0(X(0; x, t))$ as before, as long as the characteristics exist.

The method of characteristics also makes it possible to consider the case when there is a source term in the right-hand side, i.e., an equation of the form

$$\frac{\partial u}{\partial t}(x, t) + a(x, t)\frac{\partial u}{\partial x}u(x, t) = f(x, t),$$

where f is a given function. Integrating this equation along the characteristics, we obtain

$$u(x, t) = u_0(X(0; x, t)) + \int_0^t f(X(\tau; x, t), \tau)\, d\tau, \tag{10.15}$$

see Proposition 10.13 below for a proof in any dimension.

We have already mentioned nonlinear hyperbolic systems in Chap. 9. In the one-dimensional case, such systems become nonlinear transport equations,

$$\begin{cases} \dfrac{\partial u}{\partial t}(x, t) + \dfrac{\partial (f(u))}{\partial x}(x, t) = 0, x \in \mathbb{R}, t > 0 \\ u(x, 0) = u_0(x), x \in \mathbb{R}, \end{cases} \tag{10.16}$$

where f is a given function from $\mathbb{R}$ to $\mathbb{R}$, called the flux function. The linear case corresponds to flux functions of the form $f(u) = au$ with $a \in \mathbb{R}$, and we recover Eq. (10.11). The simplest nonlinear example is the classical Burgers equation, which corresponds to $f(u) = \frac{u^2}{2}$.

The method of characteristics also applies to nonlinear transport equations. In the nonlinear case, the speed of propagation depends on the solution u, which makes them much more complicated than linear transport equations. Indeed, if u is a solution of class C^1 of (10.16), we can write

$$\frac{\partial u}{\partial t} + f'(u)\frac{\partial u}{\partial x} = 0.$$

For instance, regular solutions of the Burgers equation satisfy

$$\frac{\partial u}{\partial t}(x, t) + u(x, t)\frac{\partial u}{\partial x}(x, t) = 0.$$

In this particular case, we see that when $u > 0$, the propagation goes to the right with a speed that increases with u. If we imagine an initial wave profile u_0 with a crest to the left of a trough, the crest will ride faster than the trough and eventually catch up with it, which leads to serious difficulties.

In the general case, the propagation speed is thus $f'(u)$, which is not known a priori. In the present context, the characteristics are defined by the ordinary differential equation

$$\frac{dX}{dt}(t) = f'(u(X(t), t)).$$

It is not too difficult to show that, as long as it remains regular, the solution u is constant along the characteristics and that the characteristics are therefore straight lines again. However, the speed of propagation along such a line, i.e., its slope in space–time, depends on the value of u on this particular characteristic, which in turn is equal to the value of u_0 at the point where the characteristic was at time 0. Consequently, the characteristics are in general no longer parallel to each other. Difficulties arise when these lines cross, see the crest-trough discussion above. We are not going to study this problem, which involves so-called weak solutions, or solutions in the distributional sense, and shock waves, any further here. Nevertheless, as long as the characteristics do not intersect each other, the regular solution is still given by the same formula

$$u(x, t) = u_0(X(0; x, t)).$$

10.3 Finite Volumes for the Transport Equation

We now describe the finite volume method for the transport equation (10.11) in the linear case with a constant for simplicity. We first cover the whole space $\mathbb{R}$ by cells $C_n = \left[x_{n-1/2}, x_{n+1/2}\right]$, with $x_{n-1/2} < x_{n+1/2}$, $n \in \mathbb{Z}$, and we look for a piecewise constant approximation of the solution u, which is constant on each cell C_n, at some discretized instants.

We set $h_n = x_{n+1/2} - x_{n-1/2}$. As in the elliptic case, h_n is a local cell size, it is not necessarily a constant. Let x_n denote the middle of C_n. We introduce the mean value $\bar{u}_n(t)$ of $u(\cdot, t)$ on cell C_n, i.e.,

$$\bar{u}_n(t) = \frac{1}{h_n} \int_{C_n} u(x, t)\, dx.$$

In a first step, we want to approximate $\bar{u}_n(t)$ by a quantity denoted by $u_n(t) \in \mathbb{R}$ for each t and n. Time will be discretized later on. We first integrate the equation on each cell C_n

$$\int_{C_n} \left(\frac{\partial u}{\partial t} + \frac{\partial q}{\partial x}\right)(x, t)\, dx = 0,$$

where $q = au$ is the flux, then compute both terms. The first term is

$$\int_{C_n} \frac{\partial u}{\partial t}(x,t)\,dx = \frac{d}{dt}\Big(\int_{C_n} u(x,t)\,dx\Big) = h_n \frac{d\bar{u}_n}{dt}(t)$$

by differentiation under the integral sign, and the second term reads

$$\int_{C_n} \frac{\partial q}{\partial x}(x,t)\,dx = q\big(x_{n+\frac{1}{2}},t\big) - q\big(x_{n-\frac{1}{2}},t\big),$$

by direct integration, so that we have

$$h_n \frac{d\bar{u}_n}{dt}(t) + q\big(x_{n+\frac{1}{2}},t\big) - q\big(x_{n-\frac{1}{2}},t\big) = 0.$$

So far, there is no approximation, the above relation is an exact consequence of the transport equation.

Since the semi-discrete unknowns $u_n(t)$ we are looking for are supposed to approximate the mean values $\bar{u}_n(t)$, and not the values of u at the interfaces between cells, we need an approximation of the fluxes $q\big(x_{n+1/2},t\big) = au\big(x_{n+1/2},t\big)$ that is expressed only in terms of the semi-discrete unknowns. This is achieved by means of an approximated flux or numerical flux, a function g in two variables which is used to compute this approximation as follows

$$g_{n+\frac{1}{2}}(t) = g(u_n(t), u_{n+1}(t)) \approx q\big(x_{n+\frac{1}{2}},t\big).$$

Of course, the numerical flux g must be chosen in such a way that the above approximation makes reasonable sense. This yields a semi-discrete scheme

$$h_n \frac{du_n}{dt}(t) + g(u_n(t), u_{n+1}(t)) - g(u_{n-1}(t), u_n(t)) = 0, \tag{10.17}$$

for $t > 0$ and $n \in \mathbb{Z}$, which is an infinite system of ordinary differential equations.

The properties of the scheme partly result from the choice of the numerical flux g. We will give a few examples below. If $u_n(t) = u_{n+1}(t)$ for some n, and t, there is one natural flux between the two cells, namely $au_n(t) = au_{n+1}(t)$, which should naturally be retained. This requirement is fulfilled as soon as

$$\forall v \in \mathbb{R},\ \ g(v,v) = av. \tag{10.18}$$

From now on, we will make the above assumption on the numerical flux, which is related to consistency, see Proposition 10.8 below.

We now consider the discretization with respect to time. We use a classical method of approximation for ordinary differential equations, for example the Euler explicit scheme. Let $M \in \mathbb{N}$ be an integer and $k = \frac{T}{M+1}$ be the time step. We let $t_j = jk$,

$0 \le j \le M+1$. We denote by u_n^j the approximation of $\bar{u}_n^j = \frac{1}{h_n}\int_{C_n} u(x, t_j)dx$ that we want to compute. Discretizing system (10.17) in time accordingly, we thus obtain the following explicit scheme:

$$h_n \frac{u_n^{j+1} - u_n^j}{k} + g(u_n^j, u_{n+1}^j) - g(u_{n-1}^j, u_n^j) = 0, \tag{10.19}$$

for $j \le M$. Starting from an initial data u_n^0 given by

$$u_n^0 = \frac{1}{h_n}\int_{C_n} u_0(x)\, dx = \bar{u}_n(0), \tag{10.20}$$

we can thus compute $u_n^j, n \in \mathbb{Z}$ for all $j \le M+1$. Other ordinary differential equation schemes for the discretization with respect to the time variable can also be used, for example the implicit Euler scheme, the leapfrog scheme, or any Runge–Kutta scheme. Of course, in practical computations, the set of indices n involved must somehow be restricted to a finite set, but we do not pursue this here.

Let us discuss the case of the *upwind scheme*, a commonly used scheme. In all the sequel, we assume that $h = \sup_{n\in\mathbb{Z}} h_n < +\infty$, $\inf_{n\in\mathbb{Z}} h_n > 0$ and that u is sufficiently regular.

We start from scheme (10.19)–(10.20). We just have to specify the choice of the numerical flux g, which is used to approximate $au(x_{n+1/2}, t_j)$. The simplest choice consists in approximating $u(x_{n+1/2}, t_j)$ by one of the neighboring approximated mean values u_n^j or u_{n+1}^j. In the case $a > 0$, the upwind scheme corresponds to the choice u_n^j on the left and in the case $a < 0$, to the choice u_{n+1}^j on the right.

We thus set, depending on the sign of a,

$$g(u_n^j, u_{n+1}^j) = au_n^j \text{ if } a > 0, \text{ or } g(u_n^j, u_{n+1}^j) = au_{n+1}^j \text{ if } a < 0.$$

The case $a = 0$ is not interesting. This gives the upwind scheme

$$\frac{u_n^{j+1} - u_n^j}{k} + a\frac{u_n^j - u_{n-1}^j}{h_n} = 0, \tag{10.21}$$

when $a > 0$ and

$$\frac{u_n^{j+1} - u_n^j}{k} + a\frac{u_{n+1}^j - u_n^j}{h_n} = 0, \tag{10.22}$$

when $a < 0$. In both cases, we enforce the initial condition (10.20). Both choices satisfy the consistency condition (10.18). The scheme is called *upwind*, in the sense that we take into account the direction of the wind, i.e., the exact flux coming from the left when $a > 0$ and from the right when $a < 0$.

Without loss of generality, we focus on the case $a > 0$ and consider the upwind scheme in the form (10.21). We first remark that, in the general case of a non constant cell size, this finite volume upwind scheme differs from the corresponding finite difference scheme. In fact, the finite difference upwind scheme reads

$$\frac{v_n^{j+1} - v_n^j}{k} + a\frac{v_n^j - v_{n-1}^j}{h_{n-\frac{1}{2}}} = 0, \tag{10.23}$$

$$v_n^0 = u_0(x_n), \tag{10.24}$$

where $h_{n-1/2} = x_n - x_{n-1} = (h_n + h_{n-1})/2$ and v_n^j represents an approximation of $u(x_n, t_j)$. There are two differences between the finite volume and finite difference schemes. The first difference is that the factor h_n in (10.21) is replaced by $h_{n-1/2}$ in (10.23) in order to write a difference quotient. The second difference lies in the initial conditions, (10.20) versus (10.24). However, in the case of a constant cell size, i.e., if $h_n = h$ for all $n \in \mathbb{Z}$, then the only difference comes from the discretization of the initial condition. Moreover, as in the elliptic case, we can remark that the ratio $\frac{u_n^j - u_{n-1}^j}{h_n}$ is not a consistent approximation, in the finite difference sense, of $\frac{\partial u}{\partial x}(x_n, t_j)$. Indeed, if we perform the usual Taylor expansions, we obtain

$$u(x_{n-1}, t_j) = u(x_n, t_j) - h_{n-\frac{1}{2}}\frac{\partial u}{\partial x}(x_n, t_j) + O(h^2),$$

so that

$$\frac{u(x_n, t_j) - u(x_{n-1}, t_j)}{h_n} = \frac{h_{n-\frac{1}{2}}}{h_n}\frac{\partial u}{\partial x}(x_n, t_j) + O(h),$$

and $h_{n-1/2}/h_n \not\to 1$ in general.

Let us now study the convergence of the finite volume scheme (10.21)–(10.20). We first have the following finite difference-like convergence result.

Proposition 10.4 *We suppose that u is sufficiently regular and that there exists a constant $C \geq 0$ such that*

$$\sup_{x\in\mathbb{R}, t\in\mathbb{R}_+}\left|\frac{\partial u}{\partial x}(x,t)\right| \leq C, \quad \sup_{x\in\mathbb{R}, t\in\mathbb{R}_+}\left|\frac{\partial^2 u}{\partial x^2}(x,t)\right| \leq C, \quad \sup_{x\in\mathbb{R}, t\in\mathbb{R}_+}\left|\frac{\partial^2 u}{\partial t^2}(x,t)\right| \leq C.$$

Then, under the condition

$$a\frac{k}{\inf_{n\in\mathbb{Z}} h_n} \leq 1, \tag{10.25}$$

there exists a constant $C(T) \geq 0$ such that, for any j such that $j \leq M+1$, we have

$$\sup_{n\in\mathbb{Z}} \left|u_n^j - u\big(x_{n+\frac{1}{2}}, t_j\big)\right| \leq C(T)(h+k). \tag{10.26}$$

Proof Let us compare u_n^j to $u(x_{n+1/2}, t_j)$. For this purpose, we introduce

$$r_n^j = \frac{u\big(x_{n+\frac{1}{2}}, t_{j+1}\big) - u\big(x_{n+\frac{1}{2}}, t_j\big)}{k} + a\frac{u\big(x_{n+\frac{1}{2}}, t_j\big) - u\big(x_{n-\frac{1}{2}}, t_j\big)}{h_n}. \tag{10.27}$$

We use Taylor expansions, first with respect to the space variable and then with respect to the time variable. We first have

$$u\big(x_{n-\frac{1}{2}}, t_j\big) = u\big(x_{n+\frac{1}{2}}, t_j\big) - h_n\frac{\partial u}{\partial x}\big(x_{n+\frac{1}{2}}, t_j\big) + \frac{h_n^2}{2}\frac{\partial^2 u}{\partial x^2}(\xi_n, t_j),$$

where $\xi_n \in]x_{n-1/2}, x_{n+1/2}[$, and secondly

$$u\big(x_{n+\frac{1}{2}}, t_{j+1}\big) = u\big(x_{n+\frac{1}{2}}, t_j\big) + k\frac{\partial u}{\partial t}\big(x_{n+\frac{1}{2}}, t_j\big) + \frac{k^2}{2}\frac{\partial^2 u}{\partial t^2}\big(x_{n+\frac{1}{2}}, \tau_j\big),$$

where $\tau_j \in]t_j, t_{j+1}[$. As u satisfies the transport equation (10.11), we have

$$\frac{\partial u}{\partial t}\big(x_{n+\frac{1}{2}}, t_j\big) + a\frac{\partial u}{\partial x}\big(x_{n+\frac{1}{2}}, t_j\big) = 0,$$

so that

$$r_n^j = -a\frac{h_n}{2}\frac{\partial^2 u}{\partial x^2}(\xi_n, t_j) + \frac{k}{2}\frac{\partial^2 u}{\partial t^2}\big(x_{n+\frac{1}{2}}, \tau_j\big).$$

We deduce that

$$\|r^j\|_{\infty,h} = \sup_{n\in\mathbb{Z}} |r_n^j| \leq C'(h+k), \tag{10.28}$$

where $C' = \frac{C}{2}\max(a, 1)$.

Let us now introduce the error $e^j = (e_n^j)_{n\in\mathbb{Z}}$ defined by $e_n^j = u_n^j - u(x_{n+1/2}, t_j)$. Comparing the scheme (10.21) with the definition (10.27) of r_n^j, we obtain

$$\frac{e_n^{j+1} - e_n^j}{k} + a\frac{e_n^j - e_{n-1}^j}{h_n} = -r_n^j,$$

or equivalently

$$e_n^{j+1} = \Big(1 - a\frac{k}{h_n}\Big)e_n^j + a\frac{k}{h_n}e_{n-1}^j - kr_n^j.$$

Under condition (10.25), we have $1 - a\frac{k}{h_n} \geq 0$ for all n, and $a\frac{k}{h_n} \geq 0$ by hypothesis on a, so that the first two terms form a convex combination of e_n^j and e_{n-1}^j. Therefore, by the triangle inequality

$$\|e^{j+1}\|_{\infty,h} \leq \|e^j\|_{\infty,h} + k\|r^j\|_{\infty,h} \leq \|e^j\|_{\infty,h} + C'k(h+k),$$

according to (10.28). By an immediate induction argument, we obtain, for any $j \leq M+1$

$$\|e^j\|_{\infty,h} \leq \|e^0\|_{\infty,h} + C'jk(h+k) \leq \|e^0\|_{\infty,h} + C'T(h+k).$$

To conclude, we need to estimate $\|e^0\|_{\infty,h}$. We have

$$e_n^0 = v_n^0 - u(x_{n+\frac{1}{2}}, 0) = \frac{1}{h_n}\int_{C_n} u_0(x)dx - u_0\big(x_{n+\frac{1}{2}}\big).$$

Now, still using a Taylor expansion (see (10.5)), we have

$$\left|\frac{1}{h_n}\int_{C_n} u_0(x)dx - u_0\big(x_{n+\frac{1}{2}}\big)\right| \leq C\frac{h_n}{2},$$

so that

$$\|e^0\|_{\infty,h} \leq \frac{C}{2}h.$$

This gives estimate (10.26) with $C(T) = \frac{C}{2} + C'T$. □

Remark 10.3 The hypotheses of Proposition 10.4 are satisfied provided u_0 is regular and u_0' and u_0'' are uniformly bounded on $\mathbb{R}$.

We also deduce the following finite volume convergence result.

Corollary 10.2 *Under the assumptions of Proposition 10.4, there exists a non negative constant $C(T)$ such that, for any $j \leq M+1$, we have*

$$\sup_{n\in\mathbb{Z}}\left|u_n^j - \frac{1}{h_n}\int_{C_n} u(x,t_j)dx\right| \leq C(T)(h+k).$$

Proof This results from the triangle inequality combined on the one hand with estimate (10.26) and, on the other hand, with the Taylor expansion (10.5) applied to $u(\cdot, t_j)$, from which we deduce

$$\left|u\big(x_{n+\frac{1}{2}}, t_j\big) - \frac{1}{h_n}\int_{C_n} u(x,t_j)dx\right| \leq \frac{C}{2}h,$$

hence the result. □

Remark 10.4 Condition (10.25) is in fact a CFL stability condition. We will come back to that later.

Remark 10.5 Proposition 10.4 remains true when $a < 0$ under the CFL condition

$$|a|\frac{k}{\inf_{n\in\mathbb{Z}} h_n} \leq 1,$$

which is thus valid for all a, provided that we change $x_{n+1/2}$ into $x_{n-1/2}$ in estimate (10.26) and in the proof when $a < 0$. Corollary 10.2 remains true without changes.

We can remark that, thanks to the explicit expression (10.12), the exact solution u of the transport problem satisfies a maximum principle, in the sense that if there exists two constants m_* and m^* such that $m_* \leq u_0 \leq m^*$, then the same result holds at any time t, i.e., we have $m_* \leq u(\cdot, t) \leq m^*$. Let us show that, under the stability condition (10.25), this property remains valid at the discrete level.

Proposition 10.5 *We suppose that there exists two constants m_* and m^* such that $m_* \leq u_0 \leq m^*$. Then, under the stability condition (10.25), we have*

$$\textit{for all } n, \quad m_* \leq u_n^j \leq m^*, \tag{10.29}$$

for all $j \leq M+1$. Moreover, setting $u^j = (u_n^j)_{n\in\mathbb{Z}}$, we have

$$\|u^{j+1}\|_{\infty,h} \leq \|u^j\|_{\infty,h}. \tag{10.30}$$

Proof The proof is by induction on j. The double inequality (10.29) is true for $j = 0$ by hypothesis. Let us suppose that it is true for the index j. We have

$$u_n^{j+1} = \Big(1 - a\frac{k}{h_n}\Big)u_n^j + a\frac{k}{h_n}u_{n-1}^j.$$

By condition (10.25), we have $1 - a\frac{k}{h_n} \geq 0$ for all n, and $a\frac{k}{h_n} \geq 0$ by hypothesis on a, so that u_n^{j+1} is a convex combination of u_n^j and u_{n-1}^j. It trivially follows that

$$\min(u_{n-1}^j, u_n^j) \leq u_n^{j+1} \leq \max(u_{n-1}^j, u_n^j),$$

from which $m_* \leq u_n^{j+1} \leq m^*$ for all n ensues. Likewise, we obtain (10.30). □

A natural question that arises is what would happen if we had chosen a downwind approximation instead of the upwind approximation, i.e., formula (10.22) when $a > 0$. Intuitively, the scheme is looking for information in the opposite direction to where it is coming from in the continuous case. So we can expect such a scheme to fail. Let us give a simple numerical example. We computed the solution on $[-2, 2]$ with a compactly supported initial data u_0 with both schemes. The upwind scheme performs quite well, see Fig. 10.2, whereas the downwind scheme gives rise to extreme

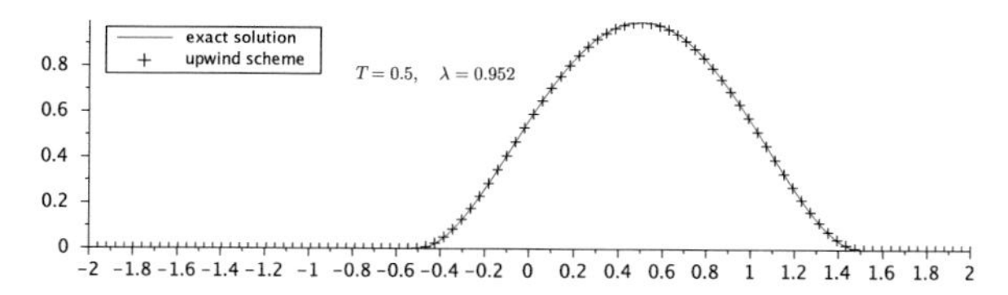

Fig. 10.2 Exact solution and upwind scheme for $u_0(x) = \big((1-x^2)_+\big)^2$, $a = 1$

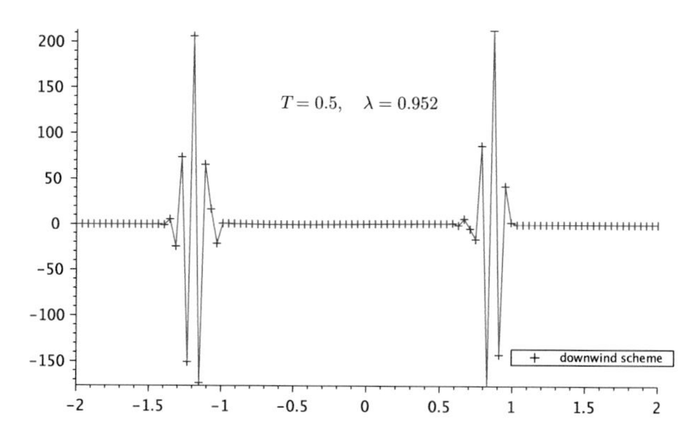

Fig. 10.3 Downwind scheme with the same data

oscillations, see Fig. 10.3. This is an instability issue, and we will go back to it later. Note that if $h_n = h$ for all n, if finite difference initial conditions $u_n^0 = u_0(x_n)$ are chosen and if $a\frac{k}{h} = 1$, then the upwind scheme computes exact values. Indeed, we have then

$$u_n^j = u_{n-1}^{j-1} = \cdots = u_{n-j}^0 = u_0(x_{n-j}) = u_0(x_n - jh) = u_0(x_n - at_j).$$

The computations shown are performed using finite volume initial data.

10.4 Explicit Three Point Schemes

Let us come back to the general form (10.19)–(10.20) of the scheme. For simplicity, we suppose from now on that the mesh is uniform, i.e., $h_n = h$ for all $n \in \mathbb{Z}$. We set

$$\lambda = \frac{k}{h},$$

and the scheme (10.19) then reads

$$u_n^{j+1} = u_n^j - \lambda\big(g(u_n^j, u_{n+1}^j) - g(u_{n-1}^j, u_n^j)\big).$$

This is a special case of a more general formulation

$$u_n^{j+1} = H(u_{n-1}^j, u_n^j, u_{n+1}^j), \tag{10.31}$$

where $H: \mathbb{R}^3 \to \mathbb{R}$ is a given function. This function H must satisfy

$$H(v, v, v) = v, \tag{10.32}$$

for all $v \in \mathbb{R}$, in order for constant states to be preserved by the scheme. Indeed, if $u_0(x) = u_0 \in \mathbb{R}$ for all $x \in \mathbb{R}$, then obviously $u(x, t) = u_0$ for all x, t.

The scheme (10.31) expresses u_n^{j+1} explicitly in terms of three values, namely u_{n-1}^j, u_n^j and u_{n+1}^j. We say it is an *explicit three point scheme*.

Definition 10.1 The scheme (10.31) can be put in *conservation form*, or is *conservative*, if there exists a function $g: \mathbb{R}^2 \to \mathbb{R}$, such that

$$H(v_{-1}, v_0, v_1) = v_0 - \lambda[g(v_0, v_1) - g(v_{-1}, v_0)].$$

The function g is called the *numerical flux*, it is defined up to an additive constant.

Note that when a three point scheme is conservative, then $H(v, v, v) = v$, for any $v \in \mathbb{R}$.

Remark 10.6 More generally, we can consider $(2\ell + 1)$ point schemes, where ℓ is a positive integer, of the form

$$u_n^{j+1} = H(u_{n-\ell}^j, \dots, u_n^j, \dots, u_{n+\ell}^j), \tag{10.33}$$

where $H: \mathbb{R}^{2\ell+1} \to \mathbb{R}$ is a given function. The definition of conservative scheme extends to this case with numerical flux $g: \mathbb{R}^{2\ell} \to \mathbb{R}$. □

Remark 10.7 This definition can be extended to nonlinear equations with a nonlinear flux $q = f(u)$. In this case, it is very important to be able to write scheme (10.33) in conservation form, see [42] for example. □

In the linear case $f(u) = au$, the function H is often taken as a linear combination of values v_j, i.e., it is of the form

$$H(v_{-1}, v_0, v_1) = \sum_{l=-1}^{1} c_l v_l, \tag{10.34}$$

where the coefficients c_l only depend on λ and a. In this case, we say that the scheme itself is linear. According to condition (10.32), we will thus always have

$$\sum_{l=-1}^{1} c_l = 1. \tag{10.35}$$

Let us now study the properties of linear three point schemes. Since the cell size is constant, these schemes can be interpreted as finite difference schemes with initial data given by the average values of u_0 on the cells. Therefore, we only concentrate on the properties of consistency, order and stability in spaces $\ell^p(\mathbb{Z})$.

We introduce the discrete norms on $\ell^p(\mathbb{Z})$,

$$\|v\|_{p,h} = h^{1/p}\Big(\sum_{n\in\mathbb{Z}} |v_n|^p\Big)^{1/p} \text{ if } 1 \leq p < +\infty, \quad \|v\|_{\infty,h} = \sup_{n\in\mathbb{Z}} |v_n|,$$

already used for $p = 2$ in Chaps. 8 and 9.[2] For simplicity, we adopt the following definition of stability in the present context.

Definition 10.2 We say that a scheme (10.31) is *stable* for the norm $\|\cdot\|_{p,h}$, $1 \leq p \leq \infty$, if the sequence $u^j = (u_n^j)_{n\in\mathbb{Z}}$ satisfies, for all $j \geq 0$,

$$\|u^{j+1}\|_{p,h} \leq \|u^j\|_{p,h},$$

for all initial data $u^0 \in \ell^p(\mathbb{Z})$.

This definition of stability is significantly more demanding than the one used before in Chaps. 8 and 9, see Definition 33. In practice, only the cases $p = 2$ and $p = +\infty$ are used.

There are sufficient conditions for ℓ^∞ stability which are very simple to check.

Proposition 10.6 *Let us consider a linear scheme* (10.34)–(10.35). *The scheme is stable in* ℓ^∞ *if and only if the coefficients* c_l *are all nonnegative.*

Proof Let us assume that $c_l \geq 0$ for all l. Then by condition (10.35), u_n^{j+1} is a convex combination of u_{n-1}^j, u_n^j, and u_{n+1}^j so that

$$\min_{l=-1,0,1} u_{n+l}^j \leq u_n^{j+1} \leq \max_{l=-1,0,1} u_{n+l}^j$$

Now we have

$$-\|u^j\|_{\infty,h} \leq \min_{l=-1,0,1} u_{n+l}^j \quad \text{and} \quad \max_{l=-1,0,1} u_{n+l}^j \leq \|u^j\|_{\infty,h},$$

therefore $-\|u^j\|_{\infty,h} \leq u_n^{j+1} \leq \|u^j\|_{\infty,h}$ for all $n \in \mathbb{Z}$. This clearly implies the stability in ℓ^∞.

Assume now that $c_0 < 0$. We thus have $c_{-1} - c_0 + c_1 = 1 - 2c_0 > 1$. The initial data $u_n^0 = (-1)^{n+1}$ is such that $u_0^1 = c_{-1} - c_0 + c_1$, hence ℓ^∞ stability is violated. Likewise, if $c_{-1} < 0$, then $-c_{-1} + c_0 + c_1 = 1 - 2c_{-1} > 1$, and an initial data such that $u_{-1}^0 = -1$, $u_0^0 = u_1^0 = 1$ with $\|u^0\|_{\infty,h} = 1$ yields $u_0^1 = -c_{-1} + c_0 + c_1$, and similarly for the case $c_1 < 0$. ☐

---

[2] Even though the infinity norm does not involve h, we keep the h subscript for notational consistency.

Recall that condition (10.35) implies that constant states are preserved by the scheme. Moreover, in this case, there exists an associated numerical flux, which is also linear.

Lemma 10.1 *Any linear scheme (10.34)–(10.35) admits a numerical flux of the form*

$$g(u, v) = \frac{1}{\lambda}(c_{-1}u - c_1 v). \tag{10.36}$$

Proof The three point scheme reads

$$u_n^{j+1} = c_{-1}u_{n-1}^j + c_0 u_n^j + c_1 u_{n+1}^j.$$

By (10.35), it follows that

$$u_n^{j+1} = u_n^j - \left((c_{-1}u_n^j - c_1 u_{n+1}^j) - (c_{-1}u_{n-1}^j - c_1 u_n^j)\right).$$

Hence g is given by (10.36). □

Definition 10.3 A linear scheme is said to be *monotone* if given two initial conditions u_0 and w_0 such that $u_n^0 \geq w_n^0$ for all n, we have $u_n^1 \geq w_n^1$ for all n.

Obviously, we also have $u_n^j \geq w_n^j$, for all j and n. Schemes that satisfy the hypotheses of Proposition 10.6 are monotone. We have already seen one example of such schemes, the upwind scheme, under the stability condition (10.25). We will see another example below, the Lax–Friedrichs scheme.

As before, ℓ^2 stability is studied via Fourier series.

Proposition 10.7 *A linear scheme (10.34) is stable in ℓ^2 if and only if the function*

$$m(s) = \sum_{l=-1}^{1} c_l e^{-ils} \tag{10.37}$$

satisfies

$$\forall s \in [0, 2\pi], \quad |m(s)| \leq 1. \tag{10.38}$$

The function m is the amplification coefficient *of the scheme and condition (10.38) is the* von Neumann condition.

Proof We use the Fourier series approach, see Sect. 8.7 of Chap. 8. For any $v \in \ell^2$, defining $\mathscr{F}(v)(s) = \sum_{n\in\mathbb{Z}} v_n e^{ins} \in L^2(0, 2\pi)$, we obtain

$$\mathscr{F}(u^{j+1})(s) = m(s)\mathscr{F}(u^j)(s),$$

where m is defined by (10.37). We know that the von Neumann condition (10.38) is then equivalent to

$$\|\mathscr{F}(u^{j+1})\|_{L^2(0,2\pi)} \le \|\mathscr{F}(u^j)\|_{L^2(0,2\pi)}, \tag{10.39}$$

for all initial conditions. The Fourier transform is an isometry, thus

$$\|u^{j+1}\|_{2,h} \le \|u^j\|_{2,h}$$

and the scheme is stable in ℓ^2. Conversely, if the scheme is stable in ℓ^2, then inequality (10.39) holds, which implies the von Neumann condition (10.38). □

We now consider consistency and order issues. Recall that $\lambda = \frac{k}{h}$. To simplify the analysis, we assume from now on that λ is constant, thus linking the time step and the cell size. Consistency will thus be studied in the limit $k \to 0$.

Definition 10.4 Let u be a bounded, regular solution of Eq. (10.11). The truncation error of the scheme is the sequence $\varepsilon_k(u)^j$, defined by

$$\varepsilon_k(u)_n^j = \frac{1}{k}\Big(u(x_n, t_{j+1}) - H(u(x_{n-1}, t_j), u(x_n, t_j), u(x_{n+1}, t_j))\Big),$$

for all $n \in \mathbb{Z}$.

Definition 10.5 We say that the three point scheme (10.31) is consistent in the ℓ^∞ norm, if for any regular solution u of Eq. (10.11) with uniformly bounded derivatives, we have

$$\sup_{j \le T/k} \|\varepsilon_k(u)^j\|_{\infty,h} \to 0 \text{ when } k \to 0.$$

It is of order p if

$$\sup_{j \le T/k} \|\varepsilon_k(u)^j\|_{\infty,h} = O(k^p).$$

There are similar definitions for $2\ell + 1$ schemes. We now give conditions on the coefficients of a linear scheme for it to be consistent. Let us introduce the *Courant number*

$$c = \lambda a = \frac{ak}{h},$$

which also plays a crucial role in stability, see condition (10.25) above.

Proposition 10.8 *A linear scheme (10.34)–(10.35) is consistent and at least of order one, if and only if*

$$\sum_{l=-1}^{1} l c_l = -c,$$

or equivalently

$$g(v, v) = av, \text{ for all } v \in \mathbb{R}, \tag{10.40}$$

with g defined by Eq. (10.36). If in addition we have $\sum_{l=-1}^{1} l^2 c_l = c^2$, *then the scheme is at least of order two.*

Proof We use Taylor expansions, assuming that u is regular enough. There exists $\tau_j, \xi_{n,l}$ such that

$$u(x_n, t_{j+1}) = u(x_n, t_j) + k\frac{\partial u}{\partial t}(x_n, t_j) + \frac{k^2}{2}\frac{\partial^2 u}{\partial t^2}(x_n, t_j) + \frac{k^3}{6}\frac{\partial^3 u}{\partial t^3}(x_n, \tau_j)$$

and

$$u(x_{n+l}, t_j) = u(x_n, t_j) + lh\frac{\partial u}{\partial x}(x_n, t_j) + \frac{(lh)^2}{2}\frac{\partial^2 u}{\partial x^2}(x_n, t_j) + \frac{(lh)^3}{6}\frac{\partial^3 u}{\partial x^3}(\xi_{n,l}, t_j),$$

for $l = \pm 1$. Since u is a regular solution of (10.11), we have $\frac{\partial u}{\partial t} + a\frac{\partial u}{\partial x} = 0$, but also

$$\frac{\partial^2 u}{\partial t^2} = -a\frac{\partial}{\partial t}\left(\frac{\partial u}{\partial x}\right) = -a\frac{\partial}{\partial x}\left(\frac{\partial u}{\partial t}\right) = a^2\frac{\partial^2 u}{\partial x^2}.$$

Since $\sum_{l=-1}^{1} c_l = 1$, the truncation error is therefore such that

$$\varepsilon_k(u)_n^j = -\left(a + \sum_{l=-1}^{1}\frac{lc_l}{\lambda}\right)\frac{\partial u}{\partial x}(x_n, t_j) + \frac{k}{2}\left(a^2 - \sum_{l=-1}^{1}\frac{l^2 c_l}{\lambda^2}\right)\frac{\partial^2 u}{\partial x^2}(x_n, t_j) + R_n^j(k),$$

where $|R_n^j(k)| \le Ck^2$ and C does not depend on n, j and k. This gives the expected results. Relation (10.40) then follows from Lemma 10.1. □

Remark 10.8 The previous result can be generalized to $2\ell + 1$ point schemes. A conservative scheme of the form (10.33) is consistent if $g(v, \dots, v) = av$ for all $v \in \mathbb{R}$, and is then at least of order one. In the nonlinear case, consistency reads $g(v, \dots, v) = f(v)$ for all v. □

We thus see that consistency or order at least one is equivalent to $c_{-1} - c_1 = c$ and order at least two to $c_{-1} + c_1 = c^2$. In fact, we can describe all conservative, consistent three point schemes as follows.

Corollary 10.3 *Any conservative, consistent, three point linear scheme is of the form*

$$u_n^{j+1} = u_n^j - \frac{\lambda a}{2}(u_{n+1}^j - u_{n-1}^j) + \frac{q}{2}(u_{n+1}^j - 2u_n^j + u_{n-1}^j). \tag{10.41}$$

for some $q \in \mathbb{R}$, called the viscosity coefficient *of the scheme. Furthermore, the scheme is of order two if and only if $q = c^2 = \lambda^2 a^2$. The numerical flux is given by*

$$g(u, v) = \frac{a}{2}(u + v) - \frac{q}{2\lambda}(v - u). \tag{10.42}$$

Proof We start from

$$u_n^{j+1} = c_{-1}u_{n-1}^j + c_0 u_n^j + c_1 u_{n+1}^j.$$

with $c_{-1} + c_0 + c_1 = 1$ and $c_{-1} - c_1 = c = \lambda a$. Letting $q = c_{-1} + c_1 = 1 - c_0$, then $c_{-1} = (q + c)/2$, $c_1 = (q - c)/2$ and the scheme reads

$$u_n^{j+1} = \frac{q + c}{2}u_{n-1}^j + (1 - q)u_n^j + \frac{q - c}{2}u_{n+1}^j, \tag{10.43}$$

from which (10.41) immediately follows. The scheme is of order two if moreover $q = c_{-1} + c_1 = c^2$. Pursuing the Taylor expansions one step further, we check that the resulting unique scheme is not of order three. Finally, Eq. (10.42) is just a rewriting of Eq. (10.36) in terms of the new parameters. □

The scheme obtained for $q = c^2$ is the Lax–Wendroff scheme, that we have already encountered in the context of the wave equation, see Chap. 9, Sect. 9.6. The name viscosity coefficient comes from the fact that the factor $u_{n+1}^j - 2u_n^j + u_{n-1}^j$ is directly linked to the three point finite difference approximation of $\frac{\partial^2 u}{\partial x^2}(x_n, t_j)$. The corresponding term in the scheme thus acts as some kind of numerical viscosity or dissipation.

Remark 10.9 Since we have assumed λ to be a constant, and the constants c_l and thus q only depend on λ and a, there is no discretization parameter left in the expression of the scheme (10.41). This can be unsettling since how then can we talk about convergence when h, and thus k, goes to 0, since the scheme apparently depends on neither one of the two? Two factors are at play here. First of all, we are working on $\mathbb{R}$ with an infinite number of discrete values. If we were working on a bounded interval with boundary conditions, the number of discrete values would be finite and depend on h and k. It would of course go to infinity when h, and thus k, goes to 0. Secondly, the discretization parameter h is still hidden in the initial data of the scheme. In a sense, it is the dependence of the initial data on the discretization parameter that drives the convergence of the scheme when h goes to 0, see also the proof of Theorem 10.1 below. □

Formula (10.41) is also useful for stability.

Proposition 10.9 *A three point linear scheme (10.41) is stable in ℓ^2 if and only if its viscosity coefficient q and Courant number c satisfy*

$$c^2 \leq q \leq 1.$$

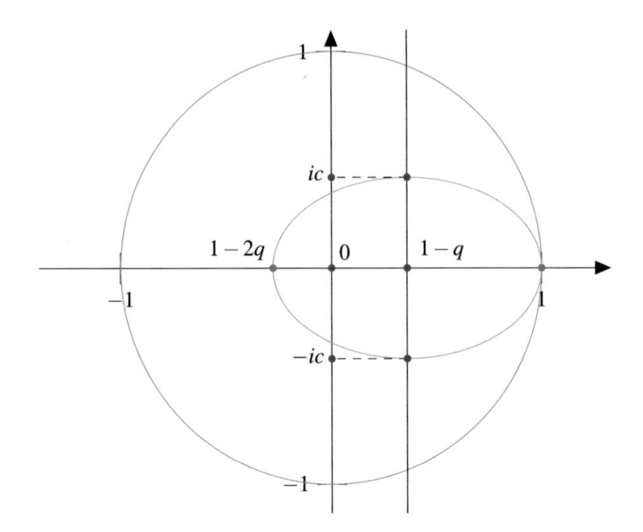

Fig. 10.4 ℓ^2 stability

Stability in ℓ^∞ *is equivalent to*

$$q \geq 0 \text{ and } c^2 \leq q^2 \leq 1.$$

Proof The amplification coefficient of the scheme is given by

$$m(s) = 1 - \frac{c}{2}(e^{is} - e^{-is}) + \frac{q}{2}(e^{is} - 2 + e^{-is}) = 1 - ic\sin s + q(\cos s - 1).$$

The image of the mapping $s \mapsto m(s)$ in $\mathbb{C}$ is an ellipse with vertices 1, $1 - 2q$, $1 - q \pm ic$. It is included in the unit disk if and only if it is included in the unit disk in a neighborhood of 1 and that $1 - 2q$ belongs to the unit disk, see Fig. 10.4. The second condition implies that $1 - 2q \geq -1$, i.e., $q \leq 1$. For the first condition, we notice that $m(s) = 1 - q\frac{s^2}{2} - ics + O(s^3)$ when $s \to 0$, so that $|m(s)|^2 = 1 - qs^2 + c^2s^2 + O(s^3)$, hence the condition $c^2 - q \leq 0$.

In view of Eq. (10.43), stability in ℓ^∞ is ensured if and only if

$$q \leq 1 \quad \text{and} \quad -q \leq c \leq q,$$

by Proposition 10.6. The double inequality implies that $q \geq 0$, it is thus is equivalent to $c^2 \leq q^2$. □

The ℓ^2 stability condition is thus $(\lambda a)^2 \leq q \leq 1$, whereas the ℓ^∞ stability condition reads $(\lambda a)^2 \leq q^2 \leq 1$ with $q \geq 0$. The latter is more stringent than the former since $q^2 \leq q$. In particular, it is not satisfied by the Lax–Wendroff scheme unless $|c| = q = 1$.

In any case, the stability of an explicit scheme is linked to a *CFL stability condition*, see also Chap. 9, Proposition 9.4, of the form

$$|a|\frac{k}{h} \leq \mu,$$

with $\mu \leq 1$ for a three point scheme, $\mu \leq \ell$ for a $2\ell+1$ point scheme. The scalar $|a|k$ is the distance covered by the transport phenomenon during one time step at speed a: the exact solution satisfies $u(x_n, t_{j+1}) = u(x_n - ak, t_j)$. The stability condition $|c| \leq 1$ says that this distance cannot exceed the length of one cell for a three point scheme (or two cells for a five point scheme, and so on). This expresses the fact that the discrete cone of influence must contain the backward characteristic passing through (x_n, t_{j+1}) (see Fig. 9.6 in Chap. 9 in the case of the wave equation).

Let us now say a few words about convergence in this context, in ℓ^∞ for simplicity. Convergence thus clearly means (recall that $\lambda = k/h$ is constant)

$$\sup_{n\in\mathbb{Z}, j\leq T/k} |u_n^j - u(x_n, t_j)| \to 0 \text{ when } k \to 0,$$

when u, i.e., u_0, is regular enough. As can be expected, we also have a Lax theorem.

Theorem 10.1 *A three point linear scheme that is stable and consistent in ℓ^∞ is convergent in ℓ^∞.*

Proof Let us set $v_n^j = u(x_n, t_j)$ and $e_n^j = u_n^j - v_n^j$. By definition of the truncation error, we have $v_n^{j+1} = H(v_{n-1}^j, v_n^j, v_{n+1}^j) + k\varepsilon_k(u)_n^j$ so that

$$e_n^{j+1} = H(e_{n-1}^j, e_n^j, e_{n+1}^j) - k\varepsilon_k(u)_n^j, \quad e_n^0 = u_n^0 - u_0(x_n).$$

Now the scheme is stable, therefore

$$\left\| \left(H(e_{n-1}^j, e_n^j, e_{n+1}^j)\right)_{n\in\mathbb{Z}} \right\|_{\infty,h} \leq \|e^j\|_{\infty,h},$$

so that

$$\|e^j\|_{\infty,h} \leq \|e^0\|_{\infty,h} + k\sum_{l=0}^{j-1} \left\|\varepsilon_k(u)^l\right\|_{\infty,h} \leq \|e^0\|_{\infty,h} + T \sup_{l\leq T/k} \|\varepsilon_k(u)^l\|_{\infty,h},$$

for all $j \leq T/k$ by a repeated application of the triangle inequality. When $k \to 0$ then $h \to 0$, and the first term in the right-hand side tends to 0 if u_0 is regular enough. The second term tends to 0 by consistency of the scheme in ℓ^∞, hence the convergence result. □

Naturally, if the scheme is of order p, we have an error estimate

$$\sup_{n\in\mathbb{Z}, j\leq T/k} |u_n^j - u(x_n, t_j)| \leq Ck^p,$$

provided u_0 is regular enough, if we use finite difference initial data or an order p approximation thereof.

Let us now give a few more examples of schemes that can be used for the transport equation, in either the finite volume or the finite difference contexts.

10.5 Examples of Linear Schemes

For all these schemes, we assume that $h_j = h$, λ constant and $u_n^0 = \frac{1}{h}\int_{C_n} u_0(x)\,dx$ is given.

Example 10.1 **The decentered scheme.**

This is the upwind scheme we already studied in the general case h_j depending on j. Let us just give a few more remarks in the light of the previous developments.

First, we have a general expression that is valid whatever the sign of a. We can write

$$u_n^{j+1} = u_n^j - \lambda(a^-(u_{n+1}^j - u_n^j) + a^+(u_n^j - u_{n-1}^j)), \tag{10.44}$$

where $a^- = \min(a, 0) = a$ if $a \leq 0$, 0 otherwise and $a^+ = \max(a, 0) = a$ if $a \geq 0$, 0 otherwise.[3] This corresponds to the numerical flux

$$g(u, v) = a^+u + a^-v.$$

Formula (10.44) thus combines the two cases (10.21) and (10.22) into one. Since $a = a^+ + a^-$, $|a| = a^+ - a^-$, rewriting the scheme in the form (10.41), we obtain

$$u_n^{j+1} = u_n^j - \frac{\lambda a}{2}(u_{n+1}^j - u_{n-1}^j) + \frac{\lambda|a|}{2}(u_{n+1}^j - 2u_n^j + u_{n-1}^j),$$

hence a viscosity coefficient $q = \lambda|a|$. We summarize the main properties of this scheme.

Proposition 10.10 *The CFL condition*

$$\lambda|a| \leq 1, \tag{10.45}$$

is a necessary and sufficient condition of stability in both ℓ^∞ and ℓ^2. The scheme is of order one in ℓ^∞, it is thus convergent in ℓ^∞. It also satisfies the discrete maximum principle.

[3]Note that this is the opposite convention to the one used for negative parts, which is $a_- = -\min(a, 0)$.

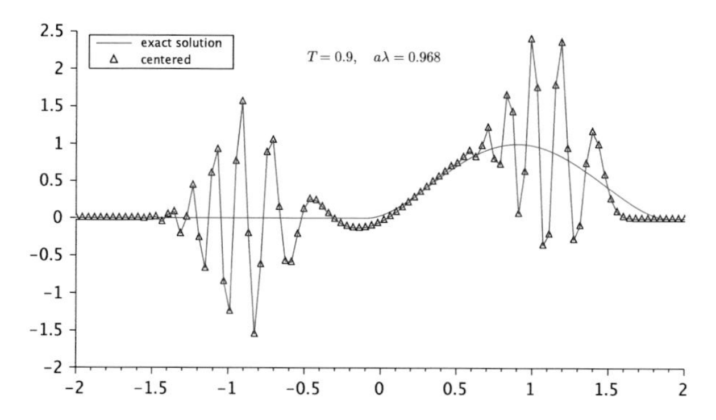

Fig. 10.5 Centered scheme divergence, same data as before

Example 10.2 **The centered scheme**.
The centered scheme is defined by

$$u_n^{j+1} = u_n^j - \frac{\lambda a}{2}(u_{n+1}^j - u_{n-1}^j).$$

This corresponds to the numerical flux

$$g(u, v) = \frac{a}{2}(u + v),$$

i.e., the mean value of the exact fluxes at the interfaces, and viscosity coefficient $q = 0$. The scheme is thus unstable in ℓ^2 and ℓ^∞ as soon as $a \neq 0$. The scheme is not used since it is not convergent, see Fig. 10.5.

Example 10.3 **The Lax–Friedrichs scheme**.
We have already encountered this scheme in the context of the wave equation in Chap. 9. It is given by the following formula

$$u_n^{j+1} = \frac{u_{n+1}^j + u_{n-1}^j}{2} - \frac{\lambda a}{2}(u_{n+1}^j - u_{n-1}^j).$$

This scheme is associated with the numerical flux

$$g(u, v) = \frac{a}{2}(u + v) - \frac{1}{2\lambda}(v - u),$$

and viscosity coefficient $q = 1$.

Proposition 10.11 *The Lax–Friedrichs scheme is stable both in ℓ^∞ and in ℓ^2 if and only if the CFL condition (10.45) is satisfied. It is then monotone. It is of order one in ℓ^∞.*

Example 10.4 **The Lax–Wendroff scheme.**

For completeness, we also record the Lax–Wendroff scheme, that was already discussed earlier and in Chap. 9. The scheme reads

$$u_n^{j+1} = u_n^j - \frac{\lambda a}{2}(u_{n+1}^j - u_{n-1}^j) + \frac{\lambda^2 a^2}{2}(u_{n+1}^j - 2u_n^j + u_{n-1}^j).$$

This scheme is associated with the numerical flux

$$g(u, v) = \frac{a}{2}(u + v) - \lambda \frac{a^2}{2}(v - u),$$

and viscosity coefficient $q = c^2$.

Proposition 10.12 *The Lax–Wendroff scheme is stable in ℓ^2 if and only if the CFL condition (10.45) is satisfied. It is unstable in ℓ^∞ unless $c = 1$. It is of order two in ℓ^∞.*

Recall that the Lax–Wendroff scheme is the only linear three point scheme of order two.

10.6 Generalizations to Discontinuous Data and Nonlinear Problems

The above schemes can also be applied when the initial data is discontinuous, see Figs. 10.6 and 10.7. In this case, the exact solution is still given by formula (10.12). It is thus also discontinuous and must be understood in the distributional sense. As a consequence, considerations of order of schemes do not apply. The question is more how accurately do the various numerical schemes capture the jumps in the solution. We show a computation in which $a\lambda$ is close to 1 but strictly lower, see Fig. 10.6. The same computation performed with $a\lambda = 1$ shows a perfect fit of all three schemes with the exact solution in Fig. 10.7. Indeed, it is easily checked that this should be the case as the three schemes coincide, in particular with the upwind scheme, which we

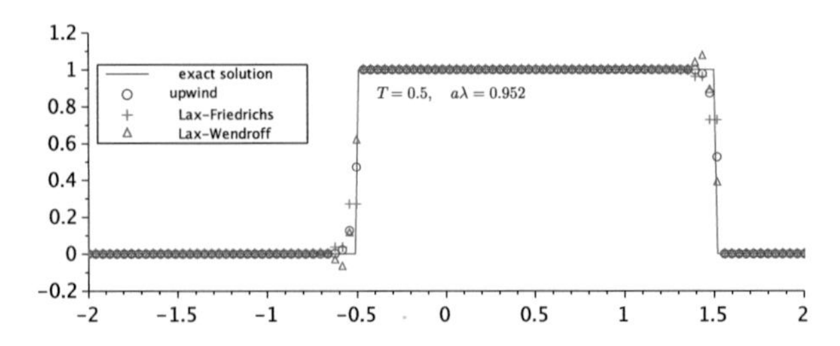

Fig. 10.6 Various schemes for a discontinuous initial data $u_0 = \mathbf{1}_{[-1,1]}$, $a\lambda < 1$

Fig. 10.7 Same with $a\lambda = 1$

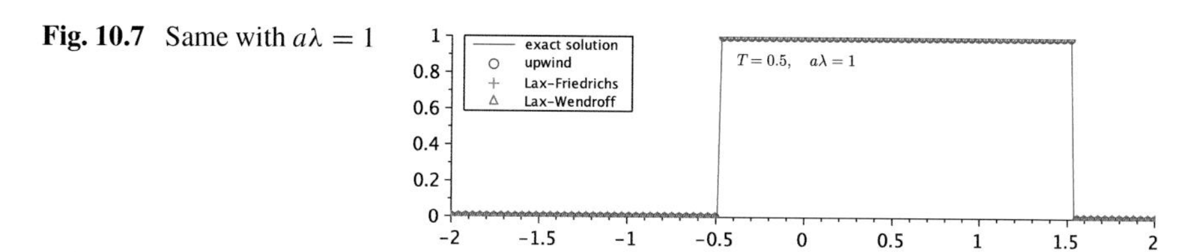

already know computes exact values in this case when finite difference initial values are used (which we did here for simplicity).

The latter remark could cast legitimate doubts on why bother with all the previous developments. The reason is that all problems are not one-dimensional, scalar and linear (anyway, there is an explicit formula for the solution in this instance). The simple one-dimensional, scalar and linear situation is a preparation for more challenging problems. We will talk about the two-dimensional case in the next section. We have already seen a few vector-valued problems in Chap. 9. Let us say a couple of words about the nonlinear case.

All the above schemes have nonlinear versions, the analysis of which is much more complicated than that of their linear version, see for example [42]. To illustrate the nonlinear case, we take the example of the Burgers equation, $\frac{\partial u}{\partial t} + u\frac{\partial u}{\partial x} = 0$ or $\frac{\partial u}{\partial t} + \frac{\partial}{\partial x}\left(\frac{u^2}{2}\right) = 0$ in conservation form. We consider two initial conditions, $u_0(x) = \left((1-x^2)_+\right)^2$ (compactly supported in $[-2, 2]$) and $u_0(x) = 1 + \sin(\pi x)/2$ (periodic on $[-2, 2]$). We plot the exact solution by using the characteristics at some time T small enough so that they have not crossed yet, and the corresponding results of the upwind, Lax–Friedrichs and Lax–Wendroff schemes, see Figs. 10.8 and 10.9.

The first two schemes assume the same form as in the linear case, with the linear flux $f(u) = au$ replaced by the nonlinear flux $f(u) = \frac{u^2}{2}$, namely,

$$u_n^{j+1} = u_n^j - \lambda\left(f(u_n^j) - f(u_{n-1}^j)\right),$$

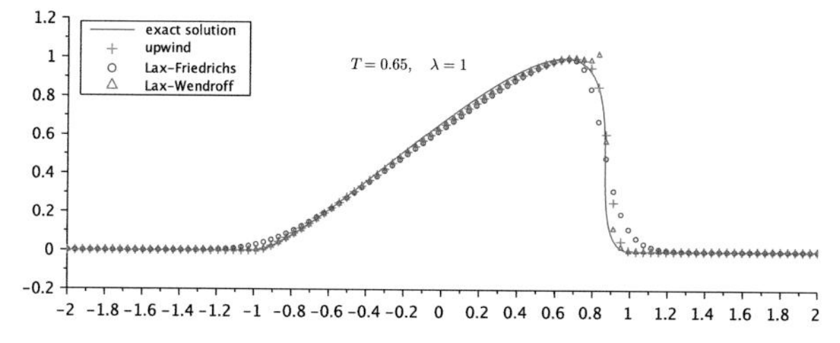

Fig. 10.8 Burgers equation, with initial condition $u_0(x) = \left((1-x^2)_+\right)^2$

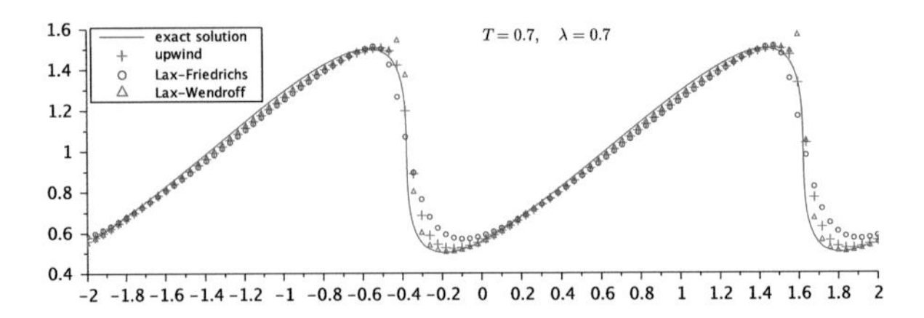

Fig. 10.9 Burgers equation, with periodic initial condition $u_0(x) = 1 + \sin(\pi x)/2$, hence periodic solution

for the upwind scheme (the initial conditions are nonnegative, hence the speeds are nonnegative too), and

$$u_n^{j+1} = \frac{u_{n+1}^j + u_{n-1}^j}{2} - \frac{\lambda}{2}(f(u_{n+1}^j) - f(u_{n-1}^j)),$$

for the Lax–Friedrichs scheme. The Lax–Wendroff scheme is of the form

$$u_n^{j+1} = u_n^j - \frac{\lambda}{2}\big(f(u_{n+1}^j) - f(u_{n-1}^j)\big) + \frac{\lambda^2}{2}\big(A(u_{n+1}^j, u_n^j)\big(f(u_{n+1}^j) - f(u_n^j)\big) - A(u_n^j, u_{n-1}^j)\big(f(u_n^j) - f(u_{n-1}^j)\big)\big),$$

where $A(u, v) = \frac{u+v}{2}$ in the particular case of the Burgers equation.

We also give space–time pictures of the corresponding characteristics before and after they cross, see Figs. 10.10 and 10.11.

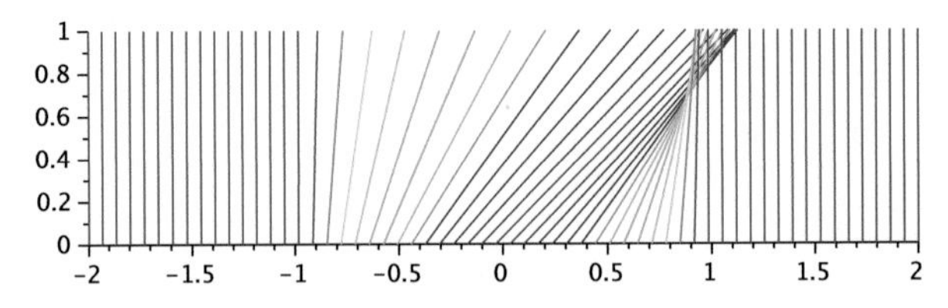

Fig. 10.10 Characteristics for $u_0(x) = \big((1 - x^2)_+\big)^2$

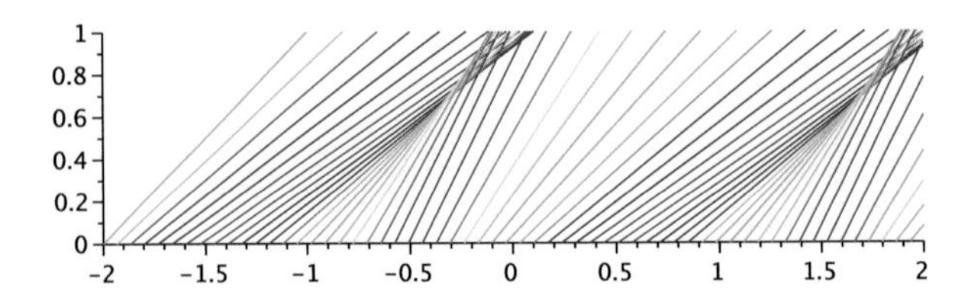

Fig. 10.11 Characteristics for $u_0(x) = 1 + \sin(\pi x)/2$

Note in both cases the already mentioned fact that the crest of the wave travels faster than the trough (which does not travel at all in the first case). One of the main issues from the theoretical side is what sense to give to a solution after the characteristics have crossed, a situation which results in the formation of a shock wave, i.e., a jump discontinuity, even if the initial condition is smooth. On the numerical side, the question is how to reliably compute the shock wave. We leave these difficult questions aside, see for example [42].

10.7 The Transport Equation in an Open Set in Higher Dimensions

In the sequel, Ω denotes a bounded open set of $\mathbb{R}^d$, with sufficiently regular boundary $\Gamma = \partial\Omega$. We here denote by ν the normal unit exterior vector. In this d-dimensional context, the advection velocity a is a given vector field defined on $\bar{Q} = \bar{\Omega} \times [0, T]$ of class C^1, with values in $\mathbb{R}^d$. Without loss of generality, we can assume that a is the restriction of a vector field defined on $\mathbb{R}^d \times \mathbb{R}$ with bounded derivatives, still denoted a, to $\bar{Q}$. We again denote by $t \to X(t; y, s)$ the characteristic passing through point $y \in \mathbb{R}^d$ at time $s \in \mathbb{R}$, still defined by (10.14).

For any $t \in \mathbb{R}$, we decompose Γ into three different parts

$$\Gamma^-(t) = \{x \in \Gamma, a(x,t) \cdot \nu(x) < 0\},$$
$$\Gamma^+(t) = \{x \in \Gamma, a(x,t) \cdot \nu(x) > 0\},$$
$$\Gamma^0(t) = \{x \in \Gamma, a(x,t) \cdot \nu(x) = 0\}.$$

The part $\Gamma^-(t)$ is called the *incoming part* of the boundary at time t, i.e., the part of the boundary where the advection velocity a points inwards and characteristics coming from outside Ω enter Ω. Likewise, $\Gamma^+(t)$ is the *outgoing part* where a points outwards at time t and characteristics leave Ω. Finally, $\Gamma^0(t)$ is called the *characteristic part*. The velocity and characteristics are tangential to Γ on the characteristic part at time t. When the velocity a is stationary, i.e., does not depend on t, the above parts also are independent of t and are then denoted by Γ^-, Γ^+ and Γ^0. For simplicity, we assume that any characteristic passing through a point y at time s with $(y, s) \in \Gamma^0(s)$ does not enter Ω, see Fig. 10.12. We consider the following general transport problem

$$\begin{cases} \dfrac{\partial u}{\partial t}(x,t) + a(x,t) \cdot \nabla u(x,t) = f(x,t) \text{ in } \Omega \times]0, T[, \\ \qquad u(x,t) = g(x,t) \text{ on } \partial Q^- \\ \qquad u(x,0) = u_0(x) \text{ in } \Omega, \end{cases} \tag{10.46}$$

where f is a given source term defined on Q, g a given Dirichlet boundary condition defined on the set $\partial Q^- = \{(x,t), x \in \Gamma^-(t), 0 \le t < T\} \subset \Gamma \times [0, T[$ and

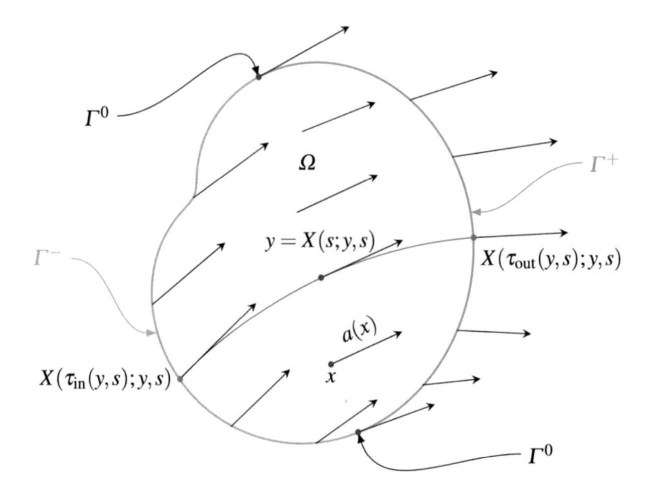

Fig. 10.12 The characteristic passing through y at time s in the case $\tau_{in}(y, s) > 0$ and $\tau_{out}(y, s) < T$. This particular drawing only makes sense if a does not depend on t

u_0 a given initial data defined on Ω. This is an initial-boundary value problem. As already mentioned in Chap. 1 in one dimension, no Dirichlet boundary condition is needed or can even be a priori imposed on the outgoing and characteristic parts of the boundary.

As in the one-dimensional case, we can express the solution of problem (10.46) by means of the characteristics, assuming all functions are regular enough. For all $(y, s) \in \bar{\Omega} \times [0, T]$, we let $[\tau_{in}(y, s), \tau_{out}(y, s)]$ denote the connected component containing s of the set of times t such that $X(t; y, s) \in \bar{\Omega}$. We remark that $X(\tau_{in}(y, s); y, s) \in \Gamma^-(\tau_{in}(y, s))$ if $\tau_{in}(y, s) > 0$, due to the hypothesis made on characteristics leaving Γ^0.

Problem (10.46) thus has an explicit solution, assuming the characteristics are known. This solution is obtained by integration along the characteristics. More precisely, we have the following result which generalizes formula (10.15).

Proposition 10.13 *If u is a regular solution of (10.46), then*

$$u(x, t) = \begin{cases} u_0(X(0; x, t)) + \int_0^t f(X(\tau; x, t), \tau)\, d\tau, & \text{if } \tau_{in}(x, t) = 0, \\ g(X(\tau_{in}(x, t); x, t), \tau_{in}(x, t)) + \int_{\tau_{in}(x,t)}^t f(X(\tau; x, t), \tau)\, d\tau, & \text{if } \tau_{in}(x, t) > 0. \end{cases} \tag{10.47}$$

Proof Let u be a regular solution of (10.46). Let us pick $(x, t) \in Q$. For $\tau \in]\tau_{in}(x, t), \tau_{out}(x, t)[$, we set $y(\tau) = X(\tau; x, t)$ and $v(\tau) = u(y(\tau), \tau)$, so that $x = y(t)$ and $v(t) = u(x, t)$. Using the first equation in (10.14) and the first equation in (10.46), we have

$$v'(\tau) = \frac{\partial u}{\partial t}(y(\tau), \tau) + y'(\tau) \cdot \nabla u(y(\tau), \tau) = \Big(\frac{\partial u}{\partial t} + a \cdot \nabla u\Big)(y(\tau), \tau) = f(y(\tau), \tau).$$

There are now two cases, depending on $\tau_{\text{in}}(x, t)$. If $\tau_{\text{in}}(x, t) = 0$, we can integrate the equation between 0 and t. We obtain

$$\begin{aligned} u(x, t) = v(t) &= v(0) + \int_0^t f(y(\tau), \tau)\, d\tau \\ &= u(y(0), 0) + \int_0^t f(y(\tau), \tau)\, d\tau = u_0(y(0)) + \int_0^t f(y(\tau), \tau)\, d\tau, \end{aligned}$$

using the initial condition. Given the definition of $y(\tau)$, this is the first expression in (10.47).

Let us now consider the second case $\tau_{\text{in}}(x, t) > 0$. In this case, we integrate between $\tau_{\text{in}}(x, t)$ and t and obtain

$$\begin{aligned} u(x, t) = v(t) &= v(\tau_{\text{in}}(x, t)) + \int_{\tau_{\text{in}}(x,t)}^t f(y(\tau), \tau)\, d\tau \\ &= u(y(\tau_{\text{in}}(x, t)), \tau_{\text{in}}(x, t)) + \int_{\tau_{\text{in}}(x,t)}^t f(y(\tau), \tau)\, d\tau \\ &= g(y(\tau_{\text{in}}(x, t)), \tau_{\text{in}}(x, t)) + \int_{\tau_{\text{in}}(x,t)}^t f(y(\tau), \tau)\, d\tau, \end{aligned}$$

on account of the incoming boundary condition (second equation in (10.46)). We thus obtain the second expression in (10.47). □

Remark 10.10 Conversely, if the function u given by formulas (10.47) is sufficiently regular, it is a solution of the transport equation. Indeed, the transport equation is clearly satisfied along the characteristics issuing from Γ^-, which cover all of Q by the Cauchy–Lipschitz theorem, and the initial and boundary conditions are clearly satisfied. Whether or not u is regular is a question of compatibility between the functions u_0 and g. □

10.8 Finite Volumes for the Transport Equation in Two Dimensions

For simplicity, we suppose that $f = 0$ and that a does not depend on x and t, so that Γ^-, Γ^+ and Γ^0 are time independent. We also suppose that Ω is polygonal. We first cover $\bar{\Omega}$ by N cells C_n, which are closed polygons such that $\bar{\Omega} = \cup_{n=1}^N C_n$ whose pairwise intersections are of zero measure. Note that these cells do not need to be triangles or rectangles, as in the case of the finite element method. They also do not need to be all of the same type: we can have triangles, quadrilaterals, and so on, in the same mesh. For simplicity, we suppose that the mesh is admissible in the

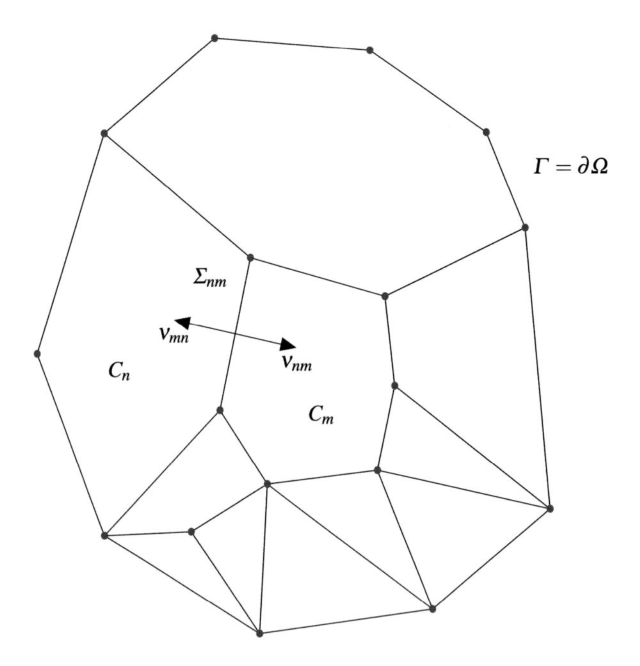

Fig. 10.13 An admissible finite volume mesh of Ω

finite element sense, i.e., the intersection between two cells is either empty, reduced to one vertex, or an entire edge, see Fig. 10.13.

We denote by ν_n the normal unit exterior vector to C_n and by $\mathscr{A}_n$ the area of C_n, so that the area of Ω is $\sum_{n=1}^{N} \mathscr{A}_n$. If two cells C_n and C_m have a common edge, this interface is denoted by $\Sigma_{nm} = \Sigma_{mn}$, and ν_{nm} denotes the restriction of ν_n to Σ_{nm} (it thus points into C_m). The length of the interface is $l_{nm} = l_{mn}$. By construction, we have $\nu_{nm} + \nu_{mn} = 0$. The boundary of C_n is the union of segments

$$\partial C_n = \bigcup_{m=1}^{M(n)} \Sigma_{nm} \bigcup_{p=1}^{M^-(n)} \Gamma_{np}^- \bigcup_{q=1}^{M^+(n)} \Gamma_{nq}^+ \bigcup_{r=1}^{M^0(n)} \Gamma_{nr}^0,$$

where $M(n)$ denotes the number of internal edges Σ_{nm} of C_n (i.e., the edges which are inside Ω), while $M^-(n)$ (resp. $M^+(n)$, $M^0(n)$) denotes the number of edges of C_n which are on Γ^- (resp. on Γ^+, Γ^0).[4] The sets

$$\bigcup_{p=1}^{M^-(n)} \Gamma_{np}^- = \partial C_n \cap \Gamma^-, \bigcup_{q=1}^{M^+(n)} \Gamma_{nq}^+ = \partial C_n \cap \Gamma^+, \bigcup_{r=1}^{M^0(n)} \Gamma_{nr}^0 = \partial C_n \cap \Gamma^0$$

[4]If one of these integers is 0, the corresponding union is empty.

represent all the boundary edges of C_n depending on whether they are on Γ^-, Γ^+ or Γ^0. We denote by l^-_{np} (resp. l^+_{nq}, l^0_{nr}) the length of Γ^-_{np} (resp. Γ^+_{nq}, Γ^0_{nr}). We also denote by ν^-_{np} (resp. ν^+_{nq}, ν^0_{nr}) the restriction of ν_n to Γ^-_{np} (resp. Γ^+_{nq}, Γ^0_{nr}).

As in the 1d-case, we first integrate the transport equation on C_n at time t and obtain

$$\frac{d}{dt}\Big(\int_{C_n} u(x,t)\,dx\Big) + \int_{C_n} (a\cdot\nabla u)(x,t)\,dx = 0. \tag{10.48}$$

We introduce the mean value $\bar{u}_n(t)$ of $u(.,t)$ on the cell C_n, i.e.,

$$\bar{u}_n(t) = \frac{1}{\mathscr{A}_n}\int_{C_n} u(x,t)\,dx.$$

The first term in (10.48) thus becomes

$$\frac{d}{dt}\Big(\int_{C_n} u(x,t)dx\Big) = \mathscr{A}_n\frac{d\bar{u}_n}{dt}(t).$$

We can use the Stokes formula (3.6) of Chap. 3 with $U(x,t) = u(x,t)a$ since a is constant. The second term in (10.48) then reads

$$\int_{C_n} (a\cdot\nabla u)(x,t)dx = a\cdot\Big(\int_{\partial C_n} u(x,t)\nu_n(x)\,d\Gamma\Big).$$

Since each Σ_{nm}, Γ^-_{np}, Γ^+_{nq} and Γ^0_{nr} is a segment, we have, taking into account the incoming boundary condition

$$\begin{aligned}\int_{C_n} (a\cdot\nabla u)(x,t)\,dx &= \sum_{m=1}^{M(n)} (a\cdot\nu_{nm})\int_{\Sigma_{nm}} u(x,t)\,d\Gamma\\ &+ \sum_{p=1}^{M^-(n)} (a\cdot\nu^-_{np})\int_{\Gamma^-_{np}} g(x,t)\,d\Gamma + \sum_{q=1}^{M^+(n)} (a\cdot\nu^+_{nq})\int_{\Gamma^+_{nq}} u(x,t)\,d\Gamma,\end{aligned}$$

with the convention that, if $M^-(n) = 0$ or $M^+(n) = 0$, the corresponding sums are equal to 0. Integrating Eq. (10.48) between t_j and t_{j+1}, we have for all $n = 1,\ldots,N$

$$\begin{aligned}&\mathscr{A}_n[\bar{u}_n(t_{j+1}) - \bar{u}_n(t_j)] + \sum_{m=1}^{M(n)} (a\cdot\nu_{nm})\int_{t_j}^{t_{j+1}}\int_{\Sigma_{nm}} u(x,t)\,d\Gamma dt\\ &+ k\sum_{p=1}^{M^-(n)} (a\cdot\nu^-_{np})\,l^-_{np}\,g^j_{np} + \sum_{q=1}^{M^+(n)} (a\cdot\nu^+_{nq})\int_{t_j}^{t_{j+1}}\int_{\Gamma^+_{nq}} u(x,t)\,d\Gamma dt = 0,\end{aligned} \tag{10.49}$$

where we have set

$$g_{np}^j = \frac{1}{k\, l_{np}^-} \int_{t_j}^{t_{j+1}} \int_{\Gamma_{np}^-} g(x,t)\, d\Gamma dt.$$

These relations are exactly satisfied, there is no approximation so far.

First, we approximate each integral with respect to time as

$$\int_{t_j}^{t_{j+1}} \phi(t) dt \approx k\phi(t_j),$$

which corresponds to the explicit Euler scheme in time. We next have to approximate the fluxes on Σ_{nm} and Γ_{nq}^+ at time t_j in terms of discrete unknowns u_n^j, where u_n^j is meant to be an approximation of $\bar{u}_n(t_j)$. Let us explain how to proceed for the computation on Σ_{nm}. We still use the fact that the information is transported by the characteristic lines. If $a \cdot \nu_{nm} = 0$, the corresponding term in (10.49) vanishes, hence there are two cases left depending on the sign of $a \cdot \nu_{nm}$ (see Fig. 10.14). More precisely, we consider the following upwind approximation:

$$\int_{\Sigma_{nm}} u(x,t_j)\, d\Gamma \approx \begin{cases} l_{nm} u_n^j, & \text{if } a \cdot \nu_{nm} > 0, \\ l_{nm} u_m^j, & \text{if } a \cdot \nu_{nm} < 0. \end{cases}$$

Using the same idea for the flux on Γ_{nq}^+, we obtain

$$\int_{\Gamma_{nq}^+} u(x,t_j)\, d\Gamma \approx l_{nq}^+ u_n^j.$$

Note that there is only one possible choice for the numerical flux in this case, and this choice is consistent with the upwind philosophy, since we are on the outgoing part of the boundary.

Putting all the above approximations together, we obtain the explicit finite volume scheme

$$\begin{aligned} \mathscr{A}_n \frac{u_n^{j+1} - u_n^j}{k} &+ \sum_{\substack{1\le m\le M(n) \\ a\cdot\nu_{nm}>0}} (a \cdot \nu_{nm}) l_{nm}\, u_n^j + \sum_{\substack{1\le m\le M(n) \\ a\cdot\nu_{nm}<0}} (a \cdot \nu_{nm}) l_{nm}\, u_m^j \\ &+ \sum_{p=1}^{M^-(n)} (a \cdot \nu_{np}^-) l_{np}^-\, g_{np}^j + \sum_{q=1}^{M^+(n)} (a \cdot \nu_{nq}^+) l_{nq}^+\, u_n^j = 0. \end{aligned} \tag{10.50}$$

We complement the scheme with the initial condition

$$u_n^0 = \frac{1}{\mathscr{A}_n} \int_{C_n} u_0(x)\, dx, \tag{10.51}$$

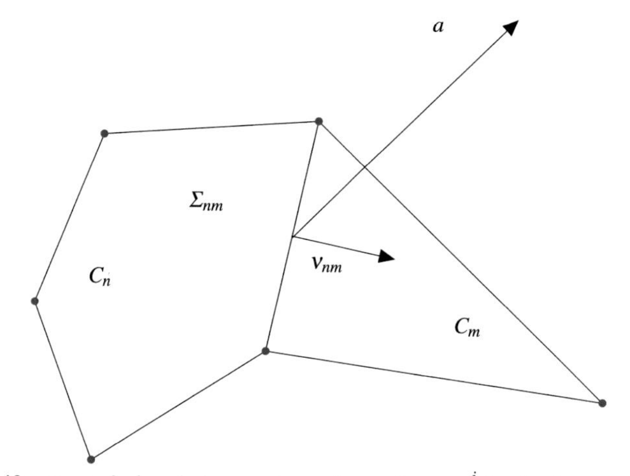

if $a \cdot \nu_{nm} > 0$, the wind comes from C_n, we choose u_n^j for the flux.

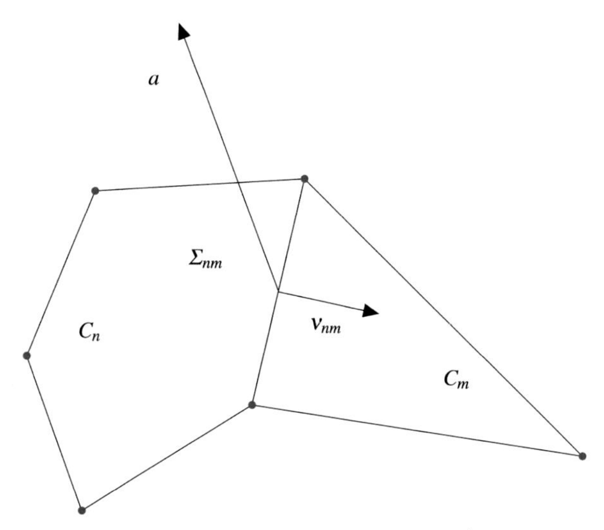

if $a \cdot \nu_{nm} < 0$, the wind comes from C_m, we choose u_m^j for the flux.

Fig. 10.14 The two cases depending on the sign of $a \cdot \nu_{nm}$

or an approximation thereof. Once a numbering of the cells is chosen, we can introduce the sequence of vectors $U^j \in \mathbb{R}^N$ with components u_n^j, and the scheme (10.50)–(10.51) takes the usual form $U^{j+1} = AU^j + G^j$, U^0 given, where A is an $N \times N$ sparse matrix and G^j is a given vector in $\mathbb{R}^N$ corresponding to the Dirichlet condition on the incoming part of the boundary.

We have the following important property:

Proposition 10.14 *The scheme is conservative, i.e., the variation of the total mass is due to what flows in and out at the boundary of* Ω

$$\sum_{n=1}^{N} \mathscr{A}_n \frac{u_n^{j+1} - u_n^j}{k} + \sum_{n=1}^{N} \sum_{p=1}^{M^-(n)} (a \cdot \nu_{np}^-) l_{np}^- \, g_{np}^j + \sum_{n=1}^{N} \sum_{q=1}^{M^+(n)} (a \cdot \nu_{nq}^+) l_{nq}^+ \, u_n^j = 0.$$

Proof The proof is straightforward. We have in fact

$$\sum_{n=1}^{N} \sum_{\substack{m \in \{1,\dots,M(n)\} \\ a \cdot \nu_{nm} > 0}} (a \cdot \nu_{nm}) l_{nm} \, u_n^j + \sum_{n=1}^{N} \sum_{\substack{m \in \{1,\dots,M(n)\} \\ a \cdot \nu_{nm} < 0}} (a \cdot \nu_{nm}) l_{nm} \, u_m^j = 0.$$

Indeed, due to the assumption that the finite volume mesh is admissible, to each term in the first sum corresponding to an edge Σ_{nm} of C_n, there corresponds one and only one term in the second sum corresponding to the edge Σ_{mn} of C_m that is shared with C_n, with $l_{mn} = l_{nm}$, $\nu_{mn} = -\nu_{nm}$ and the same value u_n^j due to the upwind approximation. The above pairing clearly exhausts all the terms of the second sum. □

We do not pursue here the numerical analysis of the above finite volume scheme and refer to [30, 36, 56] for a much more comprehensive mathematical analysis of the finite volume method. This requires more advanced mathematical techniques, in particular the use of spaces of functions with bounded variation, which are functions whose distributional derivatives are not functions but measures, to accommodate piecewise constant functions.

References

1. R.A. Adams, *Sobolev Spaces*, vol. 65, Pure and Applied Mathematics (Academic Press, New York, 1975). A subsidiary of Harcourt Brace Jovanovich, Publishers
2. R.A. Adams, J.J.F. Fournier, *Sobolev Spaces*, vol. 140, 2nd edn., Pure and Applied Mathematics (Amsterdam) (Elsevier/Academic Press, Amsterdam, 2003)
3. S. Agmon, A. Douglis, L. Nirenberg, Estimates near the boundary for solutions of elliptic partial differential equations satisfying general boundary conditions. I. Commun. Pure Appl. Math. **12**, 623–727 (1959)
4. S. Agmon, A. Douglis, L. Nirenberg, Estimates near the boundary for solutions of elliptic partial differential equations satisfying general boundary conditions. II. Commun. Pure Appl. Math. **17**, 35–92 (1964)
5. G. Allaire, *Numerical Analysis and Optimization. An Introduction to Mathematical Modelling and Numerical Simulation*, Numerical Mathematics and Scientific Computation (Oxford University Press, Oxford, 2007)
6. G. Allaire, S.M. Kaber, *Numerical Linear Algebra*, vol. 55, Texts in Applied Mathematics (Springer, New York, 2008)
7. K.E. Atkinson, W. Han, *Theoretical Numerical Analysis. A Functional Analysis Framework*, vol. 39, 3rd edn., Texts in Applied Mathematics (Springer, Dordrecht, 2009)
8. K.E. Atkinson, W. Han, D. Stewart, *Numerical Solution of Ordinary Differential Equations, Pure and Applied Mathematics* (Wiley, New York, 2009)
9. J.-P. Aubin, *Applied Functional Analysis*, 2nd edn. Pure and Applied Mathematics (New York) (Wiley-Interscience, New York, 2000)
10. C. Bernardi, Y. Maday, Spectral methods, in *Handbook of Numerical Analysis*, vol. V, ed. by P.G. Ciarlet, J.-L. Lions (North-Holland, Amsterdam, 1997), pp. 209–485
11. C. Bernardi, Y. Maday, F. Rapetti, *Discrétisations variationnelles de problèmes aux limites elliptiques*, vol. 45, Mathématiques & Applications (Berlin) [Mathematics & Applications] (Springer, Berlin, 2004)
12. E. Bohl, *Finite Modelle gewöhnlicher Randwertaufgaben*, vol. 51, Leitfäden der Angewandten Mathematik und Mechanik [Guides to Applied Mathematics and Mechanics] (B. G. Teubner, Stuttgart, 1981). Teubner Studienbücher: Mathematik. [Teubner Study Books: Mathematics]
13. N. Bourbaki, *Topological Vector Spaces. Chapters 1–5*, Elements of Mathematics (Berlin) (Springer, Berlin, 1987). Translated from the French by H.G. Eggleston, S. Madan
14. S.C. Brenner, L.R. Scott, *The Mathematical Theory of Finite Element Methods*, vol. 15, 3rd edn., Texts in Applied Mathematics (Springer, New York, 2008)
15. H. Brezis, *Functional Analysis, Sobolev Spaces and Partial Differential Equations, Universitext* (Springer, New York, 2011)
16. J.C. Butcher, *Numerical Methods for Ordinary Differential Equations*, 2nd edn. (Wiley, New York, 2008)

H. Le Dret and B. Lucquin, *Partial Differential Equations: Modeling, Analysis and Numerical Approximation*, International Series of Numerical Mathematics 168, DOI 10.1007/978-3-319-27067-8

17. C. Canuto, M.Y. Hussaini, A. Quarteroni, T.A. Zang, *Spectral Methods. Fundamentals in Single Domains*, Scientific Computation (Springer, Berlin, 2006)
18. P.G. Ciarlet, *Introduction to Numerical Linear Algebra and Optimisation*, Cambridge Texts in Applied Mathematics (Cambridge University Press, Cambridge, 1989)
19. P.G. Ciarlet, *The Finite Element Method for Elliptic Problems*, Classics in Applied Mathematics (Society for Industrial and Applied Mathematics (SIAM), Philadelphia, 2002). Reprint of the 1978 original [North-Holland, Amsterdam; MR0520174 (58 #25001)]
20. P.G. Ciarlet, *Linear and Nonlinear Functional Analysis With Applications* (Society for Industrial and Applied Mathematics, Philadelphia, 2013)
21. P.G. Ciarlet, J.-L. Lions (eds.), *Handbook of Numerical Analysis. Finite Element Methods. Part I*, vol. II (North-Holland, Amsterdam, 1991)
22. R. Courant, D. Hilbert, *Methods of Mathematical Physics. Partial Differential Equations*, vol. II, Wiley Classics Library (Wiley, New York, 1989). Reprint of the 1962 original, A Wiley-Interscience Publication
23. M. Crouzeix, A.L. Mignot, *Analyse numérique des équations différentielles*, Collection Mathématiques Appliquées pour la Maîtrise. [Collection of Applied Mathematics for the Master's Degree] (Masson, Paris, 1984)
24. R. Dautray, J.-L. Lions, *Mathematical Analysis and Numerical Methods for Science and Technology. Physical Origins and Classical Methods*, vol. 1 (Springer, Berlin, 1990). With the collaboration of P. Bénilan, M. Cessenat, A. Gervat, A. Kavenoky, H. Lanchon
25. R. Dautray, J.-L. Lions, *Mathematical Analysis and Numerical Methods for Science and Technology. Functional and variational methods*, vol. 2 (Springer, Berlin, 1988). With the collaboration of M. Artola, M. Authier, P. Bénilan, M. Cessenat, J.-M. Combes, H. Lanchon, B. Mercier, C. Wild, C. Zuily
26. R. Dautray, J.-L. Lions, *Mathematical Analysis, Numerical Methods, for Science and Technology. Spectral Theory and Applications*, vol. 3 (Springer, Berlin, 1990). With the collaboration of Michel Artola and Michel Cessenat (Translated from the French by J.C. Amson)
27. R. Dautray, J.-L. Lions, *Mathematical Analysis and Numerical Methods for Science and Technology. Integral Equations and Numerical Methods*, vol. 4 (Springer, Berlin, 1990). With the collaboration of M. Artola, P. Bénilan, M. Bernadou, M. Cessenat, J.-C. Nédélec, J. Planchard, B. Scheurer
28. R. Dautray, J.-L. Lions, *Mathematical Analysis and Numerical Methods for Science and Technology. Evolution Problems. I*, vol. 5 (Springer, Berlin, 1992). With the collaboration of M. Artola, M. Cessenat, H. Lanchon
29. R. Dautray, J.-L. Lions, *Mathematical Analysis and Numerical Methods for Science and Technology. Evolution Problems. II*, vol. 6 (Springer, Berlin, 1993). With the collaboration of C. Bardos, M. Cessenat, A. Kavenoky, P. Lascaux, B. Mercier, O. Pironneau, B. Scheurer, R. Sentis
30. B. Després, *Lois de conservations eulériennes, lagrangiennes et méthodes numériques*, vol. 68, Mathématiques & Applications (Berlin) [Mathematics & Applications] (Springer, Berlin, 2010)
31. D.A. Di Pietro, A. Ern, *Mathematical Aspects of Discontinuous Galerkin Methods*, vol. 69, Mathématiques & Applications (Berlin) [Mathematics & Applications] (Springer, Heidelberg, 2012)
32. J. Dieudonné, *Foundations of Modern Analysis*, vol. 10-I, Pure and Applied Mathematics (Academic Press, New York, 1969). Enlarged and corrected printing
33. N. Dunford, J .T. Schwartz, *Linear Operators. Part I. General Theory*, Wiley Classics Library (Wiley, New York, 1988). With the assistance of W. G. Bade, R. G. Bartle. Reprint of the 1958 original, A Wiley-Interscience Publication
34. G. Duvaut, J.-L. Lions, *Inequalities in Mechanics and Physics*, vol. 219, Grundlehren der Mathematischen Wissenschaften (Springer, Berlin, 1976)
35. L.C. Evans, *Partial Differential Equations*, vol. 19, 2nd edn., Graduate Studies in Mathematics (American Mathematical Society, Providence, 2010)

36. R. Eymard, R. Herbin, T. Gallouët, Finite volume methods, in *Handbook of Numerical Analysis. Solution of Equations in $\mathbb{R}^n$. Part 3. Techniques of Scientific Computing*, vol. VII, ed. by P.G. Ciarlet, J.-L. Lions (North-Holland, Amsterdam, 2000), pp. 713–1020
37. H. Federer, *Geometric Measure Theory*, vol. 153, Die Grundlehren der mathematischen Wissenschaften, Band (Springer, New York, 1969)
38. J.-B.J. Fourier, *Théorie analytique de la chaleur*, Cambridge Library Collection (Cambridge University Press, Cambridge, 2009). Reprint of the 1822 original, Previously published by Éditions Jacques Gabay, Paris, 1988 [MR1414430]
39. P.-L. George, Automatic mesh generation and finite element computation, in *Handbook of Numerical Analysis*, ed. by P.G. Ciarlet, J.-L. Lions (North-Holland, Amsterdam, 1996), pp. 69–190
40. D. Gilbarg, N.S. Trudinger, *Elliptic Partial Differential Equations of Second Order*, Classics in Mathematics (Springer, Berlin, 2001). Reprint of the 1998 edition
41. V. Girault, P.-A. Raviart., *Finite Element Methods for Navier-Stokes Equations. Theory and Algorithms*, vol. 5, Springer Series in Computational Mathematics (Springer, Berlin, 1986)
42. E. Godlewski, P.-A. Raviart, *Numerical Approximation of Hyperbolic Systems of Conservation Laws*, vol. 118, Applied Mathematical Sciences (Springer, New York, 1996)
43. D. Gottlieb, S.A. Orszag, *Numerical Analysis of Spectral Methods: Theory and Applications*, CBMS-NSF Regional Conference Series in Applied Mathematics (Society for Industrial and Applied Mathematics, Philadelphia, 1977)
44. P. Grisvard, *Elliptic Problems in Nonsmooth Domains*, vol. 24, Monographs and Studies in Mathematics (Pitman (Advanced Publishing Program), Boston, 1985)
45. C. Grossmann, H.-G. Roos, *Numerical Treatment of Partial Differential Equations*, Universitext (Springer, Berlin, 2007). Translated and revised from the 3rd (2005) German edition by Martin Stynes
46. W. Hackbusch, *Elliptic Differential Equations. Theory and Numerical Treatment*, vol. 18, Springer Series in Computational Mathematics (Springer, Berlin, 2010). Translated from the 1986 corrected German edition by R. Fadiman, P.D.F. Ion
47. W. Hackbusch, *The Concept of Stability in Numerical Mathematics*, vol. 45, Springer Series in Computational Mathematics (Springer, Heidelberg, 2014)
48. E. Hairer, S.P. Nørsett, G. Wanner, *Solving Ordinary Differential Equations. I. Nonstiff Problems*, vol. 8, 2nd edn., Springer Series in Computational Mathematics (Springer, Berlin, 1993)
49. E. Hairer, G. Wanner, *Solving Ordinary Differential Equations. II. Stiff and Differential-Algebraic Problems*, vol. 14, 2nd edn., Springer Series in Computational Mathematics (Springer, Berlin, 1996)
50. J.S. Hesthaven, S. Gottlieb, D. Gottlieb, *Spectral Methods for Time-Dependent Problems*, vol. 21, Cambridge Monographs on Applied and Computational Mathematics (Cambridge University Press, Cambridge, 2007)
51. F. Hirsch, G. Lacombe, *Elements of Functional Analysis*, vol. 192, Graduate Texts in Mathematics (Springer, New York, 1999)
52. C. Johnson, *Numerical Solution of Partial Differential Equations by the Finite Element Method* (Dover Publications Inc., Mineola, 2009). Reprint of the 1987 edition
53. R. Kress, *Numerical Analysis*, vol. 181, Graduate Texts in Mathematics (Springer, New York, 1998)
54. P. Lancaster, M. Tismenetsky, *The Theory of Matrices*, 2nd edn., Computer Science and Applied Mathematics (Academic Press Inc, Orlando, 1985)
55. P.D. Lax, A.N. Milgram, Parabolic Equations, *Contributions to the Theory of Partial Differential Equations*, vol. 33, Annals of Mathematics Studies (Princeton University Press, Princeton, 1954), pp. 167–190
56. R.J. LeVeque, *Finite Volume Methods for Hyperbolic Problems*, Cambridge Texts in Applied Mathematics (Cambridge University Press, Cambridge, 2002)
57. J.-L. Lions, *Quelques méthodes de résolution des problèmes aux limites non linéaires* (Dunod, Paris, 1969)

58. J.-L. Lions, E. Magenes, *Non-homogeneous Boundary Value Problems and Applications*, vol. 181, Die Grundlehren der mathematischen Wissenschaften (Springer, New York, 1972)
59. B. Lucquin, O. Pironneau, *Introduction to Scientific Computing* (Wiley, New York, 1998)
60. R. Mattheij, J. Molenaar, *Ordinary Differential Equations in Theory and Practice*, vol. 43, Classics in Applied Mathematics (Society for Industrial and Applied Mathematics (SIAM), Philadelphia, 2002). Reprint of the 1996 original
61. J. Nečas, *Les méthodes directes en théorie des équations elliptiques* (Masson et Cie, Éditeurs, Paris, 1967)
62. O. Pironneau, *Finite Element Methods for Fluids* (Wiley, New York, 1989)
63. A. Quarteroni, *Numerical Models for Differential Problems*, vol. 2, MS&A. Modeling, Simulation and Applications (Springer, Milan, 2009)
64. A. Quarteroni, R. Sacco, F. Saleri, *Numerical Mathematics*, vol. 37, 2nd edn., Texts in Applied Mathematics (Springer, Berlin, 2007)
65. A. Quarteroni, A. Valli, *Numerical Approximation of Partial Differential Equations*, vol. 23, Springer Series in Computational Mathematics (Springer, Berlin, 1994)
66. P.-A. Raviart, J.-M. Thomas, *Introduction à l'analyse numérique des équations aux dérivées partielles*, Collection Mathématiques Appliquées pour la Maîtrise. [Collection of Applied Mathematics for the Master's Degree] (Masson, Paris, 1983)
67. R.D. Richtmyer, K.W. Morton, *Difference Methods for Initial-value Problems*, vol. 4, 2nd edn., Interscience Tracts in Pure and Applied Mathematics (Wiley, New York, 1967)
68. W. Rudin, *Real and Complex Analysis*, 3rd edn. (McGraw-Hill Book Co., New York, 1987)
69. W. Rudin, *Functional Analysis*, 2nd edn., International Series in Pure and Applied Mathematics (McGraw-Hill Inc., New York, 1991)
70. A. Samarski, V. Andréev, *Méthodes aux différences pour équations elliptiques*, Traduit du russe par Djilali Embarek (Éditions Mir, Moscow, 1978)
71. M. Schatzman, *Numerical Analysis: A Mathematical Introduction* (Oxford University Press, Oxford, 2002). Translated from the French by J. Taylor
72. L. Schwartz, *Théorie des distributions. Tome I*, Actualités Sci. Ind., no. 1091 = Publ. Inst. Math. Univ. Strasbourg 9 (Hermann & Cie, Paris, 1950)
73. L. Schwartz, *Théorie des distributions. Tome II*, Actualités Sci. Ind., no. 1122 = Publ. Inst. Math. Univ. Strasbourg 10 (Hermann & Cie., Paris, 1951)
74. L. Schwartz, *Mathematics for the Physical Sciences* (Hermann, Paris, 1966)
75. J.C. Strikwerda, *Finite Difference Schemes and Partial Differential Equations*, 2nd edn. (Society for Industrial and Applied Mathematics (SIAM), Philadelphia, 2004)
76. V. Thomée, Finite difference methods for linear parabolic equations, in *Handbook of Numerical Analysis, I*, ed. by P.G. Ciarlet, J.-L. Lions (North-Holland, Amsterdam, 1990), pp. 5–196
77. V. Thomée, *Galerkin Finite Element Methods for Parabolic Problems*, vol. 25, 2nd edn., Springer Series in Computational Mathematics (Springer, Berlin, 2006)
78. F. Trèves, *Basic Linear Partial Differential Equations* (Dover Publications Inc, Mineola, 2006). Reprint of the 1975 original
79. G. Windisch, *M-Matrices In Numerical Analysis*, vol. 115, Teubner-Texte zur Mathematik [Teubner Texts in Mathematics] (BSB B. G. Teubner Verlagsgesellschaft, Leipzig, 1989). With German, French and Russian summaries
80. K. Yosida, *Functional Analysis*, Classics in Mathematics (Springer, Berlin, 1995). Reprint of the sixth (1980) edition
81. E. Zeidler, *Nonlinear Functional Analysis and its Applications. II/A. Linear Monotone Operators* (Springer, New York, 1990)
82. O.C. Zienkiewicz, R.L. Taylor, *The Finite Element Method*, vol. 1, 5th edn. (Butterworth-Heinemann, Oxford, 2000)

Index

H. Le Dret and B. Lucquin, *Partial Differential Equations: Modeling, Analysis and Numerical Approximation*, International Series of Numerical Mathematics 168, DOI 10.1007/978-3-319-27067-8

T

U

Printed in the United States
By Bookmasters